Food Process Engineering Operations

Contemporary Food Engineering

Series Editor

Professor Da-Wen Sun, Director

Food Refrigeration & Computerized Food Technology
National University of Ireland, Dublin
(University College Dublin)
Dublin, Ireland
http://www.ucd.ie/sun/

Emerging Technologies for Food Quality and Food Safety Evaluation, *edited by Yong-Jin Cho and Sukwon Kang* (2011)

Food Process Engineering Operations, *edited by George D. Saravacos and Zacharias B. Maroulis* (2011)

Biosensors in Food Processing, Safety, and Quality Control, *edited by Mehmet Mutlu* (2011)

Physicochemical Aspects of Food Engineering and Processing, *edited by Sakamon Devahastin* (2010)

Infrared Heating for Food and Agricultural Processing, *edited by Zhongli Pan and Griffiths Gregory Atungulu* (2010)

Mathematical Modeling of Food Processing, *edited by Mohammed M. Farid* (2009)

Engineering Aspects of Milk and Dairy Products, *edited by Jane Sélia dos Reis Coimbra and José A. Teixeira* (2009)

Innovation in Food Engineering: New Techniques and Products, *edited by Maria Laura Passos and Claudio P. Ribeiro* (2009)

Processing Effects on Safety and Quality of Foods, *edited by Enrique Ortega-Rivas* (2009)

Engineering Aspects of Thermal Food Processing, *edited by Ricardo Simpson* (2009)

Ultraviolet Light in Food Technology: Principles and Applications, *Tatiana N. Koutchma, Larry J. Forney, and Carmen I. Moraru* (2009)

Advances in Deep-Fat Frying of Foods, *edited by Serpil Sahin and Servet Gülüm Sumnu* (2009)

Extracting Bioactive Compounds for Food Products: Theory and Applications, *edited by M. Angela A. Meireles* (2009)

Advances in Food Dehydration, *edited by Cristina Ratti* (2009)

Optimization in Food Engineering, *edited by Ferruh Erdoğdu* (2009)

Optical Monitoring of Fresh and Processed Agricultural Crops, *edited by Manuela Zude* (2009)

Food Engineering Aspects of Baking Sweet Goods, *edited by Servet Gülüm Sumnu and Serpil Sahin* (2008)

Computational Fluid Dynamics in Food Processing, *edited by Da-Wen Sun* (2007)

Food Process Engineering Operations

George D. Saravacos
Zacharias B. Maroulis

CRC Press is an imprint of the
Taylor & Francis Group, an **informa** business

CRC Press
Taylor & Francis Group
6000 Broken Sound Parkway NW, Suite 300
Boca Raton, FL 33487-2742

Version Date: 2011908

International Standard Book Number: 978-1-4200-8353-8 (Hardback)

Library of Congress Cataloging-in-Publication Data

Saravacos, George D., 1928-
Food process engineering operations / George D. Saravacos, Zacharias B. Maroulis.
p. cm. -- (Contemporary food engineering)
Includes bibliographical references and index.
ISBN 978-1-4200-8353-8 (hardcover : alk. paper)
1. Food industry and trade. 2. Processed foods. 3. Food handling. 4. Agricultural processing. I. Maroulis, Zacharias B., 1957- II. Title.

TP370.S2184 2011
338.4'764795--dc22 2011016991

Visit the Taylor & Francis Web site at
http://www.taylorandfrancis.com

and the CRC Press Web site at
http://www.crcpress.com

To

Katie Saravacos

Sincerely yours,

George Emer

Professor George D. Saravacos

Contents

Series Preface

CONTEMPORARY FOOD ENGINEERING

Food engineering is the multidisciplinary field of applied physical sciences combined with the knowledge of product properties. Food engineers provide the technological knowledge transfer essential to the cost-effective production and commercialization of food products and services. In particular, food engineers develop and design processes and equipment in order to convert raw agricultural materials and ingredients into safe, convenient, and nutritious consumer food products. However, food engineering topics are continuously undergoing changes to meet diverse consumer demands, and the subject is being rapidly developed to reflect market needs.

In the development of food engineering, one of the many challenges is to employ modern tools and knowledge, such as computational materials science and nanotechnology, to develop new products and processes. Simultaneously, improving food quality, safety, and security continue to be critical issues in food engineering study. New packaging materials and techniques are being developed to provide more protection to foods, and novel preservation technologies are emerging to enhance food security and defense. Additionally, process control and automation regularly appear among the top priorities identified in food engineering. Advanced monitoring and control systems are developed to facilitate automation and flexible food manufacturing. Furthermore, energy saving and minimization of environmental problems continue to be important food engineering issues, and significant progress is being made in waste management, the efficient utilization of energy, and the reduction of effluents and emissions in food production.

The *Contemporary Food Engineering Series*, consisting of edited books, attempts to address some of the recent developments in food engineering. Advances in classical unit operations in engineering applied to food manufacturing are covered as well as such topics as progress in the transport and storage of liquid and solid foods; heating, chilling, and freezing of foods; mass transfer in foods; chemical and biochemical aspects of food engineering and the use of kinetic analysis; dehydration, thermal processing, nonthermal processing, extrusion, liquid food concentration, membrane processes, and applications of membranes in food processing; shelf life, electronic indicators in inventory management; sustainable technologies in food processing; and packaging, cleaning, and sanitation. The books are aimed at professional food scientists, academics researching food engineering problems, and graduate-level students.

The books' editors are leading engineers and scientists from many parts of the world. All the editors were asked to present their books to address the market's need and pinpoint the cutting-edge technologies in food engineering.

All contributions are written by internationally renowned experts who have both academic and professional credentials. All authors have attempted to provide critical, comprehensive, and readily accessible information on the art and science of a relevant topic in each chapter, with reference lists for further information. Therefore, each book can serve as an essential reference source to students and researchers in universities and research institutions.

Da-Wen Sun
Series Editor

Series Editor

Professor Da-Wen Sun, PhD, was born in Southern China and is a world authority on food engineering research and education. His main research activities include cooling, drying, and refrigeration processes and systems; quality and safety of food products; bioprocess simulation and optimization; and computer vision technology. His innovative studies on vacuum cooling of cooked meats, pizza quality inspection by computer vision, and edible films for shelf-life extension of fruits and vegetables have been widely reported in the national and international media.

Dr. Sun received first-class BSc honors and an MSc in mechanical engineering, and a PhD in chemical engineering in China before working at various universities in Europe. He became the first Chinese national to be permanently employed in an Irish university when he was appointed college lecturer at the National University of Ireland, Dublin (University College Dublin) in 1995, and was then continuously promoted in the shortest possible time to senior lecturer, associate professor, and full professor. Dr. Sun is now professor of food and biosystems engineering and director of the Food Refrigeration and Computerized Food Technology Research Group at the University College Dublin.

As a leading educator in food engineering, Dr. Sun has contributed significantly to the field of food engineering. He has trained many PhD students who have made their own contributions to the industry and academia. He has also, on a regular basis, given lectures on the advances in food engineering at academic institutions internationally and delivered keynote speeches at international conferences. As a recognized authority in food engineering, Dr. Sun has been conferred adjunct/visiting/consulting professorships from 10 top universities in China including Zhejiang University, Shanghai Jiaotong University, Harbin Institute of Technology, China Agricultural University, South China University of Technology, and Jiangnan University. In recognition of his significant contribution to food engineering worldwide and for his outstanding leadership in the field, the International Commission of Agricultural and Biosystems Engineering (CIGR) awarded him the CIGR Merit Award in 2000 and again in 2006; the Institution of Mechanical Engineers based in the United Kingdom named him Food Engineer of the Year 2004; in 2008 he was awarded the CIGR Recognition Award in recognition of his distinguished achievements as the top 1% of agricultural engineering scientists around the world; in 2007, Dr. Sun was presented with the AFST(I) Fellow Award by the Association of Food Scientists and Technologists (India); and in 2010, he was presented with the CIGR Fellow Award; the title of "Fellow" is the highest honor in CIGR, and is conferred to individuals who have made sustained, outstanding contributions worldwide.

Dr. Sun is a fellow of the Institution of Agricultural Engineers and a fellow of the Institution of Engineers of Ireland. He has also received numerous awards for teaching and research excellence, including the President's Research Fellowship, and has received the President's Research Award from the University College Dublin on two occasions. He is editor-in-chief of *Food and Bioprocess Technology—An International Journal* (Springer); series editor of the *Contemporary Food Engineering Series* (CRC Press/Taylor & Francis); former editor of the *Journal of Food Engineering* (Elsevier); and an editorial board member for the *Journal of Food Engineering* (Elsevier), the *Journal of Food Process Engineering* (Blackwell), *Sensing and Instrumentation for Food Quality and Safety* (Springer), and the *Czech Journal of Food Sciences*. Dr. Sun is also a chartered engineer.

On May 28, 2010, Dr. Sun was awarded membership to the Royal Irish Academy (RIA), which is the highest honor that can be attained by scholars and scientists working in Ireland. At the 51st CIGR General Assembly held during the CIGR World Congress in Quebec City, Canada, in June 2010, he was elected as incoming president of CIGR, and will become CIGR president in 2013–2014; the term of the presidency is six years, two years each for serving as incoming president, president, and past president.

Preface

Food engineering is an interdisciplinary field of major concern to university departments of food science, and chemical and biological engineering. It is an applied area of importance to engineers and scientists working in various food processing industries.

Food Process Engineering Operations deals with the application of chemical engineering unit operations to the handling, processing, packaging, and distribution of food products. In addition to basic process engineering considerations, such as material and energy balances, food processing should meet the special requirements of food acceptance, human nutrition, and food safety.

This book is concerned with the applications of process engineering fundamentals to food processing technology. It can be used primarily as a textbook for one or two semesters for food science students. It can also serve as an important reference for students of chemical and biological engineering interested in food engineering, and for scientists, engineers, and technologists working in food processing industries.

Chapters 1 through 5 review the fundamentals of process engineering and food processing technology, with typical examples of food process applications. Chapters 6 through 15 (the main part of the book) cover food process engineering operations in detail, including theory, process equipment, engineering operations, and application examples and problems.

Chapter 1 provides an introduction to food process engineering with brief discussions on process diagrams, material and energy balances, and engineering calculations. It also presents typical flow sheets of food preservation, food manufacturing, and food ingredient processes.

Chapter 2 deals with the transport phenomena in process engineering, including steady- and unsteady-state transport of heat, mass, and momentum. It presents the simplified solutions of the generalized diffusion equation and introduces interface transfer, transfer coefficients, and dimensionless numbers.

Chapter 3 presents an outline of thermodynamics and kinetics in relation to food processing. It introduces the laws of thermodynamics and covers several thermodynamic processes. It also discusses phase equilibria, with an emphasis on water equilibria, and introduces food kinetics, with an emphasis on microbial growth and inactivation.

Chapter 4 presents a brief overview of food processing technology with typical examples from the food preservation, food manufacturing, and food ingredient industries.

Chapter 5 reviews the engineering properties of foods, including thermophysical, mechanical, and transport properties. It also presents realistic engineering property data, which can be used in food engineering calculations and in food process designs.

Chapter 6 covers the fluid transport operations of foods, including rheology, mechanical balances, piping systems, and pumping equipment.

Chapter 7 describes the several mechanical processing operations used in the preservation and manufacturing of food products. It also discusses particle characteristics, size reduction and enlargement, mixing and forming, mechanical separations, and removal of food parts.

Chapters 8 and 9 discuss the heat transfer operations used in food processing. The heat transfer coefficients are defined and applied to the design of various heat exchangers. Thermal preservation processes (sterilization and pasteurization) are discussed in detail and thermal treatment processes (baking, roasting, and frying) are outlined.

Chapters 10 and 11 deal with the evaporation and drying operations of food processing. The design of evaporators and dryers is based on heat and mass transfer and on water activity. Energy savings are important for economic design. These chapters describe several types of drying equipment.

Chapter 12 covers refrigeration and freezing operations applied to food systems.

Chapter 13 discusses the food applications of mass transfer, including distillation, extraction, and crystallization.

Chapter 14 reviews novel food process operations, such as membrane processing, supercritical fluid extraction, and high-pressure processing.

Chapter 15 provides an overview of food packaging engineering, including packaging materials, packaging equipment, and package processing operations.

Chapter 16 introduces spreadsheet software in food process engineering. It presents examples of Excel® and Visual Basic® applications for heat exchanger, psychrometric, and rheological calculations.

Simple diagrams are used to illustrate the mechanism of each operation and the main components of the process equipment. Simplified calculations, based on elementary calculus, are used in the book. Realistic values of food engineering properties are taken from the literature and the authors' experience. Detailed data and pictures of process equipment can be obtained from suppliers of equipment. The appendix contains useful engineering data for process calculations, such as steam tables, engineering properties, engineering diagrams, and suppliers of process equipment. The book is based on our long teaching and research experiences both in the United States and in Greece.

We wish to acknowledge the contributions of our colleagues, associates, and students at the National Technical University of Athens, Rutgers University, and Cornell University. We appreciate the useful information provided by technical people in the food processing industry and the suppliers of processing equipment.

Special thanks are due to Magda Krokida for her help in preparing the numerical examples in this book. We also appreciate the help of N. Oikonomou, V. Oikonomopoulou, C. Boukouvalas, P. Eleni, I. Katsavou, A. Lazou, K. Kavvadias, and P. Michailidis for preparing the various diagrams in the book.

We hope that this book will contribute to the advancement of teaching and applications of food process engineering operations worldwide. We welcome any comments, criticisms, and corrections by the readers of this book.

George D. Saravacos
Zacharias B. Maroulis

Authors

George D. Saravacos is currently an emeritus professor at the National Technical University of Athens (NTUA), Athens, Greece. He has been a professor of chemical engineering at the NTUA, and a professor of food engineering at Cornell and Rutgers Universities. He has taught courses on unit operations and food engineering both in the United States and in Greece.

Dr. Saravacos received his diploma engineering degree in chemical engineering from NTU, his MS degree in food science from the University of California, and his ScD degree from MIT. He was awarded an honorary doctorate by the Agricultural University of Athens.

Professor Saravacos' research interests include transport properties of foods, heat and mass transfer, evaporation, dehydration, and distillation. He has participated in several fundamental and applied research projects in food engineering in Greece and in the United States, cooperating with the NYSAES (Cornell) and CAFT (Rutgers). He has published more than 150 papers in technical journals and proceedings of national and international meetings, and has authored 10 books/book chapters and 5 textbooks (in Greek). He is also the recipient of the P&G Award for Excellence in Drying Research.

Dr. Saravacos is a member of the IFT, the AIChE, and the Engineering Society of Greece. He is on the editorial board of the *Journal of Food Engineering.* Currently (2008–2011), he is serving as the president of the International Association of Engineering and Food (IAEF).

Zacharias B. Maroulis is currently a professor of chemical engineering at the National Technical University of Athens (NTUA), Athens, Greece, where he teaches and does research on unit operations and process design.

He received his diploma engineering degree and his doctorate from the NTUA and pursued his postdoctoral research at Rutgers University. He is the author or coauthor of more than 120 journal articles, 7 books and book chapters, and professional publications, including *Food Plant Economics* (CRC, 2007), *Food Process Design* (Dekker, 2003), and *Transport Properties of Foods* (Dekker, 2001).

Dr. Maroulis also serves on the editorial board of the journal *Drying Technology*; his research interests include transport properties of foods, drying science and technology, and process economics.

1 Introduction

1.1 INTRODUCTION

The process operations used in the food processing industry were adopted from the unit operations of chemical engineering, which have been applied successfully over the years in chemical process industries. The principles of unit operations are based on the transport phenomena and thermodynamics, while their industrial application requires practical knowledge and experience on process technology, process equipment, and process economics.

The engineering analysis and design of unit operations is well developed in the processing of gases and liquids, for which engineering property data are available or can be estimated with good accuracy, using theoretical and semi-theoretical models. Fluid flow and heat and mass transfer operations in gas/liquid processing can be designed and scaled up accurately. Mechanical processing operations, involving solids and semisolids are more difficult to analyze and design, and semiempirical methods are used (Saravacos and Kostaropoulos, 2002).

The food engineering operations are more difficult to analyze than simple gases and liquids because of the complexity of foods and their sensitivity to processing. Food safety, nutrient retention, and organoleptic quality should be considered carefully in addition to classical engineering analysis, economic design, and optimization. Of particular importance is the prevention of contamination and growth of food pathogenic microorganisms in the food processing and storage system. Food contamination can also occur by toxic chemicals entering the food or produced in foods during processing and storage.

Food engineering operations involving fluid foods are analyzed by methods used in chemical engineering, provided that engineering property data are available or can be estimated from semiempirical models. Experimental or approximate empirical data are used for complex fluids, semifluids, and pastes.

Most food processing operations involve solid or semisolid foods, for which limited engineering property data are available. Approximate semi-theoretical or empirical data are used in food-engineering design and process applications. In solids and semisolids processing, specialized equipment is often used, which has been tested and applied successfully in the pilot plant or similar food processing plants. Equipment manufacturers and food processors have developed various processing units, based on industrial experience with similar food products. Engineering analysis of such empirical equipment, utilizing the physical, chemical, and microbiological principles involved, can improve the design and industrial application and the quality of processed food (Maroulis and Saravacos, 2001).

Food processing is a very large industry producing large quantities of food products which are distributed to the consumers at relatively low prices. In order to be profitable, the food industry must utilize fully the available process technology and engineering (Maroulis and Saravacos, 2007).

This book is designed to be used primarily as a textbook for undergraduate and introductory graduate students of food science and technology departments as well as students of developing fields of bioengineering and biotechnology. It will be useful to engineers and technologists working in the food industry. The book emphasizes the application of fundamentals of process engineering to the design and operation of the process equipment and industrial plants applied in the various food processes. Several numerical application examples and problems of typical food process engineering applications are presented for a thorough understanding of the material of each chapter (Saravacos and Maroulis, 2001; Singh and Heldman, 2001).

1.2 PROCESS DIAGRAMS

Process engineering operations are presented usually in the form of various diagrams, which are useful for understanding the flow of materials and energy, and for visualizing the process equipment involved. They are essential in process and plant design, process control, and economic analysis of the process plants.

Two basic diagrams useful in preliminary design and plant operation calculations are the process block diagram (PBD), used mainly in material and energy balances, and the process flowsheet diagram (PFD), used for representation of the equipment sequence and the flow of materials and energy. In the PBD, each block may include more than one engineering operation and equipment. For example, a block labeled "Evaporation" may include two or more evaporator units connected in a system plus the auxiliary equipment of condensers, process pumps, piping, and controllers.

Additional diagrams used in process engineering include the three-dimensional (3D) diagram, used for space representation of the process equipment, the plant layout diagram (PLD), showing the location of the process equipment on the plant floor, the process control diagram (PCD), and the process instrumentation and piping diagram (PID), used for showing the layout of process piping and the process control system.

Figure 1.1 shows a simplified PBD for the production of orange juice concentrate (OJC) from oranges. The PFD for the same process is shown in Figure 1.2.

Industrial citrus processing plants may be more complex than the simplified versions shown in Figures 1.1 and 1.2. For example, the orange peels may be utilized by extraction of the peel oil, followed by air-drying of the peel wastes to produce animal feed.

Throughout this book, standardized symbols are used to represent various processing equipment and operations, which are explained in the Appendix.

1.3 MATERIAL AND ENERGY BALANCES

Material and energy balances are essential in the design of food processes, processing equipment, process utilities, and waste treatment facilities. They are important in process optimization and process control, and in economic (profitability) analysis of processes and food processing plants.

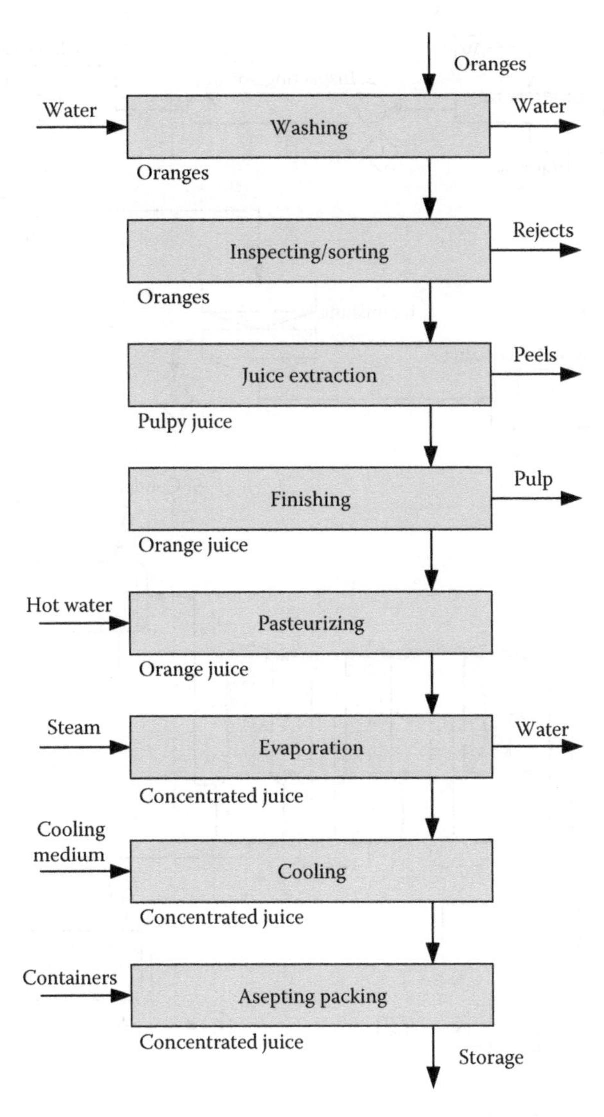

FIGURE 1.1 PBD of OJC.

Material balance calculations in food processing are based on a key food component, such as total solids, water, sugars, or oil.

Energy balances in food processing involve mainly heat energy. Mechanical and electrical energy are of secondary importance in most food processing operations.

1.3.1 Material Balances

The principles and techniques of material and energy balances used in chemical engineering can be applied to most food process engineering calculations. Food processes require special attention due to the complexity and sensitivity of food materials

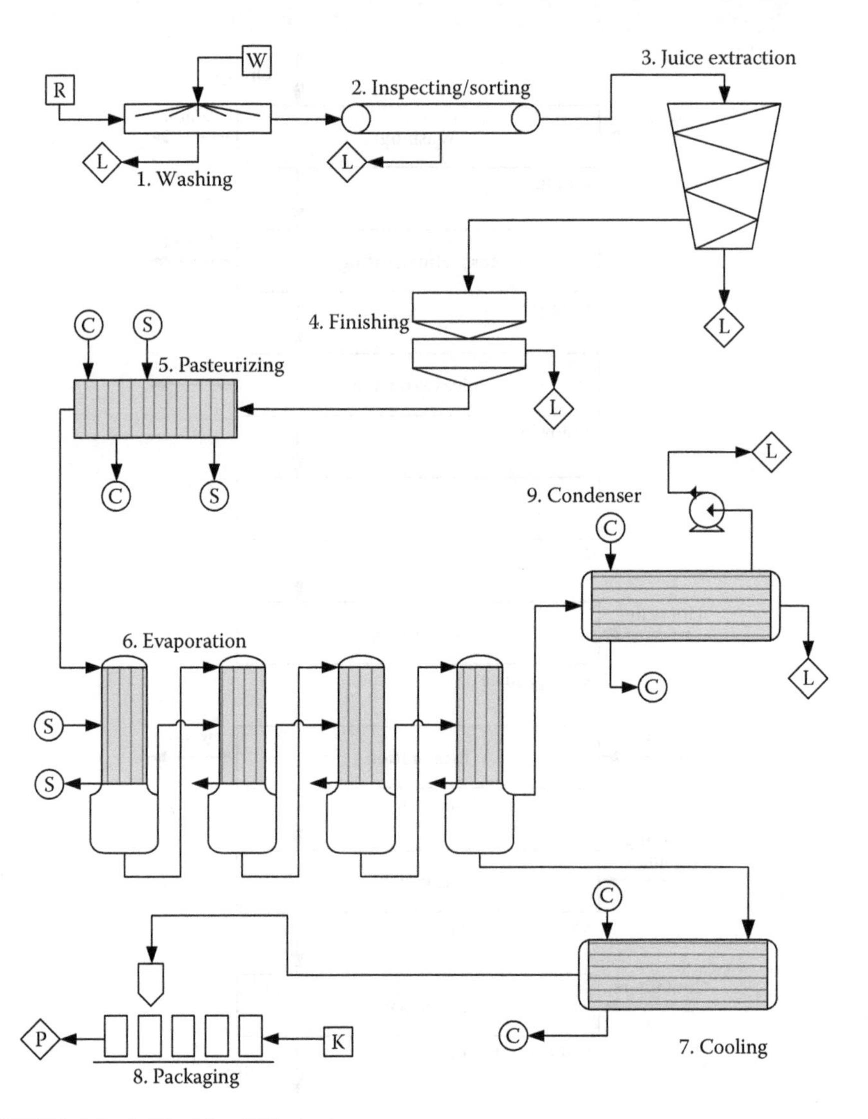

FIGURE 1.2 PFD of the OJC plant.

in storage and processing. In material balances, accurate food composition data are difficult to obtain, because of the variability of food materials in variety, growing conditions, and age of the raw materials. Reliable experimental data of a given food material are preferable to approximate composition values of the literature.

Simple material and energy balances can be performed on mechanical and heat processing operations. Simultaneous heat and mass transfer operations, such as drying, blanching, baking, and steam injection, may need more detailed analysis and experimental verification of the assumptions on food composition and energy requirements. Due to variability of raw food materials, material and energy balances may be required periodically during the operation of the food processing plants.

Two types of material balances are normally used in engineering: (1) the overall material balance of the total material in the system and (2) the component balance of a characteristic food component, such as total solids (TS), water, and sugars.

Material balances are calculated at the boundaries of a food process, from the mass conservation equation in the system:

$$(\text{Total mass in}) - (\text{total mass out}) = (\text{Total mass accumulated}) \quad (1.1)$$

$$(\text{Total component in}) - (\text{total component out}) = (\text{Total component accumulated}) \quad (1.2)$$

For continuous operations, the accumulated materials (total and component) are zero.

The component material balance (Equation 1.2) can be written for one or more food components which are important in a given processing operation. Typical components involved in food processing are water (moisture), TS, soluble solids (SS), fat, oil, salt, and protein. The SS are usually expressed as °Brix (% sucrose by weight), measured with refractometers, which are used widely in the laboratory and in the processing plant. The concentration of a components is expressed as mass or weight fraction, x_i = %(weight)/100.

Volumetric flows, e.g., (L/h or m^3/h), should be converted to mass flows, e.g., (kg/h or ton/h), using the density of the material (kg/L or kg/m^3).

1.3.2 Energy Balances

Energy balances are calculated at the boundaries of a food process from the energy conservation equation (first law of Thermodynamics):

$$(\text{Total energy in}) - (\text{total energy out}) = (\text{Total energy accumulated}) \quad (1.3)$$

For preliminary design calculations and equipment sizing, the main energy form considered in Food Process Engineering is heat, and only heat balances are calculated. The mechanical and electrical requirements for pumping, transportation, refrigeration, and operation of the various pieces of process and utility equipment are considered in the detailed process, equipment, and plant design.

The total heat of a stream (Q) is equal to the sensible (Q_s) and the latent (Q_l) heats:

$$Q = Q_s + Q_l = \sum m_l C_{pi} \Delta T + \sum m_j \Delta H_j \quad (1.4)$$

where components (i) participate in sensible heating or cooling by (ΔT) degrees (°C or K), and components (j) are involved in evaporation (condensation) or freezing (fusion); m (kg) is the mass of each component, and ΔH (kJ/kg) is the enthalpy of phase change (evaporation or freezing).

The specific heat of water (C_p) is normally taken as equal to 4.18 kJ/kg, while all food materials have lower values. The heat of evaporation or condensation of water depends on the pressure. Thus, at atmospheric pressure (P = 1.013 bar and

$T = 100°C$), $\Delta H = 2.26$ MJ/kg. The heat of freezing or fusion of water is taken as $\Delta H_f = 0.333$ MJ/kg.

Typical thermophysical and thermodynamic data for foods are given in Chapters 3 and 5 (Rahman, 1996; Rao et al, 2006).

1.4 ENGINEERING UNITS

Engineering units are very important for any calculation and application. Since Food Engineering is an international discipline, the units used should be more universal than the traditional national units. This book uses mainly the units of the International System (Système International, SI), which is applied worldwide in scientific and engineering publications. However, whenever necessary, the English units are mentioned, since this system is still used in the United States and partly in the United Kingdom.

In Food Engineering applications, the most common SI units are meter (m, length), kilogram (kg, mass), mole (mol, amount of substance), Newton (N, force), joule (J, energy), second (s, time), Kelvin (K, temperature), ampere (A, electric current), and candela (cd, luminous intensity). Table 1.1 shows the conversion of English units to SI units.

In industrial applications, a convenient mass unit is the ton (t), which is equal to 1000 kg. The common ton is sometimes written as "metric ton" or "tonne." Note that in the United States the "ton" or "short ton" is sometimes used, which is equal to 2000 lb or 908 kg.

Derived units, used extensively in engineering, are mostly expressed in SI units. Some common derived units are used instead of the basic SI units: pressure (Pa, pascal) instead of (N/m^2), power (W, watt) instead of (J/s), and viscosity (Pa s) instead of (kg/m s). Table 1.2 shows the multiplying factors for converting some common derived English units to SI units.

In engineering applications, the pressure is expressed mostly in units of (bar) (1 bar = 10^5 Pa, where 1 Pa = 1 N/m^2). Table 1.3 shows the common units of pressure.

The vacuum in process equipment is measured as absolute pressure. A vacuum of 10 Torr means that the absolute pressure in the vessel is 10 Torr, that is, the vessel is at a pressure of 760 − 10 = 750 Torr below atmospheric pressure.

TABLE 1.1
Conversion of English Basic Units to SI units

Quantity	English Unit	Conversion Factor	SI Unit
Length	Ft	0.305	m
Mass	lb	0.454	kg
Force	lb	4.45	N
Energy	Btu	1.055	kJ
Time	s	1	s
Temperature	°F	(°F − 32)/1.8	°C = K − 273

TABLE 1.2
Conversion of English to SI-Derived Units

Quantity	British Unit	Conversion Factor	SI Unit
Surface	sqft	0.093	m^2
Volume	cuft	0.0284	m^3
Volume	gal (U.S.)	3.78	L (10^{-3} m^3)
Density	lb/cuft	16.2	kg/m^3
Concentration	mole/cuft	16.2	mol/m^3
Velocity	ft/s	0.305	m/s
Volumetric flow	gal/min (GPM)	0.063	L/s (10^{-3} m^3/s)
Volumetric flow	cuft/min (CFM)	0.5	L/s (10^{-3} m^3/s)
Pressure	lb/sqin (psi)	0.0689	bar (10^5 Pa)
Enthalpy (*H*)	Btu/lb	2.326	kJ/kg
Power	Btu/h	0.293	W
Power	HP	0.745	kW
Specific heat	Btu/lb °F	1.294	kJ/(kg K)
Viscosity	cP	1	mPa s
Thermal conductivity	Btu/h ft °F	1.729	W/(m K)
Heat transfer coefficient	Btu/h sqft °F	5.678	W/(m^2 K)

TABLE 1.3
Common Units of Pressure

Common Unit	Equivalent SI Unit
Pa	N/m^2
bar	10^5 Pa
mbar	100 Pa
atm (760 Torr)	1.013 bar
Torr (mm Hg)	133 Pa
mm water	9.81 Pa

In enthalpy (heat content) calculations, two units are used: 1 kWh/kg (often written as kWh/kg) and 3.6 MJ/kg. For example, the heat of evaporation of water at 100°C is 2.3 MJ/kg or 0.639 kWh/kg, and the heat of fusion (melting) of ice is 0.333 MJ/kg or 0.093 kWh/kg.

Conversion of units from the English to the SI system is necessary in any application.

In engineering calculations, it is recommended that all given data be converted to SI units, which fit readily into the various equations and algorithms. The results of the calculations, if needed, can be easily converted to the English or other technical units, which are often used in various applications.

1.5 ENGINEERING CALCULATIONS

1.5.1 Engineering Symbols

The symbols of various quantities used in equations and numerical calculations in this book are used internationally in engineering applications. There are a few different symbols used traditionally in the United States. For example, the symbol of viscosity is often (μ) instead of the international (η), while thermal conductivity may be denoted as (k) instead of (λ). The symbols used in each chapter are listed before the References.

The symbol (/) denotes a division operator, and it is used preferably in this book instead of the exponent notation, e.g., the density (ρ) is written as kg/m^3 instead of kg m^{-3}, and the thermal conductivity (λ) is written as W/(m K) instead of W m^{-1} K^{-1} or J s m^{-1} K^{-1}.

1.5.2 Dimensions and Dimensionless Numbers

The dimensions of the various engineering quantities are very important in equations and numerical calculations.

Equations must be dimensionally correct, that is, the overall dimensions of the left side must be equal to the dimensions of the right side of the equation. For example, in the integrated heat conduction equation $Q = (\lambda/L)\, A\, \Delta T\, t$, the dimensions of heat quantity (Q) are energy (J), while the dimensions of the right side quantities are thermal conductivity [λ, W/(m K)], length (L, m), temperature difference (ΔT, K), area (A, m^2), and time (t, s). The overall dimensions of the right side will be [W/(m K)] (1/m) (K) (m^2) (s) = (W) (s) = J, which are the units of heat quantity (Q).

The dimensional consistency of complex equations should be checked before any calculation. Dimensional analysis can be used to find empirical equations describing a process, based on empirical or experimental data.

Dimensionless ratios of units are sometimes used to simplify calculations. Thus, the specific gravity (SG) of a material is defined as the ratio of the density of the material (ρ) to the density of water at the same temperature (ρ_w), $SG = \rho/\rho_w$. For example, if the SG of an edible oil at 20°C is 0.875, then its density will be $\rho = 0.875 \times 1000 = 875$ kg/m^3.

Dimensionless numbers are used extensively in engineering calculations. They are derived from various equations, when several quantities can be grouped together in a dimensionless fraction, which has a quantitative meaning of the process described (Chapter 2). Consistent SI units should be used for each quantity, resulting in a dimensionless number.

A widely used dimensionless number in fluid flow, and heat and mass transfer is the Reynolds number (Re), which for circular pipes is defined as $Re = (d\, u\, \rho)/\eta$, where d (m) is the pipe diameter, u (m/s) is the linear fluid velocity, ρ (kg/m^3) is the fluid density, and η (kg m/s) is the fluid viscosity [1 kg/(m s)] = 1 Pa s.

Table 1.4 shows commonly used dimensionless numbers in food engineering.

1.5.3 Notation of Numbers

A consistent notation of numbers, based on the SI and US systems, is used in this book. The point (.) is used in this book for decimal numbers, while the comma (,) is used in the European Union (EU). The figures of a number are grouped in three, followed by a comma (,) or an empty space (SI system), while the point (.) is used in the EU. For example, the number 1,234,567.55 (US) may be written as 1.234.567,55 (EU), or 1 234 567.550 (SI system). Numbers 1000 to 10,000 may be written without grouping of figures, e.g., 2011 and 9999.990. The writing of large numbers is simplified by using the multiplier prefixes of SI units, e.g., 100,000 t = 100 kt, 1,000,000 $ = 1 M$, 0.000 001 m = 1 μm.

In process engineering calculations, a precision of two or three figures is considered satisfactory. Several decimal figures, obtained in electronic calculations, are not

TABLE 1.4
Common Dimensionless Numbers

Dimensionless Number	Definition	Applications
Reynolds	$Re = (du\rho)/\eta$	Fluid flow
Biot	$Bi = (hL)/\lambda$	Heat transfer
Biot	$Bi = (k_cL)/D$	Mass transfer
Fourier	$F_o = (\alpha t)/L^2$	Heat transfer
Fourier	$F_o = (Dt)/L^2$	Mass transfer
Nusselt	$Nu = (hd)/\lambda$	Heat transfer
Prandtl	$Pr = (C_p\eta)/\lambda$	Heat transfer
Schmidt	$Sc = \eta/(\rho D)$	Mass transfer
Sherwood	$Sh = (k_cd)/D$	Mass transfer

an indication of precision, since errors are multiplied during calculation. The excessive number of figures is reduced by rounding off. For example, the number 3.238762 should be rounded off to 3.239 or 3.24.

1.5.4 Engineering Calculations

Classic engineering calculations, based on elementary calculus, are used throughout this book. The step-by-step procedure has the advantage of checking all calculations easily and making the necessary corrections or changes. It also gives an engineering picture to the size of the process and equipment under consideration.

Computer programs can be used in complicated calculations, provided that they fit the problem in question (Hartel et al, 2008; Sun, 2007). Simplified computer programs, such as Microsoft Excel® spreadsheets, are suitable for typical process engineering calculations (Chapter 16).

APPLICATION EXAMPLES

Example 1.1

Milk is heated in a heat exchanger at the rate of 10 GPM (gallons per minute) from 40°F to 170°F, using condensing steam at 15 pounds per square inch (psi) gauge pressure. Calculate the heat required in SI units.

Before starting any calculation, all units must be converted to the SI system (Appendix). In engineering calculations, volumetric units should always be converted to mass units using the appropriate density values.

The volume unit of (U.S.) gallon is equal to 3.785 L (liters) and the flowrate of 10 GPM corresponds to 10 × 3.785 × 60 = 2271 L/h = 2.27 m³/h. The density (ρ) of the milk at 20°C is 1030 kg/m³, and the mass flowrate will be 2.27 × 1030 = 2338 kg/h = 2.34 t/h.

In SI units, the temperature is expressed in degrees Celsius (°C) or Kelvin (K = °C + 273). It should be noted that the temperature difference in °C and K is the same. The English units of temperature (°F) are converted to SI units by the equation

°F = 1.8 (°C) + 32, while Δ°F = 1.8 (Δ°C). In this example, 40°F = 4.4°C and 170°F = 76.7°C. The heat absorbed by the heating of milk will be

$$Q = m\, C_p\, \Delta T = 2238 \times 3.8 \times (76.7 - 4.4) = 614.87 \text{ MJ/h}$$

$$= 0.171 \text{ MJ/s} = 171 \text{ kJ/s} = 171 \text{ kW}$$

The amount of steam required for heating the milk will be $m = Q/\Delta H$, where the heat of evaporation of water (ΔH) is taken from a steam table (Appendix) at pressure 15 psi. Usually the pressure is given as gauge pressure, i.e., above atmospheric pressure. From the Appendix, the gauge pressure 15 psi is equal to 15 × 0.0689 = 1.03 bar. The absolute pressure of the saturated steam will be 1.03 + 1.013 = 2.043 bar or 2.043×10^5 Pa.

From the steam table, the corresponding saturation temperature is 121°C and the heat of evaporation is $\Delta H = 2.2$ MJ/kg. Thus, the mass of the required saturated steam will be $m = 614.87/2.2 = 279$ kg/h.

Example 1.2

Orange juice (OJ) 12% TS is concentrated in a continuous evaporation system to OJC at 65% TS. A four-effect falling film evaporator is used, operated with a parallel flow of heating medium (steam) and product. Saturated steam at 1.98 bar pressure is used for heating the first effect. The water vapors from the first effect are used to heat the second effect and so on, while the vapors from the last (fourth) effect are condensed in a surface heat exchanger at a saturation pressure of 0.312 bar.

Calculate the amount of OJC produced, and the required steam and cooling water for OJ feed rate of 1 t/h.

For simplification, the boiling point elevation of the OJ can be neglected, and so the water vapors are condensed at the saturation temperature of each effect.

The steam and water vapor condensates are removed at the corresponding saturation pressure (their boiling points).

Solution

From the steam table (Appendix) saturation $T = 120$°C for $P = 1.98$ bar, and $T = 70$°C for $P = 0.312$ bar. The pressure drop between two successive effects is assumed to be the same, i.e., (1.98 − 0.312)/4 = 0.42 bar. Thus, the saturation pressure (P) and corresponding temperature (T) in each effect will be as follows.

First effect: $P = 1.57$ bar, $T = 113$°C; second effect: $P = 1.15$ bar, $T = 104$°C; third effect: $P = 0.732$ bar, $T = 91$°C; fourth effect: $P = 0.312$ bar, $T = 70$°C.

The heat of evaporation (and condensation) of water in the temperature range of 70°C–120°C varies from 2.3 to 2.2 MJ/kg, and a mean value of 2.25 MJ/kg can be assumed for overall calculations.

In evaporation, energy balance calculations are usually confined to heating and cooling requirements. The mechanical (electrical) requirements are relatively small (less than 10% of the total) and can be neglected.

The material and heat balances of the entire system, based on 100 kg of feed (OJ), are shown in Figure 1.3. The overall material balance equation yields the amount (y) of product (OJC) 65% TS: 0.12 × 100 = (0.65)y, from which $y = 18.46$ kg product.

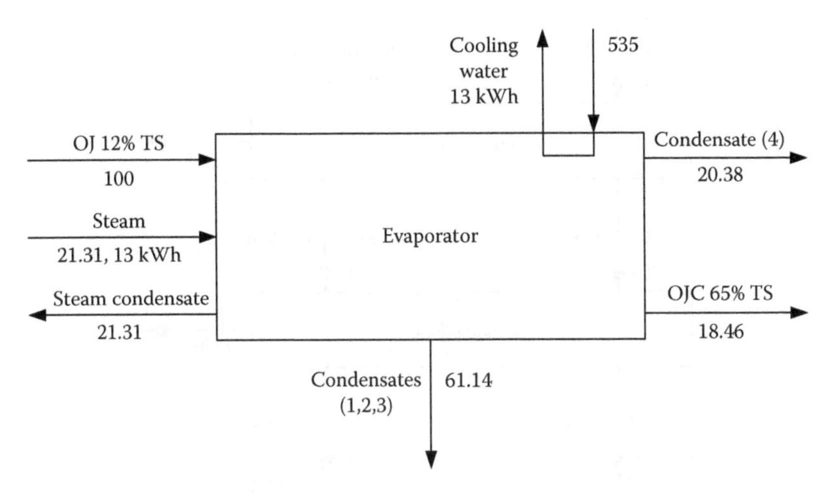

FIGURE 1.3 Overall material and energy balances of the orange juice evaporator (OJ, orange juice; OJC, orange juice concentrate).

Therefore, the amount of water vapors produced in the evaporation system will be 100 − 18.46 = 81.54 kg.

Assuming equal evaporative capacity, each effect will evaporate 81.54/4 = 20.38 kg water. The amount of steam (y) needed in the first effect can be calculated from the heat balance equation:

$$(2.2)y = 2.3 \times 20.38, \text{ from which } y = 21.31 \text{ kg steam/100 kg OJ.}$$

Condensation of the water vapors from the fourth effect at the saturation temperature of 70°C will require the removal of 20.38 × 2.2 = 44.74 MJ of heat. Assuming that cooling water is available at 20°C which can be heated up to 40°C before it is discharged into the environment, its quantity (y) will be (40°C − 20°C) × 4.18y = 44.74 × 1000, from which y = 535 kg water/100 kg OJ, where the specific heat of water is taken as 4.18 kJ/(kg K).

It must be noted that the condensate from the last effect is usually cooled to near ambient temperature, requiring an additional amount of water.

Figure 1.4 shows the material and energy balances within the evaporation system of the OJ.

Material and energy balances around each effect will result in the flows of juice, steam, water vapors, and condensate. The following assumptions simplify the solution of the underlying equations, resulting in acceptable approximate data: negligible heat losses to the environment by radiation and air convection; negligible boiling point elevation of the juice, which may be the case in the first three effects but not in the fourth effect of 65% TS concentrated juice; constant heat of evaporation of water in the temperature range of 70°C–100°C, taken as approximately equal to 2.25 MJ/kg; and removal of condensates at the saturation temperature in each effect, without further cooling (which would increase the efficiency of the system). Material balances resulted in the following flows of OJ and TS concentrations out of each effect: first effect, 79.62 kg 15.1% TS; second effect, 59.24 kg 20.3% TS; third effect, 38.38 kg 31.3% TS; fourth effect, 18.46 kg 65% TS.

The material and heat flows shown in Figure 1.4 were calculated for a feed of 100 kg of OJ 12% TS. For a feed rate of 1 t/h of OJ 12% TS the following flows can

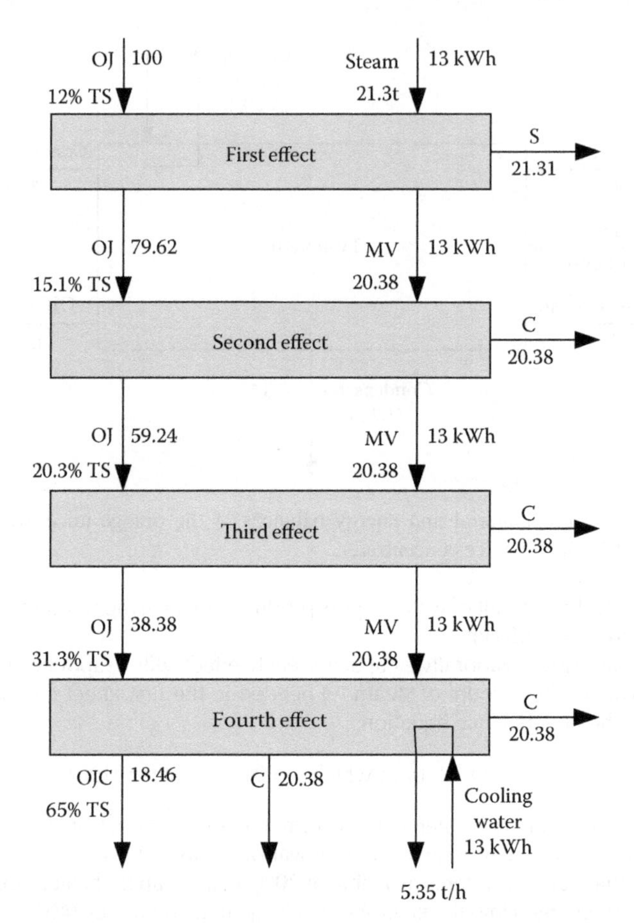

FIGURE 1.4 Material and energy balances in the orange juice evaporator (OJ, orange juice; OJC, orange juice concentrate; S, steam condensate; C, water vapor condensate).

be calculated easily: OJC 184.6 kg; steam 213 kg/h, or 130 kW; cooling water for condensing the water vapors 5.35 t/h; total water evaporated in the four effects 815.4 kg/h, approximately equal to the total steam condensates.

Example 1.3

Figure 1.5 shows the material and energy balances of the simplified citrus processing plant of Figures 1.1 and 1.2. It should be noted that citrus processing plants often include a peel oil extraction unit, and an essence recovery section, which recovers the citrus aroma components by distillation.

The material and energy balances are based on 100 kg of raw material (oranges). In some applications, the basis of calculation is 1 kg of main product.

Material balances are performed on each unit of the PBD. Simple equations are written based on the balance of entering and exiting materials. The basis of calculation is the mass faction (x) of TS of the material. The raw material (oranges) is assumed to contain 13.6% TS or $x = 0.136$.

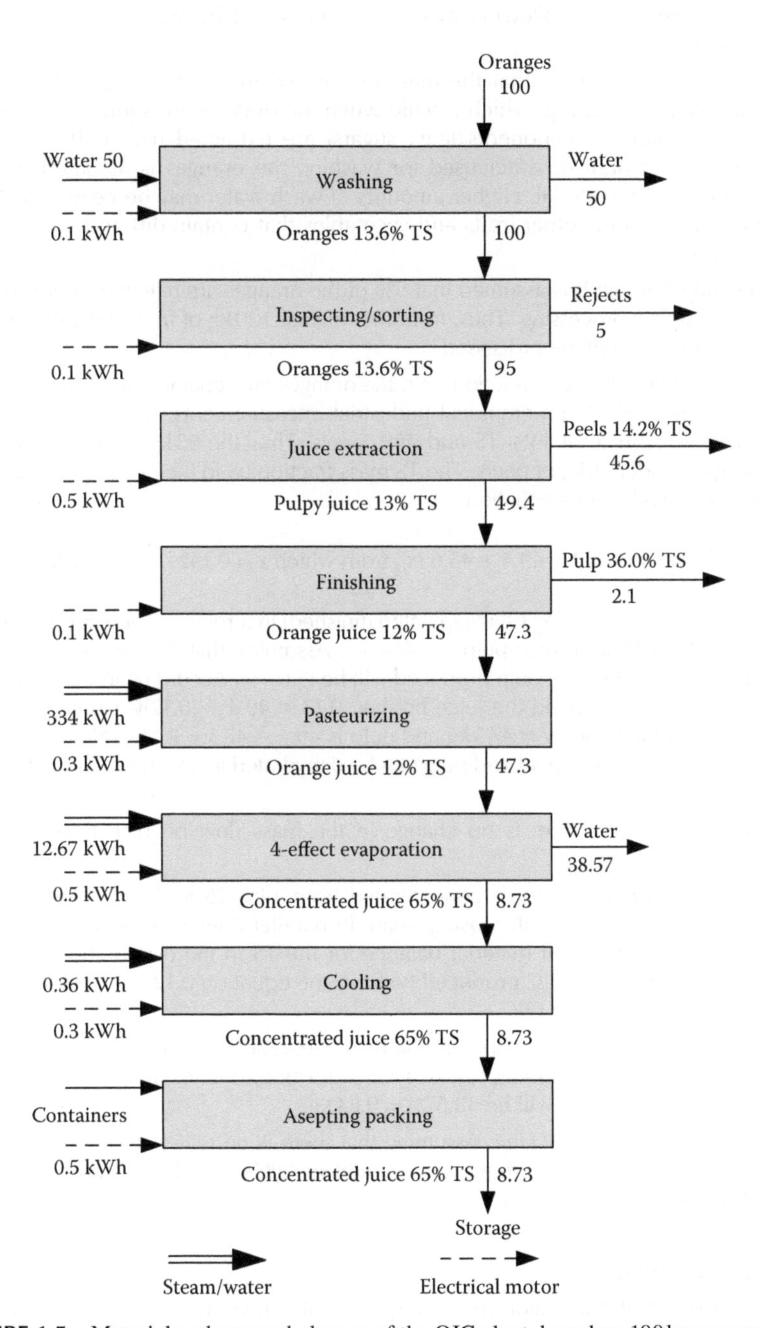

FIGURE 1.5 Material and energy balances of the OJC plant, based on 100 kg oranges.

Material Balance

From Figure 1.1 the following material balances can be written around each operation.

Washing: It is assumed that the mass of the raw material (oranges) does not change during washing, which is valid when the oranges are sound, solid, and no water-soluble components (e.g., sugars) are extracted (leached) into the water. The amount of water used for washing the oranges is assumed to be 50% of the raw material. Higher amounts of wash water may be needed in the processing of some other fruits and vegetables that contain dirt and unwanted materials.

Inspecting/Sorting: It is assumed that 5% of the oranges are rejected as non-suitable for further processing. Thus, from the original 100 kg of fruit, 95 kg of sound washed oranges will be processed.

Juice Extraction: In the extraction unit, the oranges are separated into pulpy juice and orange peels. From empirical (industrial) measurements, oranges 13.6% TS produce 52% juice of 13% TS and 48% peels. Thus the 95 kg of oranges yield 49.4 kg OJ and 45.6 kg of peels. The TS mass fraction (x) in the peels is calculated from the mass balance equation

$$0.136 \times 95 = 0.13 \times 49.4 + 45.6\,(x), \text{ from which } x = 0.142 \quad \text{or} \quad 14.2\%\ \text{TS}$$

Juice Finishing: The pulpy OJ is separated (finished) in a rotary screen into finished juice of 12% TS and juice pulp of 36% TS. Assuming that the amount of juice produced is (y) then the pulp produced will be ($52 - y$). According to the material balance equation around the juice finisher, $0.13 \times 49.4 = (0.12)y + 0.36 \times (49.4 - y)$, from which juice $y = 47.3$ kg and pulp is $49.4 - 47.3 = 2.1$ kg. Note that the significant amounts of peel and pulp can be dehydrated to a valuable animal feed by-product.

Juice Pasteurizing: There is no change in the mass flow of juice through the pasteurizer.

Juice Evaporation: The OJ is concentrated from 12% TS to 65% TS in a four-effect falling film evaporator, using steam in parallel flow as a heating medium (Figure 1.2). The overall material balance for the TS in the evaporation system yields the amount of OJC produced (y) from the equation $0.12 \times 47.3 = (0.65)y$, from which $y = 8.73$ kg OJC.

The total water vapor evaporated in the evaporation system will be $y = 47.30 - 8.73 = 38.57$ kg. Assuming similar evaporator bodies (effects), the evaporative capacity of each effect will be $38.57/4 = 9.64$ kg.

Cooling and Aseptic Packing: Assuming that there is no material loss during the cooling and packing operations, the amount of OJC packed and stored will be 8.73 kg/100 kg raw material (oranges).

Energy Balances

In the OJC plant, the major energy requirement is heat, which is used for heating and cooling operations and for evaporation of water in concentrating the OJ. Relatively smaller amounts of electrical energy are required for the motors, process control, and illumination.

Energy balances require the material flow data obtained in the material balance calculations, and thermophysical property data for the examined system. Some typical property data are given in Chapter 5.

The energy changes in the PBD of Figure 1.5 are expressed in kWh/100 kg raw material. However, in various engineering and economic calculations, the energy changes are expressed in kWh/kg or MJ/kg of product (1 kWh = 3.6 MJ).

Washing: It is assumed that washing of oranges is carried out at ambient (room) temperature and, therefore, no heating or cooling is required. The only energy needed will be for the motors that pump the water in and out of the washer, which is assumed to be 0.1 kWh/100 kg oranges.

Inspecting/Sorting: The motor of the inspection belt is assumed to require 0.1 kWh/100 kg oranges.

Juice Extraction: The juice extractor operates at ambient (room) temperature and, therefore, there is no energy change in the system. The mechanical energy required for the extraction (expression) of the juice from the oranges is assumed to be 0.5 kWh/100 kg raw material.

Juice Finishing: The mechanical finisher (screen) is assumed to require 0.3 kWh/100 kg oranges.

Juice Pasteurization: The OJ is heated from room temperature (assume 20°C) to the pasteurizing temperature (85°C for 5 s) and it is introduced into the evaporator without cooling. The mean specific heat of the OJ (12% TS) in this temperature range is approximately 3.8 kJ/kg K (Chapter 5). The thermal energy needed will be

$47.3 \times 3.8 \times (85 - 20) = 11.7$ MJ or $11.7/3.6 = 3.25$ kWh per 100 kg raw material

In addition to the heat energy, the mechanical motor requires approximately 0.3 kWh/100 kg oranges.

Juice Evaporation: The OJ is concentrated in a four-effect falling film evaporator from 12% TS to 65% TS, using saturated steam as a heating medium of the first effect. The other effects are heated by the water vapors of the previous effect. As estimated in the material balance calculations, each evaporator effect produces 9.64 kg water vapor/100 kg oranges.

Detailed analysis of the four-effect evaporator involves material and energy balances around each effect, as well as around the entire evaporation system. A simplified energy balance can be performed by considering the first and fourth effects of the evaporation system. External heat supply is needed only for the first effect to evaporate the 9.64 kg of water, while cooling water is needed to condense the 9.64 kg of water vapors of the fourth effect. The intermediate effects are heated by the vapors produced in the previous effects.

Assuming that the boiling point in the first effect is 100°C and neglecting the boiling point elevation, the heat of evaporation of water is 2.3 MJ/kg (Appendix), and the heat required will be $9.64 \times 2.3 = 22.25$ MJ or $22.2/3.6 = 6.17$ kWh/100 kg oranges.

The water vapors (9.64 kg) of the last effect will condense at 70°C, releasing heat energy of $9.64 \times 2.33 = 22.25$ MJ or $22.25/3.6 = 6.17$ kWh/100 kg oranges. Assuming that the condensate (specific heat of water $C_p = 4.18$ kJ/kg K) is cooled down to 40°C, the total heat removed from the last effect will be

$9.64 \times 4.18 \times (70 - 40)/1000 + 22.25 = 23.45$ MJ or 6.5 kWh/100 kg oranges

The mechanical energy of the evaporator motors is 0.5 kWh/100 kg oranges.

Cooling of Orange Juice Concentrate: The OJC is cooled in a heat exchanger from 70°C to 10°C. The specific heat of OJC is 2.5 kJ/kg K. The amount of heat removed per 100 kg oranges will be

$$8.73 \times 2.5 \times (70 - 10)/1000 = 131\ \text{MJ} \quad \text{or} \quad 1.31/36 = 0.36\ \text{kWh}$$

In addition, the mechanical energy of the motor is 0.3 kWh/100 kg oranges.

Aseptic Packing: The mechanical energy of the packing machine is 0.5 kWh/100 kg oranges.

The packed OJC (8.73 kg) is stored at 0°C.

Total Energy Requirements

Heat requirements $Q(\text{h}) = 3.34 + 6.17 + 6.50 + 0.36 = 16.37\ \text{kWh}$

Mechanical energy $Q(\text{m}) = 0.1 + 0.1 + 0.5 + 0.1 + 0.3 + 0.5 + 0.3 + 0.5 = 2.4\ \text{kWh}$

Total energy $Q = Q(\text{h}) + Q(\text{m}) = 16.37 + 2.4 = 18.77\ \text{kWh/100 kg oranges}$

The energy requirements are often expressed for 1 kg of product. In this application example, we have

$$Q = 18.77/8.73 = 2.15\ \text{kWh/kg} \quad \text{or} \quad 2.15 \times 3.6 = 7.74\ \text{MJ/kg product}$$

It is noted that heat amounts to 16.37 × (100/18.77) = 87% of the total energy requirements.

Actual Energy Requirements

The calculated energy requirements refer to theoretical values based on thermodynamic principles. The actual energy requirements of any industrial process can be estimated from the theoretical values if the efficiency of each process involved is known. For preliminary estimations, a mean energy efficiency of 75% can be used. Thus, in the present example, the estimated total energy requirement is 2.15/0.75 = 2.87 kWh or 10.33 MJ/kg product.

PROBLEMS

1.1 List the units used for measuring steam pressure in food engineering. Convert all units to the SI pressure unit.

1.2 Convert to the SI pressure unit (Pa) the units of vacuum used in food engineering: mm Hg (Torr), mbar, psi (lb/sqin).
Estimate the saturation temperature of water for typical vacuum operations: (a) vacuum drying at 10 Torr; (b) vacuum evaporation at 100, 300, and 600 Torr.

1.3 Convert to the basic SI units of mass flow (kg/s) the following flowrates: GPM (U.S. gallons per minute) of water, CFM (cubic feet per minute) of air at 20°C, lb/h, t/h. Use density data from Chapter 5 and the Appendix.

1.4 Calculate the material balances in a simplified PBD of canning peach halves, based on 1 t/h of raw peaches. Washing with 50% water; inspection/sorting, 5% rejects; halving/pitting, 15% pits; peeling, 5% skins; filling/syruping, 25% syrup, can content 0.9 kg; closing cans; steam sterilization; water cooling; labeling; casing, 24 cans per case.

1.5 Calculate the material and heat balances in a simplified PBD of potato granule dehydration, based on 1 t/h of raw material. Do not consider mechanical energy balances. The potatoes, containing 20% TS (total solids), are washed with water, steam-peeled (10% waste), diced, and blanched in steam. The blanched potato is dehydrated to 3% moisture content in an air dryer, operating with hot air, which can remove 0.05 kg moisture/kg air. The dehydrated product is bulk-packed.

1.6 Calculate the material and energy balances in a simplified block diagram of dried skimmed milk production, based on 1 t/h of milk. Consider both heat and mechanical energy balances. The milk is assumed to contain 9% non-fat TS and 3.0% fat. The milk fat is removed in a centrifuge and pasteurized using the high temperature/short time (HTST) process (77°C for 15 s). It is subsequently concentrated to 36% TS in a three-effect falling film evaporator, and the milk concentrate is spray-dried to a final 3% moisture content. The dried milk is bulk-packed in moisture-proof containers.

The pasteurizer is heated by pressurized hot water, which is heated from 110°C to 120°C by saturated steam at 140°C. The first effect of the evaporator is heated with saturated steam at 2 bar absolute pressure, while the third effect is operated at 60°C. The water vapors of the last effect are condensed with cold water and removed at 60°C. Assume equal evaporation per effect.

The spray dryer dehydrates the milk to 3% moisture, and operates with hot air, removing 0.2 kg moisture/kg air.

LIST OF SYMBOLS

A	m^2	Surface area
C_p	kJ/(kg K)	Specific heat
d	m	Diameter
D	m^2/s	Diffusivity
h	W/(m^2 K)	Heat transfer coefficient
k_c	m/s	Mass transfer coefficient
L	m	Length
m	kg	Mass
Q	kJ	Heat quantity
t	s	Time
T	°C or K	Temperature
TS	%	Total solids
u	m/s	Velocity

x	—	Mass fraction
α	m^2/s	Thermal diffusivity
ΔH	kJ	Enthalpy difference
ΔT	K or °C	Temperature difference
η	Pa s	Viscosity
λ	W/(m K)	Thermal conductivity
ρ	kg/m^3	Density

REFERENCES

Hartel, R.W., Connely, R.K., Hyslop, D.B., and Howell, T.A. Jr. 2008. *Math Concepts for Food Engineering*. CRC Press, New York.

Maroulis, Z.B. and Saravacos, G.D. 2003. *Food Process Design*. Marcel Dekker, New York.

Maroulis, Z.B. and Saravacos, G.D. 2007. *Food Plant Economics*. CRC Press, New York.

Rahman, S. 1996. *Food Properties Handbook*. CRC Press, Boca Raton, FL.

Rao, M.A., Rizvi, S.S.H., and Datta, A.K. 2006. *Engineering Properties of Foods*, 3rd edn. CRC Press, New York.

Saravacos, G.D. and Maroulis, Z.B. 2001. *Transport Properties of Foods*. Marcel Dekker, New York.

Saravacos, G.D. and Kostaropoulos, A.E. 2002. *Handbook of Food Processing Equipment*. Springer, New York.

Singh, R.P. and Heldman, D.B. 2001. *Introduction to Food Engineering*, 3rd edn. Elsevier, New York.

Sun, D.-W. ed., 2007. *Computational Fluid Dynamics in Food Processing*. CRC Press, New York.

2 Transport Phenomena

2.1 INTRODUCTION

Food process engineering is based mainly on transport phenomena, which were developed through chemical engineering and are applied widely in chemical process industries. Transport phenomena involve the transport of three basic engineering quantities, i.e., momentum (flow), heat, and mass (Bird et al, 1960).

The application of transport phenomena concepts to food and biological systems must consider the unique properties of the complex materials and the special requirements of quality and safety to the consumers. Process economics should also be considered in food processing, since food products are produced in large quantities and they should be available at affordable prices to the consumers (Gekas, 1992; Maroulis and Saravacos, 2003).

Transport phenomena are of fundamental importance in analyzing and designing physical unit operations and unit processes, which may also involve chemical and biochemical reactions. Most food processing operations are considered as unit processes, although they are controlled by physical unit operations. Chemical and biochemical reactions are very important to the quality and safety of food products (Barrer, 1955; Geankoplis, 1983).

Transport phenomena in gases and liquids have been modeled in mathematical terms, and they are applied widely in chemical engineering to predict and analyze various unit operations and unit processes of chemical process industries. Application of this approach to foods, which consist mostly of complex solid and semisolid materials, is more difficult, but it is very useful in understanding and controlling the food processes (Brodkey and Hershey, 1988).

Recent advances in process modeling and in food engineering properties are very useful. In food process engineering, more emphasis is placed on heat and mass transport than on fluid flow, which is considered as a supporting operation. Most food processes are controlled by heat and/or mass transfer operations.

In process engineering, transport phenomena and transport operations can take place either at steady-state or unsteady-state conditions. In steady-state operations, the driving potential for the transport remains constant, and a characteristic transport property (viscosity η, thermal conductivity λ, or mass diffusivity D) is needed to calculate the transport rate. In unsteady-state conditions, the driving potential changes with time, and complex mathematical analysis is needed to calculate the transport rate. Simplified semitheoretical or empirical models can be used to estimate the unsteady transport rate (Saravacos and Maroulis, 2001; Maroulis and Saravacos, 2003).

The term "transport" is a more general term than the term "transfer" that is used for transport between phases.

2.2 STEADY-STATE TRANSPORT

Transport of momentum (flow), heat, and mass at steady-state is analyzed readily using simple rate equations that contain the characteristic transport properties of viscosity η, thermal conductivity λ, and mass diffusivity D. The transport properties of foods are discussed in Chapters 5 and 6.

2.2.1 TRANSPORT RATE EQUATIONS

The transport rate of engineering operations is expressed by the generalized relation

$$\text{Rate} = \frac{\text{Driving potential}}{\text{Resistance}} \tag{2.1}$$

which, for one-dimensional (x-direction) transport becomes

$$\text{Rate} = -\alpha\left(\frac{\delta Y}{\delta x}\right) \tag{2.2}$$

where

Y is the transported quantity
α is the transport coefficient

The rate is the transported quantity per unit time, such as (J/s) = W in heat transfer, and (kg/s) in mass transfer. The transport flux is normally used in heat and mass transfer, defined as the transport rate per unit transfer area, (W/m^2) and [kg/(m^2 s)], respectively.

2.2.1.1 Heat Transport

The heat flux in the x-direction $(q/A)_x$ is given by the Fourier equation

$$\left(\frac{q}{A}\right)_x = -\lambda\left(\frac{\delta T}{\delta x}\right) \tag{2.3}$$

where

$\delta T/\delta x$ is the transport potential or temperature gradient in the x-direction
λ is the thermal conductivity with units W/(m K)

The Fourier equation 2.3 can be written in the form of the generalized transport equation as

$$\left(\frac{q}{A}\right)_x = -\alpha\frac{\delta(\rho C_p T)}{\delta x} \tag{2.4}$$

where

$\alpha(\lambda/\rho C_p)$, (m^2/s) is the thermal diffusivity
ρ (kg/m^3) is the density
C_p [kJ/(kg K)] is the specific heat of the material

2.2.1.2 Mass Transport

The mass flux in the x-direction $(J/A)_x$ is given by the Fick equation

$$\left(\frac{J}{A}\right)_x = -D\left(\frac{\delta C}{\delta x}\right) \tag{2.5}$$

where

$\delta C/\delta x$ is the transport potential or concentration gradient in the x-direction
D (m^2/s) is the mass diffusivity

In food engineering, the concentration of a material is expressed usually as (kg/m^3), while in chemical engineering, the units (kmol/m^3) are normally used.

2.2.1.3 Momentum Transport

The momentum transport in the x-direction is given by the Newton equation

$$\tau_{yx} = -\eta\left(\frac{\delta u_x}{\delta y}\right) \tag{2.6}$$

where

τ_{yx} (Pa) is the transport flux of momentum
$\tau_{yx} = (F_x/A)$, caused by the applied shear force F_x (N) in the x-direction
A (m^2) is the shear surface (wall) area

The driving potential or shear rate $\gamma = (\delta u_x/\delta y)$ is the gradient of velocity u_x in the y-direction of flow, and η (Pa s) is the transport coefficient viscosity.

The shear stress τ_{yx} is equivalent to the momentum flux, as shown by the units involved. Momentum is defined as the product of (mass) × (velocity) or (mu_x) with units [(kg m)/s] or (N s). Momentum flux will be $(mu_x)/(At)$ with units [(kg m)/s]/(m^2 s) = [(kg m)/s^2]/(m^2) with units of shear stress, $(mu_x)/(At) = (F_x/A) = \tau_{yx}$.

The Newton equation 2.6 can be written in the form of momentum flux as

$$\tau_{yx} = -\nu\left[\frac{\delta(u_x\rho)}{\delta y}\right] \tag{2.7}$$

where

$\nu = (\eta/\rho)$ (m^2/s) is the momentum diffusivity or kinematic viscosity
$[\delta(u_x\rho)/\delta y]$ is the momentum driving force
ρ (kg/m^3) is the density of the fluid

The quantity $(u_x\rho)$ is equivalent to the momentum per unit volume [(kg m/s)]/m^3.

The generalized transport equation describes mathematically, in an analogous form, the transport of the three quantities (momentum, heat, and mass). However, the physical mechanisms of transport are quite different, which is shown by the large differences in the transport coefficients. Thus, the transport properties of liquid water at 0°C are, momentum $\nu = 1.8 \times 10^{-6}$ m²/s, heat $\alpha = 1.4 \times 10^{-9}$ m²/s, and mass $D = 1.4 \times 10^{-10}$ m²/s (Saravacos and Maroulis, 2001).

2.3 UNSTEADY-STATE TRANSPORT

2.3.1 MATHEMATICAL MODELING

The transport of the important engineering quantities—heat, mass, and momentum—is caused by a difference in the potential of microscopic matter (molecules, small particles), and is expressed mathematically by a generalized diffusion equation (DE). Mathematical modeling and analysis is simplified by assuming that diffusion is an overall process, which includes transport by any other process, such as natural or forced convection.

In this chapter, transport of heat and mass is analyzed in more detail because it is more important in food process engineering than momentum transport.

Theoretically, diffusion takes place in three dimensions (x, y, z), but in engineering applications one-dimensional operations are considered, as expressed by Equation 2.8 (Crank, 1975).

$$\left(\frac{\delta Y}{\delta t}\right) = \alpha\left(\frac{\delta^2 Y}{\delta x^2}\right) \tag{2.8}$$

where Y is the quantity being transported, t (s) is the time, x (m) is the transport distance, and α (m²/s) is the (overall) transport property (diffusivity). For heat transport, Y (K) is the temperature and α (m²/s) is the thermal diffusivity. For mass transfer, Y (kg/m³ or –) is the mass fraction and $\alpha = D$ (m²/s) is the mass diffusivity.

In Equation 2.1, the partial derivative ($\delta Y/\delta t$) represents the accumulation of quantity Y (heat or mass) for a differential time (δt), while the second derivative ($\delta^2 Y/\delta x^2$) = $\delta(\delta Y/\delta x)/\delta x$ represents the driving potential of the transport operation. According to Thermodynamics (Chapter 3), the accumulation is positive (+) when the quantity Y enters the system, and negative (–) when Y is removed. In most engineering applications, heat or mass is transported (removed) from the system, meaning that the accumulation is negative $-(\delta Y/\mathrm{d}t)$.

Equation 2.1 expresses mathematically, the unsteady-state transport of species Y, since the driving potential $\delta(\delta Y/\delta x)/\delta x$ changes (is reduced) with increasing time. The solution of the unsteady-state diffusion equation means finding the values of Y at various distances x as a function of time t.

2.3.1.1 Analytical Solution

An exact analytical solution of the unsteady-state diffusion equation is difficult because of the complex variation of the driving potential over time. Approximate analytical solutions can be obtained, using some simplifications, which are acceptable for engineering calculations.

Reduced (dimensionless) quantities can simplify the solution of the diffusion equation 2.1. They are obtained by dividing the value of a quantity by its initial value. Thus, the reduced length x' is defined as $x' = (x/x_0)$, where x is a length $0 < x < x_0$, and x_0 is the initial length at time $t = 0$ (Gekas, 1992).

The reduced time is defined by the equation

$$t' = \frac{(\alpha t)}{b^2} \tag{2.9}$$

where

t is the diffusion time
b is the half-thickness of a slab or the radius of a sphere or infinite cylinder
α is the transport coefficient

Analytical solutions of the diffusion equation are available in the literature for the basic shapes of solid or semisolid materials, i.e., the slab, the sphere, and the infinite cylinder, when the surface resistance to transport is negligible.

The slab is a flat plate of thickness ($2b$), the other two dimensions being much larger than ($2b$). The sphere is characterized by its radius (b), and the infinite cylinder has a finite diameter ($2b$), which is much smaller than the (infinite) length. Diffusion in finite shapes, e.g., cube and finite cylinder, is analyzed readily by numerical or computer methods.

The reduced time is known as the Fourier number = $(\alpha t)/b^2$ in heat transport or as the Fick number = $(Dt)/b^2$ in mass transport.

The dimensionless (reduced) form of Equation 2.8 is

$$\left(\frac{\delta Y'}{\delta t'}\right) = \alpha\left(\frac{\delta^2 Y'}{\delta x'^2}\right) \tag{2.10}$$

where the reduced quantity Y' is defined as $Y' = (Y - Y_e)/(Y_0 - Y_e)$.

Y_0, Y, and Y_e are, respectively, the values of quantity Y at the beginning, after time t, and at equilibrium (infinite time).

The reduced diffusion equation 2.10 can be solved analytically by the separation of variables, provided that the initial (IC) and boundary (BC) conditions are known

$$\begin{array}{lll} \text{IC} & \text{for } t' = 0 & Y'(x') = 1 \\ \text{BC} & \text{for } t' > 0 & Y'(x' = +1 \text{ or } -1) = 0 \end{array} \tag{2.11}$$

The IC (2.11) means that at the beginning of the transport operation all points of the solid have the same initial quantity (concentration or temperature), $Y = Y_0$ ($Y' = 1$).

The BC (2.11) means that the surface quantity remains constant throughout the transport operation ($Y = Y_e$). This condition assumes that there is no surface resistance, a simplification for the mathematical solution of the transport equation. In practical applications, there is a surface resistance controlling the overall transport, which is considered in the numerical/computer and in generalized chart solutions.

Neglecting surface thermal resistance to transport, the solution for a slab, a sphere, or an infinite cylinder is an infinite (Fourier) series of exponential terms of the reduced time $t' = (\alpha t)/b^2$. For transport of long time ($t' > 0.1$), the solution can be approximated by the first term of the series, resulting in the following equations for heat transport:

$$\text{Slab,}\quad t = \left(\frac{1}{\alpha}\right)\left(\frac{2b}{\pi}\right)^2 \ln\left[\frac{\left(\frac{8}{\pi^2}\right)(T_0 - T_e)}{(T - T_e)}\right] \tag{2.12}$$

$$\text{Sphere,}\quad t = \left(\frac{1}{\alpha}\right)\left(\frac{b}{\pi}\right)^2 \ln\left[\frac{\left(\frac{6}{\pi^2}\right)(T_0 - T_e)}{(T - T_e)}\right] \tag{2.13}$$

$$\text{Cylinder,}\quad t = \left(\frac{1}{\alpha}\right)\left(\frac{b^2}{5.8}\right) \ln\left[\frac{0.69(T_0 - T_e)}{(T - T_e)}\right] \tag{2.14}$$

From these equations, the time t, required to reach a mean temperature T in a slab (half-thickness, b), a sphere (radius, b), or an infinite cylinder (radius, b), can be calculated if the initial (T_0) and the equilibrium (T_e) temperatures are known, together with the thermal diffusivity α.

The same Equations 2.12 through 2.14 can be used for mass transport, substituting the temperature T with the concentration C or mass fraction X, and the thermal diffusivity α with the mass diffusivity D.

For very short time of transport, the solution of the transport equation is approximated by the following error function:

$$\frac{(Y_0 - Y)}{(Y_0 - Y_e)} = 1 - \text{erf}\left[\frac{x}{(4\alpha t)^{1/2}}\right] \tag{2.15}$$

The error (or Gauss error) function erf(x) of x is defined by the integral,

$$\text{erf}(x) = 2/(\pi)^{1/2}\ \text{integral}(0, x)\exp(-t^2)\text{d}t \tag{2.16}$$

The complementary error function erfc(x) is defined as

$$\text{erfc}(x) = 1 - \text{erf}(x) \tag{2.17}$$

TABLE 2.1
Error Functions of Parameter (x)

X	erf(x)	erfc(x)
0.0	0.000	1.000
0.1	0.112	0.888
0.2	0.223	0.777
0.3	0.328	0.672
0.4	0.428	0.572
0.5	0.520	0.480
0.6	0.604	0.396
0.7	0.678	0.322
0.8	0.742	0.258
0.9	0.797	0.203
1.0	0.843	0.157
1.5	0.966	0.034
2.0	0.995	0.005
Infinity	1.000	0.000

The erf(x) is calculated using computer programs. Table 2.1 shows the typical values of erf(x) and erfc(x), of importance to transport phenomena.

The penetration depth B_p, defined as the depth within a solid, at which the transported quantity is reduced to 1% of its initial value, can be estimated from Equation 2.15

$$\frac{(Y_0 - Y)}{(Y_0 - Y_e)} = 0.01\,\mathrm{erf}\left[\frac{x}{(4\alpha t)^{1/2}}\right] = 0.99$$

From Table 2.1 erf(0.99) = 2, and

$$B_p = 4(\alpha t)^{1/2} \tag{2.18}$$

Thus, for short time transport, the penetration depth of mass (diffusion) or temperature (heat conduction) is a function of the square root of the diffusivity and time.

2.3.1.2 Generalized Chart Solution

Approximate solutions of the diffusion, useful in engineering calculations, can be obtained from the generalized charts of the literature. The simplified Gurney–Lurie charts, originally applied to heat transfer, can be used in applications of both heat and mass transfer, discussed in this book.

The charts are plots of the reduced quantity $Y' = (Y_0 - Y)/(Y_0 - Y_e)$ versus the reduced time t', which can be the Fourier number for heat transfer $t' = (\alpha t)/b^2$ or the Fick number for mass transfer $t' = (Dt)/b^2$. The surface resistance to transport m is expressed by the inverse of the Biot number Bi, which is defined as the ratio of the internal to surface resistance (Table 2.2).

TABLE 2.2
Biot Numbers in Heat and Mass Transfer

Transport Operation	Biot Number Bi	Transport Resistance $m = 1/Bi$
Heat transfer	$(hx)/\lambda$	$\lambda/(hx)$
Mass transfer	$(k_c x)/D$	$D/(k_c x)$

h [W/(m^2 K)] heat transfer coefficient; k_c (m/s) mass transfer coefficient, λ [W/(m K)] thermal conductivity, D (m^2/s) mass diffusivity; x (m) distance.

Figures 2.1 through 2.3 show simplified charts for slab, spheres, and cylinders. The charts show solutions at the center ($x' = 0$) and near the surface ($x' = 1.0$ or 0.8), where $x' = x/x_0$, and $x_0 = b$. Three resistance factors m are considered: 0, 0.5, and 1.0.

Figures 2.1 through 2.3 show that diffusion in spheres is faster than in cylinders and slabs of the some transport dimension (half-thickness or radius). Diffusion at the center of the solid ($x' = 0$) is slower than at its surface ($x' = 1$). Increasing the surface resistance (m) or decreasing the Biot (Bi) number will decrease the diffusion rate markedly.

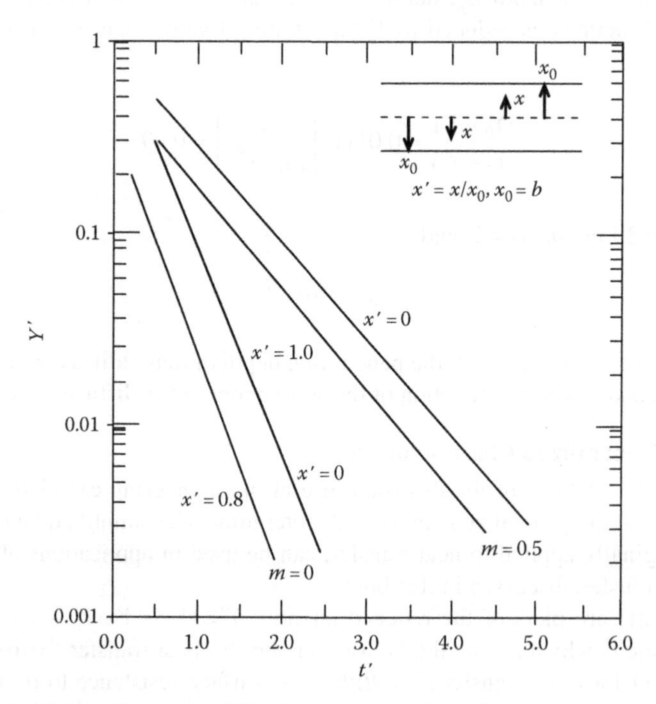

FIGURE 2.1 Reduced quantity (Y') versus reduced time (t') for slab; b, half-thickness; $m = 1/Bi$, surface resistance.

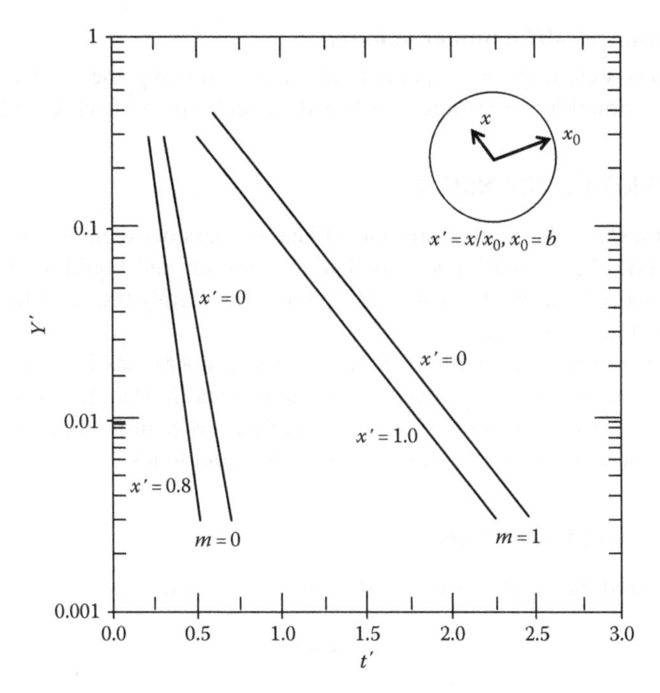

FIGURE 2.2 Reduced quantity (Y') versus reduced time (t') for sphere; b, radius; $m = 1/Bi$, surface resistance.

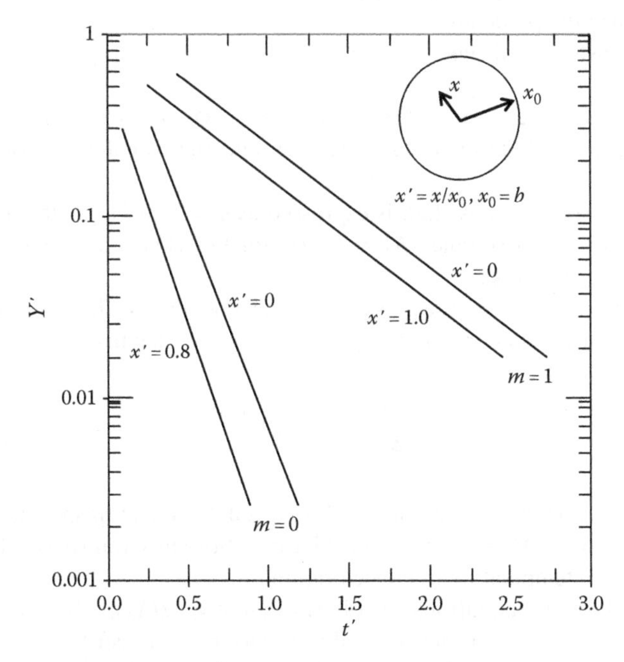

FIGURE 2.3 Reduced quantity (Y') versus reduced time (t') for infinite cylinder; b, radius; $m = 1/Bi$, surface resistance.

2.3.1.3 Numerical/Computer Solution

The unsteady-state diffusion equation can be solved using specialized computer programs or simplified procedures, such as the Excel® spreadsheet (Chapter 16).

2.4 INTERFACE TRANSFER

Interface transfer refers to the transport of heat or mass between any two surfaces, such as solid/solid, liquid/liquid, gas/liquid, gas/solid, and liquid/solid. Interface transfer includes interphase transfer, which concerns mainly transport between two phases (gas, liquid, or solid).

Interface transfer is essential in food processing operations involving heat and mass transport, such as drying, evaporation, and crystallization. It is also important in slow operations, such as moisture and volatile transfer in food products, and in absorption and desorption of gases and vapors in liquid foods.

2.4.1 TRANSFER COEFFICIENTS

The generalized transport equation at the interface is written as

$$\varphi = \kappa\, \Delta Y \tag{2.19}$$

where

- φ is the transfer flux (transfer rate per unit area)
- ΔY is the driving potential
- κ is the transfer coefficient

In heat transfer, the thermal flux is expressed as $\varphi = q$ (W/m^2), the driving force is the temperature difference $\Delta Y = \Delta T$ (K), and the transfer coefficient is $\kappa = h$ (W/m^2 K).

In mass transfer, the mass flux is expressed as $\varphi = J$ (kg/m^2 s), the driving force is the concentration difference $\Delta Y = \Delta C$ (kg/m^3) or (kg/kg db), and the transfer coefficient is $\kappa = k_c$ (m^2/s).

The transfer coefficient κ expresses the transport flux at the interface and it is related to the steady-state transport equation 2.2, which can be written as

$$\varphi = \alpha\left(\frac{\Delta Y}{\Delta x}\right) \quad \text{or} \quad \varphi = \left(\frac{\alpha}{\Delta x}\right)\Delta Y \tag{2.20}$$

where Δx is the thickness of the surface film through which transport takes place.

The driving force ΔY is taken as a positive quantity, since transport is from higher to lower Y value (temperature or concentration).

From the last two equations, it follows that $\kappa = (\alpha/\Delta x)$, where the transport coefficient is $\alpha = \lambda$ for heat transfer, and $\alpha = D$ for mass transfer.

Therefore, the heat and mass transfer coefficients (h, k_c) are given by the equations

$$h = \frac{\lambda}{\Delta x}, \quad k_c = \frac{D}{\Delta x} \tag{2.21}$$

A calculation of the transfer coefficients from these equations is difficult because the surface film thickness (Δx) cannot be measured or estimated precisely. For this reason, the transfer coefficients for a given transport system are measured experimentally or estimated from empirical correlations, as discussed in the next section.

2.4.2 Dimensionless Numbers

Empirical models, based on dimensionless numbers, are used in several food engineering applications involving transport of momentum, heat, and mass.

The dimensionless numbers are ratios of physical and transport quantities of the system, as defined in the following summary.

2.4.2.1 Momentum Transport (*Re, Fr, f*)

The Reynolds number *Re* is defined as the ratio (inertial forces)/(viscous forces). It is calculated from the relation $Re = (u\rho d)/\eta = (ud)/\nu$, where u (m/s) is the velocity, d (m) is the diameter of the flow duct, η (Pa s) is the viscosity, and ν (m^2/s) is the kinematic viscosity of the fluid.

The Froude number *Fr* is the ratio of (inertial forces)/(gravity forces). It is calculated from the relation $Fr = u^3/(dg)$, where g (9.81 m/s^2) is the acceleration of gravity.

The friction factor f is the ratio (wall shear stress)/(inertial forces). It is calculated from the relation $Fr = \tau/(\rho u^2)$, where τ (Pa) is the shear stress at the wall.

2.4.2.2 Heat Transfer (*Bi, Fo, Gr, Gz, Le, Nu, Pe, Pr, St*)

The Biot *Bi* number in heat transfer is defined as the ratio Bi = (heat transfer at the boundary)/(heat conduction within the solid). It is calculated from the relation $Bi = (hb)/\lambda$, where h [W/(m^2 K)] is the convective heat transfer coefficient, b (m) is the half-thickness or radius of the solid, and λ [W/(m K)] is the thermal conductivity of the solid.

The Fourier number *Fo* is the ratio of (heat conduction rate)/(heat storage rate). It is a characteristic dimensionless time in unsteady state transport. It is calculated from the relation $Fo = (\alpha t)/b^2$, where $\alpha = \lambda/(\rho C_p)$ (m^2) is the thermal diffusivity, t (s) is the transport time, b (m) is the transport distance, and C_p [kJ/(kg K)] is the specific heat of the solid.

The Grashof number *Gr* is the ratio of (buoyancy forces)/(viscous forces). It is calculated from the relation $Gr = (g\beta\Delta T d^3)/\nu^2$, where g (m/s^2) is the gravity constant, $\beta = (1/T)$ is the volumetric expansion coefficient, ΔT (K) is the temperature difference, d (m) is the characteristic dimension (diameter), and ν (m^2/s) is the kinematic viscosity of the fluid.

The Graetz number *Gz* characterizes heat transfer in the laminar flow. It is calculated from the relation $Gz = RePr(\pi d/4L) = [(u\rho d)/\eta]\,[(C_p\eta)/\lambda](d/L) = (\pi\, u\rho d^2 C_p)/(4\lambda L)$.

The Lewis number *Le* is the ratio of (thermal diffusivity)/(mass diffusivity). It is calculated from the relation $Le = (\alpha/D)$.

The Nusselt number Nu is the ratio of (convective heat transfer)/(conduction heat transfer). It is calculated from the relation $Nu = (hb)/\lambda$, where h [W/(m^2 K)] is the convective heat transfer coefficient, b (m) is a characteristic transport dimension (diameter), and λ (W/m K) is the thermal conductivity of the fluid.

The Peclet number Pe in heat transfer is the ratio of (convection transfer)/(diffusion transport) and it is calculated from the relation $Pe = (ud)/\alpha$.

The Prandtl number Pr is the ratio of (momentum diffusivity)/(heat diffusivity). It is calculated from the relation $Pr = (C_p\eta)/\lambda = (\nu/\alpha)$, where C_p [kJ/(kg K)] is the specific heat.

The Stanton number St is the ratio of (heat transferred into a fluid)/(thermal capacity of the fluid). It is calculated from the relation $St = h/(u\rho C_p)$, and it is related to the Re and Pr numbers by the relation $St = Nu/(RePr)$.

2.4.2.3 Mass Transfer (*Bi, Fi, Le, Pe, Sc, Sh*)

The Biot Bi number in mass transfer is defined as the ratio Bi = (mass transfer at the boundary)/(mass within the solid). It is calculated from the relation $Bi = (k_c b)/D$, where k_c (m/s) is the convective mass transfer coefficient, b (m) is the half-thickness or radius of the solid, and D (m^2/s) is the mass diffusivity in the solid.

The Fick number Fi is a characteristic dimensionless time in unsteady-state mass transport. It is calculated from the relation $Fi = (Dt)/b^2$, where D (m^2/s) is the mass diffusivity, t (s) is the transport time, and b (m) is the transport distance.

The Peclet Pe number in mass transfer is the ratio of (convection transfer)/(diffusion transport). It is calculated from the relation $Pe = (ud)/D$.

The Lewis number Le is the same in both heat and mass transfer. It is the ratio of (thermal diffusivity)/(mass diffusivity) and it is calculated from the relation $Le = (\alpha/D)$.

The Schmidt number Sc is the ratio of (momentum diffusivity)/(mass diffusivity). It is calculated from the relation $Sc = \nu/D$.

The Sherwood number Sh is the ratio of (convective mass transfer)/(diffusive mass transfer). It is calculated from the relation $Sh = (k_c b)/D$, where k_c (m/s) is the convective mass transfer coefficient, and b (m) is a characteristic length (diameter).

The dimensionless numbers are used in several empirical equations of momentum, heat, and mass transfer operations, as discussed in Chapters 6 through 13.

APPLICATION EXAMPLES

Example 2.1

A spherical fruit of diameter 4 cm and initial temperature 30°C is cooled in an air stream. The density of the fruit is 950 kg/m^3, the thermal conductivity is 0.520 W/(m K), and the specific heat is 3.5 kJ/(kg K).

a. Estimate the time required to reach a mean temperature of the fruit 5°C, assuming an equilibrium temperature 2°C and no surface resistance to heat transfer.
b. Repeat the estimation, assuming that there is a significant surface resistance to heat transfer with a Biot number $Bi = 1$.

Solution

a. This unsteady-state heat conduction problem can be solved by using either the approximate analytical solution of Equation 2.13 or the generalized chart for spheres (Figure 2.3). The thermal diffusivity of the fruit is $\alpha = \lambda/(\rho C_p) = 0.52/(950 \times 3500) = 1.56 \times 10^{-7}$ m^2/s. The reduced temperature of the spherical fruit is $Y' = (T - T_e)/(T_0 - T_e)$ or $Y' = (5 - 3)/(30 - 3) = 0.074$.
   The simplified equation for spheres 2.13 yields

$$t = (1/1.56) \times 10^7 \times (0.02/3.14)^2 \ln(0.6 \times 13.5)$$
$$= 0.64 \times 10^7 \times 4 \times 10^{-5} \times 2.1 = 538 \text{ s} = 9 \text{ min}$$

   The chart of Figure 2.3 (sphere) for $Y' = 0.074$ yields $t' = (\alpha t)/b^2 = 0.28$ or $t = (0.28b^2)/\alpha$ or $t = (0.28 \times 0.02^2)/(1.56 \times 10^{-7}) = 718$ s = 12 min.
   The cooling time for the food material is relatively short (about 10 min), due to the assumption of negligible surface resistance to thermal transport. The analytical equation yields a significantly less cooling time than the chart method, most likely due to the simplification of using only the first term of the Fourier series, which is acceptable only for long transport time.
b. The chart of Figure 2.3 (sphere) for $Y' = 0.074$, $m = 1$ and $x' = 0.5$ yields $t' = (\alpha t)/b^2 = 1.10$ or $t = (1.10\, b^2)/\alpha$ or $t = (1.10 \times 0.02^2)/(1.56 \times 10^{-7}) = 2800$ s = 47 min.
   The surface resistance to moisture transport sharply increases the cooling time from 12 to 47 min. The cooling rate can be increased by increasing the air flow rate, which increases the heat transfer coefficient h and the Biot number for heat transfer $[(h\, b)/\lambda]$.

Example 2.2

A spherical food product ($d = 2$ cm) is dried in an air stream, which is at a constant temperature of 60°C. The product has a moisture content of 75% wet basis at the beginning and 5% at equilibrium. Moisture diffusivity $D = 2 \times 10^{-9}$ m^2/s.

a. Estimate the time required to reach a mean moisture content of the product 10% (wet basis), assuming no surface resistance to mass transfer.
b. Repeat the estimation, assuming that there is a significant surface resistance to mass transfer with a Biot number $Bi = 1$.

Solution

The given moisture contents are converted to dry basis X (kg water/kg db) as follows. Initial $X_0 = 75/25 = 3.0$; after time $tX = 10/90 = 0.11$; at equilibrium $X_e = 5/95 = 0.053$. Reduced moisture content $Y' = (X - X_e)/(X_0 - X_e) = (0.11 - 0.053)/(3.0 - 0.053) = 0.019$.

a. The simplified equation 2.13 for spheres and negligible surface resistance yields

$$t = (1/2) \times 10^9 \times (0.01/3.14)^2 \ln(0.6 \times 52.6)$$
$$= 0.5 \times 10^9 \times 1 \times 10^{-5} \times 3.45 = 17{,}259 \text{ s} = 4.80 \text{ h}$$

The chart of Figure 2.3 (sphere) for $Y' = 0.019$, $m = 0$ and $x' = 0.5$ yields $t' = (\alpha t)/b^2 = 0.40$ or $t = (0.4\ b^2)/\alpha$ or $t = (0.4 \times 0.01^2)/(2 \times 10^{-9}) = 20{,}000\,\text{s} = 5.55\,\text{h}$.

The analytical equation yields significantly less drying time than the chart method (4.80 versus 5.55 h), most likely due to the use of only one term of the Fourier series.

b. The chart of Figure 2.3 (sphere) for $Y' = 0.019$, $m = 1$ and $x' = 0.5$ yields $t' = (\alpha t)/b^2 = 1.65$ or $t = (1.65b^2)/\alpha$ or $t = (1.65 \times 0.01^2)/(2 \times 10^{-9}) = 83{,}000\,\text{s} = 23\,\text{h}$.

The surface resistance to moisture transport sharply increases the drying time from 5.5 to 23 h. The drying rate can be increased by increasing the air flow rate or the temperature, which increases the heat and mass transfer coefficients and the Biot numbers for heat and mass transfer.

Example 2.3

A fruit is treated with gaseous sulfur dioxide (SO_2) for the purpose of preservation.

a. Estimate the penetration depth of sulfur dioxide in the fruit for 1 h exposure, if its diffusivity at the given temperature is $1 \times 10^{-8}\ \text{m}^2/\text{s}$.
b. Discuss the effect of surface resistance to mass transfer on the desorption of SO_2 from fruit juices and suggest two methods of effective industrial operation.

Solution

a. The penetration depth (B_p) is a measure of the unsteady-state transport rate of a compound into a material. At this distance, the concentration of the penetrant is reduced by 99%. It is estimated from the approximate solution of the diffusion equation 2.18

$$B_p = 4(\alpha t)^{1/2} \quad \text{where } \alpha = D = 1 \times 10^{-8}\,\text{m}^2/\text{s} \quad \text{and}$$

$$B_p = 4(1 \times 10^{-8} t)^{1/2} = 4 \times 10^{-4} t^{1/2}$$

For an exposure time of $t = 1\,\text{h} = 3600\,\text{s}$, we obtain $B_p = 0.024\,\text{m}$ or $B_p = 24\,\text{mm}$.

b. Desulfuring (removal of SO_2) of fruit juices and wine can be accomplished by two methods: (1) vacuum treatment of sprayed juices, which sharply reduces the surface resistance to mass transfer, and (2) blowing air in a spray column (high Biot numbers).

Example 2.4

Water at 20°C flows at the rate of 1 kg/s in a tube of 20 mm internal diameter and 2.5 mm wall thickness.

Estimate the dimensionless numbers *Re*, *Nu*, *Pr*, *Gz*, *Pe*, *Le*, and *St* of the system.

Engineering properties of water: density $\rho = 1000\,kg/m^3$, viscosity $\eta = 1\,mPa\ s$, specific heat $C_p = 4.2\,kJ/(kg\ K)$, thermal conductivity $\lambda = 0.62\,W/(m\ K)$, internal heat transfer coefficient $h_i = 1\,kW/(m^2\ K)$.

Solution

Velocity (*x*-direction of water in the tube, $u = (1/A)(m/\rho)$, where $A = 3.14 \times (0.02)^2/4 = 3.14 \times 10^{-4}\ m^2$, $m = 1\,kg/s$, and $\rho = 1000\,kg/m^3$. Therefore, $u = (1/3.14) \times 10^4 \times 1/1000 = 3.18\,m/s$. Kinematic viscosity of water $\nu = \eta/\rho = 1 \times 10^{-6}\ m^2/s$. Self mass diffusivity of water $D = 1.1 \times 10^{-9}\ m^2/s$. Thermal diffusivity of water $\alpha = \lambda/(\rho C_p) = 0.62/(1000 \times 4200) = 1.48 \times 10^{-7}\ m^2/s$.

Reynolds number $Re = (u\rho d)/\eta = (ud)/\nu = (2 \times 0.02) \times 1 \times 10^6 = 40{,}000$.

Nusselt number $Nu = (hd)/\lambda = (2{,}500 \times 0.02)/0.62 = 80.6$.

Prandtl number $Pr = (C_p\eta)/\lambda = (4{,}200 \times 1 \times 10^{-3})/0.62 = 6.8$.

Graetz number $Gz = RePr(\pi d/4L) = (\pi u\rho d^2 C_p)/(4\lambda L)$. The Gz number is used in the laminar flow ($Re < 2{,}100$), and it is not applicable here ($Re = 40{,}000$, turbulent flow).

Peclet number $Pe = (ud)/\alpha = (2 \times 0.02 \times 10^7)/1.48 = 2.7 \times 10^5$.

The Lewis number $Le = (\nu/\alpha) = 1 \times 10^{-6}/(1.1 \times 10^7) = 910$.

The Stanton number $St = h/(u\rho C_p) = 2{,}500/(2 \times 1{,}000 \times 4{,}200) = 3.0 \times 10^{-4}$. It can also be calculated from the relation $St = Nu/(RePr) = 80.6/(40{,}000 \times 6.8) = 3.0 \times 10^{-4}$.

Example 2.5

Atmospheric air at 20°C flows at a velocity 50 m/s in a tube of 30 cm internal diameter.

Estimate the dimensionless numbers *Re, Nu, Pe, Gz, Gr, Pr, Le,* and *St* of the system. Engineering properties of air: density $\rho = 1.2\,kg/m^3$, viscosity $\eta = 0.02\,mPa\ s$, specific heat $C_p = 1.1\,kJ/(kg\ K)$, thermal conductivity $\lambda = 0.03\,W/(m\ K)$, internal heat transfer coefficient $h_i = 0.1\,kW/(m^2\ K)$.

Solution

Kinematic viscosity of air $\nu = 0.02 \times 10^{-3}/1.2 = 1.7 \times 10^{-5}\ m^2/s$. Self mass diffusivity of air $D = 1.1 \times 10^{-9}\ m^2/s$. Thermal diffusivity of air $\alpha = \lambda/(\rho C_p) = 0.03/(1.2 \times 1{,}100) = 2.3 \times 10^{-5}\ m^2/s$. Reynolds number $Re = (50 \times 0.3 \times 1.2 \times 10^3)/0.02 = 900{,}000$.

Nusselt number $Nu = (hd)/\lambda = (100 \times 0.3)/0.03 = 1000$.

Prandtl number $Pr = (C_p\eta)/\lambda = (1100 \times 0.02 \times 10^{-3})/0.03 = 0.733$.

Graetz number $Gz = Re\ Pr\ (\pi d/4L) = (\pi u\rho d^2 C_p)/(4\lambda L)$. The Gz number is used in the laminar flow ($Re < 2{,}100$), and it is not applicable here ($Re = 652{,}000$, turbulent flow).

Peclet number $Pe = (ud)/\alpha = (50 \times 0.3 \times 10^5)/2.3 = 6.52 \times 10^5$.

Lewis number $Le = (\nu/\alpha) = 1.7 \times 10^{-5}/(2.3 \times 10v^5) = 0.78$.

The Stanton number $St = h/(u\rho C_p) = 100/(50 \times 1.2 \times 1{,}100) = 1.5 \times 10^{-3}$. It can also be calculated from the relation $St = Nu/(Re\ Pr) = 1{,}000/(900{,}000 \times 0.733) = 1.5 \times 10^{-3}$.

PROBLEMS

2.1 A meat slab of thickness $2b = 15$ cm and initial temperature 30°C is cooled at both flat sides to 5°C in an air stream, which is maintained at 0°C. Calculate the center and mean temperatures of the meat, assuming negligible thermal resistance. Assume an equilibrium temperature of 2°C. Density of beef $\rho = 1050\,\text{kg/m}^3$, thermal conductivity $\lambda = 0.50\,\text{W/(m K)}$, and specific heat $C_p = 3.5\,\text{kJ/(kg K)}$.

2.2 A sausage of 15 mm diameter and 15 cm long and initial temperature 30°C is heated in a retort (sterilizer) with steam at 120°C.

(a) Calculate the time required for the center temperature to reach 100°C, assuming negligible thermal resistance.

(b) If the diameter of the sausage is increased to 20 mm, what will be the required time to reach 100°C?

Consider the sausage an infinite cylinder. Density $\rho = 950\,\text{kg/m}^3$, thermal conductivity $\lambda = 0.48\,\text{W/(m K)}$, and specific heat $C_p = 3.7\,\text{kJ/(kg K)}$.

2.3 Show that:

(a) the Prandtl number $Pr = (C_p\eta)/\lambda$ can be written as $Pr = \nu/\alpha$, where ν is the kinematic viscosity and α is the thermal diffusivity.

(b) The Stanton number St, defined as $St = h/(u\rho C_p)$, can be written as $St = Nu/(RePr)$, where Nu, Re, and Pr are the Nusselt, Reynolds, and Prandtl numbers, respectively.

2.4 A spherical fish piece 2.5 cm diameter is salted in a brine solution containing 20% salt db (dry basis). Estimate the salt concentration in the center of the piece after 10 h. The Biot number is 1, the equilibrium concentration is 15%, and the salt diffusivity is $D = 5 \times 10^{-10}\,\text{m}^2/\text{s}$.

2.5 A fluid of density ρ flows at a velocity u in a tube of internal diameter d and length L with a pressure drop of ΔP.

(a) Show that the momentum transfer flux or shear stress τ_{yx} of the Newton equation is given by the relation $\tau_{yx} = d\,\Delta P/(4L)$. Note that $\tau_{yx} = F/A_w$, where $F = A_0\Delta P$ is the force exerted in the tube, A_0 is the cross-sectional area, and A_w is the internal wall surface of the tube. The flow is in the x-direction while the shear is in the y-direction.

(b) Calculate the driving potential or shear rate γ of flow in a tubular system of 4 cm diameter and 10 m length, if the fluid has a viscosity of 50 mPa s and it develops a pressure drop of 0.2 bar. Compare with the shear rate calculated from the relation $\gamma = (8u/d)$.

LIST OF SYMBOLS

A	m^2	Transport area
A_0	m^2	Cross-sectional area
A_w	m^2	Internal wall surface
b	m	Characteristic dimension (half-thickness, radius)
Bi	—	Biot number, heat transfer $(h\,b)/\lambda$, mass transfer $(k_c b)/D$

B_p	m	Penetration depth
C	kg/m^3	Concentration
C_p	kJ/(kg K)	Specific heat
d	m	Characteristic dimension (diameter)
d	m^2/s	Mass diffusivity
f	—	Friction factor, $\tau/(\rho u^2)$
F	N	Shear force
Fi	—	Fick number, $(Dt)/b^2$
Fo	—	Fourier number, $(\alpha t)/b^2$ $(\alpha t)/b$
Fr	—	Froude number, $u^2/(dg)$
g	m/s^2	Gravity constant, 9.81
Gz	—	Graetz number, $(ud)/\alpha$
Gr	—	Grashof number, $(g\beta\Delta Td^3)/\nu^2$
h	W/(m^2 K)	Heat transfer coefficient
J	kg/(m^2 s)	Mass flux
k_c	m/s	Mass transfer coefficient
L	m	Length
Le	—	Lewis number, (α/D)
m	kg	Mass, surface resistance to transport?
Nu	—	Nusselt number, $(hb)/\lambda$
P	Pa	Pressure
Pe	—	Peclet number, heat transfer $(ud)/\alpha$, mass transfer $(ud)/D$
Pr	—	Prandtl number, $(C_p\eta)/\lambda$ or (ν/α)
Re	—	Reynolds number, $(u\rho d)/\eta$ or $(ud)/\nu$
Sc	—	Schmidt number, (ν/D)
Sh	—	Sherwood number, $(k_c b)/D$
St	—	Stanton number, $h/(u\rho C_p)$
t	s	Time
T	K, °C	Temperature
u	m/s	Velocity
x	m	Distance
X	kg/kg db	Concentration
q	W/m^2	Heat flux
y	m	Direction perpendicular to x
Y	K, kg/m^3, kg/kg	Transport quantity
α	m^2/s	Thermal diffusivity
β	K^{-1}	Volumetric expansion coefficient
γ		Shear rate
η	Pa s	Viscosity
κ	h or k_c	Interface transfer coefficient
λ	W/(m K)	Thermal conductivity
ν	m^2/s	Kinematic viscosity (momentum diffusivity)
ρ	kg/m^3	Density
τ	Pa	Shear stress
φ	W/m^2 or kg/(m^2 s)	Transport flux (q, J)

REFERENCES

Barrer, R.M. 1951. *Diffusion in and through Solids*. University Press, Cambridge, U.K.

Bird, R.B., Stewart, W.E., and Lightfoot, E.N. 1960. *Transport Phenomena*. Wiley, New York.

Brodkey, R.S. and Hershey, H.C. 1988. *Transport Phenomena: A Unified Approach*. McGraw-Hill, New York.

Crank, J. 1975. *The Mathematics of Diffusion*, 2nd edn. Oxford University Press, London, U.K.

Geankoplis, C.J. 1983. *Transport Processes and Unit Operations*, 2nd edn. Allyn and Bacon, Boston, MA.

Gekas, V. 1992. *Transport Phenomena of Foods and Biological Materials*. CRC Press, New York.

Maroulis, Z.B. and Saravacos, G.D. 2003. *Food Process Design*. Marcel Dekker, New York.

Saravacos, G.D. and Maroulis, Z.B. 2001. *Transport Properties of Foods*. Marcel Dekker, New York.

3 Thermodynamics and Kinetics

3.1 INTRODUCTION

Thermodynamics is the science of heat and work, which is applied to several engineering operations. A brief review of applied thermodynamics is presented in this chapter, including principles of thermodynamics and phase equilibria.

Thermodynamics is applied in several engineering operations, discussed in this book, such as heating operations (Chapter 8), evaporation (Chapter 10), refrigeration (Chapter 12), and mass transfer operations (Chapter 13).

Food kinetics is concerned with physical, chemical, and biological changes in food systems. Of particular importance is the rate of the various reactions and its dependence on temperature. Applications of chemical and microbiological kinetics are discussed in Chapter 9. They are important in food preservation processes and in food storage.

3.2 PRINCIPLES OF THERMODYNAMICS

3.2.1 Definitions

Thermodynamic processes take place in a system, separated from its surroundings by a boundary.

Isolated systems do not exchange matter or energy with their surroundings.

A system is open when matter is exchanged with the surroundings, while closed systems may exchange energy but not matter.

Within an isolated system, changes of temperature, pressure, or volume may take place, but the system eventually reaches an internal equilibrium.

A closed system exchanging energy with the surroundings reaches eventually external equilibrium.

The state of a system is defined when all its properties, such as temperature, pressure, and volume, are fixed. When a system is displaced from an equilibrium state, it undergoes a process or change of state, which continues until its properties reach new equilibrium values.

A reversible process proceeds so that the system is displaced only differentially from an equilibrium state. Such a process can be reversed at any point by very small changes in external conditions, retracing the initial path in the opposite direction.

In an adiabatic process there is no heat transfer between the system and its surroundings (Smith and van Nesh, 1967; van Nesh and Abbott, 1999).

3.2.2 First Law

The first law of thermodynamics states that the total energy of various forms in a system remains constant. As a consequence, when one form of energy disappears, it appears instantly in other forms.

For a closed (but not isolated) system, the first law requires that energy changes of the system be compensated exactly by energy changes in the surroundings.

In a closed system, the change in internal energy ΔU is equal to the difference of heat Q minus the mechanical work W

$$\Delta U = Q - W \tag{3.1}$$

The SI units of the terms of Equation 3.1 are joules (J).

In an adiabatic process there is no heat transfer to or from the system, i.e., $Q = 0$.

The internal energy U is an intrinsic property of a system at internal equilibrium.

Heat Q is energy crossing the system boundary under the influence of a temperature difference or gradient. Q is positive when heat is added and negative when heat leaves the system.

Work W is energy in transit between a system and its surroundings, resulting from the displacement of an external force acting on the system. Like heat, W is positive when work is done on the system by the surroundings, and negative when work is done on the surroundings by the system.

The enthalpy of a system H (J) is expressed by the equation

$$H = \Delta U + PV \quad \text{or} \quad \mathrm{d}H = \mathrm{d}U + \mathrm{d}(PV) \tag{3.2}$$

where

P (Pa) is the pressure
V (m^3) is the volume

Note that the units of PV are Pa m^3 = (N/m^2) (m^3) = (N m) = J (joules).

The enthalpy change in a flow system ΔH is given by the equation (Chapter 6)

$$\Delta H + \left(\frac{\Delta u^2}{2}\right) + g\Delta z = Q - W \tag{3.3}$$

In thermodynamic calculations, the kinetic $\Delta u^2/2$ and the potential $g\Delta z$ energies are usually very small compared to the heat Q and the mechanical work W, and they can be neglected, resulting in the simplified equation

$$\Delta H = Q - W \tag{3.4}$$

For closed systems, the work of a reversible process dW_{rev} (J) is calculated from the equation

$$dW_{rev} = PdV \tag{3.5}$$

where
P (Pa) is the absolute pressure
V (m^3) is the volume of the system

The equation of state relates the specific volume V, temperature T, pressure P, and composition x of a given thermodynamic system. The internal energy U and the entropy S are also functions of P, V, x.

The specific (unit-mass) properties V, U, S, as well as the P, T, x, are called intensive properties and they are independent of system size. The total system properties of V, U, and S are called extensive properties and they depend on system size.

3.2.3 Second Law

The second law of thermodynamics states that the total entropy change (ΔS_{total}) of a system and its surroundings, resulting from any real process, is positive, approaching zero when the process approaches reversibility

$$\Delta S_{total} > 0 \quad \text{or} \quad \Delta S_{total} = 0 \tag{3.6}$$

The entropy S (J/K) is an intrinsic property of the system, the change of which, for reversible processes, is given by the equation

$$dS = \frac{dQ_{rev}}{T} \tag{3.7}$$

where
dQ_{rev} (J) is the reversible change in heat quantity
T (K) is the absolute temperature

Internal energy U and entropy S, together with the two laws, apply to all types of systems.

The second law is also expressed by the following statements:

1. No device can be operated so that the only effect is the conversion of heat absorbed by the system completely into work.
2. No process is possible in which heat is transferred solely from a lower to a higher temperature.

Thus, the second law limits the fraction of heat that may be converted to work in a cyclic process. From experience, the maximum conversion of heat to work is about 40%, while work can be converted into heat almost 100%.

3.2.4 Calculation of Entropy Change

The basic equation of the first law (3.1) in its differential form is written as

$$dU = dQ - dW \quad (3.8)$$

For reversible processes (Equation 3.5), Equation 3.8 is written as

$$dU = dQ_{rev} - PdV \quad (3.9)$$

The definition of enthalpy (Equation 3.2) in its differential form is written as

$$dH = dU + PdV + VdP \quad (3.10)$$

Substituting Equation 3.9 into Equation 3.10 results in

$$dH = dQ_{rev} - PdV + PdV + VdP \text{ or } dQ_{rev} = dH - VdP \quad (3.11)$$

For unit-mass ideal gases ($PV = RT$), Equation 3.11 is written as

$$dQ_{rev} = C_p dT - (RT)\left(\frac{dP}{P}\right) \quad \text{or} \quad \frac{dQ_{rev}}{T} = dS = C_p\left(\frac{dT}{T}\right) - R\left(\frac{dP}{P}\right) \quad (3.12)$$

where

C_p is the specific heat of the gas
R is the gas constant
(1, 2) are the initial and final values of P and T

Integration of Equation 3.12 leads to the equation for entropy change ΔS (J/K) per unit mass in reversible processes of ideal gases

$$\Delta S = C_p \ln\left(\frac{T_2}{T_1}\right) - R\ln\left(\frac{P_2}{P_1}\right) \quad (3.13)$$

For a unit mass of 1 mol, $R = 8.31$ J/mol K and C_p is in J/(mol K). For unit mass of 1 kg, the *mol* is divided by the molecular weight M of the gas, e.g., $M = 28$ kg for nitrogen.

3.2.5 Thermodynamic Flow Processes

Fluid flow operations in Food Process Engineering are considered approximately as constant pressure processes (Chapter 6). Compression and expansion are

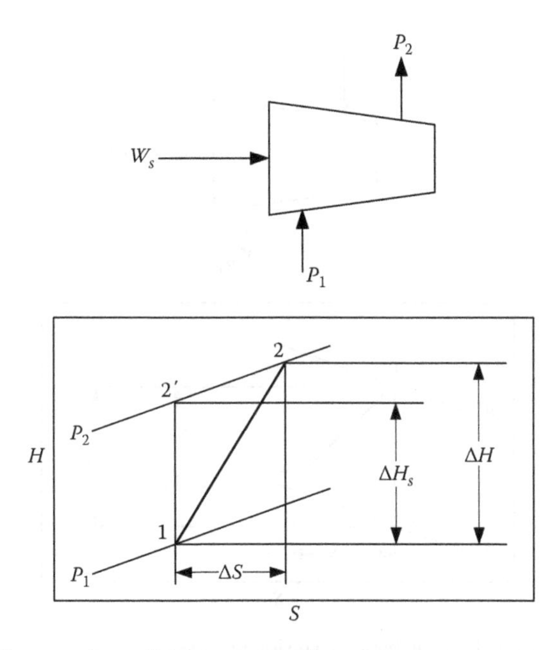

FIGURE 3.1 Diagram of an adiabatic compression process.

thermodynamic processes with several industrial applications. Compression refrigeration, because of its importance, is treated separately in Chapter 12.

3.2.5.1 Compression Processes

The compression of gases and vapors is accomplished in rotary (blades) or reciprocating (pistons) compressors. Rotary equipment is used for low pressures, while reciprocating compressors can achieve higher pressures.

Figure 3.1 shows an adiabatic compression process between pressures P_1 and P_2. In adiabatic processes there is no heat transfer between the system and its surroundings. All heat produced or absorbed in the process remains within the system boundaries. The principle of adiabatic change applies also to the expansion processes.

The process is represented in the enthalpy H–entropy S diagram by the line (1, 2). The adiabatic process is accompanied by an increase of the entropy ΔS. The enthalpy change of the process ΔH is higher than the isentropic (constant entropy) change ΔH_s, and the efficiency of the system η is

$$\eta = \frac{\Delta H_s}{\Delta H} = \frac{W_s(\text{isentropic})}{W_s} \quad (3.14)$$

The efficiency of compressors is usually in the range of 70%–80%.

3.2.5.2 Expansion Processes

Figure 3.2 shows an adiabatic expansion process between pressures P_1 and P_2.

The process is represented in the enthalpy H–entropy S diagram by the line (1, 2).

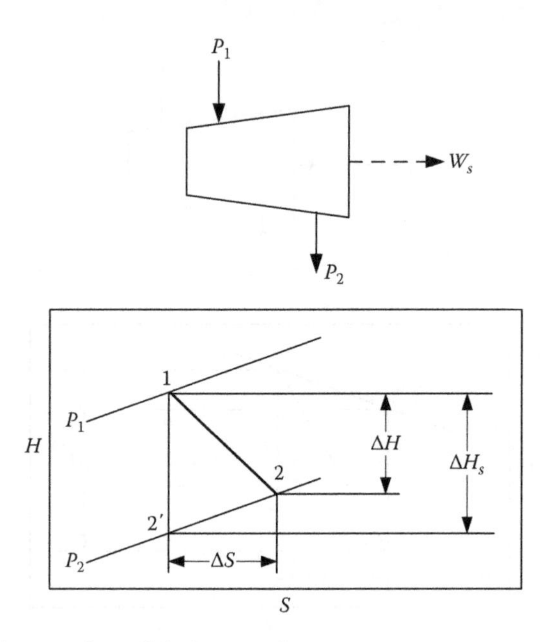

FIGURE 3.2 Diagram of an adiabatic expansion process.

The adiabatic process is accompanied by an increase of the entropy ΔS. The enthalpy change of the process ΔH is lower than the isentropic (constant entropy) change ΔH_s, and the efficiency of the system η is

$$\eta = \frac{\Delta H}{\Delta H_s} = \frac{W_s}{W_s(\text{isentropic})} \tag{3.15}$$

The efficiency of expanders is similar to compressors (range of 70%–80%).

3.2.5.3 Steam Power Cycle

Steam boilers can be used to produce power and waste heat for food processing operations. This system (cogeneration) is considered cost-effective in large plants.

Figure 3.3 shows the diagram of a steam boiler cycle. Heat is supplied to the steam boiler, producing steam in an isothermal process. A heat quantity Q_H of high-pressure steam at temperature T_H from the boiler is expanded in a turbine to produce mechanical work W_s or electricity. The exhaust low-pressure steam of thermal energy Q_C at temperature T_C can be condensed releasing heat, which can be used in process operations. The condensate is returned to the boiler by a pump, repeating the cycle.

The steam-boiler power cycle is similar to the Carnot (heat engine) cycle (Figure 3.4).

The net power output is equal to the difference between the rate of heat input and the heat rejection in the condenser, $W_s = Q_H - Q_C$.

The thermal efficiency of the cycle will be

$$\eta = \frac{W_s}{Q_H} = 1 - \left(\frac{Q_C}{Q_H}\right) = 1 - \left(\frac{T_C}{T_H}\right) \tag{3.16}$$

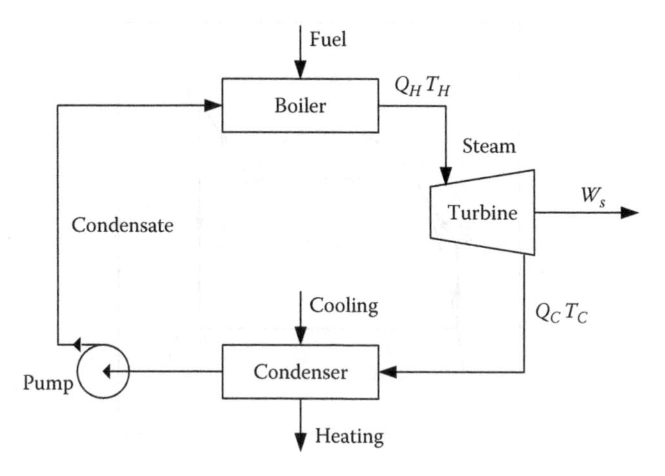

FIGURE 3.3 Diagram of a steam-boiler power cycle.

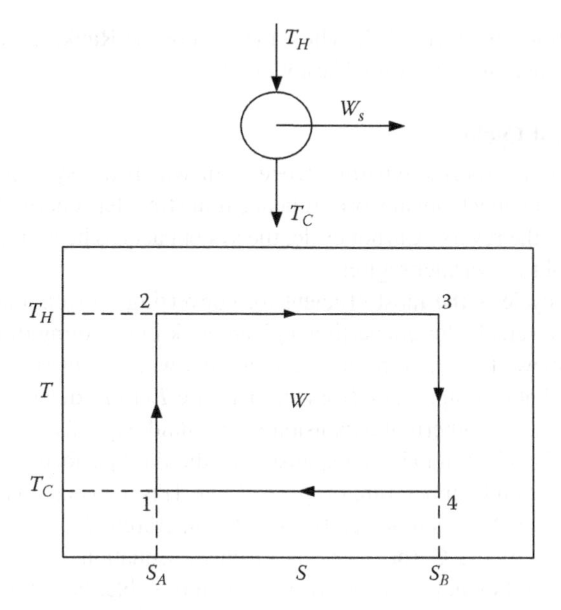

FIGURE 3.4 Diagram of a Carnot cycle.

The thermal efficiency increases as the heating temperature T_H increases or the condensation temperature T_C decreases. As an example, for $T_H = 160°C = 433$ K and $T_C = 100°C = 373$ K, $\eta = 1 - (373/433) = 0.14\%$ or 14%. Thus, the efficiency of conversion of heat to work in a steam cycle is very low.

The efficiency of steam boilers, used in power generation, can be increased by using superheated steam, which increases the heat supply Q_H and the high temperature T_H. Such a system is called the Rankine cycle (Figure 3.5). Compared to the Carnot cycle, the saturated steam (line 2–3) is superheated (line 3–3′) before it is expanded in the turbine (line 3′–4). Cooling (line 4–1′) is continued until saturation,

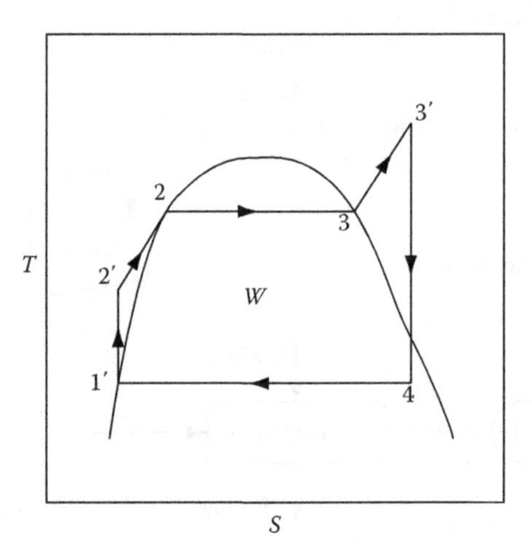

FIGURE 3.5 Diagram of a Rankine cycle.

and heating follows lines (1′–2′–2). Thus, the area of the Rankine cycle W is larger than the corresponding area of the Carnot cycle.

3.2.5.4 Carnot Cycle

A Carnot cycle is a thermodynamic process, in which the system acts as a heat engine, performing mechanical work. A heat engine transfers energy from a warm to a cool region. In the reverse Carnot cycle, the system acts as heat pump, transferring heat from a cool to a warmer region.

The Carnot cycle is the most efficient for converting a given amount of energy into work or, conversely, for converting a given work into cooling (refrigeration).

Figure 3.4 shows the diagram of a heat engine, which consists of the following steps: (1–2) adiabatic compression from temperature T_C to T_H due to work performed on the system; (2–3) isothermal expansion as the fluid expands when heat is added at temperature T_H; (3–4) adiabatic expansion as the fluid performs work during the expansion process and temperature drops to T_C; (4–1) isothermal compression as the fluid contracts when heat is removed from it at temperature T_C.

The efficiency of the Carnot cycle is given by Equation 3.16, which indicates that a maximum efficiency of 100% ($\eta = 1$) is impossible, because a cool temperature of absolute zero ($T_C = 0$) cannot be reached. The statement that "a Carnot cycle (heat engine) efficiency of 100% is impossible" is known as the third law of thermodynamics.

3.2.5.5 Refrigeration and Heat Pumps

Compression refrigeration and its applications are discussed in Chapter 12.

Heat pumps are heat engines (Figure 3.4) operated in reverse, i.e., converting mechanical work to heat and transferring heat from a lower to a higher temperature. They operate in a similar way with the compression refrigeration system, and they are used in air-conditioning systems.

The coefficient of performance (*COP*) of heat pumps is similar to the *COP* of refrigeration

$$COP = \frac{Q_C}{W} \tag{3.17}$$

where
Q_C is the heat transferred
W is the work supplied to the compressor

In heat pumps, the *COP* is approximately equal to 4, but the actual heat supplied to the heated space, due to friction, is higher (about 5 W).

3.2.6 Gas Compression

Normal food processing operations, involving gases, such as air, nitrogen, oxygen, and carbon dioxide, are carried out at low pressures (near atmospheric) and the prefect gas law is applied

$$PV = nRT \tag{3.18}$$

where
P (Pa) is the pressure
V (m^3), n (mol) is the amount of substance
$R = 8.314$ kJ/(kmol K) is the gas constant
T (K) is the temperature

It should be noted that $n = m/MW$, where m (kg) is the gas mass and MW is its molecular weight. Thus, for $m = 100$ kg of atmospheric air of molecular weight $MW = 28.8$ the amount of air will be $n = 100/28.8 = 3.47$ kmol.

The perfect gas law neglects the molecular size and the intermolecular forces, which are significant in real gases. The pressure–volume relations of nonideal gases are described by semiempirical equations of state, such as the van der Waals equation

$$\left[P + a\left(\frac{n}{V}\right)^2\right](V - nb) \tag{3.19}$$

where a, b are specific empirical constants. For air $a = 0.136$ J m^3/mol^2 and $b = 3.64 \times 10^{-5}$ m^3/mol and for water vapor $a = 0.55$ J m^3/mol^2 and $b = 3.04 \times 10^{-5}$ m^3/mol.

In the van der Waals equation, parameter a is related to the size of the molecules, while b is related to the intermolecular forces. At low gas densities, as near atmospheric pressure, correction factors (nb) and (a/V) are small and the van der Waals equation reduces to the perfect gas law (Equation 3.18).

The perfect gas law can be considered as an isothermal (constant temperature) process

$$PV = \text{constant} \quad \text{or} \quad P_1V_1 = P_2V_2 \tag{3.20}$$

In adiabatic (isothermal) processes both temperature and volume change very fast, so that heat transfer to the environment is negligible, and the following equation is applied:

$$PV^{\gamma} = \text{constant} \quad \text{or} \quad P_1V_1^{\gamma} = P_2V_2^{\gamma} \tag{3.21}$$

where

$\gamma = C_p/C_v$ is the adiabatic index

(C_p, C_v) are the specific heats of the gas at constant pressure and constant volume, respectively [kJ/(kg K)]

Note that $C_p/C_v = C_p/(C_p - R)$.

The adiabatic index γ varies in the range 1–2, and for air $\gamma = 1.4$.

In practice, most processes are polytropic (intermediate rates of changes), given by the equation

$$PV^{n} = \text{constant} \quad \text{or} \quad P_1V_1^{n} = P_2V_2^{n} \tag{3.22}$$

where the empirical polytropic index n is smaller than the adiabatic index. The index n varies from 1 to 1.4, and for air $n = 1.2$.

The work W of an adiabatic compression process from pressure P_1 to P_2 is calculated from the following equation:

$$W = \left[\frac{\gamma}{\gamma - 1}\right] P_1V_1 \left[\left(\frac{P_2}{P_1}\right)^{(\gamma-1)/\gamma} - 1\right] \tag{3.23}$$

The critical point of a gas refers to the critical pressure P_c and critical temperature T_C above which a liquid phase of the substance cannot exist. The liquid state can be formed below the critical point by reducing the temperature or the pressure.

Critical points of some important gases and vapors: air, $P_c = 39.5$ bar, $T_C = -140°C$; water vapor, $P_c = 221$ bar, $T_C = 374°C$; carbon dioxide, $P_c = 78$ bar, $T_C = 31°C$.

3.3 PHASE EQUILIBRIA

Equilibrium is associated with various thermodynamic processes. The equilibrium state is reached when transfer of mass or energy between phases is completed and the system is stable, i.e., the composition and enthalpy are constant.

Phase equilibria must obey the Phase Rule, which is expressed by the equation

$$F = 2 - P + N \tag{3.24}$$

where
F is the degrees of freedom
P is the number of phases
N is the number of components of the system

The degrees of freedom F is the number of independent variables that must be fixed arbitrarily to define the system. Thus, for vapor–liquid equilibria of a two-component system between two phases, $F = 2$. The two degrees of freedom in this case are temperature and pressure, meaning that if pressure is fixed, the only variable that can be varied is temperature.

3.3.1 Vapor–Liquid Equilibria

Vapor–liquid equilibria are discussed in Chapter 13 with emphasis on distillation, as applied to food processing. Gas–liquid equilibria are treated in a similar manner, and they are applied to gas absorption and gas desorption operations.

3.3.2 Vapor–Solid Equilibria

Vapor–solid equilibria are of importance to food drying and food storage, as discussed in Chapter 11 (Jowitt et al, 1983; Iglesias and Chirife, 1987).

Moisture sorption isotherms represent equilibrium data of water vapor and solid food materials. The water activity (α_w) is defined as the ratio of water vapor pressure of the material in the system (p) to the vapor pressure of pure water (p_0) at the given temperature

$$\alpha_w = \left(\frac{p}{p_0}\right) \tag{3.25}$$

The water activity of a material at equilibrium is equal to the water activity of the surrounding atmosphere, which is expressed as relative humidity

$$\alpha_w = \frac{\%RH}{100} \tag{3.26}$$

where $(\%RH) = 100(p/p_0)$, (p) is the partial pressure of water vapor in the atmosphere.

The equilibrium moisture content of a food material (X_e), corresponding to a given water activity and temperature, is determined experimentally and is usually presented as the water or moisture sorption isotherm of the material. The gravimetric static method is normally used, in which the food samples were equilibrated in closed jars of constant relative humidity maintained by saturated salt solutions.

The equilibrium data can be fitted to semiempirical relations, such as the Brunauer–Emmett–Tetter (BET), and the Guggenheim–Anderson and de Boer (GAB) equations.

The BET equation is written as

$$\frac{\alpha_w}{(1-\alpha_w)X_e} = \frac{1}{X_m C} + \frac{\alpha_w(C-1)}{X_m C} \tag{3.27}$$

where

α_w is the water activity
X_m (mass fraction) is the moisture content when a monomolecular layer is adsorbed
C is an empirical constant

The GAB equation is written as

$$\frac{\alpha_w}{X_e} = a\alpha_w^2 + b\alpha_w + c \tag{3.28}$$

where

$$a = \frac{k}{X_m}\left(\frac{1}{C}-1\right), \quad b = \frac{1}{X_m}\left(1-\frac{2}{C}\right), \quad c = \frac{1}{X_m Ck}$$

The constants C and k are related to the heat of sorption, the heat of condensation of water, and the temperature of the system. X_m is the monomolecular moisture content.

The water activity and sorption isotherms are affected by the composition of the food and the temperature of the system. In general, polymers sorb more water than sugars and other soluble components at low water activities. However, the soluble components sorb more water above a certain water activity, e.g., above 0.6 for sugars. Temperature has a negative effect on the equilibrium moisture content at low water activities. At high values of α_w, the soluble components, such as sugars, sorb more water, and temperature has a positive effect because of the dissolving effect of water.

Figure 3.6 shows two typical sorption isotherms of a food biopolymer (e.g., starch) and a dried food material, which can be considered a mixture of polymeric and soluble materials, such as sugars. At high water activities, the food material sorbs more water than the biopolymer, because of the presence of water-soluble components.

Water activity at a given moisture content increases normally with the temperature, according to the Clausius–Clapeyron equation

$$\ln\left(\frac{\alpha_{w2}}{\alpha_{w1}}\right) = \left(\frac{\Delta H}{R}\right)\left(\frac{1}{T_1}-\frac{1}{T_2}\right) \tag{3.29}$$

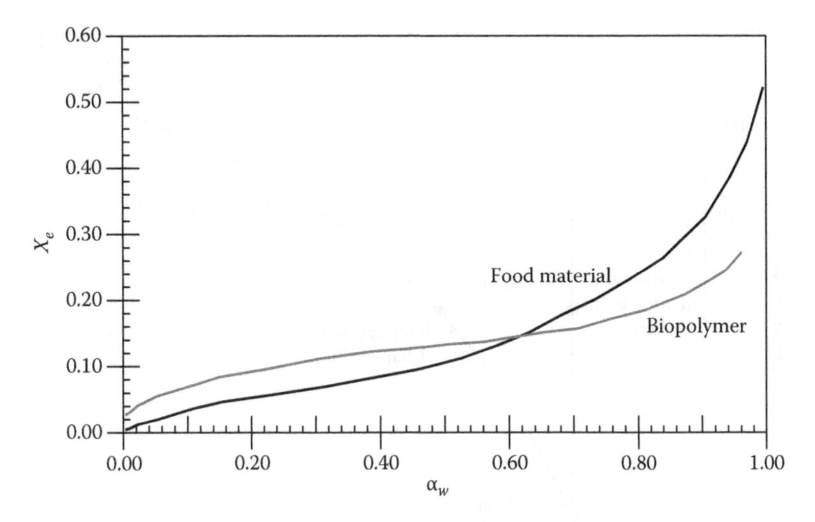

FIGURE 3.6 Moisture sorption isotherms of a biopolymer and a food material.

where

α_{w1}, α_{w2} are the water activities at temperatures T_1, T_2, respectively
ΔH (J/mol) is the heat of sorption of water at the given moisture content

The temperature is in (K) and $R = 8.31$ J/(mol K).

3.3.3 Water Phase Equilibria

The water phase equilibria are important in food process engineering operations such as evaporation and drying.

The saturated vapor pressure of water P_s as a function of the temperature is given by the empirical Antoine equation

$$\ln(P_s) = \frac{\alpha_1 - \alpha_2}{\alpha_3 + T} \tag{3.30}$$

where P_s is in bar and T in °C.

The Antoine constants for water are, $\alpha_1 = 1.19 \times 10$, $\alpha_2 = 3.95 \times 10^3$, and $\alpha_3 = 2.32 \times 10^2$, and the critical points, $P_C = 221$ bar and $T_C = 374$°C.

The triple point of water is $T = 0.01$°C and $P = 6$ mbar (Figure 3.7). It refers to the state of three coexisting phases of vapor, liquid, and solid (ice). In freeze-drying, the operating pressure and temperature must be below the triple point, so that the ice will evaporate directly into vapor without melting (not going through the liquid state).

The melting point of ice decreases nonlinearly as the pressure is increased, according to the Clapeyron equation. Thus, the melting point decreases from 0°C at 1 bar to −7°C at 1 kbar and −20°C at 2 kbar.

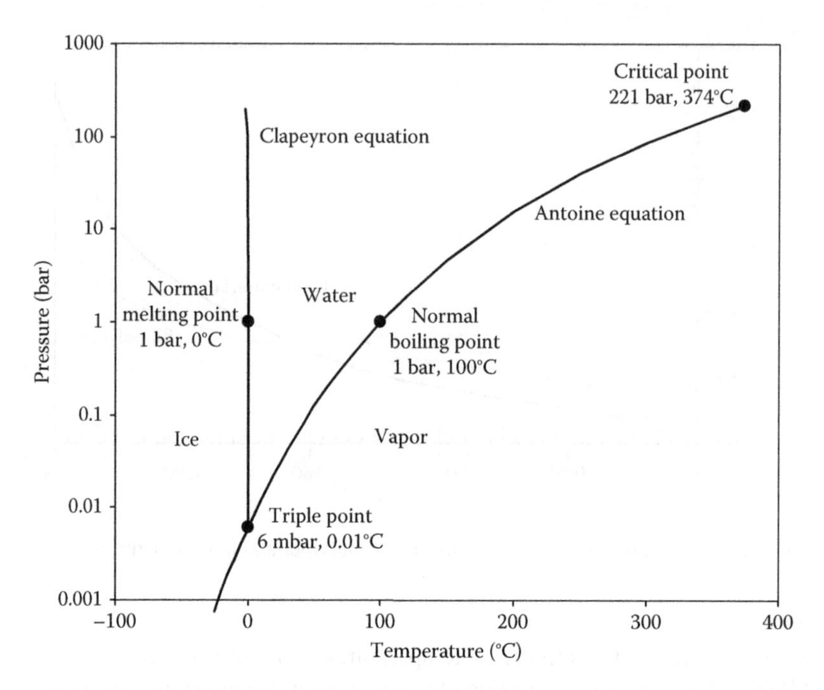

FIGURE 3.7 Phase diagram of water.

3.4 FOOD KINETICS

Food Kinetics is primarily concerned with quality changes of food materials during processing and storage. Kinetics is also important in process calculations of various food engineering operations (van Bockel, 2009).

3.4.1 Chemical Reactions

Chemical and biochemical reactions in food systems have a very complex mechanism, due to the several ingredients and complex conditions involved.

Mathematical models are used to describe the reaction mechanism and their kinetics. The rate of chemical reactions expresses the change of concentration over time and its dependence on concentration, temperature, and other environmental factors.

At equilibrium in Kinetics, the rate of the net reaction is zero, meaning that the rate of the forward reaction is equal to the reverse reaction in the system.

The order of a chemical reaction n (–) is usually estimated from experimental data, using the generalized equation (Villota and Hawkes, 1992)

$$\left(-\frac{dC}{dt}\right) = kC^n \tag{3.31}$$

where

C (kg/m³ or kg/kg)
t (s) is the time
k (1/s) is the reaction rate

Equation 3.31 is also written as

$$\ln\left(-\frac{dC}{dt}\right) = \ln(k) + n\ln(C) \tag{3.32}$$

The constants k and n can be estimated from experimental data, by a nonlinear regression of the last equation. A graphical estimation is obtained by plotting $\ln(-dC/dt)$ versus $\ln(C)$, obtaining a straight line with slope n and ordinate intercept $\ln(k)$.

In food process engineering operations, food kinetics is simplified by considering primarily first-order and sometimes zero-order reactions, expressed by Equations 3.33

$$\text{Zero-order}\left(-\frac{dC}{dt}\right) = k, \quad \text{first-order}\left(-\frac{dC}{dt}\right) = kC \tag{3.33}$$

The zero-order reaction is integrated as

$$C = C_0 - kt \tag{3.34}$$

In this case, the concentration of the reactants decreases linearly with time.

The integrated first-order reaction is written as

$$\ln\left(\frac{C}{C_0}\right) = -kt \tag{3.35}$$

Thus, in first-order reactions, the concentration of reactants decreases logarithmically with the time.

Temperature has a very significant effect on the reaction rate. In most cases, the reaction rate constant k (1/s) increases nonlinearly with the temperature, following the Arrhenius equation

$$k = k_0 \exp\left[\frac{-E_a}{RT}\right] \tag{3.36}$$

where

E_a (kJ/mol) is the activation energy
R [8.31 J/(mol K)] is the gas constant
T (K) is the absolute temperature
k_0 (1/s) is a characteristic constant

The Arrhenius equation can also be written in the following forms

$$\frac{d\ln k}{d\left(\frac{1}{T}\right)} = -\left(\frac{E_a}{R}\right) \tag{3.37}$$

TABLE 3.1
Activation Energies (E_a) of Food Reactions

Reaction	E_a (kJ/mol)
Enzymatic reactions	0–30
Ascorbic acid degradation	20–160
Anthocyanin degradation	30–125
Chlorophyll degradation	20–120
Lipid oxidation	40–100
Carotenoid oxidation	40–90
Non-enzymatic browning	40–160
Protein denaturation	300–500

$$\ln\left(\frac{k}{k_{\text{ref}}}\right) = -\left(\frac{E_a}{R}\right)\left(\frac{1}{T} - \frac{1}{T_{\text{ref}}}\right) \tag{3.38}$$

where (k_{ref}) and (T_{ref}) are the reference rate constant and temperature, respectively.

Table 3.1 shows typical activation energies of chemical and biochemical reactions in food systems.

Low activation energies ($E_a < 20$ kJ/mol) characterize high reaction rate constants $k > 1 \times 10^{-2}$ (1/s), meaning that reactions take place very fast. High activation energies ($E_a > 50$ kJ/mol) are related to slow reactions (range of k, 1×10^{-3} to 1×10^{-7} [1/s]), in which the effect of temperature is very strong.

In catalytic reactions, the catalyst does not change the chemical equilibrium, but it reduces the activation energy, increasing the reaction rate at the given temperature. Homogeneous and heterogeneous catalysts are important in food systems.

Temperature has a strong effect on degradation reactions of food components, with very high activation energy of protein denaturation.

Diffusion of the reactants in food systems may be limiting the rate of a given chemical reaction. The mass diffusivity D (m^2/s) in foods is discussed in Chapter 13.

In first-order reactions, the half-time $t_{1/2}$(s), defined as the time it takes for a reduction in concentration by 50%, is given by the equation

$$t_{1/2} = \frac{\ln 2}{k} = \frac{0.693}{k} \tag{3.39}$$

where k (1/s) is the reaction rate constant.

In Food Science, the effect of temperature on the rate of food reactions is often expressed as Q_{10} (increase of the reaction rate by a 10°C rise of the temperature) or z (temperature rise to reduce the m.o. population by 90%). Q_{10} and z are discussed in Chapter 9.

Complex food reactions that cannot be fitted to simple first-order reactions can be analyzed by the method of multiple response modeling. Such reactions include degradation of food components, such as chlorophyll, and nonenzymatic browning, where several reactants and products are involved. Experimental reaction rate data from such systems are fitted to complex equations, using special computer software.

In complex reactions, where the Arrhenius model is not applicable, empirical equations can be used to describe the effect of temperature on the reaction rate constant (k) or on related property of food, such as viscosity (η). Thus, the Williams–Landel–Ferry (WLF) equation is used in polymeric and food materials near the glass transition temperature T_g (the transition temperature from the glassy to the rubbery state) (Roos, 1992).

The WLF equation for reaction rate constant (k) is written as

$$\ln\left(\frac{k_g}{k}\right) = -\frac{C_1(T - T_g)}{C_2 + (T - T_g)} \tag{3.40}$$

where

k_g, k are the reaction rate constants at temperatures T, T_g, respectively
$C_1(-)$ and $C_2(°C)$ are empirical constants, for example, $C_1 = 17.4$ and $C_2 = 51.6°C$

Temperatures are used in °C.

A similar WLF equation is applied for the viscosity (η)

$$\ln\left(\frac{\eta}{\eta_g}\right) = -\frac{C_1(T - T_g)}{C_2 + (T - T_g)} \tag{3.41}$$

where η, η_g are the viscosities at temperatures T, T_g, respectively.

Enzymes are considered biological catalysts which reduce the activation energy and the reactions can take place at low temperatures. Enzymes are usually inactivated at temperatures above 60°C, but a few of them are very heat resistant.

Enzyme-catalyzed reactions consist of three steps: (a) binding of the enzyme with the substrate, (b) reacting with the substrate, and (c) releasing the product(s) and the enzyme to the system. Enzymatic reactions are affected by activators, inhibitors, temperature, pH, and substrate concentration.

Simple enzyme reactions can be modeled in the Michaelis–Menten equation

$$u_0 = \frac{u_{max}[S]}{k_M + [S]} \tag{3.42}$$

where

u_0 and u_{max} are the initial and maximum reaction rates, respectively
$[S]$ is the substrate concentration
k_M is the Michaelis constant

From the last equation it follows that the Michaelis constant is equal to the substrate concentration ($k_M = [S]$) when $u_0 = 0.5u_{max}$. The parameters u_{max} and k_M are determined by a regression analysis of experimental data.

3.4.2 Microbial Kinetics

Microbial Kinetics is concerned with microbial growth and microbial inactivation.

3.4.2.1 Microbial Growth

Microbial growth in food systems depends on the food composition, the chemical environment, and the enzymatic reactions of microbial cells. Kinetic models describe the growth rate of cells, while stochastic models deal with the statistical probability of growth.

Microbial growth curves show a sigmoid shape consisting of three phases, i.e., lag, exponential, and stationary (Figure 3.8).

Microbial growth can be modeled using empirical models, primarily the logistic model, and secondarily the more complicated Gompertz model. The logistic model relates the microbial growth $\log(N/N_0)$ to the time t (h), and the number of log cycles $C(-)$

$$\log\left(\frac{N}{N_0}\right) = \frac{C}{\{[1+\exp(-B(t-M))]\}} \tag{3.43}$$

where

B (1/h) is the maximum growth rate
M (h) is the time corresponding to B

Figure 3.8 shows a sigmoid microbial growth curve, fitted to the logistic model.

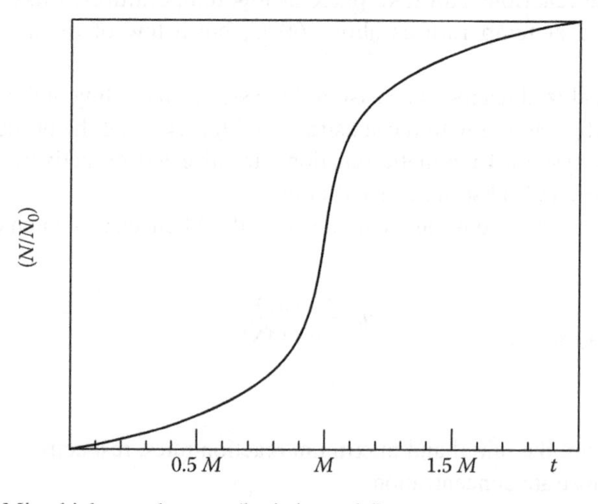

FIGURE 3.8 Microbial growth curve (logistic model).

The Arrhenius model is not applicable for microbial growth kinetics. Instead, the effect of temperature can be modeled by the empirical square root (Ratkowsky) equation

$$B^{1/2} = b(T - T_{min}) \tag{3.44}$$

where

B (1/h) is the maximum growth rate
T, T_{min} are the actual and the minimum temperatures (°C), respectively
b [1/(h$^{1/2}$ °C)] is a regression fit parameter

3.4.2.2 Microbial Inactivation

The kinetics of thermal inactivation of microorganisms (m.o.) is discussed in Chapter 9.

Simple, first-order models are used to calculate the inactivation of spoilage m.o. of importance to food processing. The reaction rate constant (k) of the rate equation 3.31 is usually expressed as the D value (min), defined by the equation

$$D = \frac{(\ln 10)}{k} = \frac{2.3}{k} \tag{3.45}$$

The effect of temperature on inactivation kinetics is expressed by the z-value (°C), which is defined as the temperature rise to reduce the m.o. population by 90%. Assuming that the Arrhenius model is applicable, the z-value (°C) can be calculated from the equation

$$z = \frac{2.3RT^2}{E_a} \tag{3.46}$$

where

T (K) is the absolute temperature
E_a (J/mol) is activation energy
R is the gas constant (8.31 J/mol K)

Although simplified first-order models are used normally in most applications, more complex probabilistic models may be more accurate for complex mechanisms. Nonlinear inactivation curves are treated by complex empirical models, such as the Weibull probability distribution. Regression analysis can estimate the constants of such models.

APPLICATION EXAMPLES

Example 3.1

A steam boiler produces steam at 10 bar absolute pressure, which is used to generate power. The steam is superheated to 400°C before it is expanded in a turbine.

Calculate

(a) The overall efficiency of the system, if the steam is expanded to an absolute pressure of 1 bar. Thermal efficiency of the turbine is 75% of the theoretical.
(b) The overall efficiency of the system, if the steam is expanded to an absolute pressure of 0.1 bar. Thermal efficiency of the turbine is 75% of the theoretical.
(c) The amount of steam required for the generation of 1 MW power, for both cases (Haar et al, 1984).

Solution

From steam tables, the temperature of saturated steam at 10 bar is $T_H = 180 + 273 = 453\,\text{K}$, at 1 bar $T_C = 100 + 273 = 373\,\text{K} = 100 + 273 = 373\,\text{K}$, and at 0.1 bar $T_C = 46 + 273 = 319\,\text{K}$. The corresponding enthalpies of the saturated steam will be 2776 kJ/kg (10 bar), 2675 kJ/kg (1 bar), and 2585 kJ/kg (0.1 bar). The superheated steam at 10 bar and $T_H = 400 + 273 = 673\,\text{K}$ has an enthalpy of $Q_H = 3264\,\text{kJ/kg}$.

(a) The theoretical efficiency of the power generation system with $T_C = 373\,\text{K}$ will be $\eta = 1 - 373/673 = 0.44$. Since the thermal efficiency of the turbine is 75%, the overall efficiency of converting heat (steam) to mechanical energy will be $\eta = 0.75 \times 0.44 = 0.33$.
(b) The theoretical efficiency of the power generation system with $T_C = 319\,\text{K}$ will be $\eta = 1 - 319/673 = 0.53$. Since the thermal efficiency of the turbine is 75%, the overall efficiency of converting heat (steam) to mechanical energy will be $\eta = 0.75 \times 0.53 = 0.40$. Thus, reducing the pressure of the expanded steam from 1 to 0.1 bar will increase the overall efficiency of power generation from 33% to 40%. However, the exhaust steam at 0.1 bar pressure cannot be utilized for heating operations, contrary to the 1 bar steam.
(c) The generated power of 1000 kJ/s requires heat power $Q_H = W_s/\eta = 1000/\eta$ kJ/s.

For the case (a), $\eta = 0.33$ and $Q_H = 1000/0.33 = 3030\,\text{kJ/s}$. Condensation of steam at 1 bar releases 2258 kJ/kg (steam tables), and steam required $m = 3030/2258 = 1.34\,\text{kg/s}$ or $m = (1.34 \times 3600)/100 = 4.83\,\text{t/h}$.

For the case (b), $\eta = 0.40$ and $Q_H = 1000/0.40 = 2599\,\text{kJ/s}$. Condensation of steam at 0.1 bar releases 2392 kJ/kg (steam tables), and steam required $m = 2500/2392 = 1.045\,\text{kg/s}$ or $m = (1.045 \times 3600)/100 = 3.76\,\text{t/h}$.

It is concluded that a significant reduction of steam requirements (22%) is achieved by reducing the steam condensation pressure from 1 to 0.1 bar.

Example 3.2

A heat pump is used to heat atmospheric air from 5°C to 25°C, utilizing a reverse Carnot cycle. Calculate the mechanical (shaft) work required to heat $1000\,\text{m}^3/\text{h}$. The system can be considered as isobaric (constant pressure), and the thermal efficiency is 75% of the theoretical. The density of air is $1.3\,\text{kg/m}^3$ and the specific heat 1.1 kJ/(kg K).

Solution

Heating of $1{,}000\,\text{m}^3/\text{h}$ of air from 5°C to 25°C requires $Q = 1{,}000 \times 1.3 \times 1.1 \times 20 = 28{,}600\,\text{kJ/h}$ or $Q = 28{,}600/3{,}600 = 7.94\,\text{kW}$.

The mechanical work (W) is related to the heat transferred in a heat pump (Q) by the equation $COP = Q/W$, where the coefficient of performance (COP) is taken usually as $COP = 5$. Thus the work rate (power) required will be $W = 7.94/5 = 1.6$ kW.

Note that the COP is discussed in detail in Chapter 12.

Example 3.3

The degradation of a food vitamin is assumed to follow first-order kinetics with an Arrhenius activation energy of 50 kJ/mol.

Estimate (a) the reaction rate constant if the percent vitamin retention is 50% after 100 days storage at 30°C, and (b) what will be the increase in vitamin retention for the same storage time, if the temperature is reduced to 20°C?

Solution

(a) The integrated first-order model $\ln(C/C_0) = -kt$ gives for $C/C_0 = 0.5$ and $t = 100$ days and 30°C gives, $\ln(0.5) = -(100 \times 24)k$, or $0.693 = 2{,}400k$, from which $k = 2.9 \times 10^{-4}$ (1/h) or $k(30) = (2.9/3{,}600) \times 10^{-4} = 0.8 \times 10^{-7}$ (1/s).

(b) The reaction rate constant at 20°C is related to the constant at 30°C by the Arrhenius equation, $\ln[k(20)/k(30)] = -(E_a/R)(1/293 - 1/303) = -(50{,}000/8.31) \times 0.1 \times 10^{-3} = -0.602$, from which $k(20)/k(30) = 0.548$ and $k(20) = 0.548 \times 0.8 \times 10^{-7} = 0.44 \times 10^{-7}$ (1/s).

Thus, when the storage temperature is reduced from 30°C to 20°C, the reaction rate constant (k) will be reduced by (0.80–0.44)/0.80 = 45%, resulting in the same % increase of storage life, i.e., the storage life for 50% vitamin retention will be 100 + 45 = 145 days.

Example 3.4

A food spoilage microorganism has a decimal reduction time $D = 5$ min and a temperature resistance factor $z = 10$°C. A first-order inactivation model is assumed.

Estimate (a) the rate constant (k), (b) the half-time ($t_{1/2}$) of the microbiological reaction, (c) the activation energy at a temperature of 120°C, assuming an Arrhenius relationship.

Solution

(a) The reaction rate constant is estimated from the equation $k = 2.3/D = 2.3/(5 \times 60) = 0.77 \times 10^{-2}$ (1/s).

(b) The half-time of the reaction will be $t_{1/2} = 0.693/k = (0.693/0.77) \times 10^2 = 90$ s or $t_{1/2} = 1.5$ min.

(c) The z-value is defined as the temperature rise in °C or K, which affects a 90% reduction of the microbial population. If the Arrhenius equation is applicable, the activation energy is given by the equation, $E_a = 2.3RT^2/z$, where $R = 8.31$ J/mol, $T = 120 + 273 = 393$ K and $z = 10$ K. Thus, $E_a = 2.3 \times 8.31 \times 393^2/10 = 295.15$ kJ/mol.

It is seen that the activation energy is very high, characteristic of microbial inactivation.

PROBLEMS

3.1 Water at 100°C and constant pressure 1 bar is heated by a heat supply Q = 2.5 MJ/kg. (a) Calculate the work produced (W), if the enthalpy change is equal to the heat of evaporation of water ΔH = 2.2 MJ/kg, and the kinetic and potential energy are neglected. (b) What will be the change of the volume (m^3/kg) in this process?
Note that 1 kJ = 0.1 bar m^3.

3.2 Calculate the entropy change ΔS (J/kg) in the reversible compression of nitrogen from pressure P_1 to P_2. Nitrogen can be considered a perfect gas and the pressure–temperature relationship is expressed by the equation $(T_2/T_1) = (P_2/P_1)^{1/4}$. R = 8.31 J/(mol K). Molecular weight of nitrogen M = 28 kg/kmol, specific heat C_p = 1.1 kJ/(kg K).

3.3 Two chemical reactions (A, B) follow the first- and second-order models, respectively, with the same rate constant (k). What will be the relative reaction rates $(-\mathrm{d}C/\mathrm{d}t)_B/(-\mathrm{d}C/\mathrm{d}t)_A$ for these reactions for the same time at a given temperature?

3.4 A food reaction has a rate constant $k = 1.0 \times 10^{-3}$ (1/s) at a temperature of 50°C. Calculate the corresponding half-time $t_{1/2}$ (min) and D-value (min) at a temperature of 90°C, if the Arrhenius activation energy is E_a = 30 kJ/mol. Given R = 8.31 J/(mol K).

LIST OF SYMBOLS

B	1/h, 1/s	Maximum growth rate
C	kg/m^3	Concentration
C_p	kJ/(kg K)	Specific heat
COP	—	Coefficient of performance
D	min	Time to reduce a population by 90%
D	m^2/s	Mass diffusivity
E_a	kJ/mol	Activation energy
F	—	Degrees of freedom
g	m/s^2	Gravity constant, 9.81
H	J	Enthalpy
N	—	Number of components
N	—	Number of microorganisms
P	Pa	Pressure
Q	J	Heat amount
R	kJ/mol	Gas constant, 8.31
S	J/K	Entropy
t	s, h	Time
T	K, °C	Temperature
U	J/kg	Internal energy
u	m/s	Velocity
V	m^3	Volume
W	J	Work (shaft)
X	kg/kg db	Moisture content

z	m	Height
z	°C	Temperature rise to reduce a population by 90%
α_w	—	Water activity
η	Pa s	Viscosity
η	—	Efficiency
ρ	kg/m^3	Density

REFERENCES

Haar, L., Galagher, J., and Kell, G. 1984. *NRC/NRC Steam Tables*. Hemisphere Publication, New York.

Iglesias, H.A. and Chirife, J. 1982. *Handbook of Food Isotherms*. Academic Press, New York.

Jowitt, R., Escher, F., Hallstrom, B., Meffert, H.F.Th., Spiess, W.E.L., and Vos, G., eds. 1983. *Physical Properties of Foods*. Applied Science, London, U.K.

Roos, Y.H. 1992. Phase transitions in food systems. In: *Handbook of Food Engineering*. D.R. Heldman and D.B. Lund, eds. Marcel Dekker, New York.

Smith, J.M. and Van Nesh, H.C. 1987. *Introduction to Chemical Engineering Thermodynamics*, 4th edn. McGraw-Hill, New York.

Van Bockel, M. 2009. *Kinetic Modeling of Reactions in Foods*. CRC Press, Boca Raton, FL.

Van Nesh, H.C. and Abbott, M.M. 1999. Thermodynamics. In: *Perry's Chemical Engineers' Handbook*, 7th edn. R.H. Perry, D.W. Green, and J.O. Maloney, eds. McGraw-Hill, New York.

Villota, R. and Hawkes, J.G. 1992. Reaction kinetics in food systems. In: *Handbook of Food Engineering*. D.R. Heldman and D.N. Lund, eds. Marcel Dekker, New York.

4 Overview of Food Process Technology

4.1 INTRODUCTION

The food processing industry uses agricultural, animal, and marine raw materials to produce processed food products, which provide the required nutrients, energy, and organoleptic (sensory) acceptance to the consumers. It is a large industry based on food science and engineering, and empirical knowledge. In order to be profitable, the food industry should utilize the principles and practice of process engineering and economics (Hui, 2006; Runken and Kill, 1993).

Food science involves principles and applications of food chemistry, food microbiology, biology, and human nutrition to the processing, storage, and utilization of food products. Food engineering involves mainly the principles and practices of process engineering. Empirical knowledge, acquired through years of practical experience, is necessary to deal with complex industrial and economic problems, which are gradually solved by applying scientific principles (Heldman and Hartel, 1997).

Computer use has improved the application of scientific and engineering principles, especially to complex food processing systems.

Food process technology can be divided according to the employed process into three categories: food preservation, food manufacturing, and food ingredients technology (Maroulis and Saravacos, 2007).

Food preservation involves microbiological, chemical, and physical processes, which inactivate or control the spoilage and/or pathogenic microorganisms and enzymes, resulting in food products which can be stored for a short or long time, without significant loss of their nutritive value and organoleptic acceptance.

Preservation of fresh foods (e.g., fruits, meat, fish) can be achieved by minimum processing, hurdle technology, or with refrigeration. Thermal processing, including blanching, pasteurization, and sterilization, inactivates the spoilage microorganisms and enzymes, and the packaged food product can be preserved without refrigeration for a long time (Rahman, 2007).

Food concentration and dehydration preserve foods by lowering the water activity, which inhibits the growth of spoilage microorganisms.

Food manufacturing involves several processing operations, including mixing, extruding, heating, drying, and fermentation to form, formulate, assemble, and preserve food products, usually packaged in consumer containers. Modern food manufacturing utilizes food materials science and food product engineering in the design and production of new food products of desired properties.

Food ingredient technology applies conventional chemical processing, based on the principles of chemical engineering. It includes the isolation of important food ingredients, such as sugar, starch, protein, and edible oils, the preparation of food powders such as wheat flour, and the production of food chemicals, such as vitamins and additives.

Both food manufacturing and food ingredient technology are high-value-added industries, where the cost of raw materials is relatively less important than in food preservation.

Food plant economics is important in processing large amounts of raw materials to produce consumer products at competitive cost. The main cost component in food processing is the raw materials, followed by labor and energy. Less important are the costs of technology and processing equipment.

Food product engineering is concerned with the industrial production of improved food products, applying the principles of food materials science.

4.2 FOOD PRESERVATION

Food preservation is an old technology developed through experience for different food products in various parts of the world. The aim of preservation is to prevent the microbial or chemical spoilage of fresh raw materials used as foods, such as fruits and vegetables, meats, and fish. In addition, preserved foods should retain the nutritive and organoleptic (sensory) quality, and they should be safe (nontoxic) to the consumers.

Modern food science and engineering have improved greatly food preservation technology, so that today large quantities of nutritious and safe processed foods are produced, which feed the expanding world population at affordable economic cost.

4.2.1 Minimal Processing and Hurdle Technology

Minimal processing is concerned with the preservation of fresh and natural foods without loss of nutritional quality, while the product is safe to the consumers. The product should have a significant shelf life to make distribution to the consumers feasible.

Minimal processing is combined with hurdle technology. Hurdles are a series of preservative factors that cannot be overcome by the spoilage microorganisms, such as temperature, water activity, pH, redox potential, preservatives, gas packaging, edible coatings, ultrahigh pressure, addition of salts, irradiation, refrigeration, and packaging.

Hurdle technology is applied to storage without refrigeration or to minimally processed and ready-to-eat meals. Natural hurdles, instead of expensive chemicals, are used in developing countries. Shelf-stable meat products and hurdle-treated fruits can be stored at ambient temperatures, while treated vegetables require refrigerated storage.

Predictive modeling of microbial growth and decay and shelf-life estimation of specific food products have been developed, using complex computer programs.

4.2.2 Refrigeration

Refrigeration is a mild preservation process applied to fresh fruits and vegetables, meat, and fish, extending the shelf life for several days or weeks. It is also used to preserve minimally processed foods. Cold storage may be combined with controlled atmosphere, resulting in improved preservation (Heldman and Lund, 2006).

Refrigerated foods are kept in bulk or in packages in cold storage rooms or cabinets, where the temperature is maintained just above the freezing point of water (0°C).

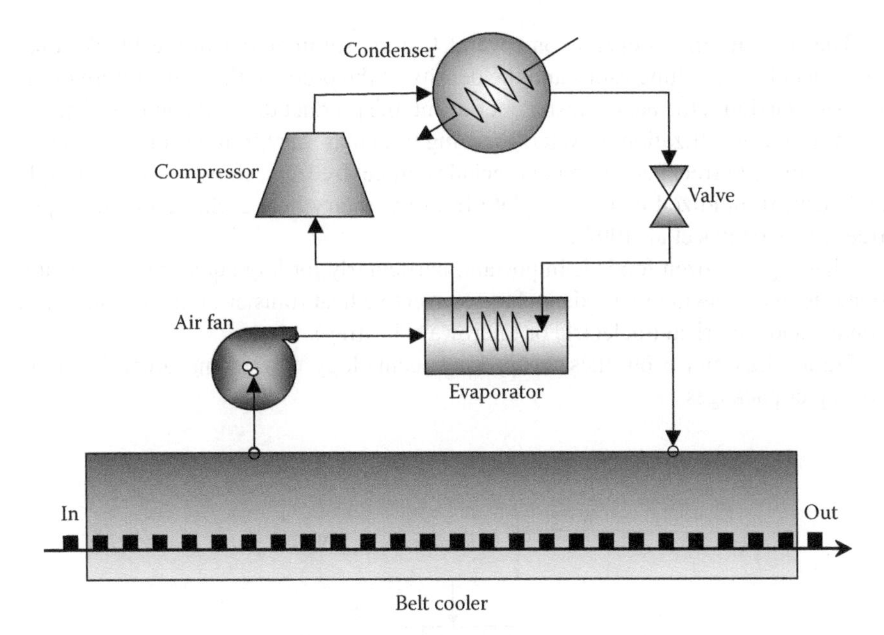

FIGURE 4.1 Diagram of a cooling belt for food refrigeration. (Modified from Maroulis, Z.B. and Saravacos, G.D. *Food Process Design*, CRC Press, May 9, 2003, fig. 5.9.)

They are transported to long distances in refrigerated trucks, tanks, or ships. Refrigeration and freezing temperatures are obtained mainly by mechanical compression of refrigerant gases, and to a minor scale by evaporation of cryogenic liquids.

Refrigerated foods require a minimum preparation before cold storage, such as washing, sorting, and packing in bulk or consumer containers and packages. Water precooling may be used to speed up cooling of the product before cold storage.

Cold storage reduces significantly food spoilage by microorganisms and biochemical or chemical reactions, preserving the nutritive and sensory quality of the products.

Figure 4.1 shows a conveyor-belt mechanical refrigeration cooler of food products, which are subsequently cold-stored or transported to the market.

4.2.3 Freezing

Freezing at temperatures below the freezing point of foods can preserve the frozen products for several months without appreciable microbial spoilage. Enzymatic deterioration in some frozen foods (vegetables) can be prevented by thermal inactivation (blanching), while chemical changes can be minimized by packaging, which prevents oxygen transfer to the frozen product. Frozen foods should be stored and distributed at low temperatures, usually near −18°C (0°F).

Fresh foods are cleaned and prepared before freezing in bulk or in packages. Freezing media include cold air at high velocity (blast freezers), cold plate freezers, and boiling of cryogenic liquids, such as liquid nitrogen and liquid carbon dioxide. Fast freezing causes formation of small crystals, which preserve the food structure, maintaining a high nutritive and sensory quality of the frozen product.

The freezing time depends on several factors, summarized in the Plank equation, including the dimensions and the density of the product, the temperature of the freezing medium, the heat-transfer coefficient, the product thermal conductivity, and the heat of crystallization of water. Freezing time may vary from minutes to hours.

Commercial freezing equipment includes tunnel freezers, conveyor belts (straight and curved), fluidized beds, cold plate freezers, rotary heat exchangers, and liquid freezers (Valentas et al, 1992).

Thawing of frozen foods is important, particularly for large quantities of institutional foods. Thawing methods include convective heat transfer of air or water, heat contact and electrical (dielectric or microwave heating).

Figure 4.2a and b outlines the process technology of freezing green beans in consumer packages.

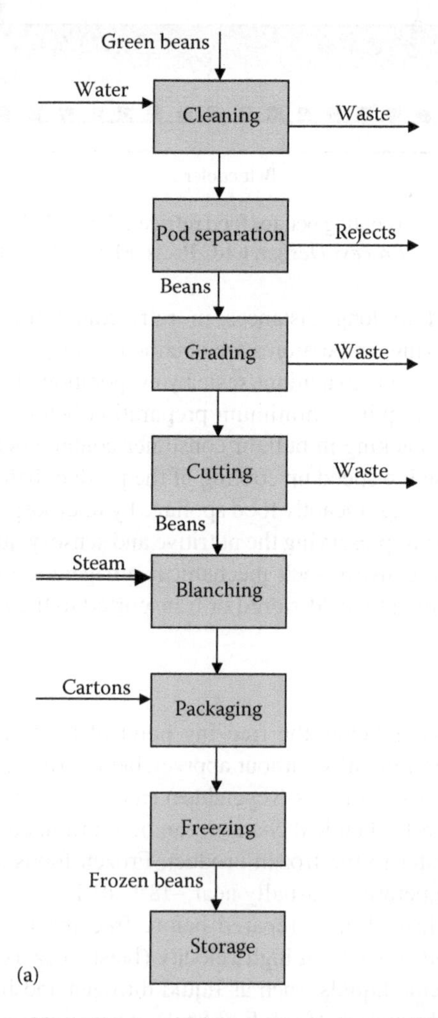

FIGURE 4.2 (a) Process block diagram of freezing green beans. (Modified from Maroulis, Z.B. and Saravacos, G.D., *Food Plant Economics*, August 2, 2007, fig. 7.5a.)

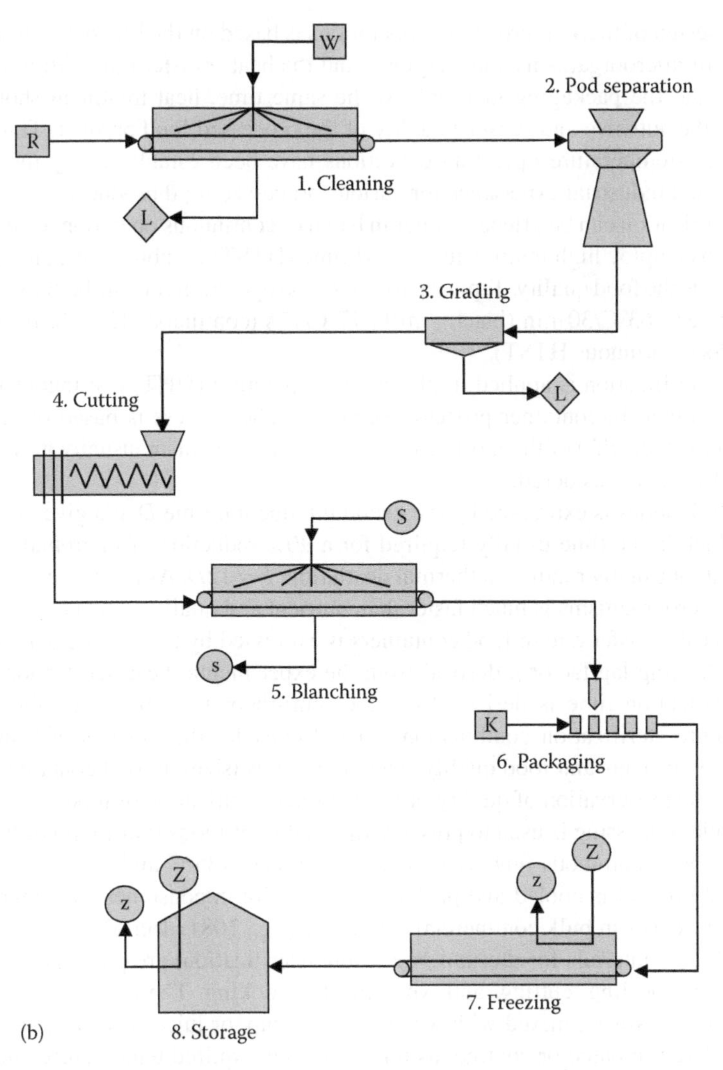

FIGURE 4.2 (continued) (b) Process flowsheet of freezing green beans. (Modified from Maroulis, Z.B. and Saravacos, G.D., *Food Plant Economics*, CRC Press, New York, August 2, 2007, fig. 7.5b.)

4.2.4 Thermal Processing

Thermal processing inactivates the spoilage microorganisms and enzymes, extending the shelf life of foods. Pasteurization inactivates all pathogenic and most spoilage microorganisms and enzymes, extending the shelf life of the food for several days. Pasteurized foods are better preserved at refrigeration temperatures. Sterilization inactivates all spoilage microorganisms and enzymes, and the packaged sterile food products can be kept at ambient temperatures for a long time. Blanching is a mild thermal treatment of fruits and vegetables before other processing (e.g., freezing), which inactivates spoilage enzymes and most microorganisms.

The design of thermal processing operations is based on the kinetics of heat inactivation of microorganisms and enzymes, and the heat-transfer rate within the food product and the packaging material. At the same time, heat treatment should not damage the nutritive and sensory quality of the processed food product. Thus, optimum time–temperature operating conditions have been established by theoretical analysis and industrial experience for various processed food products.

Pasteurization can be affected either in batch or continuous operation. Continuous operations employ high temperature–short time (HTST) combinations, and they do not damage the food quality. Typical pasteurization operations are: milk, 63°C/30min (batch); beer, 63°C/30min (batch); milk, 77°C/15s (continuous HTST); fruit juice, 85°C/15s (continuous HTST).

Food sterilization is applied in ultrahigh temperature (UHT) continuous flow or in conventional in-container process operations. The process is based on the heat inactivation ("death") of the most heat-resistant microorganism, usually its spores in the food system considered.

Death kinetics is expressed by the decimal reduction time D at a given temperature, which is the time usually required for a 90% reduction of microbial activity. Sterilization usually requires a thermal death time $F = 12D$. As a rule, heat inactivation of microorganisms is much faster than nutrient and quality damage.

The heat-transfer rate in food containers is expressed by the heating parameter f_h and the heating lag factor j, derived from the experimental heat penetration curve. The sterilization time is derived from the semi-theoretical Ball formula. Time–temperature sterilization combinations should consider the kinetics of heat damage to the nutrients and food quality. The dimensions (size) of food container affect strongly the preservation of quality of in-container sterilization of foods.

Aseptic processing is used to preserve the quality of foods (mostly fluid) by UHT sterilization. Continuous flow systems are used, e.g., 135°C and 2s for milk. The sterilized product is cooled and packaged aseptically in consumer containers (e.g., paper cartons) or in bulk commercial containers (e.g., 208L drums).

Food raw materials for thermal processing (sterilization) are prepared by washing, sorting, peeling, cutting, and when needed cooking. The prepared product is filled into containers, mixed with water, brine, syrup, or juice. The cans are sealed and sterilized in batch or continuous retorts for the required temperature and time, and then cooled to ambient temperature.

Packaged or packed sterilized foods can be stored at ambient temperatures for more than 1 year, but with some damage of the nutritive value (vitamins, protein) and the sensory quality, due to chemical and physical changes.

Figure 4.3a and b outlines the process technology of canning peach halves in syrup.

4.2.5 Concentration and Dehydration

Concentration of liquid foods, such as fruit juices and milk, can preserve the product for a substantial time. Concentrated products above 65°Brix can be stored at ambient temperatures, while lower concentrations require refrigeration or frozen storage to prevent microbial spoilage.

Concentration of foods is obtained mainly by evaporation, but also by reverse osmosis, or freeze-concentration. Evaporation of water requires large amounts of energy (e.g., 2.3 MJ/kg at 100°C) and energy-saving systems are used, such as multiple-effect evaporators. Heat damage to the food product during evaporation can be minimized by vacuum operation or high-temperature–short time systems.

Food dehydration technology has been developed though the years from an empirical art to modern industrial technology of food preservation and food product engineering. Food preservation is achieved by lowering the water activity of the product, so that microbial growth and biochemical spoilage at ambient temperatures are minimized (avoided).

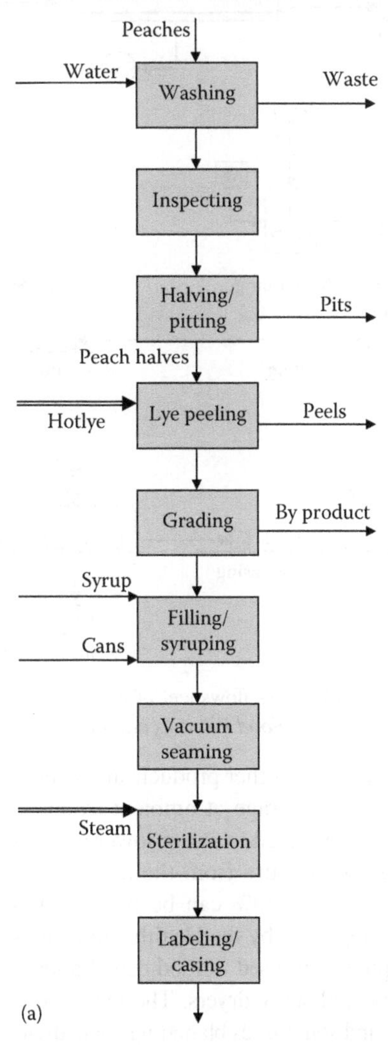

FIGURE 4.3 (a) Process flowsheet of canning peach halves. (Modified from Maroulis, Z.B. and Saravacos, G.D., *Food Plant Economics*, CRC Press, New York, August 2, 2007, fig. 7.4a.)

(*continued*)

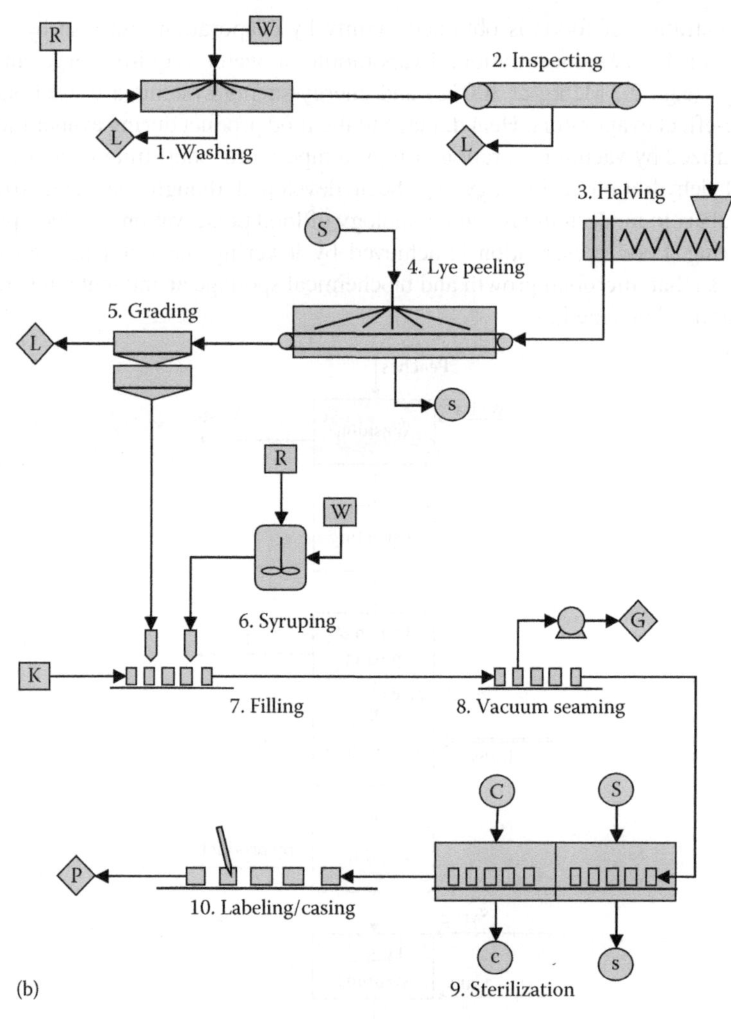

FIGURE 4.3 (continued) (b) Process flowsheet of canning peach halves. (Modified from Maroulis, Z.B. and Saravacos, G.D., *Food Plant Economics*, August 2, 2007, fig. 7.4b.)

Some fruits, cereal grains, and other products are dried to about 15%–20% moisture and can be preserved by storage at ambient (room) temperature and relative humidity. Most foods are dehydrated to 3% or lower moisture and require protective packaging from picking up moisture from the atmosphere. Intermediate moisture foods at moisture contents 20%–30% can be preserved at ambient temperatures, using special chemicals, approved by the Health Authorities.

Several types of equipment are used in food dehydration, including air dryers, vacuum dryers, freeze dryers, and spray dryers. The raw food materials are prepared by washing, sorting, cutting, and sometimes blanching. Dehydrated foods are stored in bulk or in consumer containers, and often they have to be rehydrated before use. Quality preservation may require use of specific additives, such as preservatives and antioxidants.

Figure 4.4a and b outlines the process technology of potato granule dehydration.

4.2.6 Novel Food Processes

Novel food processes are still at the research and development stages with limited industrial applications. They include food preservation and food processing methods. Irradiation (IRR), high-pressure processing (HPP), and pulsed electric field (PEF) have the potential of preserving food without significant loss of the freshness and quality of the product. Membrane separations, supercritical fluid extraction, and freeze concentration are nonthermal processing methods, which can separate or concentrate foods and food components without appreciable damage to product quality. In some cases, a combined preservation method may be more effective, e.g., combination of heat and HPP.

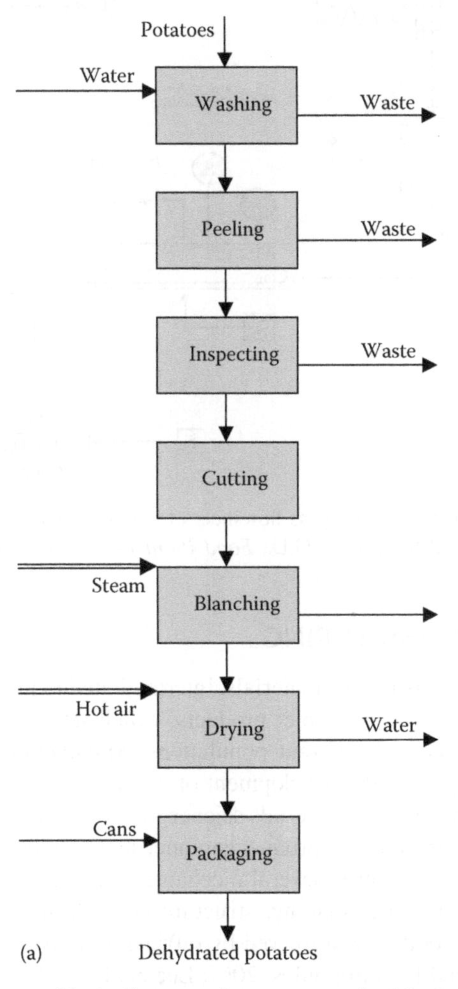

FIGURE 4.4 (a) Process block diagram of potato granule dehydration. (Modified from Maroulis, Z.B. and Saravacos, G.D., *Food Plant Economics*, CRC Press, New York, August 2, 2007, fig. 7.6a.)

(continued)

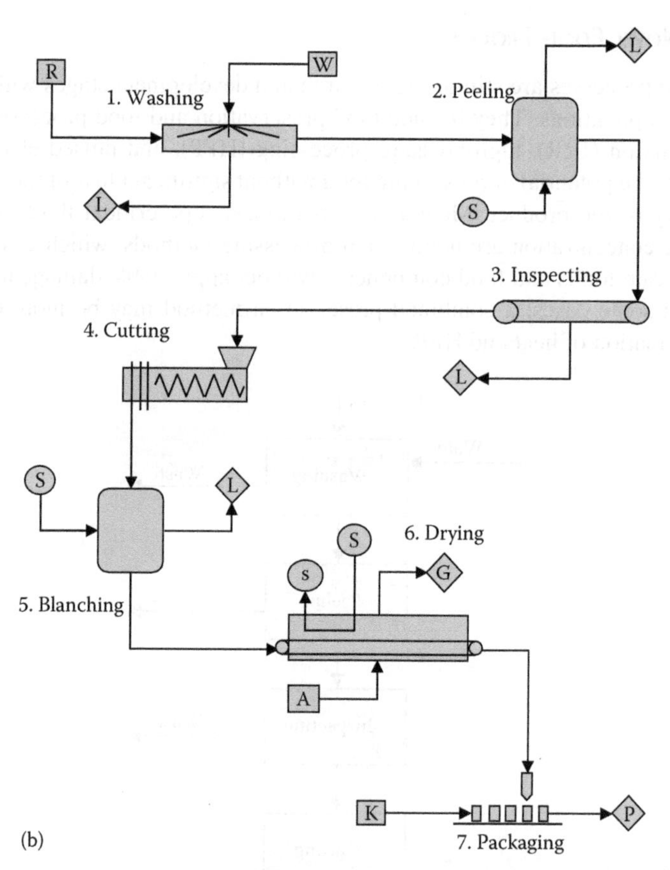

FIGURE 4.4 (continued) (b) Process flowsheet of potato granule dehydration. (Modified from Maroulis, Z.B. and Saravacos, G.D., *Food Plant Economics*, CRC Press, New York, August 2, 2007, fig. 7.6b.)

4.3 FOOD MANUFACTURING

Food manufacturing is using raw materials, intermediate food ingredients, and packaging materials to produce consumer products, which satisfy the nutritional needs and sensory preferences of different populations. Manufactured foods have been produced empirically until the development of modern food science and engineering. The fundamentals of food materials science are considered as the basis of modern food product design and food product engineering (Aguilera and Lillford, 2008).

Food manufacturing involves several mechanical operations, such as size reduction, agglomeration, mixing, shaping, structure control, and food formulation of foods. The prepared foods are preserved by refrigeration, freezing, or thermal processing (Saravacos and Kostaropoulos, 2001; Lee et al, 2008).

Typical manufacturing technologies are shown in Figures 4.5 and 4.6. Figure 4.5a and b outlines the process technology of industrial bread baking, using wheat flour as raw material. The process involves yeast fermentation of wheat dough, mechanical mixing, baking in a continuous oven, and cooling and packaging of the bread loaves.

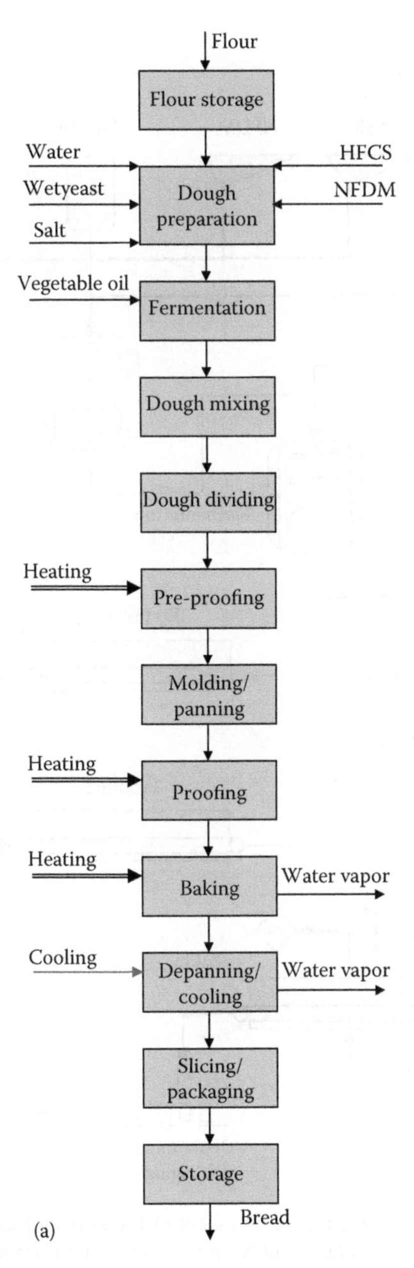

FIGURE 4.5 (a) Process block diagram of bread manufacturing. (Modified from Maroulis, Z.B. and Saravacos, G.D., *Food Plant Economics*, CRC Press, New York, August 2, 2007, fig. 8.1a.)

(*continued*)

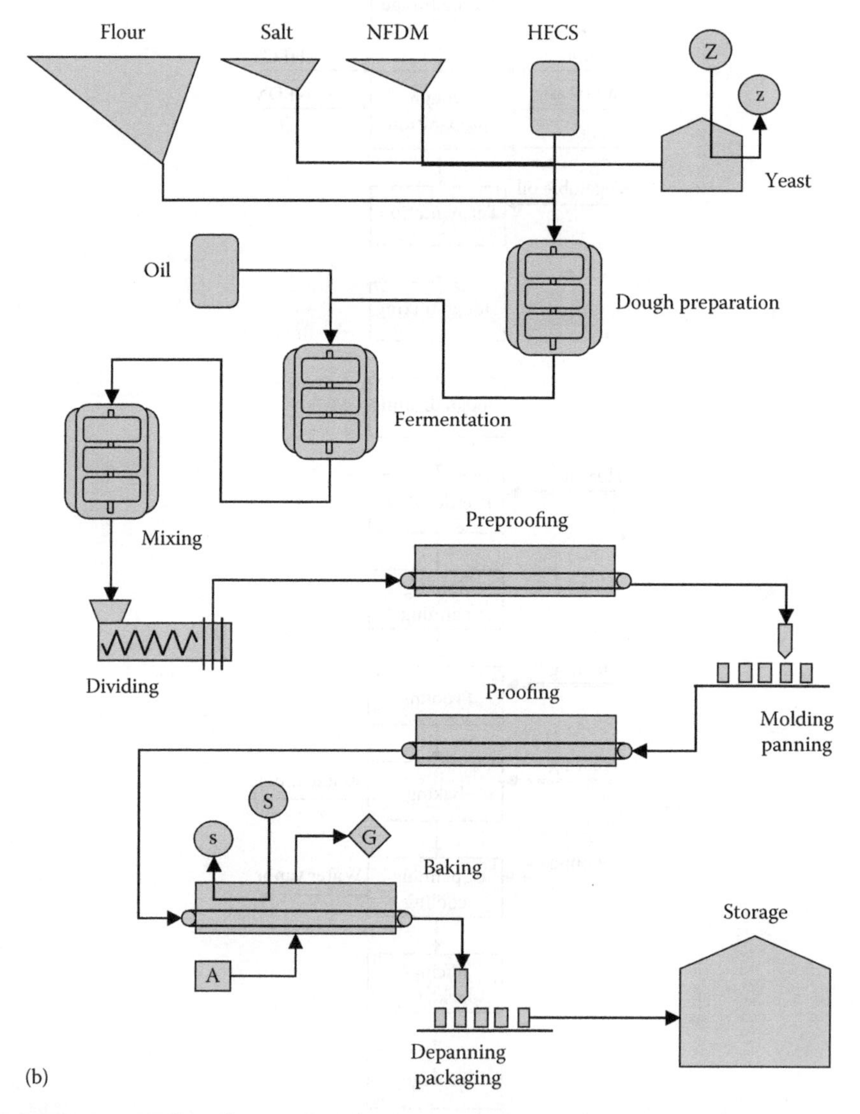

FIGURE 4.5 (continued) (b) Process flowsheet of bread manufacturing. (Modified from Maroulis, Z.B. and Saravacos, G.D., *Food Plant Economics*, CRC Press, New York, August 2, 2007, fig. 8.1b.)

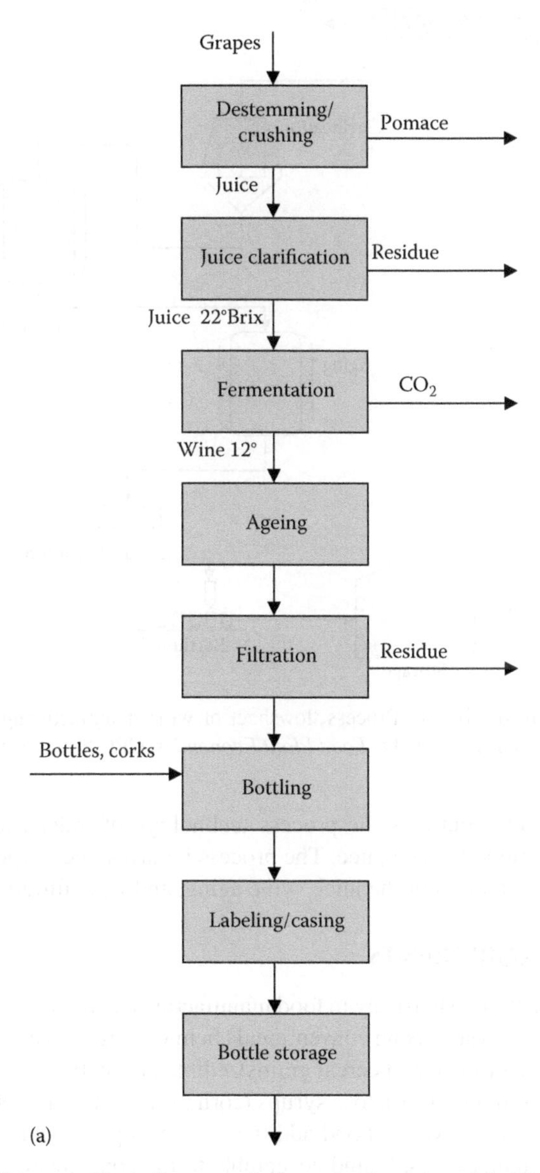

FIGURE 4.6 (a) Process block diagram of wine manufacturing. (Modified from Maroulis, Z.B. and Saravacos, G.D., *Food Plant Economics*, CRC Press, New York, August 2, 2007, fig. 8.3a.)

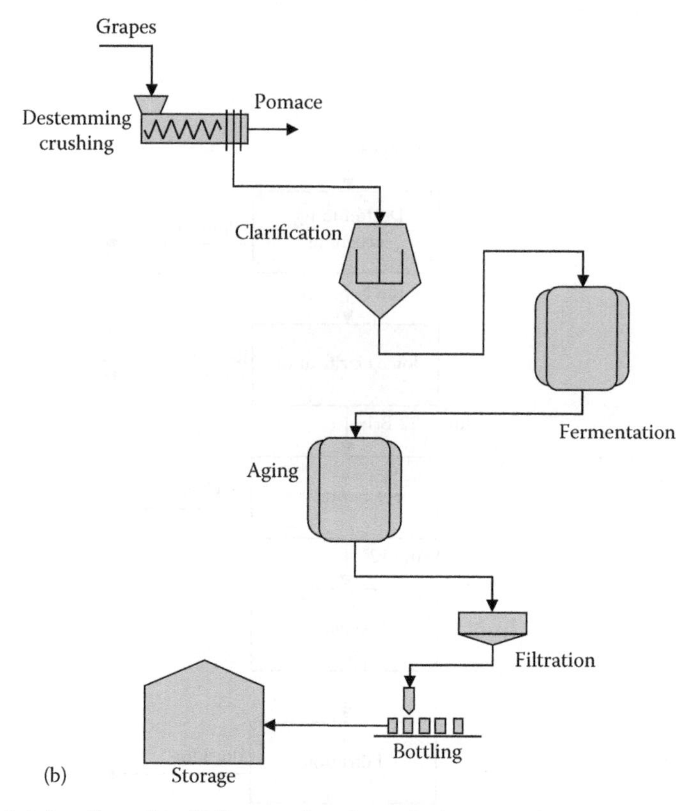

FIGURE 4.6 (continued) (b) Process flowsheet of wine manufacturing. (Modified from Maroulis, Z.B. and Saravacos, G.D., *Food Plant Economics*, CRC Press, New York, August 2, 2007, fig. 8.3b.)

Figure 4.6a and b outlines the process technology of industrial production of wine by fermentation of grape juice. The process involves juice expression from the grapes, yeast fermentation of the juice, wine aging, and wine filtration and bottling.

4.4 FOOD INGREDIENTS

Food ingredients, used extensively in food manufacturing, are produced mainly from agricultural raw materials, using conventional chemical engineering processes. They include cereal flours (milling of cereal grains), edible oils (extraction or expression of oilseeds), starches and high fructose syrups (corn processing), and protein (processing of soybeans or milk whey). Food additives, such as preservatives, antioxidants, flavor, and color materials, safe and acceptable to the consumers, are produced by specialized chemical processing technology.

The cost of food ingredients relative to the cost of raw materials is much higher than the cost of preserved or manufactured foods, mainly due to higher requirements for energy and sophisticated processing equipment.

Figure 4.7a and b outlines the process technology of crystalline sugar (sucrose) production from sugar beets. It involves several chemical engineering operations,

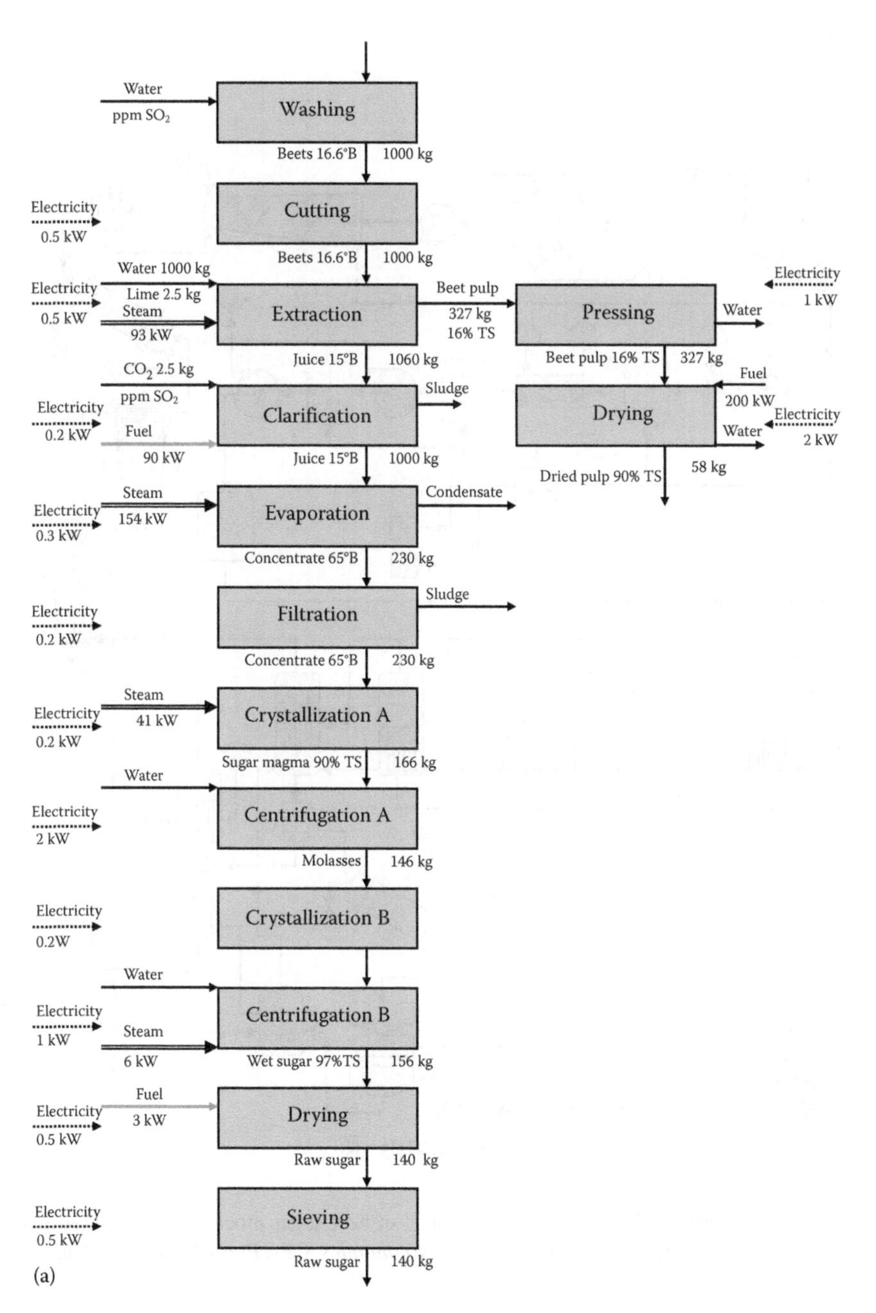

FIGURE 4.7 (a) Process block diagram of beet sugar processing. (Modified from Maroulis, Z.B. and Saravacos, G.D., *Food Plant Economics*, CRC Press, New York, August 2, 2007, fig. 9.1.)

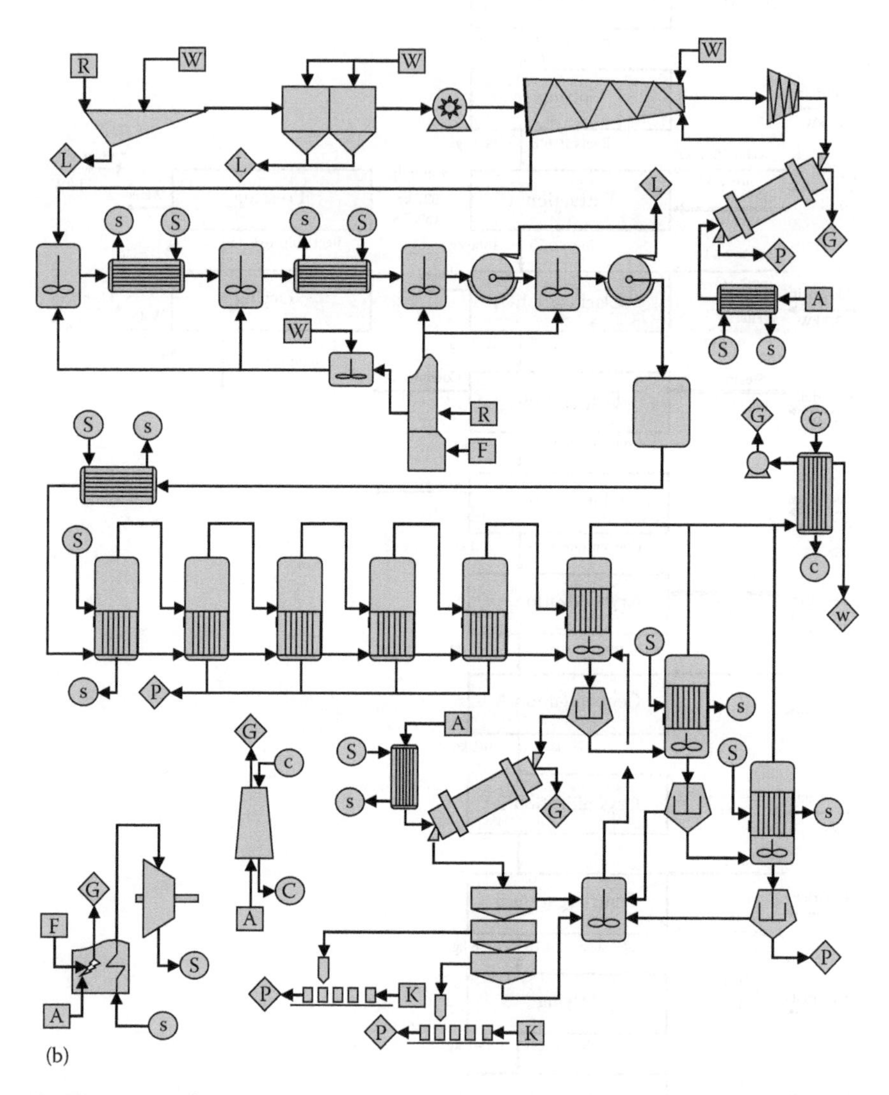

FIGURE 4.7 (continued) (b) Process flowsheet of beet sugar processing. (Modified from Maroulis, Z.B. and Saravacos, G.D., *Food Plant Economics*, CRC Press, New York, August 2, 2007, fig. 9.2.)

such as extraction (leaching), juice clarification, juice concentration, sugar crystallization, separation by centrifugation, drying, bulk storage, and drying of beet by-product.

4.5 FOOD PRODUCT ENGINEERING

Food product engineering is concerned with the molecular structure, nanostructure, microstructure, and macroscopic structure of food materials and food products, and their effect on food quality. It is based on the principles of physical chemistry, materials science, chemical engineering, food science, and food engineering.

Food product engineering operations are carried out in special equipment, developed mostly by practical experience in the food and equipment industries. Recent advances in food engineering are increasingly applied to design new food products of high quality.

Mixing equipment is used for mixing food powders, liquids, or particle suspensions. Forming equipment includes extruders, molders, sheeters, and enrobing machines. Enrobing, involving coating the food product with an edible material, such as chocolate, is used to protect the material from environmental effects (oxygen or moisture transfer), and to improve the eating quality.

PROBLEMS

4.1 Tomato paste of 32% TS is produced by concentration of tomato juice in a three-effect forced circulation evaporator. The tomatoes are washed with water sprays, inspected and cleaned from foreign matter and unfit fruit. The clean tomatoes are crushed and finished into tomato juice containing 6% TS, which is concentrated in the evaporator. The tomato paste is filled into consumer cans, which are sealed and sterilized. The canned product is packaged into carton cases.

Draw a process block diagram and a process flowsheet of the tomato paste plant, using standardized symbols.

4.2 Beer production is a complex manufacturing process, involving several engineering operations. The process technology is briefly as follows: Barley grain from storage bins is cleaned mechanically and it is steeped (immersed) in water for several hours. The wet barley is left to germinate in flat beds for a few days. The germinated barley is air-dried and it is used as the main raw material (malt) of beer manufacture. The malt is mixed with rice, water, and hops (dried flavoring plant material), and the mixture (wort) is fermented in large stainless steel tanks, using special yeast culture.

The fermentation is completed in storage tanks for several days. The beer is aged in large tanks to develop the desired flavor. The aged beer is filtered and is used either in bulk (draft beer) or it is packaged in bottles or aluminum cans. The packaged beer is often pasteurized in hot water and cooled to room temperature.

Draw a process block diagram and a process flowsheet of beer manufacture, using standardized symbols.

REFERENCES

Aguilera, J.M. and Lillford, P.J. eds. 2008. *Food Materials Science*. CRC Press, New York.
Heldman, D.R. and Hartel, R.W. 1997. *Principles of Food Processing*. Chapman and Hall, New York.
Heldman, D.R. and Lund, D.B. 2006. *Handbook of Food Engineering*, 2nd edn. CRC Press, New York.
Hui, Y.H. ed. 2006. *Hanbook of Food Science, Technology, and Engineering*, 4 volumes. CRC Press, New York.
Maroulis, Z.B. and Saravacos, G.D. 2003. *Food Process Design*. CRC Press, Boca Raton, FL.
Maroulis, Z.B. and Saravacos, G.D. 2007. *Food Plant Economics*. CRC Press, New York.
Rahman, M.S. 2007. *Handbook of Food Preservation*, 2nd edn. CRC Press, New York.
Ranken, M.D. and Kill, R.C. 1993. *Food Industries Manual*, 23rd edn. Blackie Academic and Professional, London.
Saravacos, G.D. and Kostaropoulos, A.E. 2002. *Handbook of Food Processing Equipment*. Springer, New York.
Valentas, K.J., Rotstein, E., and Singh, R.P. 1997. *Handbook of Food Engineering Practice*. CRC Press, New York.

5 Engineering Properties of Foods

5.1 INTRODUCTION

The engineering properties of foods, consisting of the thermophysical and transport properties of food materials, are essential in all food process engineering operations. Most food property data are experimental or are estimated from actual engineering data. Empirical models for particular properties are useful for approximate calculations (Rao et al, 2005; Rahman, 2009).

The thermophysical properties include the density and porosity of solid foods, whose values vary widely depending on the structure of the material. Most of the other physical properties, such as specific heat and enthalpy, radiative properties, electrical properties, acoustic, colorimetric, and surface properties, are nearly constant and characteristic for a given material.

The transport properties of foods, including the rheological properties, and the heat and mass transfer properties vary widely depending on the microstructure of the material and the processing conditions. They are reviewed in Section 5.3.

5.2 PHYSICAL PROPERTIES

5.2.1 Density and Porosity

The density of a material ρ (kg/m^3) is defined, in general, as the ratio of its mass m (kg) to the volume occupied V (m^3):

$$\rho = \frac{m}{V} \tag{5.1}$$

Three densities are used in process engineering calculations: the solids density ρ_s, the apparent density ρ_a, and the bulk density ρ_b.

The solids density ρ_s is based on the real volume of the solids, excluding the volume of the pores that can be penetrated by the volume-measuring fluid (normally pressurized gas). The solids may contain adsorbed water and any small nonpenetrated pores. The solids density is practically equal to the particle density ρ_p, used in some applications (Rahman, 2005).

The apparent density ρ_a is based on apparent volume of the solids, as measured from the geometric dimensions or from a volume-replacement fluid (organic liquid or fine sand).

The bulk density ρ_b is based on the volume of packed or stacked materials in bulk, as measured from the container volume.

The apparent porosity ε_a of a porous solid is the void (open) volume fraction, and it is given by the equation

$$\varepsilon_a = 1 - \left(\frac{\rho_a}{\rho_s}\right) \tag{5.2}$$

The bulk porosity (ε_b) of bulk solids is the void (open) volume fraction, and it is given by the equation

$$\varepsilon_b = 1 - \left(\frac{\rho_b}{\rho_s}\right) \tag{5.3}$$

In engineering applications, the simple term "porosity" refers often to either apparent or bulk porosity. Apparent porosity refers to porous individual particles or pieces, containing internal pores. Bulk porosity characterizes bulk particles or pieces packed together, containing external voids (empty spaces), in addition to any internal pores. The bulk density and porosity of individual solid particles or pieces are identical to their respective apparent properties.

Figure 5.1 shows the solids density of starch at low moisture contents. A maximum of density is observed at about 5% moisture, corresponding to the monomolecular layer, below which water is desorbed without volume change. Figure 5.2 shows the apparent porosity of granular and gelatinized starches at low moistures.

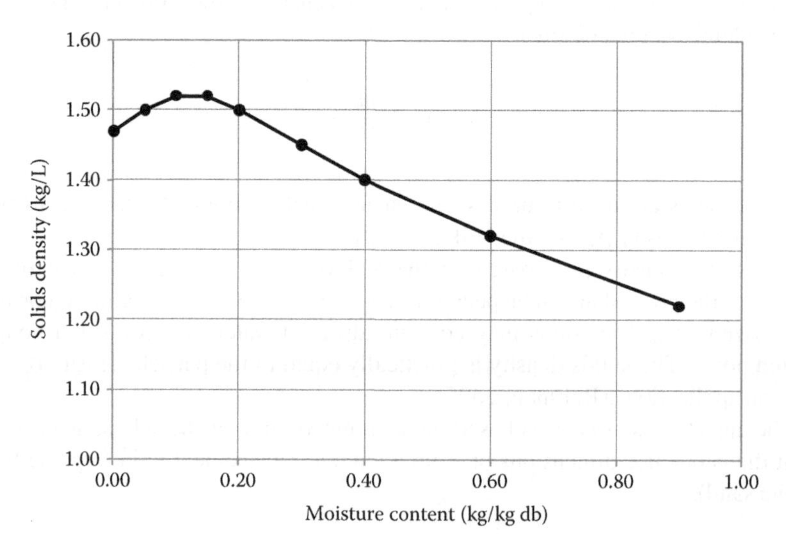

FIGURE 5.1 Solids density ρ_s of starch versus moisture content, dry basis.

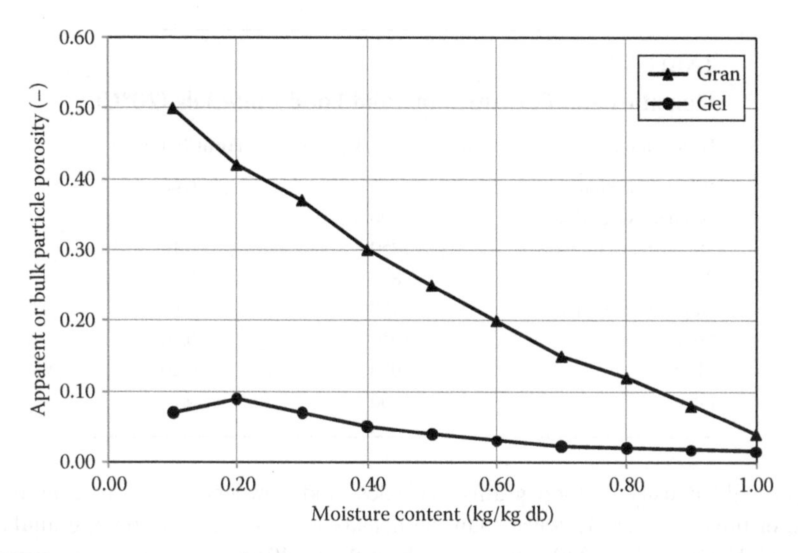

FIGURE 5.2 Apparent or bulk particle porosity ε of granulated (Gran) and gelatinized (Gel) starch versus moisture content, dry basis.

The gelatinized starch shrinks more severely during drying (low ε), contrary to the granular material (high ε).

The bulk density of a powder (mass/total volume occupied) depends on how it is packed into a container. Four types of bulk densities of powders of increasing magnitude are used based on how the particles are packed in a measurement device: (1) aerated bulk density, (2) poured bulk density, (3) tap density, and (4) compacted bulk density.

Detailed definitions and methods of measurement of density and porosity of solids are given in the literature (Rahman, 1995, 2009).

The solid density ρ_s of solid food components depends on their physicochemical structure, varying in the range of 1270–1670 kg/m^3, as shown in Table 5.1.

The density and porosity of solid foods vary widely due to their composition and physical structure, as shown in Table 5.2.

TABLE 5.1
Solids Density of Dry Food Components

Food Component	Solids Density (ρ_s) (kg/m^3)
Glucose	1544
Sucrose	1588
Starch	1500
Protein	1400
Gelatin	1270
Salt (NaCl)	2163

TABLE 5.2
Densities and Porosities of Solid Food Materials (20°C)

Food Material	Density (ρ) (kg/m³)	Porosity (ε) (–)
Wheat flour (bulk)	785	0.40
Granular sugar (bulk)	800	0.38
Starch gel (25% moist)	1250	0.10
Apple, fresh	840	0.35
Apple, freeze dried	200	0.85
Potato	1040	0.20
Meat	1070	0.20
Fish	1100	0.19

The bulk density of food grains, particles, and powders, stored in containers, silos, or bins is generally lower than the density of water, for example, granulated sugar 800 kg/m³, wheat 770 kg/m³, and wheat flour 640 kg/m³.

The shrinkage of solid foods during processing, e.g., drying, can be expressed by the apparent shrinkage coefficient S, defined by the equation

$$S = \frac{V_p}{V_o} \tag{5.4}$$

where V_o and V_p are the apparent volumes of the material before and after processing.

Shrinkage of fruit and vegetable materials is usually isotropic, i.e., uniform shrinkage in all directions. Anisotropic (nonuniform) shrinkage may be observed in some animal or fibrous foods, e.g., fish.

Table 5.3 shows typical densities of miscellaneous liquid and frozen food materials.

TABLE 5.3
Typical Densities of Miscellaneous Liquid and Frozen Food Materials

Food Material	Temperature (°C)	Density (kg/m³)
Water	20	1000
Sugar solutions	20	
15%		1060
40%		1180
65%		1280
Milk	20	1025
Vegetable oil	20	900
Ice	0	920
Frozen foods	−10	980

The density of dilute aqueous food solutions is slightly higher than that of water (1000 kg/m^3), but it increases substantially as the concentration of the solute is increased. The density of fruit and vegetable suspensions (pulps) is slightly higher than 1000 kg/m^3.

The density of vegetable oils at 20°C is about 900 kg/m^3, decreasing with the temperature. The density of semi-solid fat is 900–950 kg/m^3.

The density of milk at 20°C varies in the narrow range of 1000–1050 kg/m^3. The density of ice at 0°C is 920 kg/m^3 and of frozen foods is 980 kg/m^3.

5.2.2 Thermal Properties

The thermal properties of foods of importance to food engineering include specific heat, enthalpy, and latent heat (Nesvadba, 2005). The thermal conductivity and diffusivity, and the heat transfer properties are discussed in Section 5.3.2.

The specific heat and the related thermal properties of foods are not affected significantly by the physical structure of the material. They are affected by the chemical composition, particularly the moisture content of the food. There is a significant difference between the specific heats of nonfrozen and frozen foods, due to the different physical properties of liquid and frozen water (ice).

5.2.2.1 Nonfrozen Foods

Water has the highest specific heat of all food components, as shown in Table 5.4:

The specific heat C_p of a food can be estimated from the specific heats of its components, using the summation equation:

$$C_p = \sum x_i C_{pi} \tag{5.5}$$

where x_i and C_{pi} are the mass fraction and the specific heat of component i.

Table 5.5 gives typical values of the specific heat of food materials.

TABLE 5.4
Specific Heats (C_p) of Food Components (20°C)

Food Component	C_p (kJ/(kg K))
Water	4.18
Ice	2.06
Carbohydrate	1.52
Protein	2.99
Fat	1.98
Fiber	1.85
Food solids	1.96
Ash	1.09
Air (1 bar pressure)	1.01

TABLE 5.5
Specific Heats (C_p) of Some Food Materials (20°C)

Food Material	C_p (kJ/(kg K))
Water	4.18
Potato (73% moisture)	3.70
Apple (85% moisture)	4.00
Meat (74% moisture)	3.60
Fish (80% moisture)	3.70
Milk	3.80
Soybean oil (100°C)	2.18
Wheat flour (10% moisture)	1.83
Bread (48.5% moisture)	2.84

The enthalpy H of a food is the integral of the specific heat over a temperature range plus any latent heat ΔH:

$$H = \sum mC_p\Delta T + m\Delta H \tag{5.6}$$

A convenient reference temperature for enthalpy calculations is 0°C.

The latent heat ΔH in aqueous systems refers to either the heat of evaporation of water (ΔH_v = 2.3 MJ/kg at 100°C) or the heat of fusion (melting) of ice (ΔH_f = 0.33 MJ/kg).

The enthalpy of water and steam as a function of temperature and pressure is given in the Steam Tables (Appendix).

In fats and oils, calculation of enthalpy should include the heat of melting or crystallization, if the temperature goes through a phase change (melting). The heat of melting of lipid materials is in the range of 0.08–0.2 MJ/kg, depending on the molecular weight of the triglyceride.

5.2.2.2 Frozen Foods

The thermal properties of frozen foods are considerably different from the properties of nonfrozen food, because of the different properties of unfrozen water and ice. The specific heat of frozen foods is not constant, but it depends on the state of water (frozen–unfrozen). As the temperature is reduced, the unfrozen water decreases gradually.

The enthalpy and the unfrozen water in frozen foods can be estimated from the enthalpy–concentration diagrams, determined experimentally by Riedel.

The specific heat of ice is 1.96 kJ/(kg K) compared to 4.16 kJ/(kg K) of liquid water, while its density is 917 kg/m^3. Based on the enthalpy change of frozen food in a certain temperature range, the mean specific heat is estimated as C_p = 1.8 kJ/(kg K), which is about one-half of that of water, The specific heat of fat is C_p = 1.7–2.2 kJ/(kg K) at temperatures $T > 40$°C, and C_p = 1.5 kJ/(kg K) at temperatures $T < 0$°C.

The apparent specific heat of frozen foods is nearly C_p(ice) = 1.96 kJ/(kg K) at temperatures below −20°C. However, it increases sharply as the temperature is raised, reaching a maximum value of about 30 kJ/(kg K) just below 0°C, and dropping abruptly to nearly 4.18 kJ/(kg K) at higher temperatures.

5.2.3 Mechanical Properties

Mechanical properties are important in the design and operation of food processes involving handling and deformation of solid food materials. They are essential in determining food texture, quality, and acceptability of food products.

Fluid foods are either Newtonian or non-Newtonian and their rheological properties are expressed by the viscosity or the apparent viscosity, respectively, as discussed in Section 5.3.1.

Elastic solids follow Hooke's law, characterized by the modulus of elasticity (Young's modulus). They are deformed by the application of a force, but they return to the original condition, when the force is removed. Most solid and semisolid foods are viscoelastic, i.e., their properties lie between the viscous fluid and the elastic solid.

The mechanical properties of food materials can be evaluated by three types of tests, i.e., quasistatic, dynamic, and empirical tests (Rao et al, 2005).

5.2.3.1 Static Tests

Normal static tests are actually quasistatic, since the application of any force to a body creates dynamic conditions.

The mechanical properties of the elastic solids are characterized by the stress σ and the strain ε, defined by Equations 5.7 and 5.8:

$$\sigma = \frac{F}{A} \tag{5.7}$$

$$\varepsilon = \frac{\Delta L}{L} \tag{5.8}$$

where

- F is the applied force
- A is the cross-sectional area of the body
- ΔL is the deformation
- L is the original length of the body

Viscoelastic materials are represented by a number of springs (elastic) and dashpots (viscous). Because of their complex viscoelastic behavior, solid foods cannot be characterized accurately. They may be expressed by a combination of two basic models, i.e., the Maxwell (spring and dashpot in series) and the Kelvin model (spring and dashpot in parallel).

The viscoelastic behavior of foods can be evaluated by three simple tests, i.e., (a) the creep curve, showing the strain as a function of time at constant stress; (b) the relaxation curve, showing the stress as function of time at constant strain; and (c) the dynamic modulus test, showing the dynamic modulus as a function of frequency of the strain.

The modulus of elasticity (Young's modulus) is defined by the Hooke equation,

$$\text{Modulus of elasticity (Young's modulus)} = \frac{\sigma}{\varepsilon} \tag{5.9}$$

The Young modulus decreases slightly as the temperature is increased. The Young modulus of steel at 20°C is 2.1 Mbar and of aluminum 0.69 Mbar.

The dimensional changes of a material due to the application of a stress (or stretch) load are expressed by the Poisson ratio ν, which is the ratio of the transverse (contraction) strain, perpendicular to the applied load, to the axial (extension) strain, in the direction of the applied load (Equation 5.10),

$$\nu = -\frac{\varepsilon(\text{trans})}{\varepsilon(\text{axial})} \tag{5.10}$$

The strains ε(trans) and ε(axial) are either positive or negative, according to the stress,

$\varepsilon(\text{trans}) > 0$ for axial compression, and $\varepsilon(\text{trans}) < 0$ for axial tension
$\varepsilon(\text{axial}) > 0$ for axial tension, and $\varepsilon(\text{axial}) < 0$ for axial compression space

Most materials have positive ν values between 0.0 and 0.5. Rubber is elastic having $\nu = 0.5$. Most steels and rigid polymers have $\nu = 0.3$. A few special materials, such as foams, may have negative Poisson ratios.

5.2.3.2 Dynamic Tests

Dynamic tests are very fast methods for determining the mechanical properties (e.g., the Young modulus of elasticity) of viscoelatic materials, including foods. They apply very small strains, resulting in linear stress–strain behavior. Two dynamic tests are used mainly for food materials: direct measurement of stress and strain and resonance methods.

In the direct stress–strain method, a sinusoidally varying strain is provided to a sample, and the resulting stress is observed as a function of time. The stress σ (Pa) will vary with the same frequency as the strain (ε) but it will lag behind the strain by an angle θ, known as the phase angle, and given by the equations

$$\varepsilon = \varepsilon_0 \sin(\omega t), \quad \sigma = \sigma_0 \sin(\omega t - \theta) \tag{5.11}$$

where
ω (1/s) is the frequency
t (s) is the time

Theoretical and empirical equations are used to estimate the modulus of elasticity of a sample of known dimensions and density, utilizing measured data of the complex dynamic modulus and the phase angle of the system. A resonance dynamic tester consists of a frequency generator, transmitting a sinusoidal signal through the material, which is amplified and measured by a voltmeter.

The Young modulus and the Poisson ratio are estimated from measured frequency response and sample dimensions using semitheoretical equations.

5.2.3.3 Empirical Tests

Mechanical and texture measurements should be able to replace human sensory evaluation of food quality. Some of the empirical mechanical/texture measurement instruments used for foods are the following (Bourne, 1982):

The General Foods texturometer, which simulates the masticating action of human mouth, uses strain gauges to detect and record the applied forces. The instrument can generate a texture profile analysis curve, from which some important texture parameters can be obtained, such as hardness, fracturability, and gumminess.

The shear press, which is adaptable to the Instron universal testing machine, consists of a metal box with 10 blades, shearing the food sample. It is used for measuring the maturity and textural properties of fruits and vegetables.

Other empirical texture measuring instruments are the Warner-Bratzler shear, used in evaluating the texture of meat; the Penetrometer, used for measuring the quality of gels and jellies; the Ridgelimeter for gels; the Magness-Taylor pressure tester for fruits and vegetables; and the Baker compressimeter for measuring the firmness of bread.

5.2.4 Electrical Properties

Electrical properties are important in the development and application of novel food processing technologies, such as ohmic heating, dielectric and microwave heating, and pulsed electric field (PEF) processing.

The dielectric and microwave properties of importance to food heating operations are reviewed in Section 8.4.

Electrical conduction in foods is caused mainly by ions and second by dipole rotation of water molecules. The electrical conductivity σ of a material is defined by the equation

$$\sigma = \frac{L}{AR} \tag{5.12}$$

where

L (m) is the length of the conduction path
A (m^2) is the cross-sectional area
R (ohm) is the electrical resistance, defined by the equation

$$R = \frac{V}{I} \tag{5.13}$$

where

V (volt) is the applied electrical potential
I (ampere) is the electrical current

The units of electrical conductivity are (s/m), where s = Siemens (1 s = 1/ohm = 1 mho). The electrical conductivity of liquid foods is approximately analogous to the mass diffusivity of spherical particles (ions or colloidal particles), as discussed in Section 5.3.3. Thus, electrical conductivity σ is inversely proportional to the viscosity (η) of the liquid. Suspended solids in the solution have a strong effect on σ.

The electrical conductivity of liquid foods increases almost linearly with the temperature.

Typical values of σ in the temperature range 5°C–60°C are beer 0.08–0.26 s/m, orange juice 0.31–0.61 s/m, skim milk 0.31–0.69 s/m, and tomato juice 1.19–3.14 s/m.

The electrical conductivity of solid foods depends strongly on their microstructure (porosity), an analogous behavior to the mass transport properties (diffusivity) of small particles (molecules) in similar materials (Section 5.3.4).

5.2.5 Acoustic Properties

5.2.5.1 Sound Propagation

In process engineering applications, sound waves are considered to propagate longitudinally, i.e., parallel to the direction of transport.

Acoustic properties of materials include sonic (0.1–20 kHz) and ultrasound ($f > 20$ kHz) frequencies f. The speed of sound c (m/s) is given by the equation

$$c = f\lambda \tag{5.14}$$

where

f (Hz or 1/s) is the frequency of vibration

λ(m) is the wavelength

The speed of sound c (m/s) in solids, such as steel, is related to the Young modulus of elasticity E (Pa), according to the equation

$$c = \frac{(E)^{1/2}}{\rho} \tag{5.15}$$

where ρ (kg/m³) is the density of the material. For steel $\rho = 7800$ kg/m³.

Typical speeds and wavelengths of sound, sonic (10 kHz) and ultrasound (5 MHz), are shown in Table 5.6.

It is noted that aluminum and steel have approximately similar sound constants, although aluminum is a lighter material. This is due to the lower Young's modulus for aluminum.

5.2.5.2 Food Applications

Ultrasound waves at low power levels are used in food engineering to characterize the structure and composition of various food materials. Ultrasonic attenuation, i.e., energy loss of sound waves during transmission in a medium, can be used to characterize various properties, such as the viscosity of fluid foods.

TABLE 5.6
Typical Speeds of Sound (c) and Wavelengths (λ) at 20°C

Material	c (m/s)	λ (10 kHz) (cm)	λ (5 MHz)
Air	343	3.43	69 μm
Water	1482	14.8	298 μm
Steel	5790	58.0	1.12 mm
Aluminum	5100	51.1	1.02 mm

The speed of sound in aqueous sugar solutions and in fruit juices increases almost linearly with the sugar concentration and the temperature. A similar relationship is observed in aqueous ethanol solutions.

The speed of sound in vegetable oils decreases significantly from about 1500 to 1300 m/s, as the temperature is raised from 20°C to 70°C. In the whole milk the speed of sound increases from 1460 to 1560 m/s in the temperature range of 9°C to 50°C.

The composition of meat and fish can be estimated using empirical models, based on the different speeds of sound in the major components, fat, protein, and water.

The texture of fruits and cheese can be evaluated by measuring the speed of sound, which is a function of the Young modulus of elasticity (Equation 5.15).

5.2.5.3 Doppler Velocimetry

The Doppler effect is the change in frequency of a sound wave, caused by the motion of a fluid, in which the sound is propagated.

Ultrasound Doppler velocimetry (UDV) is used to measure the (pipe) flow velocity in various engineering applications. Thus, the velocity distribution of laminar flow in a pipe can be determined, from which the shear stress–shear rate curve and the apparent viscosity of the fluid are determined (Chapter 6).

5.2.6 Colorimetric Properties

Human vision detects colors in the range of electromagnetic radiation 390–760 nm. Empirical systems of measurement of color of a material are based on the visual comparison with a number of standard color solids, such as the Munsell system, which contains 1225 color chips, arranged for convenient visual comparison. For example, the official USDA grades of tomato juice are measured by a number of spinning disks of specific Munsell designation.

Instrumental methods of color measurement are more convenient than visual comparison. They are based on the spectrophotometric reflection and transmission curves, from which the characteristic XYZ values are calculated. The XYZ international color system Commission Internationale d' Eclairage (CIE) is based on coordinate spectra corresponding approximately to red (X), green (Y), and blue (Z) colors.

Tristimulus colorimeters use the glass filters with transmission spectra that duplicate the X, Y, Z curves, measured with a photocell. Color measurement can be used for evaluating food quality, for example, carotenoid content, during processing and storage of fruits and vegetables. Color measurement instruments have been developed for quality control in food processing. Modern instruments combine color measurement with data treatment.

5.2.7 Surface Properties

5.2.7.1 Introduction

The surface properties of food materials are important in food materials science and engineering and in processing of food products.

According to thermodynamics (Chapter 3), the Helmholtz free energy A (J) at interfaces of materials is given by the equation

$$A = U - TS \tag{5.16}$$

where

U (J) is the internal energy
S (J/K) is the entropy
T (K) is the temperature

The Helmholtz energy is identical to the Gibbs energy G (J) at constant volume and temperature.

The surface energy γ (J/m^2) of a material is given by the thermodynamic equation

$$\gamma = \left(\frac{\mathrm{d}A}{\mathrm{d}a}\right)_{T,V} \tag{5.17}$$

where a (m^2) is the interface surface area.

Systems tend to minimize the total free energy, for example, by minimizing the interfacial area or by preferential adsorption of a substance.

The work of cohesion W_c (J/m^2) is the reversible work per unit area required to separate two surfaces of a bulk material. By definition, $W_c = 2\gamma$. The work of adhesion W_a (J/m^2) is the reversible work per unit area required to separate two dissimilar surfaces.

The higher energy of surface molecules is attracted inward and a surface tension γ is created with units (N/m), which are identical to the units of surface energy γ (J/m^2). The term surface tension is used mainly for liquid systems, while surface energy is used as a more general term, involving solids/liquids, solids/solids, and liquids/liquids.

5.2.7.2 The Gibbs Equation

The Gibbs adsorption equation relates the adsorbed amount of a species Γ_i (mol/m^2) to the surface energy γ (J/m^2) and the chemical potential of the species μ_i (J/mol):

$$-\mathrm{d}\gamma = \Gamma_i \mathrm{d}\mu \tag{5.18}$$

According to this equation, adsorption results in a sharp reduction of the surface energy. The generalized Gibbs equation 5.18 can be simplified for dilute solutions, substituting the chemical species potential μ with the concentration c, resulting in the equation

$$\Gamma_l = -\left(\frac{c_i}{RT}\right)\left(\frac{d\gamma}{dc_i}\right) \tag{5.19}$$

5.2.7.3 Liquids

The surface tension (liquid surface energy) can be measured readily by the four following methods:

1. The capillary rise method, used for pure liquids. The surface tension is calculated from height of the liquid rise, the radius of the capillary tube, the (liquid–vapor) density, and the contact angle between the liquid and the capillary tube wall.
2. The drop method, in which the volume of a drop of the liquid is measured, as it is detached from the tip of a tube of known radius.
3. The Wilhelmy plate method, in which a plate is immersed in the liquid, and the surface tension is calculated from the force needed to detach the plate from the liquid.
4. The Du Nouy ring method, in which the force required to lift the ring from the liquid is related to the surface tension.

Table 5.7 shows the surface tension γ of some liquids of importance to food processing. The γ of pure liquids decreases considerably as the temperature is increased, for example, γ (water) decreases from 75.6 mN/m (0°C) to 58.9 mN/m (100°C).

5.2.7.4 Solids

Determination of surface energy in solids is more complicated than in liquids.

Contact angle analysis of liquid/solid surfaces is used to evaluate the surface energy of solids. The contact angle θ of a liquid droplet on a solid surface, measured

TABLE 5.7
Surface Tension (γ) of Some Liquids Related to Food Processing (20°C)

Liquid	γ (mN/m)
Water	71.2
Water/10% ethanol	49.8
Water/20% ethanol	40.7
Water/30% ethanol	35.4
Ethanol	22.4
Nonpolar organics	20–30

with a goniometer, is related to the surface tension of the liquid γ_L and the work of adhesion W_a, according to the equation

$$W_a = \gamma_L(1+\cos\theta) \tag{5.20}$$

In the system solid/liquid/air three surface energies are involved, (1) liquid (γ_L), (2) solid (γ_S), and (3) liquid/solid (γ_{SL}). Energy (force) balance in the three-phase system around the droplet results in the equation

$$\gamma_S = \gamma_{SL} + \gamma_L\cos\theta \tag{5.21}$$

5.3 TRANSPORT PROPERTIES

The transport properties of food materials are very important in food engineering calculations involving heat and mass transfer operations. The principal transport properties of interest to food engineering are the viscosity (momentum transport) η, the thermal conductivity (heat transport) λ, and the mass diffusivity (mass transport) D. Due to the complex structure of foods, theoretical prediction of the transport properties is not possible, and experimental values or empirical correlations may be used, when available, for the specific food material (Saravacos and Maroulis, 2001).

5.3.1 Transport Properties of Gases and Liquids

The transport properties of gases and simple (Newtonian) liquids have been investigated in molecular dynamics and thermodynamics. Property data are available in the chemical engineering literature in the form of data banks or semitheoretical correlations. Such data are useful in process design and processing operations.

The transport of momentum, heat, and mass in one direction (x) is expressed by the following analogous transport equations (Reid et al, 1987; Gekas, 1992):

$$\tau_{yx} = -\eta\left(\frac{\delta u}{\delta x}\right) \tag{5.22}$$

$$\left(\frac{q}{A}\right)_x = -\lambda\left(\frac{\delta T}{\delta x}\right) \tag{5.23}$$

$$\left(\frac{J}{A}\right)_x = -D\left(\frac{\delta C}{\delta x}\right) \tag{5.24}$$

where

τ_{yx} (Pa/m²) is the shear stress and η (Pa s) or kg/(m s) is the viscosity
$(q/A)_x$ (W/m²) is the heat flux and λ [W/(m K)] is the thermal conductivity
$(J/A)_x$ (kg/m² s) is the mass flux and D (m²/s) is the mass diffusivity
u (m/s) is the velocity in the x-direction, T (K) is the temperature
C (kg/m³) is the concentration of the species being transported
($\delta u/\delta x$), ($\delta T/\delta x$), and ($\delta C/\delta x$) are the gradients of the three transport processes

The following derived transport properties are also used for convenience in calculations,

$\nu = \eta/\rho$ (m^2/s), kinematic viscosity or momentum diffusivity

$\alpha = \lambda/(\rho C_p)$ (m^2/s), thermal diffusivity, where ρ (kg/m^3) is the density and C_p (J/kg K) is the specific heat.

The transport properties of real gases can be estimated from the following semi-theoretical equations:

$$\eta = 2.669 \times 10^{-26} \left[\frac{(MT)^{0.5}}{\sigma^2 \Omega_\eta} \right] \tag{5.25}$$

$$\lambda = 8.3224 \times 10^{-22} \left[\frac{(T/M)^{0.5}}{\sigma^2 \Omega_\eta} \right] \tag{5.26}$$

$$D = 1.883 \times 10^{-26} \left[T^{1.5} \frac{(1/M_A + 1/M_B)}{P\sigma^2 \Omega_D} \right] \tag{5.27}$$

where

M is the molecular weight

T (K) is the temperature

P (Pa) is the pressure

σ (m) is the collision diameter

Ω_η, Ω_D (–) are the collision integrals

In the diffusivity equation, M_A and M_B are the molecular weights of diffusing species and gas medium, respectively.

The collision diameter and integrals are characteristic parameters of gases, described in molecular dynamics.

Equations 5.25 and 5.26 show that η and λ increase with the square root of the temperature, while D is a function of $T^{1.5}$. Equation 5.27 indicates that at constant temperature, PD = constant, i.e., the mass diffusivity is inversely proportional to the pressure. This relationship is important in evaluating gas diffusivities at very low pressures, for example, in freeze-drying.

The viscosity η and the thermal conductivity λ of real gases are correlated with empirical equations, such as the Eucken equation for monoatomic gases:

$$\frac{(\lambda M)}{(\eta C_v)} = 2.5 \tag{5.28}$$

where

M (kg/kmol) is the molecular weight

C_v (kJ/kmol K) is the specific heat at constant volume

Note that $C_v = C_p - R$, where $R = 8.314\,\text{kJ/kmol K}$. The viscosity and thermal conductivity of liquids are determined experimentally or obtained from empirical correlations, since theoretical prediction is difficult. Temperature has a negative effect on the viscosity of liquids, which is stronger at high viscosities, for example, vegetable oils and sugar solutions.

The temperature effect on viscosity η can be described by the Arrhenius equation

$$\eta = \eta_0 \exp\left(\frac{E_0}{RT}\right) \tag{5.29}$$

where

E_0 (kJ/mol) is the energy of activation for flow
T (K) is the temperature
$R = 8.314\,\text{kJ/kmol K}$

The E_0 of water is 14.4 kJ/mol, but it increases to about 60 kJ/mol for concentrated sugar solutions or fruit juice concentrates. Pressure has a negligible effect on both viscosity η and thermal conductivity λ.

The mass diffusivity D of a molecular species (A) in a liquid (B) is estimated by the semitheoretical Wilke–Chang equation

$$D_{AB} = \frac{[1.17 \times 10^{-16}\, T(\varphi M_B)^{0.5}]}{[\eta_B V_A^{\,0.6}]}\ \text{m}^2/\text{s} \tag{5.30}$$

where

M_B is the molecular weight of the liquid
V_A (m^3/kmol) is the molar volume of (A) at the boiling point
φ (–) is an interaction parameter, for example, 2.26 for water

The D of a particle of radius r_p (m/molecule) is given by the Stokes–Einstein equation

$$D = \frac{(k_B T)}{(6\pi\eta_B r_p)} \tag{5.31}$$

where

k_B (1.38×10^{-23} J/molecule K) is the Boltzmann constant
T (K) is the temperature
η_B (Pa s) is the viscosity of the liquid

From Equation 5.30 it follows that

$$\frac{(\eta_B D)}{T} = \text{constant} \tag{5.32}$$

TABLE 5.8
Transport Properties of Air/Water at 25°C

Air	$\eta = 0.017\,\text{mPa s}$	$\lambda = 0.025\,\text{W/(m K)}$
Water	$\eta = 0.90\,\text{mPa s}$	$\lambda = 0.62\,\text{W/(m K)}$
Oxygen (A)/air (B)	$D_{AB} = 1.7 \times 10^{-5}\,\text{m}^2/\text{s}$	
Oxygen (A)/water (B)	$D_{AB} = 1.7 \times 10^{-9}\,\text{m}^2/\text{s}$	

This means that the mass diffusivity is inversely proportional to the viscosity of the liquid.

Temperature has a positive effect on the mass diffusivity D, according to the Arrhenius equation:

$$D = D_0 \exp\left(-\frac{E_D}{RT}\right) \tag{5.33}$$

where

E_D (kJ/mol) is the activation energy for diffusion
($R = 8.314\,\text{kJ/kmol K}$) is the universal gas constant
D_0 (m^2/s) is a constant

Table 5.8 shows some transport properties of air, oxygen, and water, which are useful in food engineering calculations:

Some useful analogies of transport properties of gases (G) and liquids (L),

$$\left(\frac{\eta_L}{\eta_G}\right) = \left(\frac{\lambda_L}{\lambda_G}\right) = 10 - 10^3, \quad \left(\frac{D_L}{D_G}\right) = \left(\frac{1}{10^4}\right)$$

5.3.2 Rheological Properties

Food rheology deals with the deformation and flow of food materials due to external (mechanical) forces. Viscometry is concerned with the flow of fluids, while visco-elasticity and elasticity (Section 5.2.3) characterize the solids and semisolids.

Most fluid foods are characterized either by the viscosity (Newtonian fluids) or the rheological constants (non-Newtonian fluids). Due to the complex physical and chemical structure of foods, theoretical prediction of the rheological properties is not possible, and experimental measurements and empirical correlations are used in food engineering calculations (Rao, 2007).

In Newtonian fluids, the shear rate τ (Pa/m^2) is proportional to the shear rate γ (1/s) and the proportionality constant is the viscosity η (Pa s), according to Equation 5.34, which is similar to the basic momentum transport Equation 5.6.

$$\tau = \eta\gamma \tag{5.34}$$

Most non-Newtonian fluid foods are represented by the power law model

$$\tau = K\gamma^n \tag{5.35}$$

where

K (Pa s^n) is the flow consistency coefficient or rheological constant
n (–) is the flow behavior index

Most fluid foods are pseudoplasic or shear-thinning fluids ($n < 1$), while a few are considered dilatant ($n > 1$).

The rheological constant (K) decreases as the temperature is increased, according to an Arrhenius-type equation, similar to equation for viscosity:

$$K = K_0 \exp\left(\frac{E_0}{RT}\right) \tag{5.36}$$

The flow behavior index n is not affected significantly by the temperature.

The solids concentration C in a liquid food has an exponential effect on the rheological constants (η, K) according to the equation

$$K = K_0 \exp(BC) \tag{5.37}$$

where B is a characteristic constant.

The Herschel–Bulkley model is a generalized equation for non-Newtonian fluids:

$$\tau = \tau_0 + K\gamma^n \tag{5.38}$$

where τ_0 (Pa/m^2) is the yield stress.

The Casson rheological model is used to characterize some special foods, such as chocolate:

$$\tau^{0.5} = \tau_0^{0.5} + K_c\gamma^n \tag{5.39}$$

where K_c (Pa s^n) is the flow consistency coefficient.

The apparent viscosity η_a (Pa s) of a non-Newtonian fluid is a useful property in food engineering calculations, and it is defined by the equation

$$\eta_a = K\gamma^{n-1} \tag{5.40}$$

Most non-Newtonian fluid foods are time-independent of shearing, that is, their rheological properties are not affected by the duration of (mechanical) shearing. Some fluid food may be affected by the time of shearing, behaving either as thixotropic (the apparent viscosity at constant shearing decreases with time), or as rheopectic (the apparent viscosity at constant shearing increases with time).

The rheological properties of fluid foods (η, K, n) are determined experimentally by measuring the shear stress at various shear rates applied to the material.

TABLE 5.9
Viscosities of Newtonian Food Liquids (20°C)

Material	η (mPa s)
Water	1
Sugar solution 20°Brix	2
Sugar solution 60°Brix	20
Honey	20,000
Corn oil	60
Homogenized milk	2

TABLE 5.10
Rheological Constants of Some Pseudoplastic Food Fluids

Material	K (Pa s^n)	n (–)
Orange juice concentrate 65°Brix	0.4	0.76
Apple sauce	2.6	0.30
Mayonnaise	6.1	0.50
Mustard	18.5	0.39
Tomato paste 35°Brix	120.0	0.30

The common instruments are based on the flow of the material in a long tube of small diameter or on the shearing of the fluid in a rotating system (Chapter 6).

Typical viscosities of Newtonian liquid foods are shown in Table 5.9.

The rheological constants (K, n) of some non-Newtonian (pseudoplastic) food fluids are shown in Table 5.10.

The effect of temperature on the rheological constants depends on the structure and the solids concentration of the food material. The temperature effect is expressed by the activation energy for flow (Equations 5.28 and 5.35), as shown in Table 5.11 for aqueous sugar solutions.

The mechanical properties of solids are discussed in Section 5.2.3. The elastic (Hookean) solids are analogous to the Newtonian fluids, with the Young modulus of elasticity ($Y = \sigma/\varepsilon$) corresponding to the viscosity ($\eta = \tau/\gamma$).

In a generalized expression for both solids and fluids, the shear stress σ (Pa) is a function of the applied strain γ (1/s):

$$\sigma = G\gamma \tag{5.41}$$

Solid and semisolid foods may be viscoelastic or viscoplastic materials. In viscoelastic foods, both Newton's viscosity Equation 5.33 and Hooke's elasticity law (Equation 5.9) may be applicable.

TABLE 5.11
Activation Energies for Flow (E_0) of Sugar Solutions

Sugar % Total Solids	E_0 (kJ/mol)
0 (water)	14.5
30	19
40	22
50	30
60	42
65	56

Viscoplastic solids may be described by the simple Bingham plastic model (Equation 5.42), which is derived from the Newton's viscosity model (Equation 5.18) by adding a yield stress τ_0:

$$\tau = \tau_0 + \eta\gamma \tag{5.42}$$

Thus, the Bingham plastic (solid or semisolid) foods behave as Newtonian fluids, once the yield stress τ_0 is overcome by mechanical stress.

5.3.3 Thermal Transport Properties

The thermal conductivity λ [W/(m K)] is a fundamental transport property of foods, used in many food engineering operations, such as heating/cooling, thermal processing, and dehydration of foods. It is defined by the basic heat transport Equation 5.23, which is integrated to give the steady-state heat conduction (Fourier) equation:

$$\left(\frac{q}{A}\right)_x = -\lambda\left(\frac{\Delta T}{x}\right) \tag{5.43}$$

where

$(q/A)_x$ (W/m^2) is the heat flux in the x-direction
$\Delta T = (T_1 - T_2)$ (K) is the temperature difference in the same direction
x (m) is the thickness, across which heat is conducted (Chapter 8)

Theoretical prediction of thermal conductivity is not feasible, and experimental measurements are necessary. Two methods are normally used: (1) the steady method (hot plate), based on heat conduction through a thin layer of the material, applying Equation 5.28, and (2) the unsteady state method of heat conduction, measuring the temperature with a heated probe.

TABLE 5.12
Thermal Conductivity (λ) and Thermal Diffusivity (α) of Food Materials

Food Material	λ (W/m K)	$\alpha \times 10^{-7}$ m²/s
Water	0.620	1.48
Aqueous sugar solution 15%	0.567	1.42
Tomato paste 35% TS	0.550	1.40
Corn oil	0.180	0.89
Apple 15% TS	0.513	1.62
Potato 18% TS	0.533	1.40
Beef 26% TS	0.452	1.23

The thermal diffusivity α (m²/s), used in several food engineering calculations, is defined by the equation

$$\alpha = \frac{\lambda}{(\rho C_p)} \tag{5.44}$$

where
λ (W/m K) is the thermal conductivity
ρ (kg/m³) is the density
C_p (kJ/kg K) is the specific heat of the material

The thermal diffusivity of foods is usually determined from the thermal conductivity λ, the density ρ, and the specific heat C_p, using Equation 5.44. In some cases, α is estimated from experimental data on unsteady state heat conduction in a food material of defined shape and size. The unsteady state heat conduction is discussed in Chapters 2 and 8.

Temperature and pressure have little effect on the thermal conductivity of foods, but they may be significant in the thermal diffusivity. Table 5.12 shows typical values of thermal conductivity and diffusivity of food materials.

5.3.4 Mass Transport Properties

Mass transport of water and various food components is involved in some important food processing operations, such as dehydration and storage. Mass diffusivity and permeability are the fundamental mass transport properties of interest in food engineering (Saravacos and Maroulis, 2001; Saravacos and Kostaropoulos, 2002).

5.3.4.1 Mass Diffusivity

Mass diffusivity D (m²/s) is defined by the basic transport equation of gases and liquids (Equation 5.24), which for steady state diffusion in food systems is integrated to yield the Fick's first law:

$$\left(\frac{J}{A}\right) = -D\left(\frac{\Delta C}{\Delta x}\right) \tag{5.45}$$

where

(J/A) [kg/(m²s)] is the mass flux in the x-direction
C (kg/m³) is the concentration of the diffusing species
x (m) is the diffusion path

The concentration gradient $\Delta C/\Delta x$ is considered the driving force for diffusion.

In most food applications, one-directional (x) diffusion is considered in both steady-state and unsteady-state mass transport (Chapter 13).

The mass diffusivity D in solid and semisolid foods is usually determined experimentally from unsteady state experiments, applying the Fick's second law, as discussed in Chapters 2, 11, and 13. It can also be determined indirectly from permeability measurements (Equation 5.47).

In various food processing operations, mass transport is a more complex process than simple molecular diffusion, involving hydrodynamic flow or other mechanisms. When diffusion is the main mass transport mechanism, it may be assumed that all mass transport is by molecular diffusion and the Fick's equation is applicable. Thus, the only transport property required is the apparent or effective diffusivity D_e, which in this book denoted as the diffusivity D of the system.

The mass diffusivity of water (liquid and vapor) in food system is the most important mass transport property, which is involved in basic food engineering operations, such as dehydration and storage. The diffusivities of small solute molecules, such as sugars, salts, nutrients, and flavor compounds are important in food product engineering and in food storage.

The mass diffusivity is depended strongly on the physical structure of the food material. It is much faster in porous foods, such as freeze-dried and extruded food products. Temperature has a positive effect on mass diffusivity, which is stronger in compact than in porous foods. The Arrhenius Equation 5.29 is applicable and the activation energy for diffusion E_D is higher at low porosities. Pressure has negative effect on diffusivity, especially in porous materials.

Low moisture diffusivities are observed in starchy foods containing proteins or lipids, due to the low water permeability of protein and lipid films.

Approximate values of the moisture diffusivity D (m²/s) at various bulk porosities ε (–) can be obtained from empirical correlations of experimental data, such as Equation 5.46 for granular starch:

$$D = 10^{-10}\left\{4.842 + 0.57X^{-4.34} + 34.2\left[\frac{\varepsilon^3}{(1-\varepsilon)^2}\right]\exp\left(\frac{4.5}{RT}\right)\right\} \tag{5.46}$$

Typical values of moisture diffusivity D and activation energy for diffusion E_D in food materials are shown in Table 5.13.

The mass diffusivity of various food solutes in water can be estimated from the Wilke–Chang (Equation 5.30) or the Einstein–Stokes (Equation 5.31) equations. Table 5.14 gives some typical diffusivities D of food molecules of various sizes.

TABLE 5.13
Moisture Diffusivity (D) in Food Materials (20°C, 1 bar)

Food Material	Porosity ε	D (m²/s)	Activation Energy E_D (kJ/mol)
Freeze dried food	0.90	50×10^{-10}	10
Apple	0.30	10	20
Potato	0.20	5	30
Beef meat	0.20	1	30
Starch/protein foods	0.15	0.1	50
Starch/lipid foods	0.15	0.01	60

TABLE 5.14
Diffusivity (D) of Food Solutes in Water Solutions (20°C)

Food Solute	$D \times 10^{-10}$ m²/s
Ethanol	8.40
Acetic acid	12.10
Glucose	6.90
Sucrose	5.40
Maltose	4.80
Lactoglobulin	0.70

5.3.4.2 Permeability

The permeability P [kg/(m s Pa)] of a thin layer or a film is defined by the equations

$$J = P\left(\frac{\Delta P}{z}\right) = DS\left(\frac{\Delta P}{z}\right) \tag{5.47}$$

where

J [kg/(m² s)] is the mass transfer rate
ΔP (Pa) is the pressure difference
z (m) is the thickness of the material
D (m²/s) is the mass diffusivity
S [kg/(m³ Pa)] is the solubility of the penetrant in the material

The solubility is the inverse of the Henry's constant $S = 1/H$. The Henry's law here is defined by the equation $P = HC$, where C (kg/m³) is the concentration and H [(m³ Pa)/kg] is the Henry's constant. In Chapter 13, constant H [Pa/(mole fraction)] is defined by the equation $p = Hx$, where x (mole fraction) is the concentration and p (Pa) is the partial pressure.

From Equation 5.47, it follows that $P = DS$, i.e., the permeability of a material increases as the diffusivity D and solubility S of the penetrant increase. Thus, if

TABLE 5.15
Permeability (*P*) and Diffusivity (*D*) of Water Vapor

Material	$P \times 10^{-10}$ g/(m s Pa)	$D \times 10^{-10}$ m²/s
Gluten	5.00	1.00
Corn pericarp	1.60	0.10
Cellophabe	3.70	1.00
Chocolate	0.11	0.01
HDPE	0.002	0.005
LDPE	0.004	0.010
PVC	0.041	0.050

experimental data on permeability P are available, the diffusivity D can be estimated indirectly. Table 5.15 shows some data on P and D of food and packaging materials.

The permeability P of gases (oxygen and carbon dioxide) in food films is in general much lower than the P of water vapor in the same materials. For example, the P of oxygen in HDPE is 7×10^{-13} g/(m s Pa), while that of carbon dioxide in the same material is 5.3×10^{-15} g/(m s Pa).

The permeability and diffusivity of water vapor and gases in glassy polymers and foods is lower than in crystalline or amorphous materials.

APPLICATION EXAMPLES

Example 5.1

The apparent (bulk) density ρ_b of apple slices was studied during air and freeze drying, and the experimental results are given in Table 5.16.

TABLE 5.16
Experimental Apparent Density of Apple during Drying

Apple % TS	Air Drying ρ_b (kg/m³)	Freeze Drying ρ_b (kg/m³)
20	900	500
30	850	400
40	750	350
60	710	340
70	680	330
80	665	300
85	650	250
90	620	230
95	600	200

1. Plot apparent density ρ_b as a function of apple moisture content X (dry basis) for both air and freeze drying.
2. Plot apparent (bulk) porosity ε as a function of apple moisture content X (dry basis) for both air and freeze drying.

The solids density ρ_s of dry (non porous) apple is assumed to be $\rho_s = 1500\,kg/m^3$.

Solution

The moisture content X, dry basis of apple is calculated from the apple total solids content (%TS) using the formula $X = (100 - \%TS)/(\%TS)$, resulting in Table 5.17.

Figure 5.3 shows the apparent density of apple ρ_b as a function of moisture content X. It is seen that the apparent density of apple decreases continuously as

TABLE 5.17
Experimental Apparent Density of Apple versus Moisture Content (db)

Apple %TS	Apple X (kg/kg)	Air Dried Apple ρ_b	Freeze Dried Apple ρ_b
20	4.00	900	500
30	2.33	850	400
40	1.50	750	360
60	0.67	710	340
70	0.43	680	330
80	0.25	660	300
85	0.17	650	250
90	0.11	620	230
95	0.05	600	200

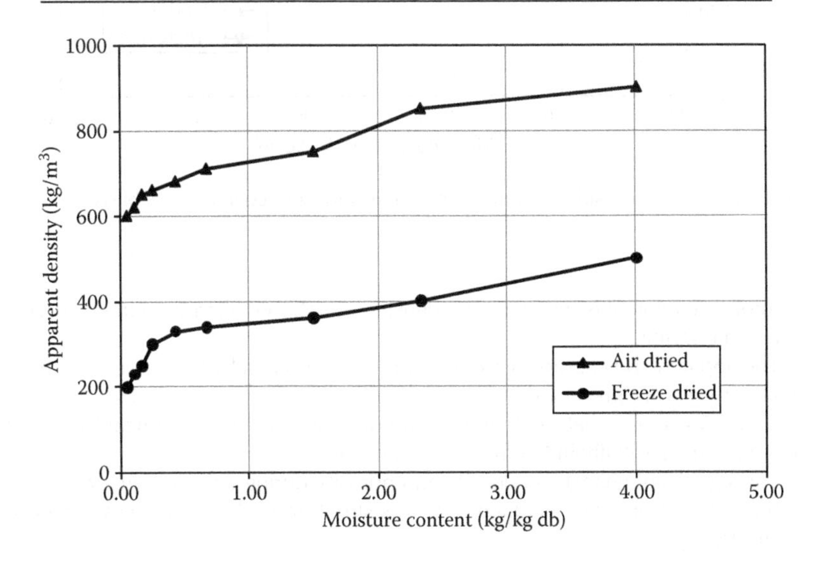

FIGURE 5.3 Bulk densities ρ_b of apple at various moisture contents, dry basis.

TABLE 5.18
Calculated Porosity of Apple versus Moisture Content (db)

Apple X (kg/kg db)	Air Dried Apple ε	Freeze Dried Apple ε
4.00	0.40	0.67
2.33	0.43	0.73
1.50	0.50	0.76
0.67	0.52	0.77
0.43	0.54	0.78
0.25	0.55	0.80
0.17	0.57	0.82
0.11	0.58	0.85
0.05	0.60	0.87

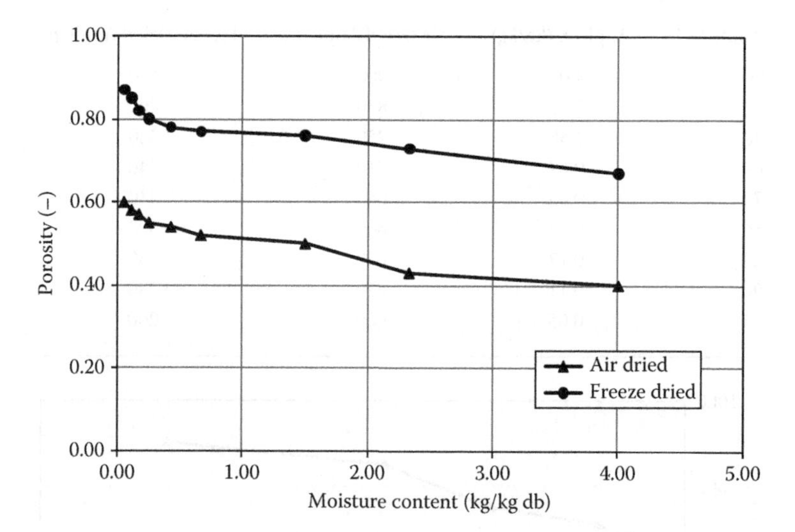

FIGURE 5.4 Bulk porosities ε of apple at various moisture contents, dry basis.

the moisture content is reduced. The density reduction is sharper in freeze drying than in air drying.

The apparent (bulk) porosity ε of the apple samples is calculated by $\varepsilon = 1 - \rho_b/\rho_s$ and the results are presented in Table 5.18.

Figure 5.4 shows that the bulk porosity of apple increases significantly during dehydration, especially in the freeze dried material.

The significant changes in density and porosity during dehydration are caused by the removal of water, creating a porous structure in dried materials, larger in freeze drying.

Example 5.2

Experimental measurements of shear stress τ–shear rate γ in a chocolate sample at 20°C gave the following data (Table 5.19):

1. Estimate the constants (τ_0, K_c, n) of the Casson rheological model, which is assumed to be applicable to chocolate.
2. At temperatures higher than 45°C, the yield stress of the Casson model disappears, and chocolate becomes a pseudoplastic, a dilatant, or a Newtonian fluid.

TABLE 5.19
Experimental Measurements of Shear Stress–Shear Rate in Chocolate at 20°C

Shear Stress τ (Pa)	Shear Rate γ (1/s)
30	0.01
29	0.1
28	1
40	10
130	40
500	100
1700	1000

Solution

The given rheological data are plotted on a logarithmic diagram as log(τ) versus log(γ), as shown in Figure 5.5.

By extrapolation of the lower nearly straight line of the data, the yield stress τ_0 is obtained as approximately $\tau_0 = 28$ Pa. Extrapolation of the higher shear rate data yields nearly a straight line with slope $n = 0.8$. Since the flow behavior index n is lower than 1, the chocolate sample at temperatures higher than 45°C behaves as a pseudoplastic fluid, following the rheological equation $\tau = K\gamma^n$, where $K = 40$ Pa s^n and $n = 0.8$.

Example 5.3

The moisture diffusivity D in granular and gelatinized starch at two temperatures is given in Table 5.20.

1. Calculate the energy of activation for diffusion of water for both starch materials, assuming that the Arrhenius equation is applicable.
2. What would be the effect of temperature on D of both starch materials, when water-soluble sugar is added to the material?

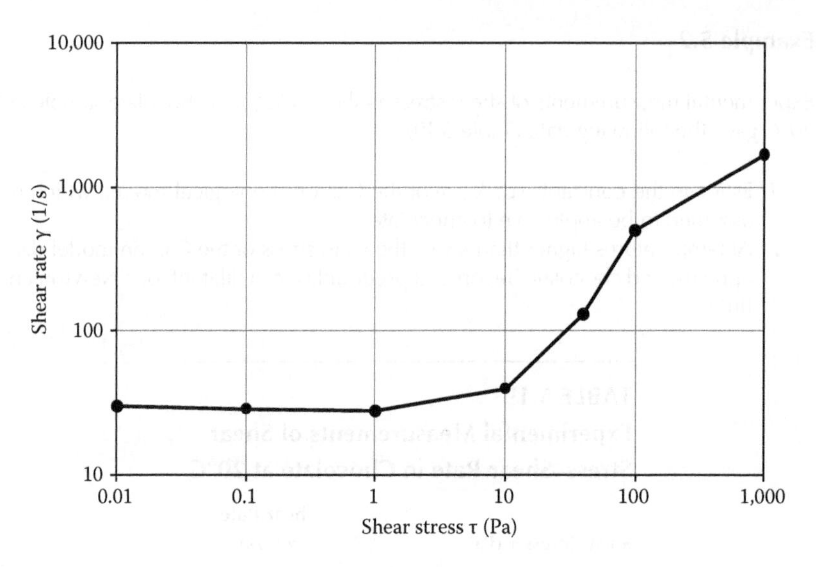

FIGURE 5.5 Rheological data of chocolate (20°C).

Solution

The Arrhenius equation for mass diffusion (Equation 5.17) at temperatures T_1 and T_2 (K) is written as follows:

$$D_1 = D_0 \exp\left(\frac{-E_D}{RT_1}\right) \quad \text{and} \quad D_2 = D_0 \exp\left(\frac{-E_D}{RT_2}\right)$$

Written in the natural logarithmic form these equations become,

$$\ln(D_1) = \ln(D_0) - \frac{E_D}{RT_1} \quad \text{and} \quad \ln(D_2) = \ln(D_0) - \frac{E_D}{RT_2}$$

By subtraction, the last equations yield

$$\ln(D_2) - \ln(D_1) = -\left(\frac{E_D}{R}\right)\left(\frac{1}{T_2} - \frac{1}{T_1}\right) \quad \text{or} \quad \ln\left(\frac{D_2}{D_1}\right) = \left(\frac{E_D}{R}\right)\left[\frac{(T_2 - T_1)}{T_1T_2}\right] \tag{5.48}$$

TABLE 5.20
Moisture Diffusivity of Starch $D \times 10^{-10}$ m²/s

Temperature (°C)	Granular Starch	Gelatinized Starch
20	10	1
50	20	6

For a temperature rise from $T_1 = 20 + 273 = 293\,\text{K}$ to $T_2 = 50 + 273 = 323\,\text{K}$, the moisture diffusivity increases by the ratio $D_2/D_1 = 2$ in air drying, and $D_2/D_1 = 6$ in freeze drying. Substituting these values in Equation 5.48, the following data are obtained:

$$\ln(2) = \left(\frac{E_D}{R}\right)\left[\frac{30}{(293)(323)}\right] \quad \text{or} \quad 0.32\times10^{-3}\left(\frac{E_D}{R}\right) = 0.69 \text{ for granular starch}$$

$$\text{and} \quad \ln(6) = \frac{E_D}{R}\left[\frac{30}{(293)(323)}\right] \quad \text{or} \quad 0.32\times10^{-3}\left(\frac{E_D}{R}\right) = 1.79 \text{ for gelatinized starch}$$

From the last equations it follows that, $(E_D/R) = (0.69/0.32) \times 10^3 = 2156$ and $E_D = 2156R$, where the universal gas constant is $R = 8.314\,\text{kJ/kmol} = 0.0083\,\text{kJ/mol}$ and $E_D = 17.9\,\text{kJ/mol}$ for granular starch. For gelatinized starch, we obtain $(E_D/R) = (1.79/0.32) \times 10^3 = 5600$ and $E_D = 5600R$, or $E_D = 5600 \times 0.0083 = 46.48\,\text{kJ/mol}$.

It is concluded that the energy of activation for water diffusion in gelatinized starch is considerably higher than in granular material. Granular starch has evidently higher open (porous) structure than the compact gelatinized material, facilitating the transfer of water molecules by a combination of gas and liquid diffusion.

Addition of water-soluble sugars to the granular starch will decrease the open porosity, resulting in a decrease of water diffusivity. A smaller effect on water diffusivity is expected when sugars are incorporated in the gelatinized material, which has a more compact structure of low open porosity. The energy of activation for diffusion of water in granular material is expected to increase significantly when soluble sugar is incorporated. A smaller effect is expected for the sugar-containing starch gel material.

PROBLEMS

5.1 Estimate the viscosity η and the thermal conductivity λ of nitrogen and carbon dioxide at 20°C, based on the thermal conductivity of oxygen $\lambda = 0.025\,\text{W/(m K)}$ at the same temperature. Molecular weight of oxygen 32, nitrogen 28, carbon dioxide 44.

Assume that the collision diameters and collision integrals of the two gases are the same.

5.2 In a food processing operation, a 60% aqueous glucose solution of (Newtonian) viscosity 50 mPa s at 20°C is to be mixed with a non-Newtonian fruit pulp in a rotary mixer. The mixing can be facilitated substantially if the apparent viscosity of the pulp approaches the viscosity of the sugar solution. The fruit pulp behaves as a pseudoplastic fluid with rheological constants $K = 1$ Pa s and $n = 0.7$.

Calculate the required rotational speed (rounds per minute, RPM) of the mixer so that the apparent viscosity of the pulp approaches 50 mPa s. The mean shear rate γ (1/s) in the mixer is given by the empirical relation $\gamma = 10 \times (\text{RPM})/60$.

5.3 Calculate the thermal diffusivity α of the following materials at 20°C: water, ethanol, aqueous sugar solution 60% TS, and vegetable oil.

Thermal conductivities λ [W/(m K)]: water 0.80, ethanol 0.17, sugar solution 65% 0.35, vegetable oil 0.18. Densities ρ (kg/m^3): water 1000, ethanol 785; sugar solution 65% 1310, vegetable oil 910. Specific heats [kJ/(kg K)]: water 4.18, ethanol 2.3, sugar solution 65% 2.65, vegetable oil 1.67.

5.4 Estimate the mass diffusivity of a protein molecule of mean diameter 50 nm in water and in a concentrated 60% sugar solution at 20°C.

Explain (qualitatively) how the diffusivity D will change when the temperature of the solution is increased substantially. Viscosities, water 1 mPa s, sugar solution 40 mPa s.

5.5 The mass diffusivity of water vapor in air at 20°C and atmospheric pressure is $D = 1.7 \times 10^{-5}$ m^2/s.

1. Estimate the thermal diffusivity of water vapor in air at the same (P,T) conditions; assuming $\rho = 1.3$ kg/m^3 and $C_p = 1.1$ kJ/kg K. The thermal conductivity of the air mixture can be taken as $\lambda = 0.022$ W/(m K).
2. Estimate the mass and thermal diffusivities of water vapor in a freeze drying chamber operating at 20°C and pressure 1 mbar. The atmospheric pressure is assumed to be 1 bar.
3. Based on the thermal and mass diffusivities of water vapor at 1 bar and 1 mbar, explain what is the controlling mechanism at the two pressures (heat or mass transfer).

5.6 The permeability P of a polymer membrane to oxygen and carbon dioxide is very important in food packaging. In general, membranes should be more permeable to carbon dioxide than to oxygen.

Explain qualitatively how the permeability of a membrane can be increased by physicochemical modification. The mass diffusivities D of oxygen and carbon dioxide are assumed to be constant in the polymer material under consideration.

5.7 The moisture diffusivity D of solid and semisolid foods during dehydration changes significantly, due to the changes of the physical structure (particularly porosity) of the material. The changes in bulk porosity ε during air and freeze drying of potato pieces were determined in Application Example 5.1.

Estimate and plot the (effective) moisture diffusivity D as a function of porosity for both drying processes, using the empirical Equation 5.46 of this chapter.

5.8 Explain why the mass diffusivity D of water in granular (porous) starch is much higher than in the gelatinized material.

LIST OF SYMBOLS

A	M^2	Surface area
A	J	Free energy
C	m/s	Velocity
C_p	kJ/(kg K)	Specific heat
d, D	m	Diameter
db	—	Dry basis

D	m^2/s	Mass diffusivity
E	Pa	Modulus of elasticity (Young's)
f	1/s	Frequency
F	N	Force
H	kJ/kg	Enthalpy
I	Ampere	Electrical current
J	kg/s	Mass flow
K	Pa s^n	Flow consistency coefficient (Rheological constant)
L	m	Length
m	kg/s	Mass flow rate
n	—	Flow behavior index
q	W	Heat flow rate
p, P	Pa	Pressure
r	m	Radius
R	kJ/(kg K)	Gas constant (8.314)
R	Ohm	Electrical resistance
S	—	Shrinkage coefficient
S	kg/m^3	Solubility
S	J/K	Entropy
t	s	Time
T	°C or K	Temperature
u	m/s	Velocity
V	m^3	Volume
V	Volt	Electrical potential
x	m	Distance
x	—	Mass fraction
x, X	kg/(kg db)	Moisture content, dry basis
W	J/m^2	Surface work (surface energy)
α	m^2/s	Thermal diffusivity
γ	1/s	Shear rate
γ	N/m	Surface tension
Γ	mol/ m^2	Adsorbed mass
Δp	Pa	Pressure difference
ΔT	K or °C	Temperature difference
ΔH	kJ/kg	Enthalpy change
ε	—	Emissivity
ε	—	Porosity
ε	—	Strain
η	Pa s	Viscosity
λ	W/(m K)	Thermal conductivity
μ	kJ/mol	Chemical potential
ν	—	Poisson ratio
ρ	kg/m^3	Density
σ	Pa	Mechanical stress
τ	Pa	Shear stress

REFERENCES

Bourne, M.C. 1982. *Food Texture and Viscosity: Concept and Measurement*. Academic Press, New York.

Gekas, V. 1992. *Transport Phenomena of Foods and Biological Materials*. CRC Press, New York.

Nesvadba, P. 2005. Thermal properties of unfrozen foods. In: *Engineering Properties of Foods*. Rao, M.A., Rizvi, S.S.H., and Datta, A.E., eds., 3rd edn. Taylor & Francis, New York.

Rahman, M.S. 2005. Mass-volume-area related properties of foods. In: *Engineering Properties of Foods*, Rao, M.A., Rizvi, S.S.H., and Datta, A.E., eds. 3rd edn. Taylor & Francis, New York.

Rahman, M.S. 2009. *Food Properties Handbook*, 2nd edn. CRC Press, New York.

Rao, M.A. 2007. *Rheology of Fluid and Semisolid Foods*, 2nd edn. Springer, New York.

Rao, M.A., Rizvi, S.S.H., and Datta, A.E. 2005. *Engineering Properties of Foods*, 3rd edn. Taylor & Francis, New York.

Reid, R.C., Prausnitz, J.M., and Poling, B.E. 1987. *The Properties of Gases and Liquids*, 4th edn. McGraw-Hill, New York.

Saravacos, G.D. and Maroulis, Z.B. 2001. *Transport Properties of Foods*. Marcel Dekker, New York.

Saravacos, G.D. and Kostaropoulos, A.E. 2002. *Handbook of Food Processing Equipment*. Springer, New York.

6 Fluid Flow Operations

6.1 INTRODUCTION

Fluid flow is an important basic operation, applied to several food processing operations, such as heat transfer, evaporation, thermal processing, and drying. Fluid flow is also involved in several food product engineering operations, such as mixing, homogenization, and emulsification. This chapter is concerned with incompressible flow, i.e., flow of liquids without changes of volume (constant density), such as fluid and semifluid food materials. The flow of air in various applications of food engineering, such as drying, cooling/freezing, and air conditioning, is usually considered as an incompressible flow (constant air density), since the pressure differences in the system are less than 10%. Compressible flow, where the volume (or density) changes significantly during the operation, is outlined in Chapter 3. Some fluid flow applications are treated in other chapters of this book, such as pneumatic and hydraulic transport, mixing (Chapter 7), heating operations (Chapter 8), evaporation (Chapter 10), refrigeration/freezing (Chapter 12), and thermal processing (Chapter 9).

6.2 FOOD RHEOLOGY

The rheological properties of foods (Chapter 5) are of primary importance in most fluid flow calculations. Some fluid foods can be considered as Newtonian, i.e., their viscosity is not affected by the rate of shearing (flow velocity). However, several fluid foods show a non-Newtonian behavior, i.e., their apparent viscosity changes, as the shear rate (fluid velocity) is changed (Saravacos and Maroulis, 2001).

6.2.1 Capillary Tube Viscometry

The capillary or tube viscometer consists of a long tube of small diameter, through which the fluid is forced by pressure to flow (Figure 6.1). In this system, the flow is laminar and the Poiseuille equation for Newtonian fluids is applicable. The pressure drop is due to the friction of the fluid within the pipe, neglecting the effects of entrance and exit end (Rao et al, 2005; Rao, 2007),

$$\frac{d\Delta P}{4} = \eta\left(\frac{8u}{d}\right) \tag{6.1}$$

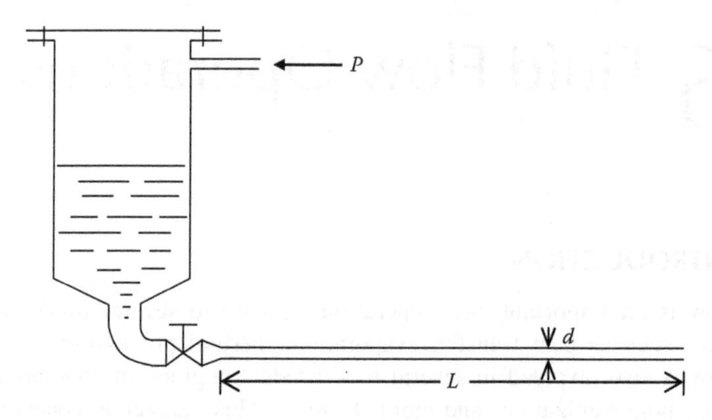

FIGURE 6.1 Capillary viscometer.

where

ΔP (Pa) is the pressure drop
d, L (m) are the interior diameter and length of the pipe, respectively
u (m/s) is the mean velocity
$d\Delta P/4L = \tau$ (Pa) is the shear stress
$8u/d = \gamma$ (1/s) is the shear rate
η [Pa s or kg/(m s)] is the viscosity of the fluid

The shear stress τ and the shear rate γ are considered at the pipe wall. The shear rate can also be calculated from the volumetric flow rate Q (m^3/s), as $\gamma = 32Q/(\pi d^3)$. The Poiseuille equation for power-law non-Newtonian fluids becomes

$$\frac{d\Delta P}{4L} = K\left[\frac{(3n+1)}{4n}\right]^n \left(\frac{8u}{d}\right)^n \tag{6.2}$$

The rheological constants K (Pa s^n) and n (dimensionless) of power-law fluids are determined from experimental data of capillary viscometry by plotting the shear stress $(d\Delta P)/4L$ versus shear rate $8u/d$ on a log–log diagram. If the power-law model is applicable, a straight line is obtained with a slope n and an intercept $K[(3n + 1)/4n]^n$ from which the constants n and K are determined.

6.2.2 Rotational Viscometry

In rotational viscometers, the shear stress is measured at various rotational speeds, corresponding to different shear rates. The coaxial narrow gap viscometer, shown in Figure 6.2, is the simplest basic instrument, in which the shear rate can be calculated precisely. Another basic rotational and more accurate instrument is the expensive Weissenberg cone and plate viscometer (rheogoniometer). Some other rotational viscometers, such as the Brookfiled and the Stormer instruments, use empirical shear stress–shear rate correlations.

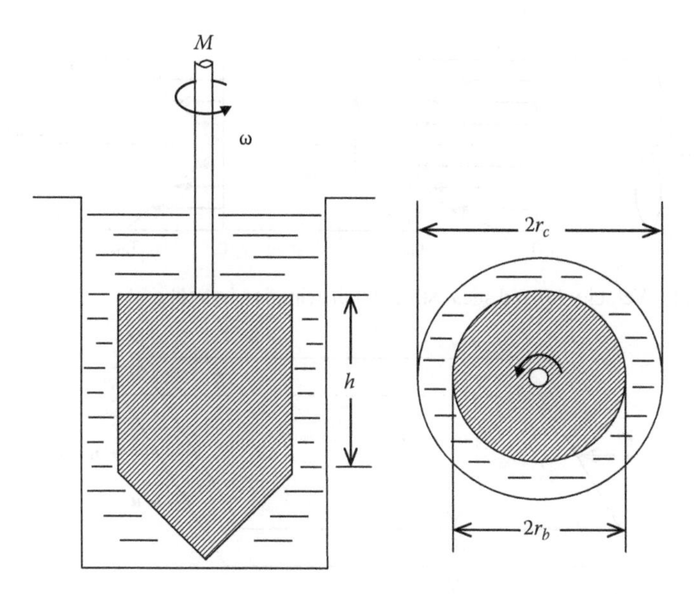

FIGURE 6.2 Rotational (coaxial) viscometer.

The viscosity η [kg/(m s)] of a Newtonian fluid in a coaxial viscometer is calculated from the following equation:

$$\eta = \left[\frac{M}{(2\pi h\omega)}\right]\left[\left(\frac{1}{r_b^2}\right) - \left(\frac{1}{r_c^2}\right)\right] \tag{6.3}$$

where

M (N m) is the mechanical moment (measured directly)

r_b, r_c (m) are the radii of the rotating (inner) and stationary (outer) cylinders, respectively

h (m) is the height of the rotating cylinder

ω (1/s) is the rotational speed of the cylinder

6.2.3 Velocity Distribution

The velocity distribution in a pipe (tube) depends on the type of fluid (Newtonian or non-Newtonian) and the type of flow (laminar or turbulent). Figure 6.3 shows the velocity and shear stress distribution in the pipe flow of a Newtonian fluid.

The fluid velocity u in the tube in the flow x-direction is changed from nearly zero at the wall to a maximum u_{max} at the center. The shear stress τ, caused by changes of velocity in the y-direction, is decreased from a maximum τ_{max} at the wall to zero at the center.

Figure 6.4 shows a diagram of developing a shear rate in a flow system. Two neighboring parallel planes of the fluid are moving at a velocity u when a parallel force F acts on one plane, causing an increase of the velocity to u + du. The shear

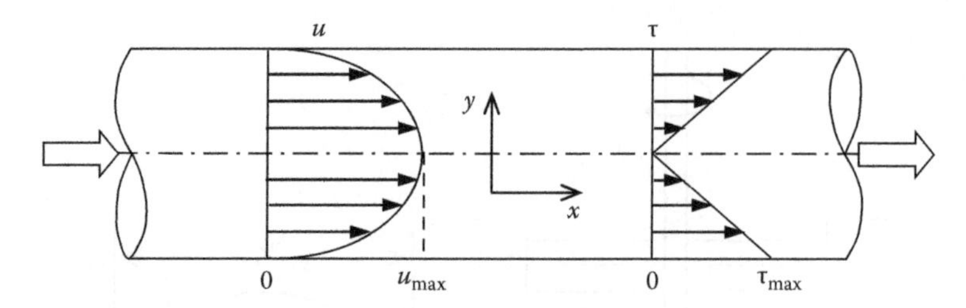

FIGURE 6.3 Velocity (u) and shear stress (τ) distribution in a tube.

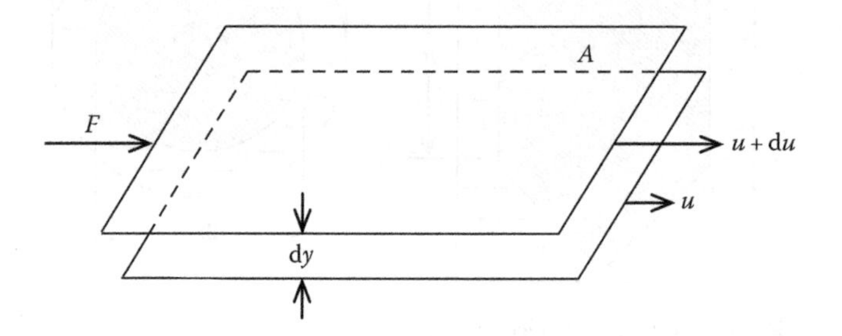

FIGURE 6.4 Shear rate between two parallel planes.

rate γ is defined as the change in velocity du over the perpendicular distance between the two planes dy:

$$\gamma = \frac{du}{dy} \tag{6.4}$$

In Newtonian fluids, the shear stress–shear rate relationship defines the viscosity η:

$$\tau = \eta\gamma \tag{6.5}$$

In the laminar (streamline) flow of Newtonian fluids in a tube, the mean linear velocity is one-half the maximum velocity $\bar{u} = 0.50\, u_{max}$, while in the turbulent flow the mean velocity is 80% of the maximum velocity $\bar{u} = 0.80\, u_{max}$, as shown in Figure 6.5.

Figure 6.6 shows the velocity distribution in the tubular flow of non-Newtonian fluids, particularly power-law and Bingham plastic fluids (see Chapter 5).

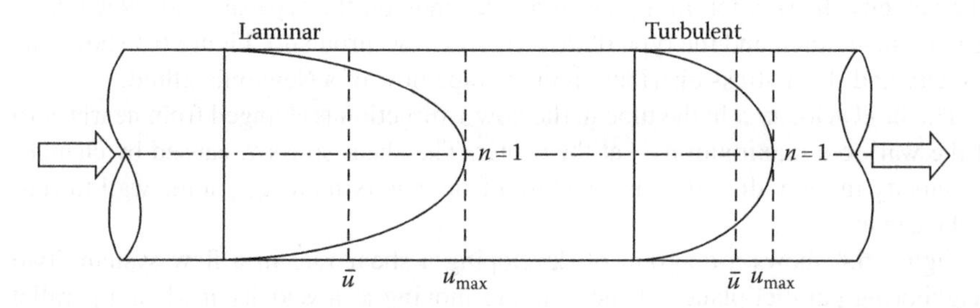

FIGURE 6.5 Velocity distribution in Newtonian fluids.

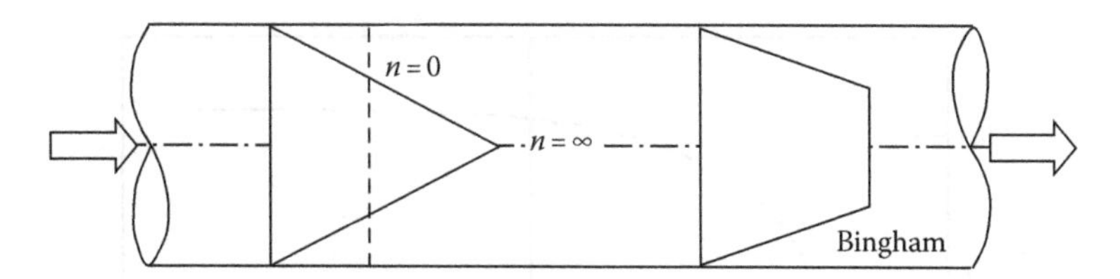

FIGURE 6.6 Velocity distribution in non-Newtonian fluids.

6.2.4 Reynolds Number

The Reynolds number *Re* (dimensionless) is a fundamental index of the type of flow in a pipe or flow system (laminar, turbulent, or intermediate). It is defined as the ratio of inertial to viscous forces, and for pipe (tube) flow it is calculated from either one of the following equations:

$$Re = \frac{(du\rho)}{\eta} \quad \text{or} \quad Re = \frac{dm}{\eta} \quad \text{or} \quad Re = \frac{dQ\rho}{A\eta} \tag{6.6}$$

where

d (m) is the interior pipe (tube) diameter
u (m/s) is the (linear) mean fluid velocity
ρ (kg/m^3) is the fluid (mass) density
η [Pa s or kg/(m s)] is the (Newtonian) fluid viscosity
m [kg/(m^2 s)] is the mass fluid velocity
Q (m^3/s) is the volumetric fluid velocity

The *Reynolds number* for flow in channels other than circular (pipes) is calculated by considering the geometry (flow cross section) of the system. The generalized Reynolds number is given by the following equation:

$$Re = \frac{4r_h u\rho}{\eta} \tag{6.7}$$

where r_h (m) is the hydraulic radius of the flow cross section, defined as r_h = (cross-sectional area of flow)/(wetted perimeter of the cross section).

For example, the hydraulic radius of a rectangular cross section is $r_h = (bh)/2\,(b + h)$, where b, h (m) are the dimensions (base, height) of the rectangular.

For non-Newtonian fluids, the *Reynolds number* is based on the apparent viscosity η_a (Pa s), which, for a power-law fluid, is estimated from Equation 6.8:

$$\eta_a = K\gamma^{n-1} \tag{6.8}$$

For pseudoplastic fluids ($n < 1$), which constitute the majority of non-Newtonian foods, the apparent viscosity decreases as the shear rate γ is increased (higher linear or rotational velocities), and the *Re* increases sharply.

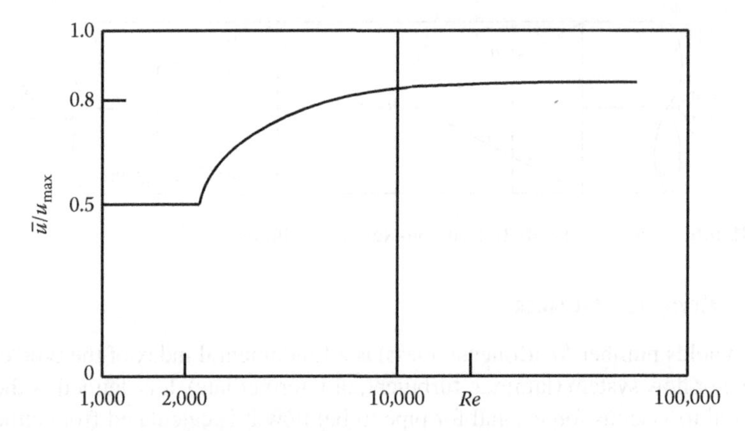

FIGURE 6.7 Velocity distribution versus Reynolds number in Newtonian fluids.

The generalized Reynolds number *Re* of power-law fluids can also be calculated from the following equation:

$$Re = \frac{d^n u^{2-n} \rho}{8^{n-1} K} \tag{6.9}$$

where *K*, *n* are the rheological constants of the non-Newtonian fluid.

It should be noted that the generalized *Re* (Equation 6.9) reduces to $Re = (du\rho)/\eta$ for Newtonian fluids, when $n = 1$ and $K = \eta$. The effect of *Reynolds number* on the velocity distribution of a Newtonian fluid is shown in Figure 6.7. The mean velocity $\bar{u}$ is 50% of the u_{max} below $Re = 2{,}100$ (laminar flow) and it approaches 80% of the u_{max} at $Re > 10{,}000$ (fully developed turbulent flow). The $\bar{u}$ is between 50% and 80% of the u_{max} at intermediate *Re* numbers.

The *Reynolds number* is affected strongly by the viscosity of the fluid. The density does not change much for a given fluid, while its velocity can be changed within a certain range. The viscosity can change manifold due to changes in composition and temperature. Thus, the *Reynolds number* of a highly viscous fluid can be very low, e.g., below 10, resulting in laminar flow, which cannot be changed to turbulent flow, even at high velocities. However, high temperatures will decrease the viscosity and may develop turbulent flow. Turbulent flow is desirable in some process operations, such as heat and mass transfer.

6.3 MECHANICAL ENERGY BALANCES

6.3.1 Mechanical Flow Systems

Fluid flow calculations and applications are based on mechanical energy balances in flow systems. The mechanical energy balance between two points (1, 2) in a flow system, consisting of piping, valves, and a pump, is expressed by the Bernoulli equation (Holland and Bragg, 1995)

$$\left(\frac{\Delta P}{\rho}\right)+\left(\frac{\Delta u^2}{2}\right)+g\Delta z+h_f-W_p=0 \tag{6.10}$$

where

$\Delta P/\rho = (P_2 - P_1)/\rho$ is the pressure energy
$\Delta u^2/2 = (u_2^2 - u_1^2)/2$ is the velocity energy
$\Delta z = z_2 - z_1$ is the elevation (height) energy
h_f is the friction energy
W_p is the pump energy
$g = 9.81\ \text{m/s}^2$ is the gravity constant
ρ is the density of the fluid (kg/m^3)

The energy units of the terms of the Bernoulli equation are (J/kg) or (N m/kg).

The units of the various quantities are, pressure P (Pa or N/m^2), velocity u (m/s), elevation z (m), friction h_f (J/kg), and pump work W_p (J/kg). It should be noted that the units of the squared velocity (m^2/s^2) are energy per unit mass, i.e., (J/kg) = (N m)/kg = (kg mm)/(kg s^2) = (m^2/s^2).

The terms of the energy balance (Bernoulli) equation can be written in the form of energy "heads" by dividing them by the gravity constant (g), resulting in Equation 6.11:

$$\frac{\Delta P}{(\rho g)}+\left(\frac{\Delta u^2}{2g}\right)+\Delta z+\frac{h_f}{g}-\frac{W_p}{g}=0 \tag{6.11}$$

The (energy) head is the equivalent energy of the liquid height in appropriate units, such as (m) in the SI system and (ft) in the English system.

The Bernoulli equation 6.10 is mostly used to calculate the pump energy W_p in a flow system when all the other energies are known. The pressures are measured with pressure gauges (sensors), the fluid velocity with flow meters, and the height with measure tape. The friction energy is calculated using semiempirical models, as explained later in this chapter.

The axial pump power P (kW) for a given flow rate m (kg/s) is calculated from the theoretical pump energy W_p, using the following equation:

$$P=\frac{mW_p}{1000\beta} \tag{6.12}$$

where β is the pump efficiency, normally $\beta = 0.70$–0.85.

The axial power refers to the power supplied by the motor to the axis (shaft) of the pump for accomplishing the transport of the fluid m (kg/s) in the given system. The power required by the motor for this system will be P/β_M, where β_M is the motor efficiency, normally $\beta_M = 0.90$. Thus, the required power will be about 25% higher than the theoretical pump work, or the efficiency of the pumping installation is about 75%. Figure 6.8 shows a diagram of a pumping system for transporting a liquid between levels z_1, z_2, pressures P_1, P_2, and fluid velocities u_1, u_2, through a pipe and a pump.

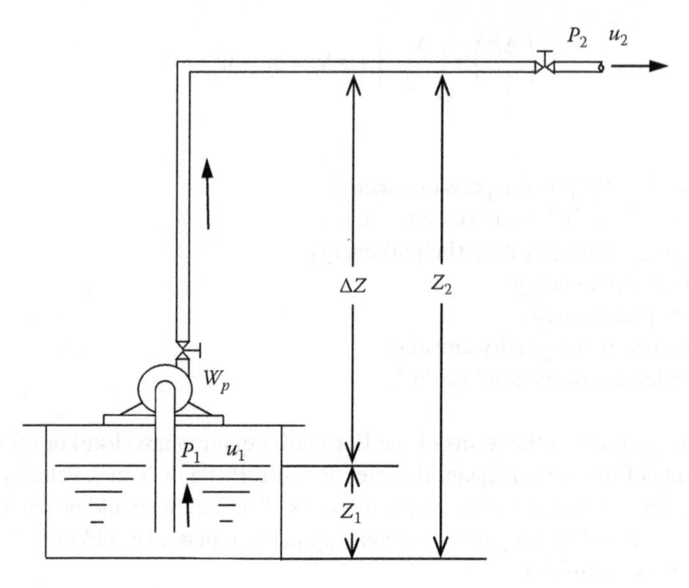

FIGURE 6.8 Diagram of a pumping system.

The Bernoulli equation of mechanical energy balance in a pipe without a mechanical pump yields the following equation:

$$\left(\frac{P_1}{\rho}\right)+\left(\frac{u_1^{\,2}}{2}\right)+gz_1=\left(\frac{P_2}{\rho}\right)+\left(\frac{u_2^{\,2}}{2}\right)+gz_2+h_f \tag{6.13}$$

Figure 6.9 shows a diagram of energy balance in a pipe, according to Equation 6.13.

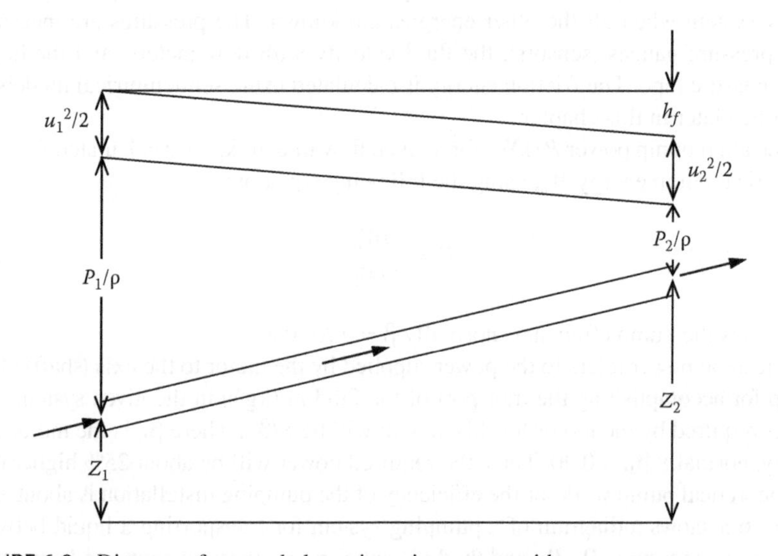

FIGURE 6.9 Diagram of energy balance in a pipe system without a pump.

6.3.2 Friction in Piping Systems

The mechanical friction of piping systems represents an important energy, which should be accounted in energy balance calculations. Empirical equations are used to estimate the friction energy as a function of fluid velocity and characteristic geometric factors of the piping.

The friction in any flow conduit (pipe, tube, channel) is caused by the contact of the flowing fluid with the solid walls, the shear between the fluid layers, or the eddies (turbulence) created by the flow.

For straight pipes, the friction energy h_f (J/kg) is estimated from the Fanning equation

$$h_f = \left(\frac{\Delta P}{\rho}\right)_f = 4f\left(\frac{L}{d}\right)\left(\frac{u^2}{2}\right) \tag{6.14}$$

where

L, d (m) are the pipe length and interior diameter, respectively
u (m/s) is the mean fluid velocity
ΔP_f (Pa) is the pressure drop
ρ (kg/m^3) is the density
f (dimensionless) is the friction factor

The Fanning friction factor f is given in the Moody (or Fanning) diagram (Figure 6.10) as a function of the Reynolds number Re. In the laminar flow region $Re < 2100$, the friction factor f is calculated from the following equation:

$$F = \frac{16}{Re} \tag{6.15}$$

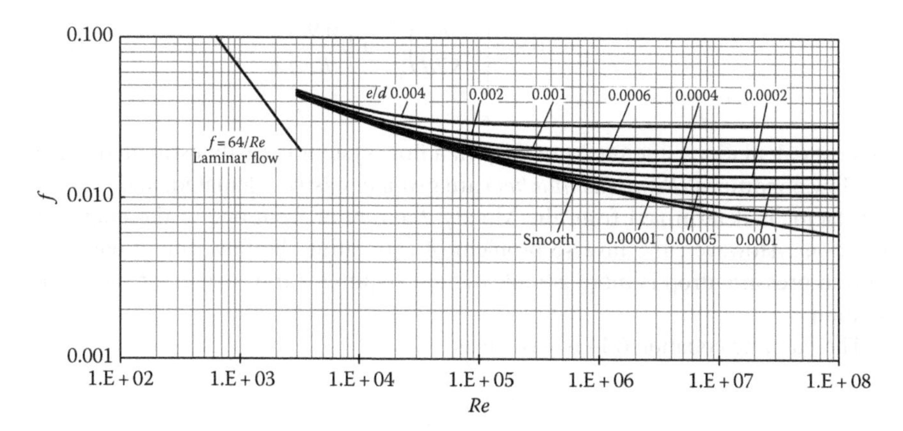

FIGURE 6.10 Moody diagram for friction factor (f) of Newtonian fluids in straight pipes, (e/d) = relative roughness factor. Abscissae in powers (E) of 10, example 1.E+3 = 10^3.

The Fanning equation 6.11 is derived by combining the Poiseuille equation 6.10 with Equation 6.6, which defines the Reynolds number. In the laminar flow, the roughness of the piping system has no effect on the friction factor.

The turbulence region in a flow system starts at about $Re = 2100$ and it is considered as fully developed at $Re > 8000$. At intermediate *Re* $2100 < Re < 8000$ the flow is considered as mixed laminar and turbulent.

In the turbulent flow region, the friction factor increases, as the relative roughness e/d is increased (Figure 6.10). Here, e (m) is the mean height of the roughness, and d (m) is the internal pipe diameter. The roughness may be caused by the type and structure of the material (e.g., smooth metal, concrete) or during construction and fitting of the pipes in the flow system (e.g., riveted steel pipe). It is may also result during the flow and heating of some fouling fluids, which form insoluble deposits on the internal surfaces of pipes and other flow conduits. Smooth metal pipes have the lowest friction factor at a given *Re*.

In the completely turbulent region $Re > 8000$, instead of the Moody diagram, the friction factor f can be estimated from certain empirical equations, such as the equations Blasius, von Karman, and Haaland.

The Blasius equation is applicable to smooth pipes:

$$F = 0.079R^{-0.25} \tag{6.16}$$

von Karman equation (smooth pipes):

$$\frac{1}{f^{0.5}} = 4.0\log(Re\, f^{0.5}) - 0.4 \tag{6.17}$$

Haaland equation (rough pipes):

$$\frac{1}{f^{0.5}} = -3.6\log\left[\left(\frac{e}{3.7d}\right)^{1.11} + \left(\frac{6.9}{Re}\right)\right] \tag{6.18}$$

In most calculations, the pressure drop in a flow system is easily estimated, when the flow conditions (*Reynolds number*) are given or easily calculated, and the friction factor is determined from the Moody diagram or from one of the empirical equations. However, the inverse problem of estimating the flow conditions from a given pressure drop is more complex and it requires a trial and error (iteration) solution.

The Fanning equation 6.14 is written as

$$f = \frac{d\Delta P_f}{2L\rho u^2} \tag{6.19}$$

Multiplying both sides of this equation by the square of $Re^2 = [(du\rho)/\eta]^2$ we obtain,

$$fRe^2 = \left[\frac{d\Delta P_f}{2L\rho u^2}\right]\left[\frac{d^2u^2\rho^2}{\eta^2}\right] \quad \text{or} \quad fRe^2 = \frac{d^3\rho\Delta P_f}{2L\eta^2} \tag{6.20}$$

Equation 6.20 is solved by an iteration method: a reasonable mean fluid velocity u is assumed, the Re is calculated, and the friction factor f is estimated from the Moody diagram; finally, the pressure drop is calculated from Equation 6.20 and compared to the given ΔP_f value. If different, the procedure is repeated until an acceptable velocity is obtained. A computer spreadsheet method can also be used to solve Equation 6.20 for the velocity u.

6.3.3 Friction in Non-Newtonian Fluids

Friction in non-Newtonian fluid foods is treated in a similar manner with the friction in Newtonian fluids, described in this section.

The analytical equations and diagrams of the Newtonian fluids are modified to account for the change of apparent viscosity due to a change of shear rate in non-Newtonians.

The Fanning friction factor f of power-law fluids for laminar flow is given by Equation 6.15 for Newtonian fluids $f = 16/Re$. In non-Newtonian fluids, the generalized Reynolds number Re (Equation 6.9) can be used. For turbulent flow of non-Newtonian fluids, the friction factor f is determined from the modified Moody diagram (Figure 6.11).

6.3.4 Rabinowitsch–Mooney Equation

In fluid flow calculations on non-Newtonian fluids, it is often necessary to estimate the shear rate at a point of the flow conduit for a given shear stress. The Rabinowitsch–Mooney equation for flow of non-Newtonian fluids in pipes relates the shear rate at the

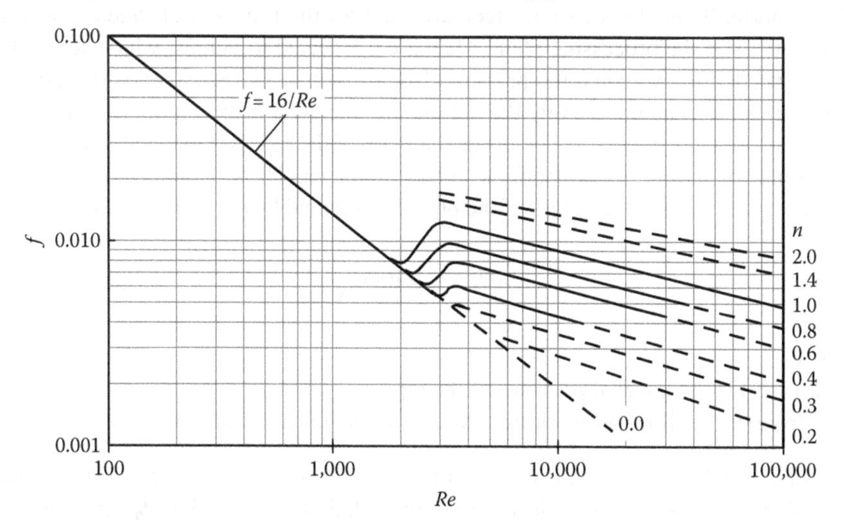

FIGURE 6.11 Modified friction diagram for non-Newtonian fluids.

wall γ_w to the shear stress τ_w and the flow characteristic $8u/d$, where u is the mean flow velocity and d is the internal diameter.

$$\gamma_w = \gamma_{wN}\left\{\frac{3}{4}+\left(\frac{1}{4}\right)\left[\frac{d\ln\left(\frac{8u}{d}\right)}{d\ln\tau_w}\right]\right\} \tag{6.21}$$

where $\gamma_{wN} = 8\ u/d$ is the shear rate at the wall for a Newtonian fluid. The gradient in the square brackets $[d\ln(8u/d)/d\ln\tau_w]$ is the slope of a logarithmic (ln–ln) plot of the flow characteristic $8u/d$ versus the shear stress at the wall τ_w. The same slope is obtained by plotting $\log(8u/d)$ versus $\log \tau_w$. The inverse of this slope is the flow behavior index n of the power-law model.

The basic rheological quantities and the rheological curve of shear stress τ_w (Pa) versus shear rate γ_w (1/s) of a power-law (non-Newtonian) fluid can be obtained from the Rabinowitsch–Mooney equation 6.21, using experimental data from a capillary (tube) viscometer (diameter d, length L) as follows:

1. Measure flow rates Q (m^3/s) at various pressure drops ($\Delta P/L$).
2. Calculate shear stress at the wall from $\tau_w = (d\Delta P/4L)$.
3. Plot $\log(8ud)$ versus $\log \tau_w$ and determine the gradient at various points on the curve.
4. Calculate the real shear rate at the wall γ_w, using the estimated gradients in Equation 6.21.

6.4 PIPING SYSTEMS

6.4.1 Pipes and Tubes

Pipes, normally made of carbon steel, are used for the transport of fluids to various destinations in the processing plant. They are connected to pumps, which supply the power needed to overcome the pressure drop caused by elevation, velocity, and friction. Pipes are characterized by the nominal diameter in inches, ranging from 1/8 to 24 in., as shown in Table 6.1. The inside diameter of the pipe is obtained by subtracting the double of wall thickness from the outside diameter. The schedule number *Sch No.* characterizes the wall thickness of the pipe. It is defined by the empirical equation

$$Sch\ No. = 1000\frac{P}{S} \tag{6.22}$$

where

P (psig) is the internal pressure of the pipe
S (psig) is the allowable working pressure, which for carbon steel is about 10,000 psig

For example, for an operating pressure of 30 bar or $P = 30 \times 14.5 = 435$ psig, and $S = 10{,}000$ psig, we obtain $Sch\ No. = (1{,}000 \times 435)/10{,}000 = 43.5$ or approximately $Sch\ No. = 40$, which is commonly used in industrial piping.

TABLE 6.1
Dimensions of Steel Pipes

d_N in.	Schedule No.	d_o in.	d_o mm	d_i mm	$A_i \times 10^{-3}$ m^2
1/8	10S	0.405	10.29	7.80	0.047
1/4	10S	0.540	12.80	10.41	0.085
3/8	10S	0.675	17.14	13.84	0.150
1/2	10S	0.840	21.34	17.12	0.230
3/4	10S	1.050	26.67	22.45	0.396
1	10S	1.315	33.40	27.86	0.610
1 ¼	40S	1.660	42.16	35.05	0.966
1 ½	40S	1.900	48.26	40.89	1.314
2	40S	2.375	60.32	52.50	2.164
2 ½	40S	2.875	73.02	62.71	3.076
3	40S	3.500	88.90	72.93	4.765
4	40S	4.500	114.30	102.26	8.213
5	40S	5.563	141.30	128.19	12.910
6	40S	6.625	168.37	154.05	18.630
8	40S	8.625	219.07	202.72	32.270
10	40S	10.750	273.05	254.51	50.860
12	40S	12.750	323.85	394.80	72.960
14	40S	14.000	355.60	333.35	87.300
16	40S	16.000	406.40	381.00	114.010
18	40S	18.000	457.20	428.35	144.300
20	40S	20.000	508.00	477.82	179.320

d_N, nominal diameter; d_o, outside diameter; d_i, inside diameter; A_i, inside cross section.

For a given nominal (and outside) diameter, there are different wall thicknesses (schedule numbers), depending on the internal (operating) pressure of the pipe (Table 6.2). Thick pipes are used when pumping liquids to higher elevations or when large pressure drops develop in the piping system. The sanitary pipes used in food processing have special dimensions (Table 6.3). National and international piping codes contain detailed information on the construction and the dimensions of various types of pipes, tubes, and fittings. Such codes are published by the American Society of Mechanical Engineers (ASME), the American Society for Testing and Materials (ASTM), the American National Standards Institute (ANSI), the American Iron and Steel Institute (AISI), the International Organization for Standardization (ISO), the Society of German Engineers VDI (Verein Deutsche Ingenieure), and the German Standards Institute DIN (Deutsches Institut fur Normung).

Smooth sanitary pipes and tubes made of stainless steel are used in food processing. They are made preferably by extrusion, instead of welding, which is used in other applications. For long permanent pipelines, stainless steel pipes are joined normally by butt welding, without overlapping.

TABLE 6.2
Schedule Numbers and Dimensions of 1–2 in. Steel Pipes

d_N in.	Schedule No.	d_o in.	mm	d_i mm	$A_i \times 10^{-3}$ m^2
1	5S	1.315	33.400	30.10	0.713
	10S	1.315	33.400	27.86	0.610
	40S	1.315	33.400	26.64	0.557
	80S	1.315	33.400	24.30	0.464
	160	1.315	33.400	20.70	0.336
1 ½	5S	1.900	48.260	44.96	1.588
	10S	1.900	48.260	42.72	1.433
	40S	1.900	48.260	40.89	1.314
	80S	1.900	48.260	38.10	1.138
	160	1.900	48.260	33.98	0.907
2	5S	2.375	60.325	57.02	2.554
	10S	2.375	60.325	54.79	2.358
	40S	2.375	60.325	52.50	2.165
	80S	2.375	60.325	49.25	1.904
	160	2.375	60.325	42.85	1.442

d_N, nominal diameter; d_o, outside diameter; d_i, inside diameter; A_i, inside cross section.

TABLE 6.3
Dimensions of Stainless Steel Sanitary Pipes

d_o in.	mm	d_i mm	$A_i \times 10^{-3}$ m^2
1	25.40	22.91	0.412
1 ½	38.10	35.61	0.995
2	50.80	47.49	1.770
2 ½	63.50	60.19	2.843
3	76.20	72.89	4.170
4	101.60	97.39	7.445

d_o, outside diameter; d_i, inside diameter; A_i, inside cross section.

In piping systems of food processing plants, occasionally the pipes must be disassembled (taken apart) for cleaning and sanitizing. For this purpose, mechanical couplings and flanged gaskets are used. Gaskets are made of food-grade synthetic elastomers, such as Neoprene and Teflon. Screwed joints of small-diameter pipes are difficult to clean and sanitize properly.

Tubes made of copper alloys, stainless steel, or aluminum are used in various types of heat exchangers, which are discussed mainly in Chapter 8. The size of tubes is generally smaller than the dimensions of pipes. The tubes are characterized by the outside diameter and the thickness of the tube wall (Table 6.4), which range from 5/8 to 2 in. (15–50 mm) and 1.24 to 3.40 mm, respectively. The wall thickness is expressed by the Birmingham Wire Gauge (*BWG*), ranging from *BWG* 10 (3.40 mm) to 18 (1.24 mm).

TABLE 6.4
Dimensions of Heat Exchanger Tubes

d_o		*BWG*			
in.	mm	No.	X mm	d_i mm	$A_i \times 10^{-4}$ m^2
5/8	15.87	12	2.77	10.34	0.839
		14	2.11	11.66	1.067
		16	1.65	12.57	1.241
		18	1.24	13.38	1.407
3/4	19.05	12	2.77	13.51	1.434
		14	2.11	14.83	1.728
		16	1.65	15.75	1.948
		18	1.24	16.56	2.154
7/8	22.22	12	2.77	16.69	2.187
		14	2.11	18.00	2.547
		16	1.65	18.92	2.812
		18	1.24	19.74	3.059
1	25.40	10	3.41	18.58	2.715
		12	2.77	19.86	3.098
		14	2.11	21.18	3.525
		16	1.65	22.10	3.836
1 ¼	31.75	10	3.41	24.94	4.887
		12	2.77	26.21	5.397
		14	2.11	27.53	5.954
		16	1.65	22.10	6.356
1 ½	38.10	10	3.41	31.29	7.690
		12	2.77	32.56	8.329
		14	2.11	33.88	9.019
2	50.80	10	3.41	43.99	15.194
		12	2.77	45.26	16.091

d_o, outside diameter; d_i, inside diameter; x wall thickness; A_i tube inside cross section.

Fluid velocities in pipes and tubes should not be too high, because of the excessive pressures developed, which may damage the piping system and require higher pump power. An economical (optimum) pipe diameter for a piping system may be calculated, using the cost data for pipes and pump power. Typical recommended fluid velocities are water and dilute aqueous solutions and suspensions 2–3 m/s, viscous and non-Newtonian fluids 1–2 m/s, air near atmospheric pressure 20 m/s, saturated steam near 1 bar gauge pressure 30 m/s, and water vapor at 100 mbar absolute pressure 100 m/s.

6.4.2 Pipe Fittings

Various types of fittings are installed in the piping systems in order to control, direct, and measure the flow of fluids. The fittings cause significant friction losses and pressure drops, which should be taken into account in designing and operating piping systems. The friction energy loss is estimated from empirical relations or diagrams (nomograms) (Walas, 1988).

The simplest method for estimating the friction in fittings is to use the empirical equivalent length L_e of straight pipe, which is often given as the ratio L_e/d_i, where d_i is the inside diameter of the pipe.

The friction losses through pipe fittings h_f (J/kg) can also be estimated from the empirical equation

$$h_f = K\left(\frac{u^2}{2}\right) \tag{6.23}$$

where

u (m/s) is the mean fluid velocity

K is a dimensionless constant, which depends on the type and dimension of the fitting, the fluid velocity, and the ratio L/d of the piping system

By comparing Equation 6.23 to the Fanning equation 6.14 it follows that $K = 4f(L/d)$, i.e., the empirical constant K is a measure of the friction in the fitting and the dimensions of the piping system. Values of the constant K for various fittings are given in the technical literature. It should be noted that the numerical value of the constant K depends on the units used in Equation 6.23. Literature values in other units, such as (ft), should be converted to the standard SI units.

6.5 FILM AND OREN CHANNEL FLOWS

6.5.1 Film Flow

The film flow of liquids is of importance to several food processing operations, such as heat exchange, evaporation, and refrigeration.

The flow rate of a liquid film flowing in an inclined surface, due to gravity forces, is given by the equation

$$\Gamma = \frac{(\delta^3 \rho^2 g \cos\theta)}{3\eta} \tag{6.24}$$

where

$\Gamma = m/b$ [kg/(ms)] is the mass flow rate per unit width of wetted surface
m (kg/s) is the mass flow rate
b (m) is the width of the wetted surface
δ (m) is the film thickness
θ is the angle of inclination of the flow surface relative to the vertical
η (Pa s) is the viscosity of the fluid

For vertical flow surfaces $\theta = 0$, $\cos\theta = 1$, Equation 6.24 yields

$$\delta = \left[\frac{3\eta\Gamma}{\rho^2 g}\right]^{1/3} \tag{6.25}$$

The Reynolds number Re of a falling film is calculated from the equation

$$Re = 4\frac{\Gamma}{\eta} \tag{6.26}$$

In film flow of liquids on vertical surfaces, the following flow regions exist:

$Re < 5$–25 laminar flow, no ripples
$25 < Re < 1000$–2000 rippled flow
$Re > 1000$–2000 turbulent flow

The ripples or waves on the liquid surface are desirable in heat and mass transfer operations, because they increase significantly the transfer coefficients.

6.5.2 Open Channel Flow

The flow in open channels is of importance to the flow of water in food plants. Open channels are often used in the hydraulic transport of fruits and vegetables and the transport of liquid wastewater.

Figure 6.12 shows the diagram of an open flow channel with length ΔL and rectangular $b \times h$ cross section.

The flow of a liquid in a sloped open-channel is caused by the atmospheric pressure acting on the free surface. The frictional energy loss Δh_f (J/kg) divided by the gravity constant $g = 9.81$ m/s^2 gives the friction head loss $\Delta h_f/g$ (m) which is related to the length of the channel ΔL (m) by the equation

$$\frac{\Delta h_f}{g} = s\Delta L \tag{6.27}$$

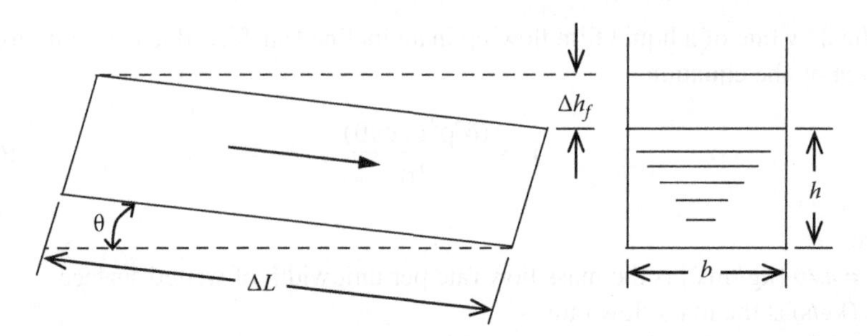

FIGURE 6.12 Diagram of an open-channel flow.

where $s = \cos\theta$ is the inclination of the free liquid surface to the horizontal.

The Fanning equation 6.14 for flow in a channel length ΔL can be written as

$$\left(\frac{\Delta h_f}{g}\right) = 4f\left(\frac{\Delta L}{d_e}\right)\left(\frac{u^2}{2g}\right) \tag{6.28}$$

where the equivalent flow diameter is $d_e = 4r_h$ and r_h is the hydraulic radius.

The hydraulic radius is defined in Equation 6.7, and for the rectangular cross section of flow channel in Figure 6.12 it is $r_h = (b \times h)/(b + h)$.

By combining Equations 6.27 and 6.28, we obtain the Chevy equation

$$u = C\left(\frac{sd_e}{2}\right)^{1/2} \tag{6.29}$$

where $C = (g/f)^{1/2}$ is the Chevy factor (constant) with SI units ($m^{1/2}s$). For flow of water in concrete (cement) channels, $C = 100\,m^{1/2}s$.

The flow of water in open channels is usually turbulent, because of its low viscosity.

6.6 FLOW MEASUREMENT

Flow measurement is important in the transport of liquids and gases used in food processing. The operation of common flow meters is based on energy balances of the Bernoulli equation. In principle, a constriction in the flow conduit increases the fluid velocity with a corresponding decrease of the pressure head (increase of pressure drop).

Mechanical and electromagnetic flow meters operate on different principles and they do not require a restriction to flow (Liptak, 1995).

6.6.1 Differential Pressure Flow Meters

The differential pressure flow meters consist of a primary element (orifice, venturi meter, nozzle, Pitot tube), which causes the pressure (head) loss, and a secondary element (U-tube, differential pressure transducer), which measures it.

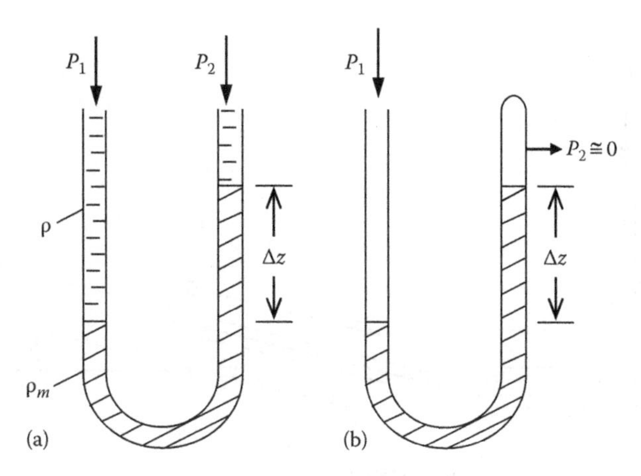

FIGURE 6.13 U-tube manometers: (a) open and (b) closed.

6.6.1.1 U-Tube Manometer

The U-tube manometers (open or closed type) measure pressure differences as shown in Figure 6.13. The U-tube contains a fluid heavier and immiscible with the flowing fluid, such as mercury for flow of liquids, and mineral oil for flow of air or gases. The two arms of the open tube are connected to the higher and lower pressures P_1, P_2, before and after the constriction of the flowing fluid. The flow velocity is proportional to the pressure difference $\Delta P = P_1 - P_2$ which is related to the difference of the level of the measuring liquid in the two arms $\Delta z = z_1 - z_2$, according to the following equation:

$$\Delta P = (\rho_m - \rho)(\Delta z)g \tag{6.30}$$

where

ρ_m, ρ (kg/m^3) are the densities of the measuring and flowing fluid, respectively
$g = 9.81$ m/s^2 is the gravity constant

In the closed U-tube, the second arm is closed and P_2 is equal to the vapor pressure of the measuring fluid at room temperature, which for mercury and oil is practically $P_2 = 0$. It is used for measuring pressures below atmospheric, while the open type manometers measures higher pressures and pressure differences.

The head energy difference Δh, corresponding to the above pressure difference ΔP is given by the equation

$$\Delta h = \frac{\Delta P}{\rho g} = \frac{[(\rho_m - \rho)(\Delta z)]}{\rho} \tag{6.31}$$

For measuring the flow of air and gases, the pressure differences are small and the measuring fluid is usually mineral oil. More accurate measurements are made using the inclined manometer (Figure 6.14), which expands the scale considerably.

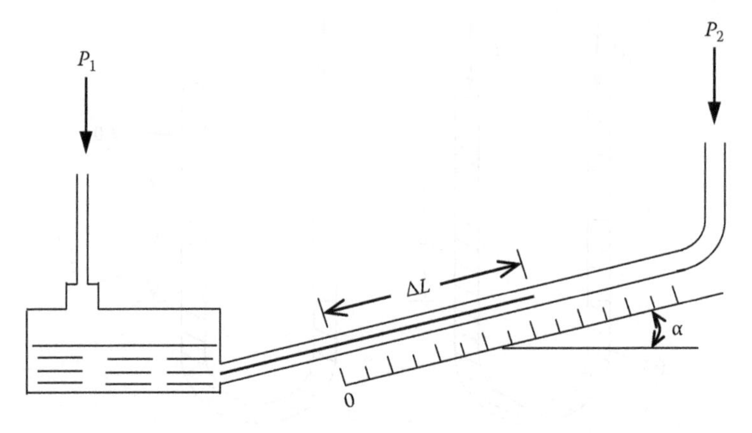

FIGURE 6.14 Inclined U-tube manometer.

The pressure difference ΔP measured in the inclined manometer is given by the equation:

$$\Delta P = (\rho_m - \rho)(\Delta L)g \sin \alpha \tag{6.32}$$

where

ΔL is the length of the measuring fluid in the inclined tube

α is the angle of inclination of the manometer scale to the horizontal, $\Delta z = \Delta L \sin \alpha$

6.6.1.2 Orifice Meter

The orifice meter measures pressure differences caused by a constriction of flow (orifice), installed in the pipe perpendicularly, as shown in the diagram of Figure 6.15. The orifice is made by drilling a concentric, eccentric, or segmental hole in a circular metallic plate. Sharp-edged orifices are preferred, giving high pressure drops. The orifice plate is installed between two sections of a pipe using the proper flange to prevent leaks.

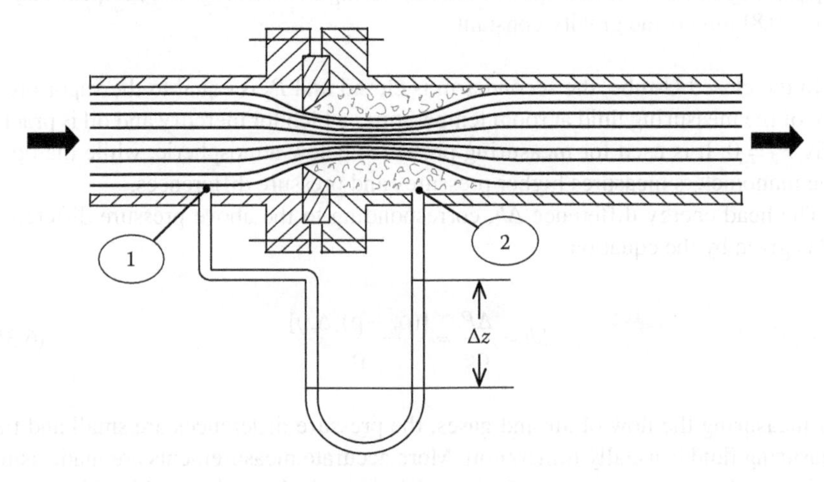

FIGURE 6.15 Diagram of an orifice meter.

The flow measurement is based on the application of the Bernoulli equation 6.13 without a pump, between the highest (1) and the lowest (2) pressures before and after the orifice, yielding the semitheoretical equation

$$u = C_o \left\{ \left(\frac{2\Delta P}{\rho} \right) \left[1 - \left(\frac{d_{or}}{d_i} \right)^4 \right] \right\}^{1/2} \tag{6.33}$$

where

u (m/s) is the mean fluid velocity in the pipe
d_{or} and d_i (m) are the orifice and inside pipe diameters, respectively
$\Delta P = P_1 - P_2$ (Pa) is the pressure difference in the orifice
ρ (kg/m^3) is the fluid density
C_o is the orifice coefficient, an empirical constant $C_o = 0.60–0.70$, the value of which depends on the type of orifice and the *Reynolds number*

For small orifice d_{or} and large pipe d_i diameters the term $(d_{or}/d_i)^4$ can be neglected and Equation 6.33 becomes

$$u = C_o \left(\frac{2\Delta P}{\rho} \right)^{1/2} \tag{6.34}$$

Equation 6.32 indicates that in an orifice system only part of the pressure energy loss $\Delta P/\rho$ is recovered as velocity energy $u^2/2$, since $C_o < 1$.

The head difference of the flowing fluid Δh (m) is often used in Equations 6.33 and 6.34, instead of the pressure difference ΔP (Pa), where $\Delta h = \Delta P/(\rho g)$, $g = 9.81$ m/s^2.

6.6.1.3 Venturi Meter

The Venturi meter is similar in principle with the orifice meter, but it is specially constructed so that the energy losses in the system are minimal (Figure 6.16). The shape of the Venturi meter is similar to the streamlines shown in the orifice meter (Figure 6.15). It is a more elaborate and expensive instrument than the simple orifice meter, in which the pressure energy is recovered almost quantitatively (about 96%) as velocity energy.

The mean fluid velocity in the pipe is given by an equation similar to Equation 6.34:

$$u = C_v \left(\frac{2\Delta P}{\rho} \right)^{1/2} \tag{6.35}$$

where C_v is the empirical Venturi coefficient, usually $C_v = 0.98$.

The flow nozzle is a less expensive flow meter similar to the orifice meter, but with a smoother flow channel, resembling the Venturi meter. The pressure energy is converted to velocity energy almost quantitatively $C = 1$.

6.6.1.4 Pitot Tube

The Pitot tube is a simple flow meter, which is used for measuring the point velocity in a pipe, while the previous flow meters measure the mean fluid velocity in the pipe.

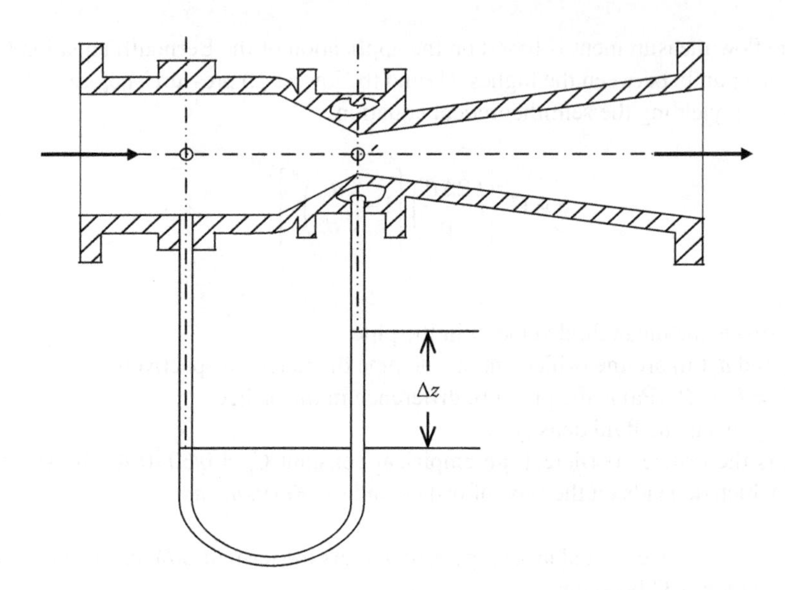

FIGURE 6.16 Venturi flow meter.

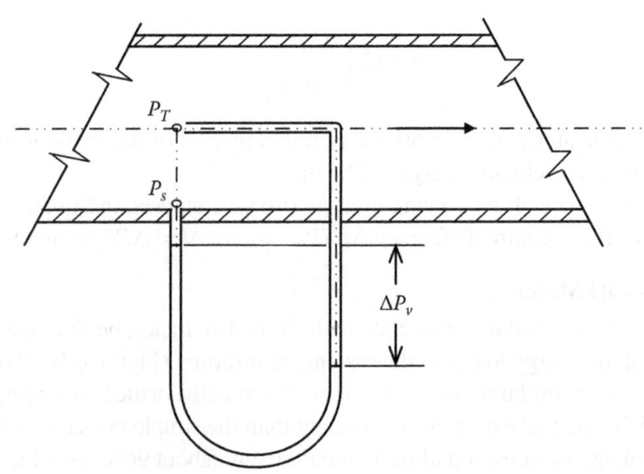

FIGURE 6.17 Pitot tube flow meter.

It is used mostly for flow measurements of air and gases (Figure 6.17). The Pitot tube consists of a U-tube, one leg of which is attached to the wall of the pipe, and the other is bent 90° against the flow direction. The energy of pressure difference $\Delta P_v/\rho$ is equal to the velocity energy $u^2/2$, resulting in the equation

$$u = \left(\frac{2\Delta P_v}{\rho}\right)^{1/2} \quad \text{or} \quad u = (2g\Delta h)^{1/2} \qquad (6.36)$$

where

$\Delta P_v = P_T - P_s$ and P_T and P_s are the total and static pressures, respectively.
$\Delta h = \Delta P/(\rho g)$ is the head loss (height of the fluid, m).

6.6.2 Mechanical Flow Meters

Mechanical flow meters consist of arrangements of moving parts, placed in the flow path, calibrated to measure directly the volumetric flow rate. They include rotameters, positive displacement (PD) flow meters, turbine flow meters, vortex and inertial flow meters.

6.6.2.1 Rotameters

Rotameters or variable area meters consist of a vertical tapered tube (glass, Plexiglas, plastic, or metal), in which the fluid flows from the bottom to the top, and a float is free to flow up and down, depending on the flow rate (Figure 6.18). The float reaches a mechanical equilibrium when the gravity force (weight) equals the force exerted by the upward moving fluid. The tapered tube is calibrated for a given fluid (usually water or air) to indicate a certain flow rate at a fixed scale reading.

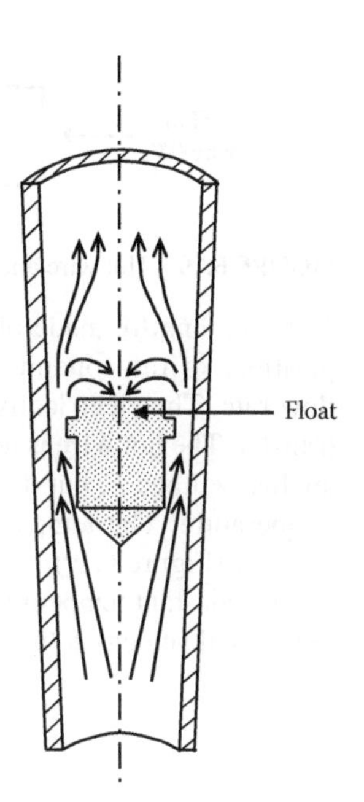

FIGURE 6.18 Rotameter flow meter.

6.6.2.2 Positive Displacement Flow Meters

The PD flow meters measure flow using an arrangement of moving parts (gears, lobes, discs) which pass isolated, known volumes of the fluid (liquid or gas) at a known rate. They operate better with clear liquids (no suspended particles). The PD flow meters resemble the positive displacement pumps, such as piston, peristaltic, gear, lobe, and diaphragm pumps (see later in this chapter). The turbine flow meters consist of a multibladed rotor mounted at right-angles to the flow, suspended in the flow stream on a free-running bearing. The diameter of the rotor is close to the inside diameter of the metering chamber, and its speed of rotation is proportional to the volumetric flow rate. Turbine rotation can be determined by solid state devices (conductance, inductance, capacitance) or mechanical (gear) sensors. Other types of rotary elements include propeller and paddlewheel devices.

The Vortex flow meters are based on the artificial creation of vortices (vortexes), or eddies in the pipe; which follow the flow pattern of the fluid. The frequency of vortex generation is related to the fluid velocity. The Inertial or Coriolis flow meters consist of two vibrating tubes through which a fluid can flow. The phase shift of tube vibration due to inertial (Coriolis) forces is proportional to the flow rate.

6.6.3 Electrical and Magnetic Flow Meters

Electrical and related flow meters utilize the electrical, magnetic, and acoustic properties of fluids, measuring the fluid velocity with electrical circuits/instruments.

6.6.3.1 Hot Wire Anemometer

The hot wire anemometer determines the velocity of air/gases and liquids by measuring changes in the electrical resistance of a wire placed in the flow path. The measuring probe consists of a very thin and short wire (0.0038 mm diameter and

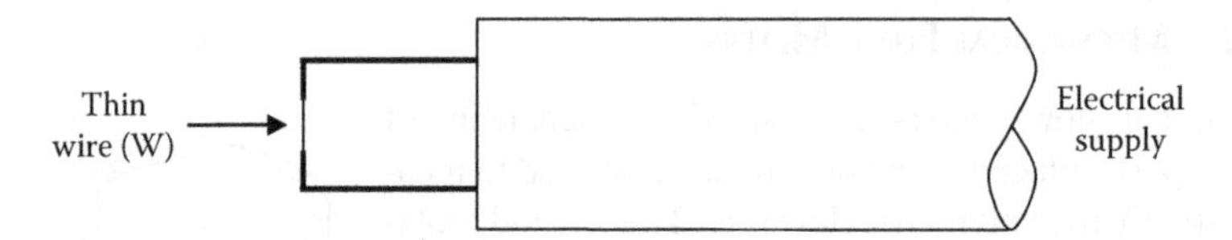

FIGURE 6.19 Hot wire anemometer.

1.25 mm length), made of platinum or tungsten, which can be placed in various positions of the conduit cross section for spot measurements of the velocity or flow rate. The electrically heated wire is cooled by the flowing fluid due to heat transfer. The hot wire is normally operated at constant temperature by varying the applied voltage E, but it can be used at constant resistance (voltage) and variable temperature. The temperature of the fluid is measured simultaneously with the velocity (Figure 6.19).

In a constant temperature wire system, the applied voltage E (volts) is a function of the fluid velocity u (m/s) according to King's equation

$$E^2 = A + Bu^{0.45} \tag{6.37}$$

where A and B are characteristic constants.

Several spot measurements can be measured quickly and the data of the electrical circuit can be transferred to a PC computer. The instrument is calibrated with a simple flow meter, such as the Pitot tube (Figure 6.17).

6.6.3.2 Magnetic and Acoustic Flow Meters

The magnetic flow meters are used for measuring volumetric flows of electrically conducting fluids, based on Faraday's law. The voltage induced across a magnetic field in the flowing fluid is proportional to the flow rate. The acoustic flow meters include the transit time, and the Doppler flow meters. The transit time flow meter is used for clear fluids and it operates by alternatively transmitting and receiving a burst of sound energy into the fluid. The time it takes for sound to travel between two transducers in the pipeline is proportional to the velocity.

The Doppler flow meter is used for particle suspensions in liquids and it employs two ultrasound transducers in the system. One transmits a continuous ultrasound wave in the flowing stream, and the other receives the ultrasound wave scattered from the suspended particles. The received wave has a frequency shift, compared to the transmitted one, which is called the Doppler frequency shift, which is proportional to the flow velocity.

6.6.4 Weir Flow Meters

The flow rate in open channels is measured using head flow meters or weirs, through which the liquid overflows by the force of gravity. The cross section of a weir can be triangular (Figure 6.20) or rectangular. The weir flow meters are based on the application of the Bernoulli equation to gravity flow. The flow rate Q in triangular

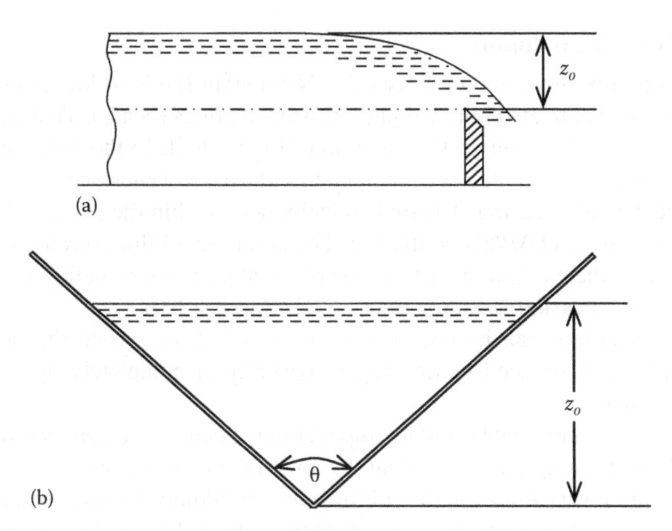

FIGURE 6.20 Weir flow meters: (a) rectangular and (b) triangular.

V-notch weirs is a function of the notch angle θ and height z_o of overflow in the weir, according to the empirical equation

$$Q = \left(\frac{8}{15}\right) C_d \left[\tan\left(\frac{\theta}{2}\right)\right] (2g)^{1/2} z_o^{5/2} \tag{6.38}$$

where C_d is the discharge coefficient (dimensionless), characterizing the given weir, typically $C_d = 0.62$ for $\theta = 20°C$ and $z_o = 0.15\,m$, $g = 9.81\,m/s^2$.

For rectangular weirs, the flow rate Q is a function of the liquid width b and height z_o of overflow in the weir, according to the empirical equation:

$$Q = \left(\frac{2}{3}\right) C_d b (2g)^{1/2} z_o^{3/2} \tag{6.39}$$

where C_d is the discharge coefficient (dimensionless), characterizing the given weir, typically $C_d = 0.65$, and $g = 9.81\,m/s^2$.

6.7 PUMPING EQUIPMENT

6.7.1 Pumps

Pumps are mechanical devices, which are used widely to transport liquids and semi-fluid materials through piping systems in food processing plants. Pumps provide the mechanical energy (or head) required to overcome the various resistances to flow (heads) in the piping system. There are various types of pumping equipment, the most common of which are centrifugal and positive displacement pumps.

6.7.1.1 Centrifugal Pumps

Centrifugal pumps are commonly used for Newtonian fluids of low viscosity, and they are operated at relatively low liquid pressure energies (heads). They are mainly of the volute type, as shown in the diagram of Figure 6.21. In the volute pump, the liquid enters the axis of a high speed impeller with curved vanes and is thrown radially into the widening casing. The high velocity head within the pump $u^2/2g$ is converted to pressure head $\Delta P/(\rho g)$ at the exit. The efficiency of this conversion depends on the design of the pump impeller and the physical properties (mainly viscosity) of the liquid (Saravacos and Kostaropoulos, 2002; Bpyce, 1999).

Centrifugal pumps can be operated while throttled, i.e., when the flow at the discharge piping is reduced or interrupted partially or completely by closing the appropriate valves.

The axial flow pumps consist of an impeller in the center of a cylinder, rotating at high speed and pushing the liquid from the low to the high pressure sides. They are used for pumping very high volumes of low-viscosity liquids (higher than 500 m^3/h) at heads less than 15 m (pressure drop less than 1.5 bar). A typical application is the recirculation of the liquid in a forced circulation evaporator (Chapter 10), where the axial pump is installed in the elbow or the recirculation loop.

6.7.1.2 Positive Displacement Pumps

PD pumps are used to transport viscous, non-Newtonian fluids, and suspensions of particles in liquids, such as several food products. They can operate at low flow rates and against high pressures (heads), and are suitable for shear-sensitive products, because of the very low shear rates developed during their operation.

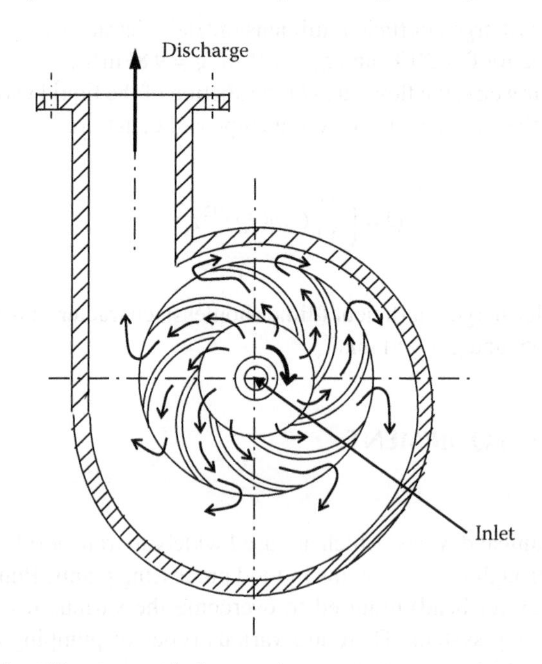

FIGURE 6.21 Centrifugal pump (volute type).

Construction and operation of PD pumps should eliminate or reduce the development of excessive slip flow, i.e., backflow of fluid past the rotors or pistons from the high pressure (discharge) to the low pressure (suction side) of the pump system.

PD pumps can be operated over a wide range of flow rates and pressures. Since they can deliver a fixed volume of fluid per rotation or stroke, they can be used as metering pumps in processes where flow rate is critical.

PD pumps must not be throttled, i.e., not be operated with closed valves in the discharge section, since the high mechanical pressures developed can damage mechanically the pump and the piping system.

PD pumps include reciprocating and rotary pumps. Reciprocating pumps are divided into piston, plunger, and diaphragm pumps, as shown in the diagrams of Figure 6.22. In reciprocating pumps, the fluid enters through a valve to a chamber, where it is pressurized mechanically and discharged through another valve to the high-pressure side of the piping system. Pumping rate is controlled by changing the frequency or the length of the stroke. In diaphragm pumps, the liquid is forced through the suction and discharge valves to a pulsating plastic or rubber diaphragm, which is compressed and expanded with a reciprocating piston. Diaphragm pumps are used when the (corrosive or sensitive) fluid should not come into contact with the metallic pump.

6.7.1.3 Rotary Pumps

Rotary pumps transport fluids through rotating gears, lobes, or screws (Figure 6.23). The external gear pumps consist of a fixed casing, which contains two meshing gears of equal

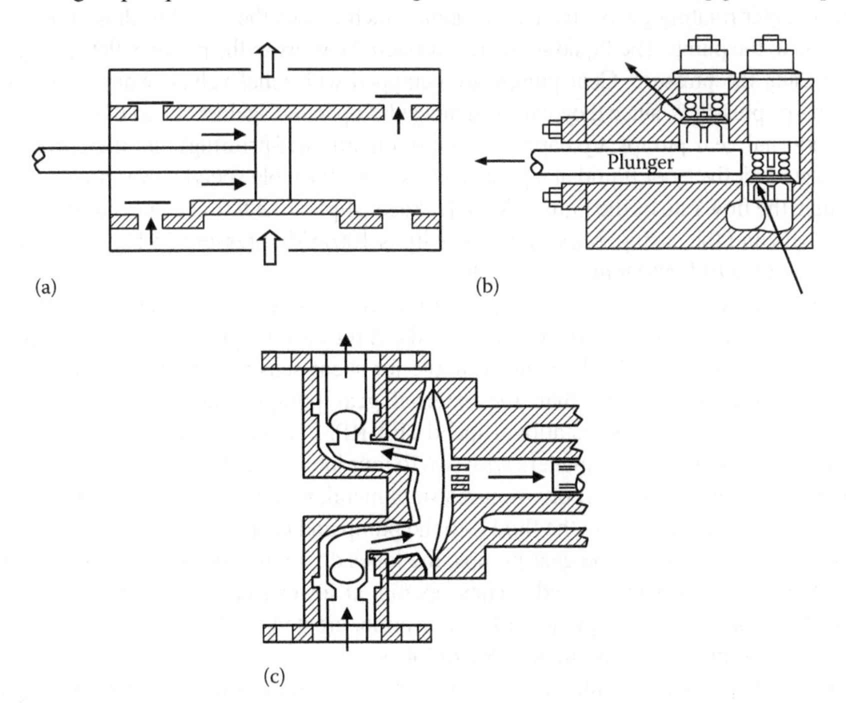

FIGURE 6.22 Reciprocating positive displacement pumps. (a) Pistol, (b) plunger, and (c) diaphragm.

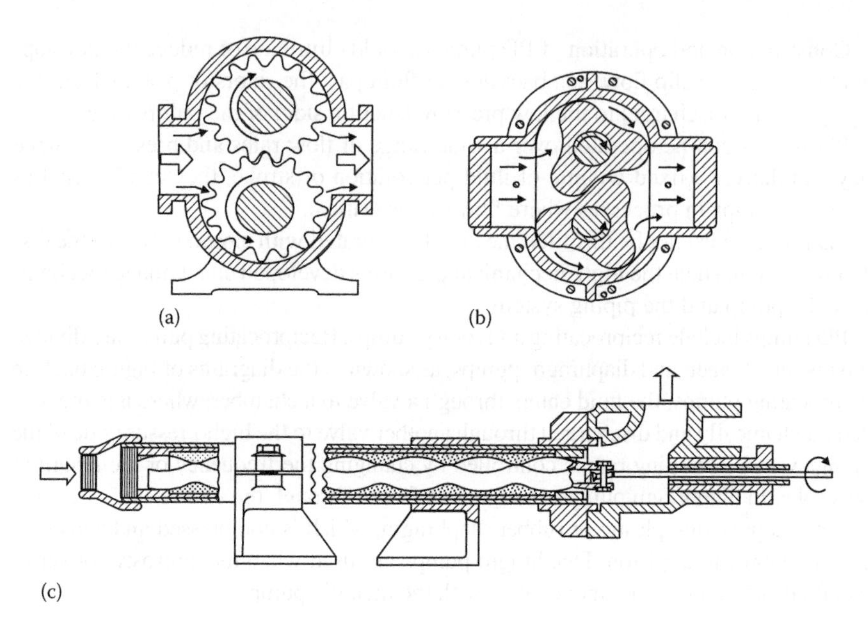

FIGURE 6.23 Rotary positive displacement pumps: (a) gear, (b) lobe, and (c) progressing cavity.

size. The driving gear is connected to the motor, while the other (idler) gear runs free. The counter-rotating gears create a vacuum, which causes the liquid to flow from the suction to the pump. The liquid is carried through the gears to the pump outlet, strongly increasing the pressure. Gear pumps are equipped with relief valves, which limit the discharge pressures, since they will be damaged if operated against closed discharge.

External gear pumps are capable of pumping air (self-priming) and they provide a constant delivery of liquid at a given rotor speed. Variable speed drives are used to change the flow rate of the liquid. Some leakage (slip) of the liquid can occur through the gears and the casing. Maximum capacity is 100 m^3/h, pressure 35 bar, rotational speed 4000 RPM, and temperature 400°C.

The internal gear pumps have a small drive gear inside a larger idler gear. As the gears rotate, a vacuum is created sucking the fluid into the pump, which is pushed to the high pressure side. They are suitable for shear-sensitive fluids, because they develop lower shear rates than the external gear pumps. Maximum capacity is 250 m^3/h, pressure 17 bar, rotational speed 1100 RPM, and temperature 340°C.

The lobe pumps have impellers with two to four lobes, which counter-rotate inside the pump casing, made of a corrosion-resistant metal, such as stainless steel 316. The counter-rotating lobes draw the fluid into the pump and push it along the casing wall to the discharge. Unlike the gear pumps, the lobes do not touch each other, enabling the flow of nonlubricating fluids. They operate at lower rotation speeds and shear rates than the rotary gear pumps. Flow pulsation, created by the rotating lobes, can be smoothed by increasing the number of lobes.

Lobe PD pumps are suitable for the transport of viscous fluids, containing large amounts of solids. Maximum is capacity 450 m^3/h, pressure 25 bar, viscosity 100 Pa s, and temperature 300°C.

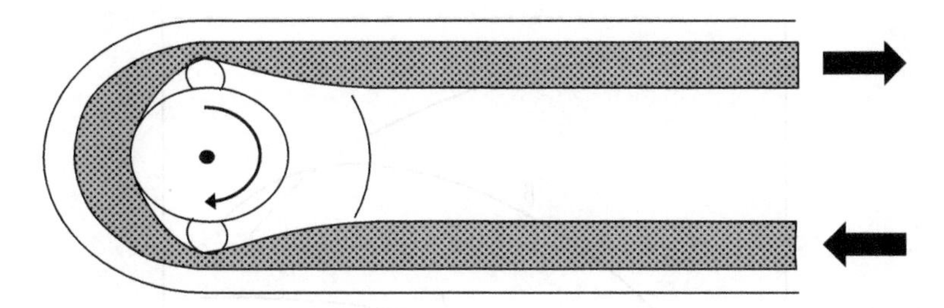

FIGURE 6.24 Diagram of a peristaltic pump.

The progressing cavity pumps are a special type of screw or helical pump, consisting of a threaded rotor turning inside a double-threaded stator. The long rotor is usually made of stainless steel, while the stator is made of an elastomer, supported by a metal casing. The process fluid is conveyed from the suction to the discharge end of the pump via a series of cavities, formed as the rotor turns within the stator. The build-up of high pressure is gradual, and pulsation of the flow rate can be minimized by using more cavities (up to 12) in the pump.

Progressing cavity pumps are used to transport thick and lumpy fluids and slurries (suspensions of particles or solid pieces in viscous fluids). They can be operated at very low flow rates, corresponding to very low shear rates. Maximum capacity is 250 m^3/h, rotational speed 1200 RPM, pressure 60 bar, and temperature 260°C.

The peristaltic pumps consist of a flexible tube inside a casing, through which the fluid is forced to move by the action of two rollers, located at the two ends of a rotor. As the rotor turns, the rollers press against and run along the tubing squeezing the fluid through the tube, and producing flow (Figure 6.24). The liquid does not contact the pump. The tube should be resilient enough to allow the rollers to collapse it and spring back to the original shape, creating a suction. Typical tubing sizes 5–12.5 mm. Peristaltic pumps are simple devices without check valves, they operate at low speeds and shear rates, and they are suitable for sensitive fluids, such as biological materials. They are used for low flow rates up to 100 L/h at pressures up to 1.8 bar.

6.7.2 Pump Characteristics

The energy for pumping W_p (J/kg) in a system involving pressure, velocity, elevation, and friction is calculated from the Bernoulli equation 6.10. The power P (kW) required to pump m (kg/s) of a fluid is given by the equation

$$P = \frac{mW_p}{1000\beta} \quad (6.12)$$

where β is the efficiency of the pump, normally $\beta = 0.75$–0.85.

6.7.2.1 Pump Curves

The characteristic curves of centrifugal pumps as functions of the volumetric flow rate Q (m^3/s) are shown in Figure 6.25. The pump head Δh (m) is calculated from the

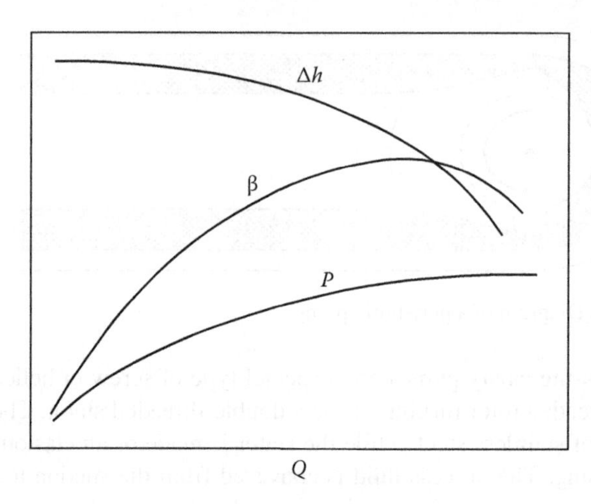

FIGURE 6.25 Characteristic curves if centrifugal pumps.

Bernoulli equation 6.10, which is modified by replacing the friction head h_f/g with the Fanning equation

$$\Delta h = \frac{\Delta P}{\rho g} + \frac{\Delta u^2}{2} + \Delta z + \left(\frac{2f}{g}\right)\left(\sum \frac{L_e}{d}\right)\left(\frac{4Q}{\pi d^2}\right)^2 \tag{6.40}$$

where

Q (m³/s) is the volumetric flow rate (note that $4Q/(\pi d^2) = u$, fluid velocity, m/s)
d (m) is the inside pipe diameter
$\sum L_e$ (m) is the total equivalent length of the pipe (including fittings)
f (dimensionless) is the Fanning friction factor

For laminar flow ($Re < 2100$), the Fanning friction factor is substituted by $f = 16/Re$:

$$\Delta h = \frac{\Delta P}{\rho g} + \frac{\Delta u^2}{2} + \Delta z + \left[\frac{32\eta}{(\rho g d)}\right]\left(\sum \frac{L_e}{d}\right)\left(\frac{4Q}{\pi d^2}\right) \tag{6.41}$$

where η (Pa s) is the viscosity (Newtonian fluids) or apparent viscosity (non-Newtonian fluids). As the flow rate Q is increased the pump head Δh decreases gradually, while the pump power P increases continuously. The pump efficiency β increases at low flow rates, it decreases at very high flows, and it reaches a maximum (optimum economic operation) at an intermediate flow rate.

According to Equations 6.40 and 6.41, the pump head is a quadratic (parabolic) function of fluid velocity in the turbulent region, becoming linear in the laminar flow.

In some pumping systems the velocity of the fluid in the pipes is nearly constant, and therefore, the difference in velocity head $\Delta u^2/2$ in both Equations 6.40 and 6.41 can be neglected.

The centrifugal pump can be operated at any point of the head Δh versus flow rate Q graph by opening or closing (throttling) the exit (control) valve. For a given piping

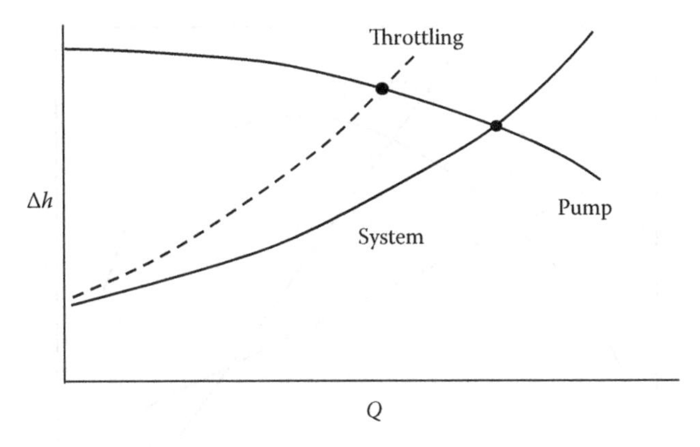

FIGURE 6.26 Centrifugal pump and pumping system curves. Pump throttling.

installation, the system curve Δh, u is given by Equation 6.40. A pumping system operates properly at the point of intersection of the pump and system curves (Figure 6.26). However, for more stability, centrifugal pumps are preferably operated with partially closed the discharge valve (throttled). The following empirical equations are useful in evaluating the performance of centrifugal pumps:

$$P = C_1 Q \Delta h = C_2 \rho N^3 D^5 \tag{6.42}$$

$$Q = C_3 N D^5 \quad \text{and} \quad \Delta h = C_4 N^2 D^2 \tag{6.43}$$

where
- P (W) is the pump power
- Q (m^3/s) is the flow rate
- Δh (m) is the pump head $\Delta h = \Delta P/(\rho g)$
- ρ (kg/m^3) is the fluid density
- D (m) is the diameter of pump impeller
- N (1/s) is the rotational speed $N = RPM/60$
- C_1, C_2, C_3, C_4 are empirical constants

From the last equations it follows that the ratio $C = (N^2 Q)/\Delta h^{2/3}$ is constant.

By a combination of Equations 6.42 and 6.43 it is possible to estimate the flow rate Q, the (manometric) head Δh, and the power P of a new pump, from the corresponding characteristics of a known similar (homologous) pump, if the impeller diameters D and the rotational speeds N are known.

If the characteristic curve ($\Delta h - Q$) of a given pump at a speed N_1 is known, the curve for another speed N_2 can be constructed, using the relation $\Delta h_2/\Delta h_1 = (N_2/N_1)^2$, which is derived from Equation 6.43.

Figure 6.27 shows the performance curves for PD pumps. All curves start at a low flow rate, but the pumps cannot be operated at $Q = 0$, contrary to the centrifugal pumps, because of excessive pressure buildup, which may damage the pump. Both energy head Δh and pump power P decrease as the flow rate is increased. The pump efficiency reaches a maximum at an intermediate flow rate.

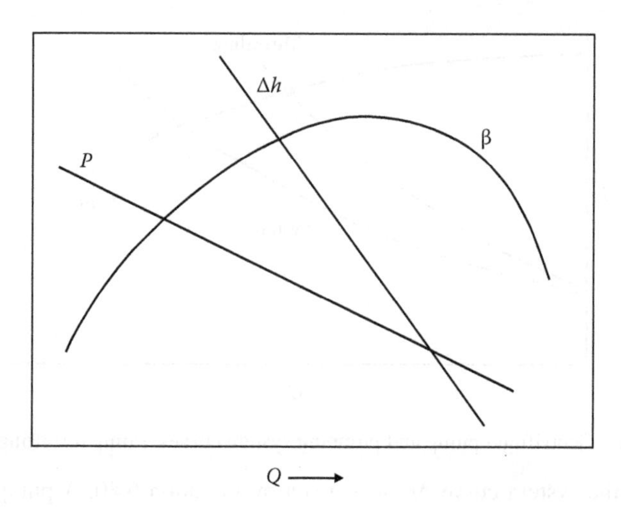

FIGURE 6.27 Characteristic curves of positive displacement pumps.

6.7.2.2 Net Positive Suction Head

Pump operation is considered normal and efficient when the flow of fluid in the piping system is continuous and constant, without interruptions or noise. The system should be so designed and operated to prevent cavitation, a condition when bubbles and vapors are formed in the suction of the pump, which collapse in the high-pressure section, causing noisy operation and loss of pumping efficiency. Cavitation may cause mechanical damage to the pump.

Cavitation conditions are avoided when the Net Positive Suction Head (NPSH) of the pump is positive. The *NPSH* is defined and estimated from the equation

$$NPSH = \frac{P_s - P_v}{\rho g} - \left(z_s + \frac{h_{fs}}{g} \right) \tag{6.44}$$

where

P_s (Pa) is the pressure on the free surface of the liquid in the suction
P_v (Pa) is the vapor pressure of the liquid at the suction temperature
ρ (kg/m³) is the density of the fluid
z_s (m) is the elevation of pump suction (entrance) above the surface
h_{fs}/g (m) is the friction head in the suction section of the system

Figure 6.28 shows a diagram of the *NPSH* components of a pumping system. In this example, the pump is located above the surface level of the liquid in the suction side.

In order to prevent cavitation, $NPSH > 0$, or $(P_s - P_v)/(\rho g) > (z_s + h_{fs}/g)$.

The vapor pressure of the liquid at the suction should be lower than the total pressure, i.e., $P_v < P_s$. This condition is obtained at low liquid temperatures. It should be noted that at the boiling point $P_v = P_s$ and $NPSH = -(z_s + h_{fs}/g)$. To prevent cavitation, $NPSH > 0$, or $z_s + h_{fs}/g < 0$. In normal pumping systems the friction head in

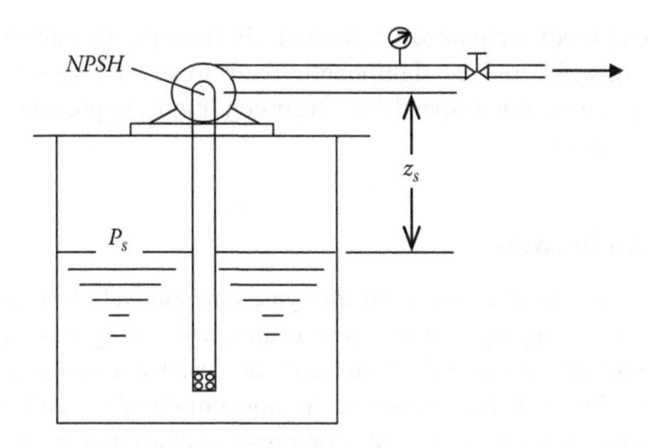

FIGURE 6.28 NPSH in a pumping system.

the suction side h_{fs}/g is relatively low, compared to the elevation head z_s, using large diameter piping, without valves. Under these conditions and boiling of the liquid, cavitation is prevented when $z_s < 0$, i.e., the pump entrance is at a lower elevation than the surface of the liquid. Such conditions exist, for example, in pumping systems of evaporators (Chapter 10).

The maximum elevation of the pump entrance above the liquid surface z_{sm} is obtained when the *NPSH* becomes zero and the friction head in the suction side is negligible, $z_{sm} = (P_s - P_v)/(\rho g)$. Reducing the temperature of the liquid will reduce its vapor pressure P_v, therefore, increasing the elevation z_s.

6.7.2.3 Pump Selection

For a given liquid (fluid) pumping application, pumps are selected on the basis of pump capacity Q against pump head Δh. Q is usually expressed in (m^3/h), which should be converted to (m^3/s) for engineering calculations in the SI units, Q (m^3/s) = Q (m^3/h)/3600. In technical U.S. units, pump capacity is usually expressed in U.S. gallons per minute (GPM), which are converted to SI units as 1 GPM = 0.063 × 10^{-3} m^3/s.

Pump head (manometric height) is expressed in (m) of liquid, while in the United States (ft) are traditionally used (1 ft = 0.305 m).

Centrifugal pumps are available at high capacities (>100 m^3/h or >500 GPM), and pump heads up to 10 bar (100 m water). PD pumps can handle up to 100 m^3/h (500 GPM), and pump heads above 10 bar (150 psig).

Centrifugal pumps are generally less expensive than PD pumps, especially in large capacities. They develop lower pump heads and they can be throttled (operated with partially closed valve at the discharge section). Contrary to the PD pumps, centrifugal pumps are not self-priming, and they must be primed before starting the operation. Priming means removal of air from the pump interior by some procedure, such as filling with liquid, evacuating, or using an auxiliary PD pump. Pumps, after a long operation should be cleaned either by the cleaning-in-place (CIP) system or by disassembling the pump and cleaning the various parts individually.

The viscosity of the fluid to be pumped is very important in pump selection. Low viscosity (Newtonian) fluids can be handled effectively by centrifugal pumps. High viscosity and non-Newtonian fluids are handled by PD (preferably rotary) pumps.

Pseudoplastic fluid foods are pumped with rotary PD pumps, operated at high speeds (high shear rates), which reduce significantly their apparent viscosity. Food emulsions should be pumped at low speed/low shear conditions, to prevent any mechanical damage of the product.

6.7.3 Fans and Blowers

Fans and blowers are used to move air and gases at relatively low pressure drops in various food processing operations, such as air-conditioning and air-drying. The terms fan and blower are equivalent and they are used interchangeably. In these applications, the flow can be considered as noncompressible, and the Bernoulli energy balance equation can be applied. Compression of air and gases above 10% is considered as compressive flow and it is treated in Thermodynamics (Chapter 3) and in Refrigeration (Chapter 12).

Most of the industrial fans are centrifugal with either backward or forward curved blades (Figure 6.29). A limited number of fans operate on the axial flow principle, similar to the axial flow pumps. Fan flow rates Q are expressed usually in in.^{3}/h or CFM (1 CFM = 1.7 in.^{3}/h). For smaller flow rates, the unit L/h may be used. For numerical applications in the SI units, all flow rates should be converted to (m^3/s).

Pressure drops (pressure heads) in air (gas) transport conduits are usually expressed in mbar or InWg (inches water gauge), which should be converted to Pa (1 mbar = 100 Pa) and (1 InWg = 250 Pa). Industrial fans or blowers can move up to 40,000 m^3/h or 23,500 CFM of air at atmospheric pressure.

The characteristic curves of centrifugal fans (backward curved) are similar to those of centrifugal pumps (Figure 6.25). However, the forward-curved centrifugal fans and the axial fans show two maxima of the Δh, Q curve, one at $Q = 0$ and the other at $Q = 0.4Q_{max}$.

Fans and blowers move air and gases through large diameter pipes or large conduits of rectangular cross section. The piping elbows should have a large curvature to reduce friction losses. Air pipes are constructed of steel or aluminum sheets, and the flow rate is controlled by dampers.

For a given centrifugal fan, the performance characteristics (flow rate Q, fan head Δh, and power P) are changed, when increasing the rotor speed from N_1 to N_2, according to the analogous (similar) Equations for centrifugal pumps 6.42 and 6.43:

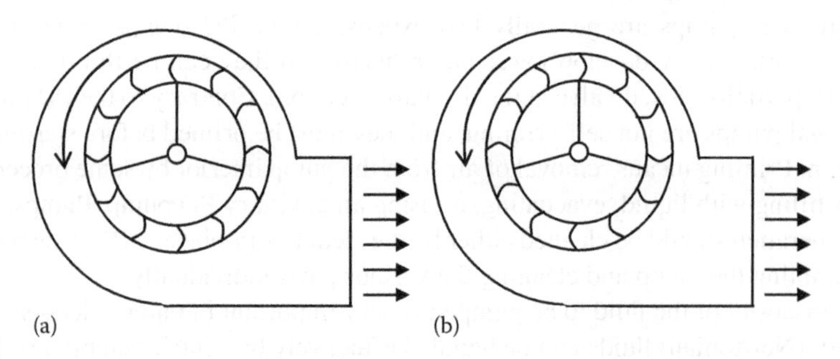

FIGURE 6.29 Centrifugal fans: (a) backward bladed and (b) forward bladed.

TABLE 6.5
Velocity Pressures and Velocities of Air

P_v		u	
inWg	mmWg	FPM	m/s
0.1	2.5	1,266	6.4
0.5	12.7	2,832	14.4
1.0	25.4	4,005	20.3
2.0	50.8	5,664	28.8
5.0	127.0	8,995	45.5
10.0	254.0	12,665	64.3
20.0	508.0	17,011	91.0
30.0	762.0	21,936	111.4
40.0	1016.0	25,330	128.7

$$Q_2 = Q_2\left(\frac{N_2}{N_2}\right) \quad \Delta h_2 = \Delta h_1\left(\frac{N_2}{N_2}\right)^2 \quad P_2 = P_1\left(\frac{N_2}{N_2}\right)^3 \tag{6.45}$$

The velocity of air or gases in a pipe or conduit u (m/s) is related to the velocity pressure P_v (Pa) according to Equation 6.36, which is also used in the Pitot tube calculations

$$u = \left(\frac{2P_v}{\rho}\right)^{1/2} \tag{6.36a}$$

where ρ (kg/m^3) is the density of air, which at 1 bar pressure and 20°C is 1.164 kg/m^3.

Table 6.5 shows typical velocity pressures P_v and velocities of air u near the atmospheric pressure.

In the technical literature, the air pressure is commonly expressed in (inches) or in (mm) of water gauge (inWg or mmWg), while the air velocity is expressed in (m/s) or FPM (ft/min).

6.7.4 Vacuum Pumps

Vacuum pumps are widely used in food engineering to create and maintain low pressures (below atmospheric), which are required in various operations, such as evaporation, drying, and distillation. In laboratory (small volume) applications, oil vacuum pumps are used, which are special rotary PD pumps, filled with mineral oil, which compress and transport the air from the low pressure chamber through the oil to the atmosphere. Oil vacuum pumps can operate down to 1 mbar (100 Pa) absolute pressure. For higher vacuum (pressures below 1 mbar), a combination of oil pump, followed by a diffusion pump, can be used. Water vapors, when present in

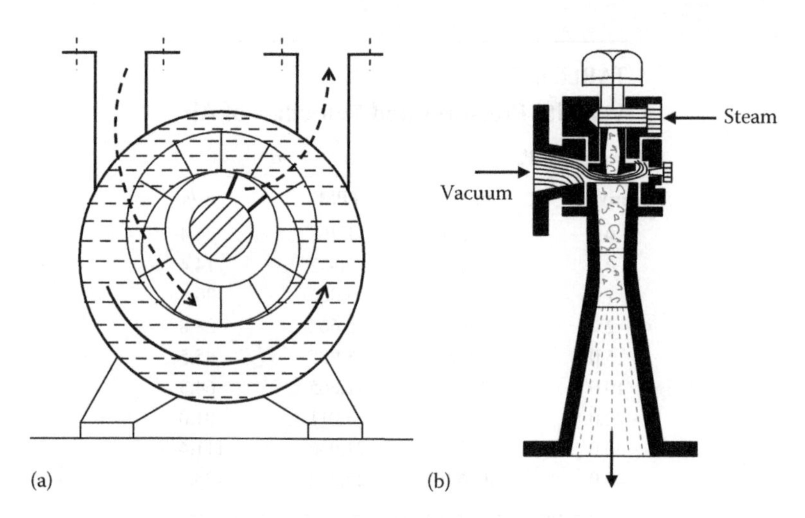

FIGURE 6.30 Industrial vacuum pumps: (a) water ring and (b) steam ejector.

considerable quantity in the vacuum system, can be removed by a cooling condenser, placed before the oil pump to prevent vapor condensation in the oil, which would spoil the oil and increase the pressure.

Industrial vacuum systems use mainly two vacuum pumps, i.e., water ring pumps and steam ejectors (Figure 6.30). The water ring pump operates on the same principle with the laboratory oil pump, using a water ring through which the air (gas) is pumped from the lower to the higher (atmospheric) pressure. The vanes rotate eccentrically in the water-containing pump casing. The vacuum obtained is limited by the vapor pressure of water at the pump temperature, e.g., 12 mbar at 10°C, 23 mbar at 20°C, and 42 mbar at 30°C. These pumps are used in vacuum evaporation or distillation of water solutions and juices, where a moderate vacuum is needed.

Steam ejectors are used in various industrial operations for moderate to high vacuum. The ejectors require high pressure steam, e.g., 10 bar, and they compress the air from the vacuum chamber to a higher pressure (atmospheric). Multistage ejectors, used in series, can create and maintain high vacuum, e.g., 3 mbar created by a 3-stage ejector, which could be used in a freeze-drying operation. The time t (min) required to evacuate a vessel is a function of the pump capacity, according to the empirical equation

$$t = C\left(\frac{V}{Q}\right) \tag{6.46}$$

where

V (m^3) is the volume of the vessel

Q (m^3/min) is the capacity of the vacuum pump

C (dimensionless) is an empirical constant, depending on the final pressure; for a vacuum (absolute pressure) of 1 Torr (133 Pa), $C = 7$

APPLICATION EXAMPLES

Example 6.1

A non-Newtonian fluid, independent of time, power-law model, flows at a mean velocity of 1.5 m/s in a straight pipe 5 m long and nominal diameter 1 ½ in. schedule No. 40 S. The density of the fluid is 980 kg/m^3 and the rheological constant $K = 4$ Pa s^n, where n (dimensionless) is the flow behavior index.

Calculate

The apparent viscosity, the *Reynolds number*, and the pressure drop of the fluid in the pipe for the following values of the flow behavior index $n = 0.5$, 1.0, and 1.5.

Solution

Table 6.1 indicates that a pipe of nominal diameter 1 ½ in. schedule No. 40S has an internal diameter $d = 40.89$ mm or $d = 4.1 \times 10^{-2}$ m.

The shear rate at the wall of the pipe γ (1/s) can be calculated from the mean fluid velocity and the diameter of the pipe as follows (Equation 6.1):

$$\gamma = \left(\frac{du}{dy}\right) = \frac{8u}{d} = \frac{(8 \times 1.5)}{(4.1 \times 10^{-2})} = 293\ 1/\mathrm{s}$$

The apparent viscosity η_a (Pa s) of the fluid at the pipe wall will be (Equation 6.8)

$$\eta_a = K\gamma^{n-1} = 4 \times (293)^{n-1}$$

The Reynolds number *Re* of the fluid in the pipe will be

$$Re = \frac{du\rho}{\eta_a} = \frac{(4.1 \times 10^{-2} \times 1.5 \times 980)}{[4 \times (293)^{n-1}]} = 15.1 \times (293)^{1-n}$$

Assuming that the flow in the pipe is laminar, the pressure drop ΔP (Pa) in the pipe can be calculated from the Poiseuille Equation for power-law fluids 6.1:

$$\Delta P = K\left(\frac{4L}{d}\right)\left(\frac{8u}{d}\right)^n = 4 \times \left[\frac{4 \times 5}{(4.1 \times 10^{-2})}\right](293)^n = 195 \times (293)^n$$

In the given application problem, the apparent viscosity, the *Re* number, and the pressure drop are functions of the flow behavior index n, as shown in Table 6.6.

The rheological constant n (flow behavior index) plays a very important role in the pipe flow of non-Newtonian fluids. Pseudoplastic materials $n < 1$ are less viscous and develop much lower pressure drops than Newtonian $n = 1$ or dilatant materials. These differences are important considerations in the design of pumping systems and in heat transfer operations.

TABLE 6.6
Effect of Flow Behavior Index (n)

n	0.5	1.0	1.5
η_a, Pa s	0.234	4.0	68.40
Re, –	258.2	15.1	0.88
ΔP, kPa	3.3	57.1	978
ΔP, bar	0.03	0.57	9.78

Example 6.2

A vegetable oil is pumped from a truck tank at the rate of 20 m³/h to a storage tank located at a distance of 30 m and a level 5 m higher than the truck. A pipe of 2 in. nominal diameter 40 schedule is used, and the piping system includes four standard elbows 90° and two gate valves, one fully open and the other half-closed.

Calculate

The pump power required for two operating temperatures, 25°C and 5°C.

Viscosity of the vegetable oil (Newtonian fluid), 60 mPa s (25°C) and 200 mPa s (5°C).

The mean density of the oil in the given temperature range is 920 kg/m³.

Solution

The flow rate of the oil in the piping system will be $m = (20 \times 920)/3600 = 5.1$ kg/s.

The inside diameter of the 2 in. pipe will be $d = 0.0525$ m and the cross-sectional area $A = 21.6 \times 10^{-4}$ m² (Table 6.2). Therefore, the mean oil velocity in the pipe will be $u = 20/(21.6 \times 10^{-4} \times 3600) = 2.6$ m/s, which is a reasonable fluid velocity in pipes. The *Reynolds number* of the fluid in the pipe at 25°C will be $Re = (0.0525 \times 920 \times 2.6)/0.06$ or $Re = 2093$, i.e., the flow is in the transition region between laminar and turbulent flow.

Assuming laminar flow, the friction factor of the oil in the piping will be $f = 16/Re$, or $f = 16/2093 = 0.0076$.

For flow of the oil at a lower temperature (5°C), the viscosity is increased considerably and the flow becomes more viscous (laminar). Reynolds number $Re = 2093 \times (60/200)$ or $Re = 628$, and friction factor $f = 16/628 = 0.0255$.

Length of the pipe 30 m and equivalent $L_e/d = 30/0.0525 = 571.42$
Equivalent length of four standard elbows 90° (Table 6.7) $4(L_e/d) = 4 \times 32 = 128.00$
Equivalent length of one gate valve open $L_e/d = 7.00$
Equivalent length of one gate valve half-closed $L_e/d = 100.00$ *spaces*
Sum of equivalent lengths $\Sigma(L_e/d) = 806.42$

The terms of the Bernoulli energy balance Equation 6.10 for the given pumping system are the following:

Pressure energy $\Delta P/\rho = 0$, since the feed and storage tanks are assumed to be open.
Velocity energy $\Delta u^2/2 = (2.6)^2/2 = 3.38$ J/k.
Elevation energy, $g\Delta z = 5 \times 9.81 = 49.95$ J/kg.

TABLE 6.7
Equivalent Pipe Lengths (L_e/d_i) of Pipe Fittings

Fitting	L_e/d_i
45° Elbow	15
90° Long sweep elbow	20
90° Standard elbow	32
90° Square elbow	60
180° Double elbow	50–75
Tee, entrance main line	60
Tee, entrance side line	90
Gate valve fully open	7
Gate valve half-open	100
Check valve fully open	75
Globe valve fully open	300
Sudden contraction (2/1)	18
Sudden expansion (1/2)	12

L_e, equivalent pipe length (m); d_i, internal pipe diameter (m).

Friction energy $h_f = 4f[\Sigma(L_e/d)](u^2/2) = 4 \times 0.0076 \times 806.42 \times 3.38 = 82.86$ J/kg at 25°C.
Friction energy $h_f = 4f[\Sigma(L_e/d)](u^2/2) = 4 \times 0.0255 \times 806.42 \times 3.38 = 278.02$ J/kg at 5°C.

The Bernoulli equation yields the required pump energy:

$$W_p = \frac{\Delta P}{\rho} + \frac{\Delta u^2}{2} + g\Delta z + h_f = 0 + 3.38 + 49.95 + 82.86 = 136.19 \text{ J/kg at } 25°\text{C}$$

$$W_p = 0 + 3.38 + 49.95 + 278.02 = 331.35 \text{ J/kg at } 5°\text{C}$$

The pump power required will be (m W_p) = 5.1 × 136.19 = 695 W at 25°C, and

$$5.1 \times 331.35 = 1690 \text{ W at } 5°\text{C}$$

The power of the motor that moves the pump, assuming an efficiency of 75%, is estimated as P = 695/0.75 = 927 W = 0.927 kW at 25°C, and P = 1690/0.75 = 2253 W = 2.253 kW at 5°C.

In this application example, the importance of viscosity in power requirement is evident: the piping system includes a long pipe with several fittings, increasing considerably the friction energy losses. Due to high viscosity, low *Reynolds numbers* are estimated, which yield higher friction factors.

When the viscosity of the liquid is increased substantially, as by low temperature of the oil, the pump power is increased considerably, in this example by 250%. A piping system should be designed for the highest possible viscosity at the operating conditions, e.g., during winter operation.

Example 6.3

A pumping system is used to pump tomato concentrate through a tubular heat sterilizer, followed by aseptic packing. The tomato product, at a concentration of 30°Brix, is fed from a storage tank through a rotary PD pump to an aseptic packaging system, consisting of a heater, a cooler, and packing equipment (Figure 6.31).

The pump is located at the bottom of the storage tank and it is connected to the heater by a sanitary pipe 2 in. outside diameter and 6 m long. The heater is a shell and tube heat exchanger, consisting of 10 parallel tubes 5/8 in. No. 12 and 3 m long, heated by saturated steam in the shell side. The cooler is a heat exchanger of the same dimensions with the heater, but it is cooled with cold water in the shell side. The piping system includes five standard elbows 90° and two half-closed gate valves. For friction calculations, all piping and fittings are assumed to be smooth. The tomato product is transported at the rate of 20 tons/h, the mean temperature in the supply piping is 20°C, in the heater 80°C and in the cooler 40°C.

Calculate

1. The mean fluid velocity in the main pipe and the tubes of the heat exchanger
2. The apparent viscosity of the tomato concentrate at the walls of the pipe and the tubes
3. The *Reynolds number* in the pipe and the tubes
4. The friction energy losses in the pipe, the heater and the cooler
5. The pump power, assuming 75% mechanical efficiency

The tomato concentrate has a density of 1100 kg/m^3, and is assumed to be a non-Newtonian power-law fluid with rheological constants $K = 20\,\text{Pa}\,\text{s}^n$ and $n = 0.5$. The rheological constant K or flow consistency coefficient decreases significantly with the temperature, following the Arrhenius equation, with an energy of activation $E_a = 20\,\text{kJ/mol}$. The flow behavior index n does not change significantly with the temperature.

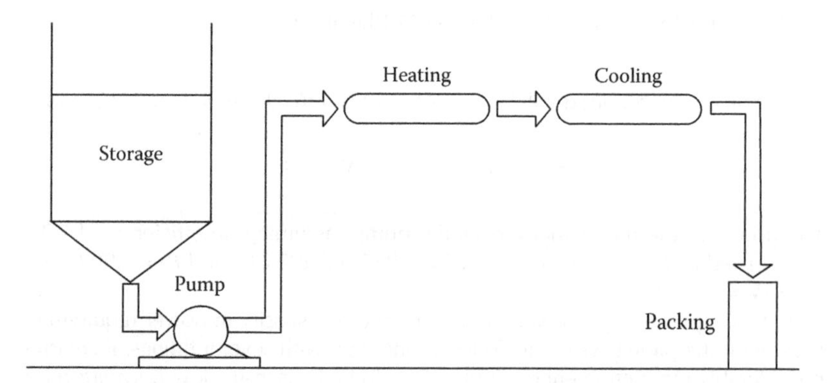

FIGURE 6.31 Pumping of tomato concentrate in aseptic packing.

Solution

1. The mean fluid velocity in the main pipe u is calculated as follows:
   Inside pipe diameter of the 2 in. sanitary pipe d = 47.49 mm (Table 6.3). Cross-sectional area $A = 3.14 \times (47.49)^2/4 = 1770\,mm^2 = 17.70 \times 10^{-4}\,m^2 = 0.00177\,m^2$. Friction energy L_e/d_i (elbow) = 32, L_e/d_i: (valve) = 100.
   The flow rate of the tomato product in the main pipe will be m = 20 tons/h or $m = 20 \times 10^3$ kg/h = 5.55 kg/s. Volumetric flow rate $Q = m/\rho = 20 \times 10^3/1100 = 18.18\,m^3/h$ or $Q = 18.18/3600 = 0.00505\,m^3/s$. The mean velocity will be $u = Q/A = 0.00505/0.00177 = 2.85$ m/s.
   The 5/8 in. No. 12 BWG tubes have an inside diameter of 10.33 mm and a cross-sectional area $A = 0.839 \times 10^{-4}\,m^2$ (Table 6.4). The total volumetric flow rate of the product $Q = 0.005\,m^3/s$ is assumed to be distributed evenly among the 10 tubes of each exchanger, i.e., $0.00505/10 = 0.000505\,m^3/s$ for each tube. The mean velocity in each tube will be $u = 0.000505/0.0000839$ = 6 m/s, which is a reasonable value for heat exchangers.
2. Tomato concentrate is a non-Newtonian fluid, following the power-law model, and its apparent viscosity η_a is estimated from Equation 6.8:

$$\eta_a = K\gamma^{n-1} \quad \text{where} \quad K = 20\ \text{Pa s}^n,\ \text{at } 20°\text{C} \quad \text{and} \quad n = 0.5\,spaces$$

The shear rate γ at the pipe or tube wall will be $\gamma = 8u/d$. In the 2 in. sanitary pipe $\gamma = (8 \times 2.85)/0.04749 = 480$ 1/s. Apparent viscosity of the tomato concentrate in the main pipe, $\eta_a = 20 \times (480)^{-0.5} = 20/(480)^{0.5} = 0.913$ Pa s = 913 mPa s, at 20°C.

The rheological constant K_2 at a temperature T_2 can be estimated from the constant K_1 at temperature T_1, using the Arrhenius equation

$$\ln\left(\frac{K_2}{K_1}\right) = \left(\frac{E_a}{R}\right)\left(\frac{1}{T_2} - \frac{1}{T_1}\right) \tag{6.47}$$

where
E_a (kJ/mol) is the energy of activation for flow
R = 8.314 J/(mol K)

The K(80°C) is calculated from the K(20°C) = 20 Pa s^n for E_a = 20 kJ/mol as $\ln[K(80)/K(20)] = (20/0.008314)(1/353 - 1/293) = -1.395$ and $K(80)/K(20) = 1/4.0345$, and $K(80) = 20/4.0345 = 4.957$ Pa s^n.

The shear rate in the 5/8 in. tubes will be $\gamma = (8 \times 6)/0.01033 = 4646$ 1/s.

Apparent viscosity of the tomato concentrate in the heater tubes at 80°C will be, $\eta_a = 4.957 \times (4646)^{-0.5} = 4.957/(4646)^{0.5} = 0.073$ Pa s = 73 mPa s.

The K(40°C) is calculated from the K(20°C) = 20 Pa s^n for E_a = 20 kJ/mol as $\ln[K(40)/K(20)] = (20/0.008314)(1/313–1/293) = -0.535$ and $K(40)/K(20) = 1/1.69$ and $K(40) = 20/1.69 = 11.834$ Pa s^n.

Apparent viscosity of the tomato concentrate in the cooler tubes at 40°C will be

$$\eta_a = 11.834 \times (4646)^{-0.5} = 11.834/(4646)^{0.5} = 0.173\ \text{Pa s} = 173\ \text{mPa s}.$$

3. The *Reynolds number* in the pipe and the tubes $Re = (du\rho)/\eta$ is calculated as follows:
   Main 2 in. pipe: $Re = (0.04749 \times 2.85 \times 1100)/0.913 = 163$
   5/8 in. tubes in the heater: $Re = (0.01033 \times 6 \times 1100)/0.073 = 934$
   5/8 in. tubes in the cooler: $Re = (0.01033 \times 6 \times 1100)/0.173 = 394$
4. Since the flow in the main pipe and the cooler tubes of this application is laminar $Re < 2100$, the friction factor f can be calculated from the equation $f = 16/Re$
   Main 2 in. pipe $f = 16/163 = 0.0982$
   5/8 in. tube in the cooler $f = 16/394 = 0.041$
   Flow in the heater 5/8 in. tubes $f = 16/934 = 0.017$
5. The pump energy W_p of the system is calculated from the Bernoulli equation 6.10, which in this case is reduced to the simplified equation $W_p = h_f$ (friction energy), since the pressure, velocity, and elevation energies are relatively negligible.
   The total friction energy loss h_f is the sum of the friction energies of the system:
   h_f (total) = h_f (main pipe) + h_f (heater) + h_f (cooler)
   h_f (main pipe) = $4f[\Sigma(L_e/d)](u^2/2)$, where $\Sigma(L_e/d) = 6/0.04749 + 5 \times 32 + 2 \times 100 = 486.3$
   h_f (main pipe) = $4 \times 0.0982 \times 486.3 \times (2.85)^2/2 = 775.8$ J/kg
   h_f (heater) = $4 \times 0.017 \times (3/0.01033) \times 6^2/2 = 355$ J/kg
   h_f (cooler) = $4 \times 0.041 \times (3/0.01033) \times 6^2/2 = 857$ J/kg
   h_f (total) = $W_p = 775.8 + 355 + 857 = 1988$ J/kg
   The required pump power will be $P = mW_p/\beta = 5.55 \times 1988/(1000 \times 0.75) = 14.7$ kW.

Notes

This application illustrates the importance of rheological constants of non-Newtonian fluids in the design of food processing operations. The flow parameters in a pumping system, such as shear rate, apparent viscosity, *Reynolds number*, and friction factor f, are affected strongly by the velocity of the fluid. The apparent viscosity of pseudoplastic fluids, such as tomato concentrate, decreases significantly at high velocities, reducing the friction energy losses and power requirements.

The significant reduction of apparent viscosity of pseudoplastic fluids at high shear rates is very important in heat transfer applications, such as heat exchangers, sterilizers, and evaporators (Chapters 8 through 10). Improved heat transfer is desirable for economic and food quality reasons.

Example 6.4

Air at atmospheric pressure and 60°C is transported through a circular conduit (pipe) of 60 cm inside diameter. The air velocity at various points of the pipe cross section was determined using a Pitot tube, which measures directly the velocity head (Δh_v), as shown in Table 6.8.

TABLE 6.8
Velocity Heads (Δh_v) in Air Flow Conduit

Distance from Pipe Wall, mm	Δh_v, mm Water
12	14
50	78
88	134
135	188
205	248

Calculate

1. The air velocity distribution in the conduit cross section
2. The mean air velocity and flow rate in the conduit

The density of air at the measurement temperature is $\rho = 1.29\,kg/m^3$ and of water $\rho = 1000\,kg/m^3$.

Solution

In the Pitot tube measurement system, the Bernoulli equation is reduced (simplified) to the equation of pressure and velocity energies, i.e., $u^2/2 = \Delta P_v/\rho$, from which the air velocity is calculated as $u = (2\Delta P_v/\rho)^{1/2}$. In this equation, the pressure is expressed in Pa, the density is in kg/m³ and the velocity is in m/s. In practice, the pressure energy is measured in mm water Δh_v, where $\Delta h_v = \Delta P_v/(\rho g)$. Therefore, the equation for velocity becomes $u = (2\Delta h_v\, g)^{1/2}$. Note that 1 mm water pressure = 9.81 Pa.

1. Table 6.9 shows the calculated air velocities at various locations in the flow cross section, based on the given pressure heads of Table 6.8. A distance x versus air velocity u diagram is shown in Figure 6.32. The velocity profile is assumed to be symmetrical to the central axis ($x = 30\,cm$).

TABLE 6.9
Calculated Air Velocities in the Flow Conduit

Distance from Pipe wall x, mm	Δh_v, mm water	Δh_v, m air	Air u, m/s
0	0	0	0
12	14	10.85	14.60
50	78	60.45	34.40
88	134	103.85	45.14
135	188	146.70	53.47
205	248	192.20	61.42

Energy is measured in mm water Δh_m, and $\Delta h_v = [\rho(\text{water})/\rho(\text{air})]\,\Delta h_w = (1000/1.29)\,\Delta h_w = 775.2\Delta h_w$.

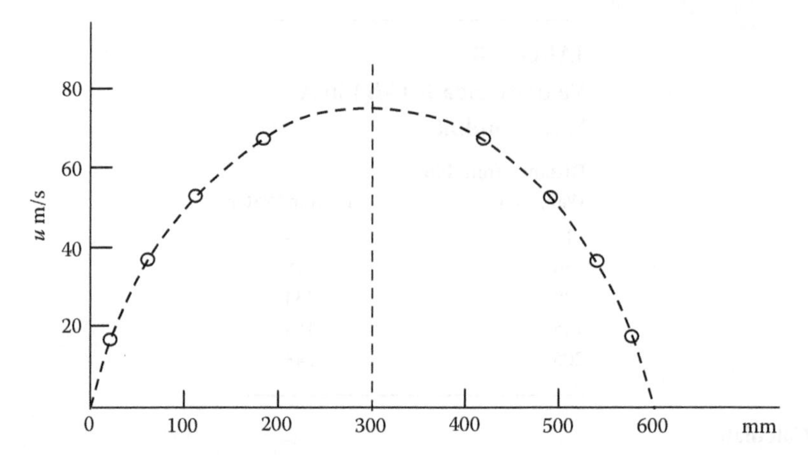

FIGURE 6.32 Air velocity distribution in a circular conduit (pipe).

2. The mean air velocity in the conduit can be estimated approximately as the mean value of the experimental velocities at various locations:

$$u = (0 + 14.60 + 34.44 + 45.13 + 53.47 + 61.42)/6 = 34.84 \text{ m/s}$$

The cross-sectional area of the circular conduit is $A = \pi d^2/4 = 3.14 \times (0.6)^2/4 = 0.283 \text{ m}^2$.

The volumetric flow rate of air will be $Q = 0.283 \times 34.84 = 9.84 \text{ m}^3/\text{s}$ or $590.4 \text{ m}^3/\text{min}$.

A more accurate calculation of the air flow rate would consider the experimental air velocities as mean values of a circular ring and sum up (integrate) all the rings to give the conduit cross-sectional area. Such an analysis gave Q and u values similar to the approximate calculation.

Example 6.5

Two pumps are operated in parallel to transport milk from a truck tank to a refrigerated storage tank. The pump capacities are Q_1 and Q_2, and the system is operated at a constant total head Δh_T.

Show in a diagram the capacities of the two pumps, the total system, and an operating curve of the system.

Solution

Figure 6.33 shows the head Δh versus capacity diagrams for pumps Q_1 and Q_2 and the total capacity Q_T. Given the pump capacities Q_1 and Q_2, the total system capacity Q_T at a given pump head Δh is calculated as $Q_T = Q_1 + Q_2$. The pump operating curve for the two-pump system intersects the total capacity at Q_T, as shown in the diagram.

Example 6.6

A vertical cylindrical tank contains a liquid, which is to be drained by gravity through a piping at the bottom of the tank. The tank internal diameter is D_T (m), the level of the liquid in the tank above the discharge point is z (m), and the drainage pipe has an internal diameter of d (m) and an equivalent length $\Sigma L_e = 10 \text{ m}$ (Figure 6.34).

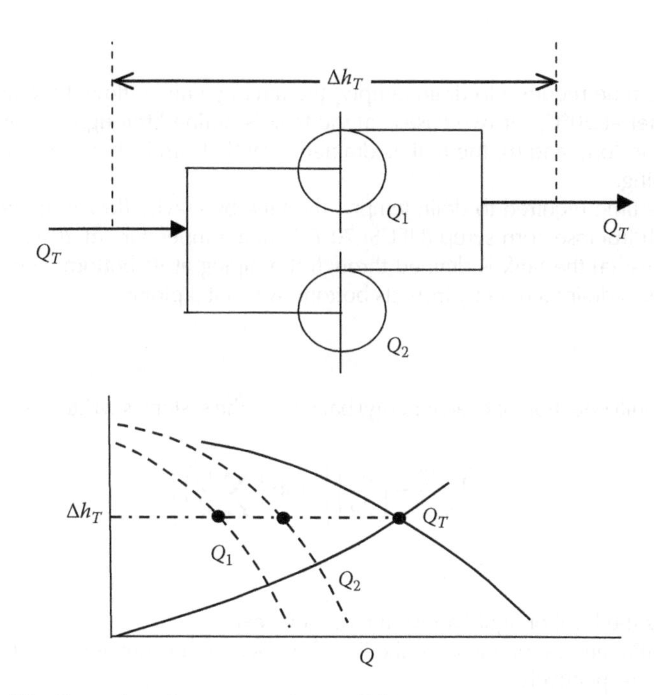

FIGURE 6.33 Operation of two pumps in parallel.

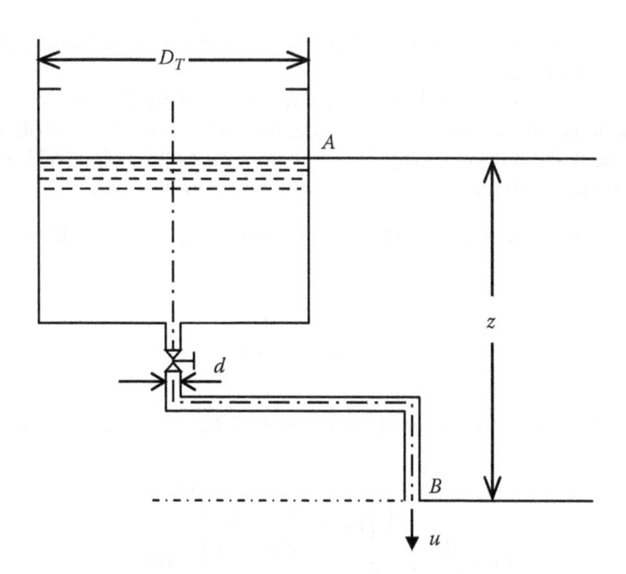

FIGURE 6.34 Flow in draining a tank.

Calculate

1. The time required to drain (empty) the tank by gravity flow if the liquid is water at 20°C, for two cases: (a) the tank is drained through the piping at its bottom, and (b) the tank is drained directly from its bottom, without a piping.
2. The time required to drain (empty) the tank by gravity flow if the liquid is high fructose corn syrup (HFCS) 70°Brix at a temperature of 20°C, for two cases: (a) the tank is drained through the piping at its bottom, and (b) the tank is drained directly from its bottom, without a piping.

Solution

The Bernoulli equation of head (energy) balance of the system is written as follows:

$$z + \frac{P_A - P_B}{\rho g} = \left(\frac{u^2}{2g}\right)\left[1 + 4f\left(\sum \frac{L_e}{d}\right)\right] \tag{6.48}$$

where
z (m) is the level of liquid above the outflow (exit)
P_A, P_B (Pa) are the pressures at the liquid surface in the container and the outflow, respectively
u (m/s) is the velocity in the outflow pipe
d (m) is the pipe inside diameter
ΣL_e (m) is the equivalent total pipe length (including fittings)
f (–) is the Fanning friction factor in the pipe

The velocity at the surface of the liquid is considered negligible, compared to the outflow velocity u.

At a differential time dt the level of the liquid will fall by dz, and volume of the liquid in the tank will decrease by $(\pi D_T^2/4)$dz, where D_T (m) is the tank diameter. At the same time dt the outflow of liquid volume from the bottom pipe will be $(\pi d^2/4)u$, which will be equal to the liquid volume removed from the tank,

$(\pi D_T^2/4)\,\mathrm{d}z = (\pi d^2/4)u$, from which the following equation is obtained:

$$dt = \left(\frac{D_T}{d}\right)^2 \left(\frac{\mathrm{d}z}{u}\right) \tag{6.49}$$

Substituting the value of velocity (u) from Equation 6.48 we obtain the following:

$$dt = \left(\frac{D_T}{\mathrm{d}}\right)^2 \left\{\frac{\left[1 + 4f\left(\sum \frac{L_e}{\mathrm{d}}\right)\right]}{\left[\frac{2g(z + (P_A - P_B)}{\rho g}\right]}\right\}^{1/2} \mathrm{d}z \tag{6.50}$$

Integrating Equation 6.50 between level (0, z) and time (0, t), and assuming that $P_A = P_B$, the time required to empty the tank is obtained

$$t = \left(\frac{z}{2g}\right)^{1/2} \left(\frac{2D_T^2}{d^2}\right) \left[1 + 4f \sum \frac{L_e}{d}\right]^{1/2} \quad (6.51)$$

In the case that the tank is drained directly from its bottom, with no piping, the discharge time is calculated from the simplified equation:

$$t = \left(\frac{z}{2g}\right)^{1/2} \left(\frac{2D_T^2}{d^2}\right) \quad (6.52)$$

The last two equations are applied to the following examples:

1. Time to empty a tank of 2 m diameter containing water at 20°C at a level of 5 m above the drain (exit) point. A drain pipe of 10 cm inside diameter is used with an equivalent length of 10 m (including an open valve). The tank system is open to the atmosphere, meaning that $P_A = P_B$. The water viscosity is assumed 1 mPa s. The only difficult parameter to estimate is the Fanning friction factor f, which is a function of the *Reynolds number*. The liquid velocity u in the drainage pipe changes from zero at the end to a maximum at the beginning of the operation. A trial and error procedure was used to estimate the initial u from Equation 6.48, yielding a mean value of $u = 5$ m/s, friction factor $f = 0.0032$, and $Re = 500{,}000$ for the water flow. For a mean value $f = 0.0032$, an equivalent pipe length $L_e = 10$ m and inside diameter $d = 0.1$ m, Equation 6.51 yields the draining time for water $t = 607$ s or 10.1 min.

   Emptying of the same tank of water directly from the bottom, without drainage piping, can be calculated from Equation 6.52 as $t = 6.7$ min. Thus, adding a drainage piping to the tank will increase the draining time by about 50%.

2(a). Time to drain the same tank, containing high fructose corn syrup (HFCS) 70°Brix at a temperature of 20°C. The syrup is considered a Newtonian fluid with a viscosity of 50 Pa s at 20°C.

   The only difference from water is the higher viscosity of the syrup, which will affect the friction factor f and the draining time. A 50-fold increase of viscosity will decrease the Re to about 10,000 and increase the average f factor to about 0.007, according to the Moody diagram and Equation 6.16. The estimated drainage time for the syrup will be $t =$ 760 s or 12.7 min, which is about 20% longer than the time to drain the water.

2(b). The drainage time of the syrup from the bottom of the tank, without any piping, is calculated from Equation 6.52, yielding the same value with water, $t = 6.7$ min. Evidently, the viscosity of the liquid is not involved, since there is no frictional flow resistance through a piping system.

PROBLEMS

TABLE 6.10
Experimental Data

Speed (RPM)	Torque M (10^{-5} N m)
40	25
100	45
160	65
230	85

6.1 A coaxial cylinder viscometer was used to determine the apparent viscosity of a food liquid. The inner and outer diameters of the cylinders were 26 and 28 mm, respectively, and the cylinder length was 40 mm. The experimental data were obtained at a constant temperature are shown in Table 6.10. Determine the values of the shear rate, shear stress, and apparent viscosity of the sample.

6.2 Determine the *Re* number of a Newtonian fluid for flow in an annular cross section between outside and inside diameters d_o and d_i, respectively. Fluid velocity u, density ρ, viscosity η.

6.3 Milk of viscosity 2.5 mPa s and density 1015 kg/m^3 flows in a 1½ in. sanitary pipe at the rate of 500 kg/h. Calculate the pressure drop in a smooth straight pipe, expressed in mbar/m pipe.

6.4 A liquid is transported through a steel pipe 50 m long and 2½ in. nominal diameter Sch No. 40, at a mean velocity of 2.5 m/s. The pressure drop due to friction is 1 bar. Estimate the mass flow rate, the *Re* number, and the viscosity of the liquid, assuming laminar flow. Density of the liquid, 1100 kg/m^3.

6.5 A Newtonian food liquid is sterilized in a sanitary pipe 2 in. nominal diameter at the flow rate of 2 tons/h. The product density is 1050 kg/m^3. Estimate the product viscosity so that the flow is laminar at a *Reynolds number* of 2000.

6.6 A centrifugal pump has the following characteristics at the maximum efficiency point: Flow rate Q = 100 m^{3}h, total head Δh = 70 m, required $NPSH$ = 15 m, initial power 15 kW, impeller speed 3000 RPM, impeller diameter D = 0.25 m. Evaluate the performance of an homologous pump which operates at an impeller speed of 1800 RPM, but which develops the same total head Δh and requires the same *NPSH*.

6.7 Calculate the available net positive suction head *NPSH* in a pumping system, if the liquid density is ρ = 1200 kg/m^3, the viscosity is η = 80 mPa s, the mean velocity u = 1 m/s, the static head on the suction side z = 3 m, the inside pipe diameter d_i = 52.5 mm, the gravity constant g = 9.81 m/s^2, and the equivalent length of the suction side ΣL_e = 5 m. The liquid is at its boiling point, and the entrance and exit losses are neglected.

6.8 Two pumps are operated in series to transport milk from a truck tank to a tall refrigerated storage tank. The pump capacities are Q_1 and Q_2, and the system is operated at a total head Δh_T. Show in a diagram the capacities of the two pumps and of the total system, and an operating system curve for constant capacity ($Q_T = Q_1 + Q_2$) and total head $\Delta h_T = \Delta h_1 + \Delta h_2$.

6.9 A centrifugal pump is operated at 1800 RPM, flow rate 100 m^3/h, and total head 50 m. Estimate the flow rate, total head, and % power increase, if the pump speed is doubled.

6.10 A Newtonian food liquid is fed as a film to the inside top of a vertical sanitary 2 in. pipe at the rate of 100 kg/h. The liquid has a viscosity of 30 mPa s. Determine the Reynolds number and the type of flow of the falling film.

LIST OF SYMBOLS

A	m^2	Area
a	m	Distance
b	m	Width
C	kg/m^3	Concentration
C_D	—	Drag coefficient
CFM	cuft/min	Cubic feet per minute
D, d	m	Diameter
E	kWh	Energy
F	—	Friction factor
g	m/s^2	Acceleration of gravity (9.81)
H, h	m	Height
h	m	Energy head
m	kg/s	Mass flow rate
K	Pa s^n	Rheological constant
L	m	Length
M	N m	Torque
N	1/s	Rotational speed
n	—	Flow behavior index
P	Pa	Pressure
P	W	Power
Q	m^3/s	Volumetric flow rate
Re	—	Reynolds number ($du\rho/\eta$)
RPM	1/min	Rounds per minute
r	m	Radius
t	s	Time
u	m/s	Velocity
V	m^3	Volume
W	J/kg	Pump energy
x, y	m	Size, distance
a	—	Angle
Γ	kg/(m s)	Film flow rate
γ	1/s	Shear rate
δ	m	Thickness
ε	—	Porosity
Δ	—	Difference
η	Pa s	Viscosity
π	3.14	
ρ	kg/m^3	Density
τ	Pa	Shear stress
ω	1/s	Rotational speed

REFERENCES

Bhatua, M.V. ed. 1982. *Process Equipment Series*, vol. 4. *Solids and Liquids Conveying Systems*. Technomic Publication, Westport, CT.

Bourne, M.C. 1982. *Food Texture and Viscosity: Concept and Measurement*. Academic Press, New York.

Boyce, M.P. 1999. Transport and storage of liquids. In: *Perry's Chemical Engineers' Handbook*, 7th edn. McGraw-Hill, New York, pp. 10-5–10-151.

Holland, F.A. and Bragg, R. 1995. *Fluid Flow for Chemical Engineers*. Edward Harold, London, U.K.

Rao, M.A. 2007. *Rheology of Fluid and Semisolid Foods*, 2nd edn. Springer, New York.

Rao, M.A., Rizvi, S.S.H., and Datta, A.E. 2005. *Engineering Properties of Foods*, 3rd edn. Taylor & Francis, New York.

Saravacos, G.D. and Kostaropoulos, A.E. 2002. *Handbook of Food Processing Equipment*. Springer, New York.

Saravacos, G.D. and Maroulis, Z.B. 2001. *Transport Properties of Foods*. Marcel Dekker, New York.

Walas, S.M. 1988. *Chemical Process Equipment*. Butterworths, London, U.K.

7 Mechanical Processing Operations

7.1 INTRODUCTION

Most foods are solid or semisolid materials and their processing involves various mechanical operations and equipment, which have been developed mostly from industrial experience. While the processing of liquids and gases is based on transport phenomena and chemical engineering, solids are converted to pieces and particles of various sizes, and their processing is based mostly on empirical technology.

In this chapter, mechanical operations, used in food processing, are described briefly, emphasizing the underlying physical and engineering principles.

7.2 PARTICLE CHARACTERISTICS

7.2.1 PARTICLE SIZE AND SHAPE

Particulate matter includes solid particles, liquid droplets, and gas bubbles. It is characterized by unique properties, which may differ from the properties of the bulk material (solid, liquid, or gas). Among them, the most important are the particle size, particle shape, and particle size distribution (Saravacos and Kostaropoulos, 2001).

Table 7.1 summarizes the size ranges of particulate materials in food processing.

The size of particles is commonly referred to as the particle diameter, which is usually measured by geometric methods (microscopic or sieving), as shown in Figure 7.1. The equivalent diameter is the diameter of a circle, having an area equal to the projection surface of the particle on the plane of maximum stability. The Feret diameter is the mean distance between two tangents of the opposite sides of the projection perimeter. The Martin diameter is the mean value of a straight line bisecting the projection diameter, which is drawn at the same direction in all measurements. Both Feret and Martin diameters are based on optical microscopy, and the straight line of constant direction corresponds to the linear scale of the eye piece (Allen, 1960).

In operations involving particle movement (settlement), the Stokes diameter d_{st} (m) is used, defined by the Stokes equation

$$u = \frac{\left[d_{st}^2(\rho_s - \rho)g\right]}{18\eta} \tag{7.1}$$

where

u (m/s) is the fall velocity due to gravity
ρ_s, ρ (kg/m^3) are the densities of solid particle and fluid, respectively
η (Pa s) is the fluid viscosity
$g = 9.81\,\text{m/s}^2$

TABLE 7.1
Size of Particulate Materials

Size Range	Particulate	Materials
1–100 nm	Molecules	Colloids
0.1–10 μm	Bacteria	Starch
10–1000 μm	Flours	Food particles
>1 mm	Food pieces	Crystals

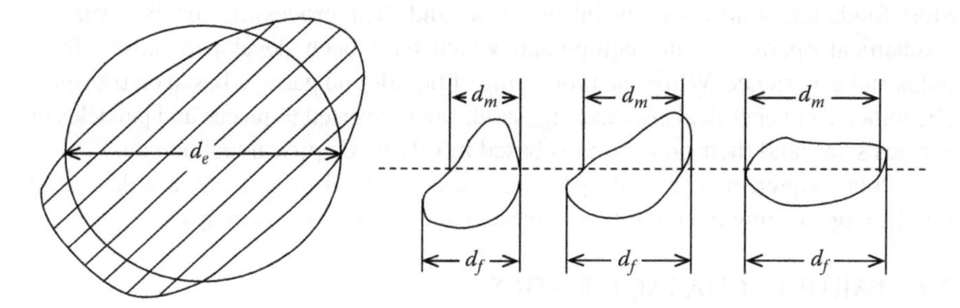

FIGURE 7.1 Particle diameters: d_e equivalent, d_f Feret, d_m Martin.

The Sauter diameter of a particle is defined as the diameter of a sphere that has the same volume to surface ratio as the particle of interest. It is usually denoted as $d_{3,2}$, where subscripts (3,2) mean the volume and surface, respectively. The mean $d_{3,2}$ in a particle size distribution is determined experimentally.

The surface f_a and volume f_v shape factors are defined by the equations

$$A = f_a\, d_e^2 \quad \text{and} \quad V = f_v\, d_e^3 \tag{7.2}$$

where

A and V are the external surface and volume of the particle, respectively
d_e is the equivalent diameter

The sphericity ψ of particles is defined as the ratio of the external surface of an equivalent (equal volume) sphere to the real surface of the particle. Table 7.2 shows the important shape factors of various particles.

TABLE 7.2
Shape Factors of Various Particles

Particle Shape	Surface, f_a	Volume, f_a	Sphericity, ψ
Spheres	3.14	0.52	1
Rounded particles	2.7–3.14	0.32–0.40	0.80
Ground particles	2.5–3.0	0.29–0.30	0.65
Flaky particles	1.5–2.5	0.05–0.15	0.30

7.2.2 Particle Size Distribution

The size of biological particles usually follows the normal size distribution

$$f(x) = \frac{1}{\left[s(2\pi)^{1/2}\right]}\exp\left[\frac{-(x-x_m)^2}{2s^2}\right] \tag{7.3}$$

where

x_m is the mean size
s is the standard deviation of the particles

Most particles, produced in industrial processing, follow the logarithmic normal distribution

$$f(\log x) = \frac{1}{\left[\log s(2\pi)^{1/2}\right]}\exp\left[\frac{-(\log x - \log x_m)}{2(\log s)^2}\right] \tag{7.4}$$

In normal distributions, the mean particle size x_m and the standard deviation s are calculated from the equations

$$x_m = \frac{\Sigma(Nx)}{N_T}, \quad \sigma = \left[\frac{\Sigma(x-x_m)^2}{N_T}\right]^{1/2} \tag{7.5}$$

In log-normal distributions, the mean particle size $\log x_m$ and the standard deviation $\log s$ are calculated from the equations

$$\log x_m = \frac{\Sigma(N\log x)}{N_T}, \quad \log s = \left[\frac{\Sigma(\log x - \log x_m)^2}{N_T}\right]^{1/2} \tag{7.6}$$

Particle size distributions are usually presented in graphical form, using special diagrams of cumulative distribution versus particle size. Normally, the residue cumulative distribution R is used, defined as $R = \int_x^{\infty} f(x)\,\mathrm{d}x$, that is, the fraction ($R < 1$) of all particles larger than x.

Normal and log-normal distributions are plotted as straight lines on special normal or log-normal diagrams of R versus x or $\log R$ versus $\log x$, respectively.

Distribution plotting paper is available in the literature (Internet).

Industrial particles may also follow the Rosin–Rammler distribution

$$R = \exp\left(\frac{-x}{x'}\right)^n \tag{7.7}$$

where

R is the cumulative distribution of particles larger than x
x' is a characteristic size
n is the uniformity index of the particles

The Rosin–Rambler distribution is represented as a straight line in a diagram of $\log[\log(1/R)]$ versus $\log x$, from which the parameters x' and n are estimated. Plotting paper for the Rosin–Rammler distribution is available in the Internet.

The Rosin–Rammler distribution is a simplified version of the generalized Weibull distribution applied widely in statistics.

7.2.3 Particle Size Analysis

7.2.3.1 Methods of Measurement

The particle size of various materials is measured by various methods, resulting in various characteristic "diameters," as shown in Table 7.3.

TABLE 7.3
Methods of Particle Size Measurement

Method	Size Range, μm	Size Measurement
1. Optical		
Microscope	1–10	Statistical diameter
Electron microscope	0.01–0.5	Projection diameter
Colorimetry	0.5–5	Volume diameter
Laser	0.1–5	Volume diameter
2. Geometric		
Sieving	>10	Particle diameter
3. Mechanical		
Gravity settlement	1–50	Stokes diameter
Air classification	5–50	Stokes diameter
Centrifuge	0.05–10	Stokes diameter
Impact	0.1–5	Impact diameter
4. Electrical		
Coulter counter	1–100	Volume diameter
5. Surface		
Gas adsorption	>1 m^2/g	Particle surface

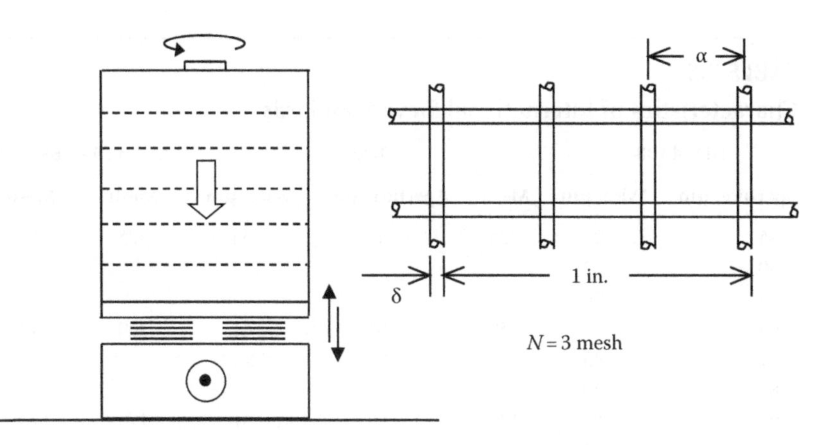

FIGURE 7.2 Sieving and mesh number.

7.2.3.2 Sieve Analysis

Sieving is widely used in the analysis of particle sizes larger than 10 μm, which are of interest in food processing, for example, various flours.

A series of standard sieves are used, which have screens of definite size of openings.

The sieves have a cylindrical shape of 10–20 cm diameter and short height, and they are arranged in a pile of decreasing aperture a from top to bottom (Figure 7.2).

The sieves are characterized by the opening of the holes, which have normally square shape of side length a (μm). In the United States, the mesh number N is used, defined as the number of openings per inch (25,400 μm) length Equation 7.8

$$N = \frac{25{,}400}{a + \delta} \tag{7.8}$$

where δ (μm) is the thickness of the screen wire.

The screens used in determining the size of particles are standardized. For particle analysis, the international standards commonly used are listed in Table 7.4.

TABLE 7.4
International Standards for Sieve Analysis

Country	Standard Designation
The United States	Tyler
The United States	ASTME11
Canada	8-GP-1d
Germany	DIN4188
The United Kingdom	BS410
France	AFNORX11–501
EU	EN933
International	ISO3310

The US Tyler series is based on a screen of 200 mesh (wire diameter 53 μm, aperture opening 74 μm). The ratio of openings of two successive sieves is $2^{0.5} = 1.41$ or $2^{0.25} = 1.19$. Note that the mesh number is not equal to the aperture, since the length of 1 in. (25,400 μm) includes the wire thickness.

The DIN (Deutsche Ingenieure Normung) series is based on the aperture (opening) and the thickness of wire, expressed in μm. The International Standards Organization (ISO) standardization is similar to the DIN series, based on the aperture (opening) and the thickness of wire (μm). The BS (British Standards) 410 and the ASTM-E11

TABLE 7.5
Characteristics of International Sieve Standards

DIN 4188		Tyler			ASTM E11	BS 410
Aperture, μm	Wire, μm	Mesh	Aperture, μm	Wire, μm	Mesh	Mesh
45	28	325	43	36	325	350
50	32					
56	36					
63	40	250	61	41	230	240
71	45					
80	50					
90	56	170	88	61	170	170
100	63					
125	80	115	124	97	120	120
160	100					
200	125					
250	160	60	246	179	60	60
—	—	—	—	—	—	—
1000	630	16	991	597	18	16
—	—	—	—	—	—	—
2000	1000	5	1981	938	10	8

standards are similar to that of the Tyler series. Table 7.5 shows the characteristics of DIN, Tyler, ASTM (American Society for Testing and Materials), and BS standards.

7.3 SIZE REDUCTION

7.3.1 INTRODUCTION

Size reduction operations are important in several food processes, where solid foods are converted to food pieces or particles as part of processing or as finished products. They are characterized by the final size of the product; the type and way the forces are applied, and the type of equipment used (Perry and Green, 1997; Rumpf, 1975).

Size reduction produces food products such as chocolate powder, sugar powder, meat slices, and fruit juices. It increases the specific surface of the product, facilitating several processes, such as heat exchange, extraction, chemical and biological reactions (e.g. blanching, sterilization, freezing, extraction of seed oil), and it enhances mixing, blending, and stability.

Size reduction is distinguished into (a) cutting, (b) crushing, and (c) grinding. Forces applied during the operation include compression, shear, bending, friction, collision, punching, and impact. Brittle materials are broken down to smaller pieces or particles easier than viscoelastic or elastic materials.

Table 7.6 shows the main size reduction operations, based on the size of the product pieces/particles.

TABLE 7.6
Size Reduction Operations

Operation	Product Size
Cutting	>1 mm
Breaking	>1.5 mm
Crushing	1.5–8 mm
Fine crushing	0.5–5 mm
Milling	50–750 μm
Fine milling, colloidal	<50 μm

In Mechanical Process Operations, food materials can be divided into hard and soft or strong and weak. The approximate ranges of these properties are illustrated in a shear–strain diagram (Figure 7.3). In this diagram, line (1) represents soft, weak, and brittle materials; line (2) is for soft, weak, and ductile; line (3) is for hard, strong, and ductile; and line (4) is for hard, strong, and brittle.

The size reduction equipment requires considerable energy for its operation, which is consumed mostly (about 95%) as mechanical friction. Three empirical equations (Rittinger, Kick, and Bond) are used to estimate the energy required E (kWh/kg) for reducing the particle size from x_1 to x_2. They are

Rittinger

$$E = K_R \left(\frac{1}{x_2} - \frac{1}{x_1} \right) \tag{7.9}$$

Kick

$$E = K_K \ln \left(\frac{x_1}{x_2} \right) \tag{7.10}$$

Bond

$$E = K_B \left[\left(\frac{1}{x_2} \right)^{1/2} - \left(\frac{1}{x_1} \right)^{1/2} \right] \tag{7.11}$$

where K_R, K_K, and K_B are the corresponding empirical constants.

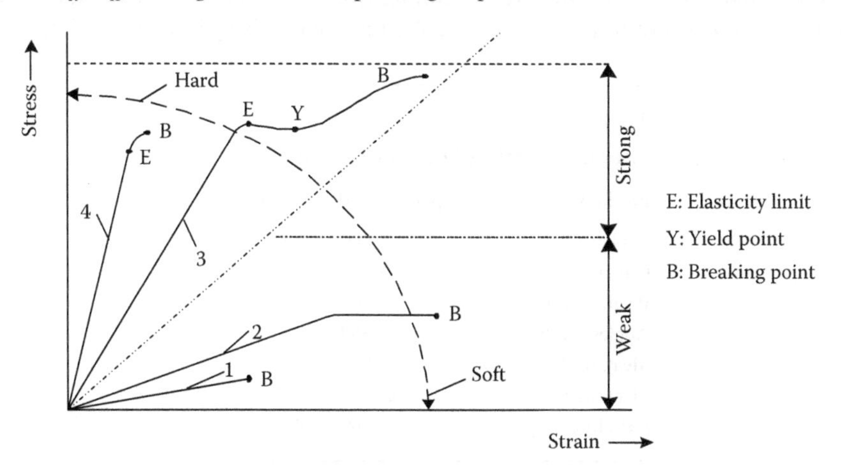

FIGURE 7.3 Stress–strain diagram of food materials.

The Bond constant can be estimated from the relation $K_B = 10W_B$, where W_B (kWh/ton) is the Bond work index, defined as the energy required for size reduction from a very large initial size $[(1/x_1)^{1/2} = 0]$ to a size that can be sieved by screens of 100 μm aperture.

7.3.2 Cutting

7.3.2.1 Introduction

Cutting is applied to ductile, viscoelastic, and elastic materials. Products of cutting are large pieces, slices, diced products, flakes, and pulps. Cutting is affected mainly by shear forces, and it is performed by knives, saws, shears, and thin wires.

The factors influencing cutting depend on the method applied. They include product to be cut, sharpness of cutting tool, cutting force, direction of force applied, and cutting speed (Stiess, 1992, 1994).

Cutting operations are applied over a wide product range (from meters to micrometers), increasing the value added, and removing undesired material that otherwise would burden processing and product.

Industrial cutting equipment is characterized by a wide variety of equipment, the cutting tools of which wear out frequently. Skilled personnel are required and productivity is relatively low.

In cutting machines, the cutting tool may be fixed, while the product moves toward the tool, or the cutting tool may move toward the fixed product. In some cases, both tool and product move in counter direction, increasing cutting effectiveness.

The effectiveness of cutting depends on the cutting edge of the knife and the "cutting angle," which is the inclination between the knife axis and perpendicular to the cutting direction. The cutting edge depends on the quality/hardness of the metal, the width of cutting blade, and the wedge angle. The finer the cutting blade and the narrower the cutting angle, the finer the cut.

Table 7.7 shows the specific cutting force of some foods.

It is noted that the specific force of cutting frozen food is much higher than that of fresh products. Fresh fruits and vegetables require relatively low cutting forces.

TABLE 7.7
Specific Cutting Forces of Foods

Product	Specific Cutting Force, N/m
Cabbage	1,000–1,200
Onions	1,700–1,800
Potatoes, fresh	600–700
Apples, fresh	330
Meat, fresh	5,000–8,000
Meat, frozen	23,000–30,000
Fish, frozen	3,200–3,700

7.3.2.2 Cutting Tools

Table 7.8 shows the basic tools (elements) of cutting equipment. Selection of the right cutting tool is based on the product, its condition (e.g. fresh or processed), and the desired quality of the product. Knives and other cutting tools can be used as portable for manual use or as part of cutting machines.

Knives are the basic cutting tools. Guillotines are composed of two knives. Only wires are basically different mechanical cutting tools.

Simple knives may be straight or curved. They can operate independently or they may be part of some other cutting mechanism, such as of shears, in which two knives swing parallel opposite to each other. The motion of simple straight and curved knives during cutting may be one way or reciprocating. Simple knives can rotate around their support at one end, while sickle knives rotate or swing. In some cases, as in meat cutters, two or more sickle knives are bound together to form a uniform knife system. In several cases, combinations of knives assembled in different ways (e.g. vertically or horizontally) cut products in certain form (e.g. cubes).

Saws are toothed knives, either straight or in the form of discs. Straight saws can reciprocate, or operate in one direction only, such as toothed band saws. These are used in cutting off relatively large pieces of fresh meat or frozen products (meat or fish), operating as fixed or portable.

Guillotines are oblique, thick blades slipping along guides, and they are used in cutting large pieces, such as blocks of frozen food, to smaller ones. Wires are mainly used in cutting sticky products. Wire reduces the influence of adhesion, since its surface is smaller than that of blades. In the range of wire thickness 0.1–1 mm, the reduction of the required cutting power is proportional to the diameter of wires. Wire cuts bread better, producing less waste and crumbs. Tube knives may be used in boring food for sucking its juice, or removing blood from fresh slaughtered animals. Spiral knives are used in special operations such as cutting the flesh around clingstone peaches.

National and international safety standards and regulations exist for various cutting machines, such as Occupational, Safety and Health Administration (OSHA) and European Organization for Standardization (CEN) or provisional standards (prEN). For example, prEN 1974 for slicing machines and prEN 12331 for mincing machines.

TABLE 7.8
Tools of Cutting Equipment

Cutting Tool	Motion	Cutting System
Knife	One direction	Straight or curved knife
Sickle	Rotation	Disk
Saw	Rotation	Disk, belt
Guillotine	Rotation	Chopper
Wire	One direction	Single or system of wires
Shears	Shear	Counter-moving knives

7.3.2.3 Portable Cutting Tools

Portable cutting tools for manual operation include knives, saws, and shears. Simple knives get their name from products they cut and operations in which they are involved (e.g. steak knife, boning knife, sticking knife). Large portable cutting tools are electrical or air-powered. They are hung from a chain, connecting them to a swinging hoist mechanism, which secures a flexible operation in any position. The weight of portable cutting tools varies between 2.5 and 20 kg, and for safety reasons, parts of the blades are covered.

Most of the portable cutting tools are used in cutting operations involving animal parts, used in food processing, such as foot shears, horn and leg shears, and pig and beef saws.

7.3.2.4 Cutting Equipment

Most of size reduction machines are product specific. Special cutting machines have been developed for meat, fish, fruit, etc. Even for the same products, different types of machines are often used, such as peeling or cutting of fruits. The basic cutting tools, mainly knives, are the basic elements of cutting equipment. In the meat industry, a variety of cutting machines has been developed. In some cases, cutting machines may also be used for other products, for example, a meat cutter for cheese.

Slicing, cutting, dicing, and shredding equipment is used widely as a preliminary step in several fruit and vegetable processing operations, such as canning, freezing, drying, cooking, and frying.

The selection of cutting equipment depends primarily on the intended use. The following general guidelines are important in selecting cutting equipment:

(1) The cutting edge must be sharp, without reducing the firmness of the cutting tool. This requires fine cutting edges and hard material for reduced wear; (2) the cutting metal must be made of hardened stainless steel, steel containing 5% chromium, and corrosion resistant tungsten carbide; (3) the cutting efficiency must be high, requiring exact cutting in relation to adequate cutting speed; (4) energy requirements must be as low as possible; (5) replacement must be easy and fast; (6) safety must be high; and (7) easy and thorough cleaning.

Cutting machines, used commonly in food processing, include the following:

7.3.2.4.1 Band Saws

Band saws are used for cutting frozen products in straight pieces, for example, cutting fish blocks into fish fingers. Bands are usually stretched between an upper and a lower wheel, which is driven electrically. Cutting is done on a processing table by the front band. The width and length of cutting bands/blades varies with the type and dimensions of machines. Bands are automatically adjusted to the right tension. The band can be automatically cleaned up during operation and easily removed for further cleaning. All parts of the band, except that in the cutting area, are covered for safety reasons. Indicative values for a band are: length, 1.5–2 m, width, 1–2 cm, and thickness, about 0.05 cm.

7.3.2.4.2 Slicers/Dicers

Slicers cut products such as meat, ham, sausages, fish, cheese, fruits, vegetables, etc. into slices, whose thickness usually varies from 0.7 mm to about 7 cm. Slicers consist of one or more rotating discs or knives, adjusted to cut products to the desired thickness. Either the knives move toward the product, or the products move toward the knives. Products are fed automatically into slots or some other devices, lying above rotating knives that slice them.

The thickness of a slice depends on the adjustment of knives, which can be controlled electronically, so that all slices maintain a more or less equal thickness, throughout the cutting of a whole piece. An automatic correction of the slicing thickness, based on the remaining piece, guarantees little end cut loss. Cutting quality is influenced by the speed of cutting, the product consistency, its temperature, and the quality of knives. The speed can be varied. Soft products have to be cooled down, if even and clean cuts are required. Sharpness of knives is very important. If knives are not sharp or well adjusted, splintering of bones may take place, when cutlets or chops are cut.

In dicers, the product is firstly cut into strips, and then chopped by bent or inclined knives, mounted around a rotor. The size and capacity of diced products (e.g. French fries) depends on the type of knives used. The size of cube dices may vary from about 2 mm to more than 50 mm.

7.3.2.4.3 Meat Mincers

Mincing machines are used for mincing different kinds of minced meat or in the preparation of processed cheese. The mincing machine consists of a structured tube in which a rotating worm forwards the material toward the other end of the cylinder, where it is compressed through a system of vertical plates and rotating knives, behind each plate. The pitch of the worm may be larger at the beginning and smaller near its end. Thus, the compression is more intensive as the product draws near the plates. In some mincers, a second worm with a large pitch, or a shaft with paddles, rotates parallel and above the main worm, mixing the products before mincing. The number of vertical plates (2–3) depends on the fineness of cutting. The aperture of perforation of the plates decreases progressively. Rotating knives behind the plate cut the product as it comes out of the machine.

7.3.2.4.4 Cutters

Cutters are versatile machines used mainly in mincing, mixing, and emulsifying meat and added fat, etc., required in the preparation of fillings of salamis and sausages. In meat processing, the size reduction lies between small pieces and colloids (30 mm to 5 μm). Cutters are also used in fine cutting of vegetables and cheese. They are made up of a circular bowl that rotates around a vertical axis. The product in the bowl is minced by 3 to 12 concentrically coupled rotating sickle knives. The knives rotate as close as possible to the surface of the bowl bottom, with speeds that can be adjusted between 40 and 5000 RPM. The higher the speed of rotation, the finer the cutting.

Double-jacketed bowls may be used for cooling or heating the bowl contents. Bowls may be hermetically closed for processing under vacuum or for using inert gases. In preparing meat mash for sausages, meat at low temperature (–2°C to 0°C) is used, so that cutting is clear and smearing is avoided. The capacity of bowl cutters is 200–550 L.

7.3.3 Crushing and Grinding Operations

7.3.3.1 Introduction

Size reduction can be divided into crushing of solids into pieces and particles of size larger than 750 μm, and grinding, which produces smaller particles and is more important in food processing operations.

Crushing and grinding are used to prepare the material for further size reduction (e.g. fine milling), preparing particles for agglomeration, and produce final products. It is an inefficient operation in terms of both energy consumption and particle size distribution.

Grindability is a measure of the facility of grinding a material. It depends on the texture, hardness, moisture content, and the breakdown mechanism of the material. Grindability is determined experimentally under controlled conditions of energy consumption and size reduction.

Solid materials break after being deformed (inelastic deformation), while elastic materials return to the original condition when stress is removed (elastic deformation). Brittle materials break readily by small deformation, while elastic materials break after significant deformation. Brittle materials must be stressed more than elastic materials up to their break, but elastic materials require more energy.

Breaking starts at points of material "defects." The larger the number or significance of defects, the easier the material breaks. Structure deformities and incorporation of foreign substances also count as material defects. The further increase and expansion of these cracks depends on the relation between energy absorption and release.

Breaking of smaller particles is more difficult than large pieces and their further size reduction requires more energy.

Size reduction equipment in food processing has been developed through experience and it is specific for the various products. Table 7.9 lists some important crushing and grinding equipment used for foods.

7.3.3.2 Crushers

Roll crushers and pulpers are used widely in food processing. Roll crushers are used in size reduction of larger pieces, in coarse grinding (e.g. grinding of corn, coffee, frozen products) and in pregrinding.

TABLE 7.9
Food Crushing and Grinding Equipment

Equipment	Size of End Product	Reduction Ratio	Main Force Applied
Roll crusher	>10 mm	4–6	Pressure/shear
Roll mill	5–100 μm	>20	Pressure/shear
Pan mill	0.05–1.0 mm	10	Pressure/shear
Hammer mill	>8 mm	5	Impact
Pin disc	2–50 μm	>50	Impact/shear
Colloidal mill	5–20 μm	>50	Shear
Rotary grinder	20–100 μm	19–25	Shear/impact
Jet mill	1–100 μm	>50	Impact

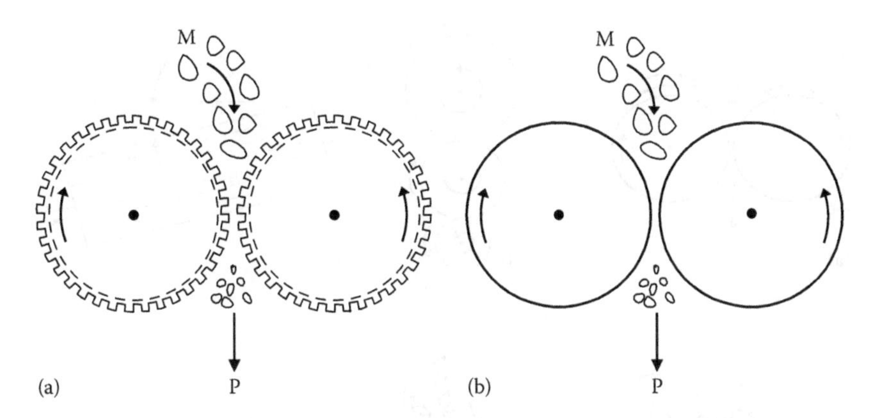

FIGURE 7.4 Roll crushers: (a) teethed; (b) smooth. M, feed material; P, product.

Roll crushers (Figure 7.4) have two counter rotating rolls that may be either smooth or have ripples, pins, or teeth. The forces applied to product crushed between smooth rolls are compression and shear. Toothed rolls exercise additional bending forces. In the case of smooth rolls, the angle of nip is important. This is defined as the largest angle between the solid piece and the rolls that will grip (grasp) the piece between the rolls. The angle of grip a is estimated from the friction coefficient f of (solid piece)/(roll surface) using the equation $\tan(a) = f$.

Roll crushers, besides brittle materials, can also crush or grind softer plastic or viscoelastic materials. Toothed roll crushers can be used for producing sizes down to 10 mm. Their advantage is relatively lower energy consumption and little dust production. The actual capacity of roll crushers is about 50% of the theoretical capacity, which can be calculated from the length of the rolls, the distance between the rolls, and the peripheral velocity.

Roll crushers can be applied to a wide range of products (brittle, ductile, dry, moist). However, they are subject to severe wear, particularly with hard products, and they consume large amounts of energy.

Pulpers and disintegrators are used in pulping fruits, in separating the flesh of fruits and vegetables, and in juice production. They consist of tools (paddles) rotating near the surface of a cylindrical or conical screen. Rotating paddles compress the product on the screen, or several rotating tools grind and compress the material through the screen. In both cases, the solid waste is rejected, while the soft part and the liquid pass through the screen. The fineness of the product depends on the screen used and the clearance between the screen and the rotating tools. This clearance is controlled by axial adjustment of the rotor. The rotation of the rotor in extractors (1,500–2,000 RPM) is slower than in disintegrators (up to 10,000 RPM).

7.3.3.3 Grinders

Grinders are used extensively in food processing to produce particles of various sizes.

7.3.3.3.1 Roll Mills

The roll mills are the main type of milling equipment used in the fine grinding of cereals (5–100 μm). They consist of counter-rotating roll pairs or in some

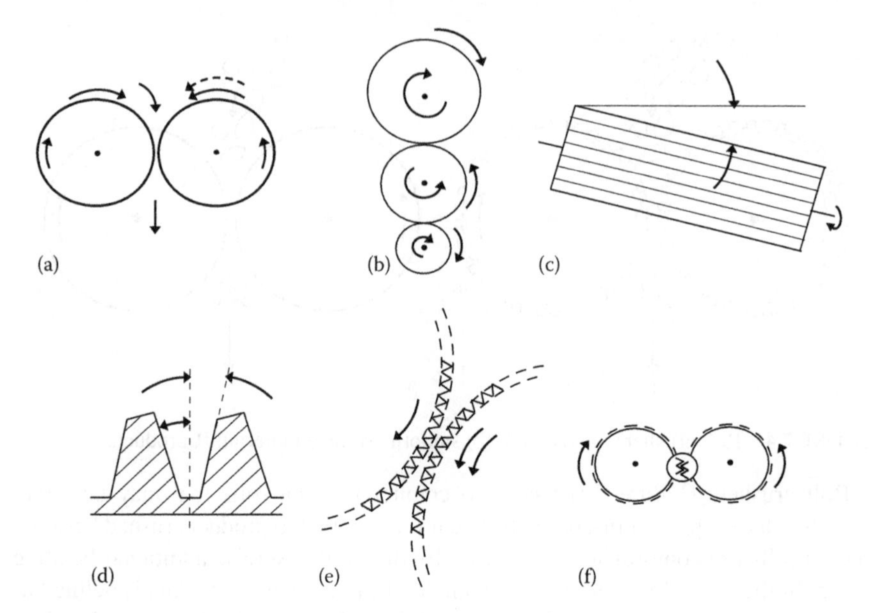

FIGURE 7.5 Roll mills: (a) two-roll; (b) multiple-roll; (c) corrugated; (d), (e), (f) grooved.

cases (e.g. wet fine milling) of a system of rolls (Figure 7.5). After each grinding operation, sifting follows, separating oversized products that are further ground. In some cases, the product is ground twice (it passes through two pairs of rolls) before it is sifted.

The efficiency of roll mills depends on the product, the dimensions, and condition of the rolls, the gap between rolls, the speed of rotation of rolls, and the moisture content of the product.

The diameter of standardized rolls is 220–315 mm and the length 315–1500 mm. The rolls may be smooth, grooved (fluted) or have corrugations. The rotation velocity of the faster roll of each pair is about 6–9 m/s.

The diameters of grinding roll pairs are equal but their speed is different. The rotation velocity of the faster roll of each pair is about 6–9 m/s. The ratio of rotation speeds of the rolls depends on product ground, for example, 1/2.5 for wheat and one-third for rye grinding. The gap between the rolls depends on the type of product and the fineness of grinding and it can be adjusted accurately. In the last stages of grinding the gap becomes narrower. Wheat grinding usually requires four to seven passes through the rolls. Table 7.10 gives typical gaps in multi-pass roll mills.

Abrasion (wear) of the rolls is especially high when the gap between rolls is small and the speed of rotation high. Wear is measured by the abrasion indicator, which is the mass of metal loss of rolls in relation to unit power used (g/kWh) and it depends on the material and method applied (wet or dry grinding).

TABLE 7.10
Typical Gaps in Multi-Pass Roll Mills

Grinding Stage	Roll Gap, mm
1	0.50
2	0.15
3	0.09
4	0.08

The forces exercised during grinding (cut, press, or shear) depend on the type, the position, and the relative speed of the rolls.

In roll milling, a high amount of energy is transformed into heat, elevating significantly the temperature of the rolls. Therefore, in some equipment, a cooling mechanism such as circulation of water in rolls may be provided.

In grain milling, two basic methods are used, that is, dry and wet grinding. In dry grinding, the moisture content of the grains is about 16%, while in wet milling it is higher. Wet grinding is especially applied to corn grinding in connection with wet processing, used in removing the oil-containing corn germ.

7.3.3.3.2 Hammer Mills

The hammer mill is a commonly used equipment for food size reduction by impact forces. It is used for producing a wide range of medium to fine particles. The ground product can be dry, moist or even lumpy, soft, brittle, crystalline or fibrous. Hammer mills consist of a rotor, including axially assembled metallic rods, rotating in a chamber (stator), whose lower part has a replaceable screen (Figure 7.6).

Depending on the product fed and its final size requirement, different hammers or beaters (chip hammers, blade beaters, etc.) are used on the rods and swing. The product fed in the chamber is hit by the quickly rotating hammers and it leaves the chamber, passing through the sieve at its bottom. The fineness of the ground product depends on the product (consistency, moisture, etc.), the rotor speed, the type of tools used, the aperture of the sieve, and the clearance between rotating tools and chamber/sieve.

The end size of particles may vary from 10 mm to 50 μm. The hammers/beaters usually rotate at 500–3000 RPM. The faster the rotation, the finer the grinding of the product, if all other parameters remain constant. In some applications, a temperature

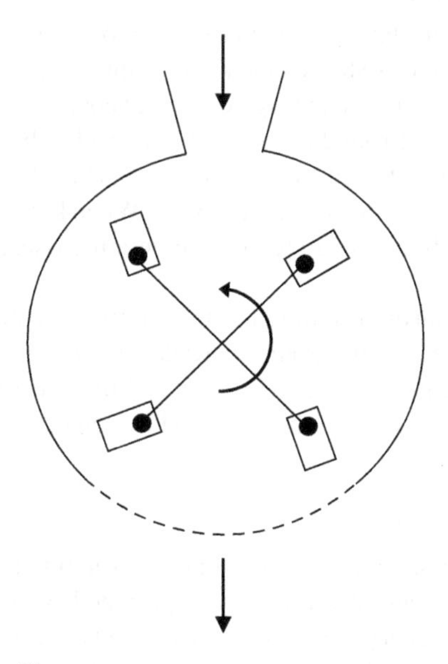

FIGURE 7.6 Hammer mill.

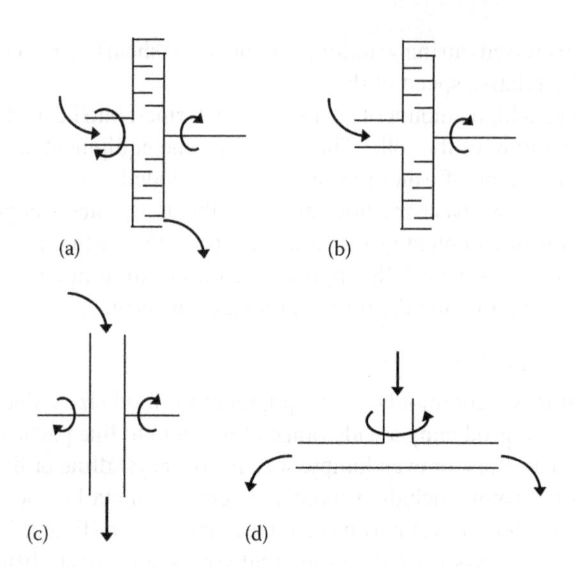

FIGURE 7.7 Disc grinders: (a), (b) pin mills; (c), (d) vertical, horizontal flat disc.

increase during grinding may be prevented by blowing air (e.g. 200–400 m^3/h) or by other cooling method (e.g. using cooled jacketed walls).

Hammer mills have relatively small volume and they can be used in the production of a wide range of particles that are of interest to food processing. They have relatively reasonable energy requirements, about 0.2–2 kWh/ton of product. However, they wear out more seriously, and they may produce more fines than mill rolls.

7.3.3.3.3 Disc Grinders

Disc grinders are used for fine grinding of soft up to medium hard brittle materials, and for dry as well as moist slurry materials. Some foods that are ground by disc grinders are starch, dry fruits, sugar, spices, and cocoa.

Disc grinders consist of round discs parallel to each other, with the material fed continuously in the gap between the rotating discs (Figure 7.7). Depending on the type of the machine, one or both discs can rotate. When both discs rotate, the second disc rotates in counter direction. To reduce temperature rise, air may be blasted during grinding.

The forces exerted during grinding in molded discs are impact and shear. When one disc rotates, the size of the ground particles is 50–5 μm. If both discs counter rotate, the particle size may be reduced down to 2 μm. Size reduction is controlled by adjustment of the gap between the discs and by screens placed on the lower part of the mill.

7.3.3.3.4 Rotary Grinders

Rotary grinders consist of concentric rolls or discs, rotating fast in cylindrical chambers or rings. The cylinders are grooved or provided with adjustable inclined or vertical knives. The periphery of the discs is engraved or curved, forming sometimes a kind of propeller. The product is fed into the gap between chamber and rotating

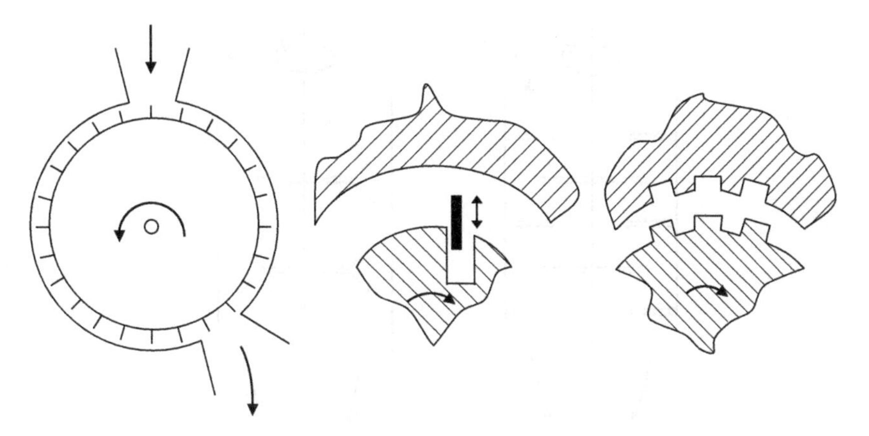

FIGURE 7.8 Rotary grinders.

devices (Figure 7.8). Size reduction is done by shear/attrition. Attrition occurs among particles and between particles and chamber surface.

Rotary grinders are used to produce small randomly shaped pieces, which thereafter are directly used in further processing (e.g. sugar production from sugar beets, or starch from potatoes). They are also used in wet size reduction of products down to 1 μm, suitable for dispersing and emulsifying. Feed is usually smaller than 5 mm. In the case of sugar beets, rotors of 2 m diameter with 20–24 knives are used. The cutting speed is 4–10 m/s and the peripheral velocity of rotors is 7 m/s.

Rotary grinders have a high capacity, a relatively low wear of moving parts due to slow rotation, and a grinding versatility through adjustment/change of knives/rotating discs.

7.3.3.3.5 Impeller Attritors

Impeller attritors (friction grinders) are used in fine grinding of various soft and medium hard products, such as chocolate crumbs, powder coatings, baking mixes, and milk–sugar mixes. They consist of a cylindrical or conical chamber and blades rotating on a perforated plate at the bottom of the chamber, while air may be blasted upward through the plate for reducing the temperature arising during grinding. The blades rotate at 600–4500 RPM and the speed can be adjusted linearly.

Size reduction is achieved through attrition among particles, by attrition of particles with the wall of the chamber, and impact with the rotating blades. In some units, the wall is perforated or screens control the final size of particles, which can be from 20 μm to more than 100 μm. Large particles (>30 μm) are produced by collision between particles and chamber wall. The size and amount of fine particles depend on their average free path in the chamber, which is a function of the rate of feeding. The larger the average free path, the more often collisions occur.

7.3.3.3.6 Disintegrators

Disintegrators consist of chambers in which several concentric tools (blades or rods) rotate with a peripheral velocity of 4–20 m/s. The tools rotate in cylinders or between bars that are part of the walls of the cylinders (Figure 7.9).

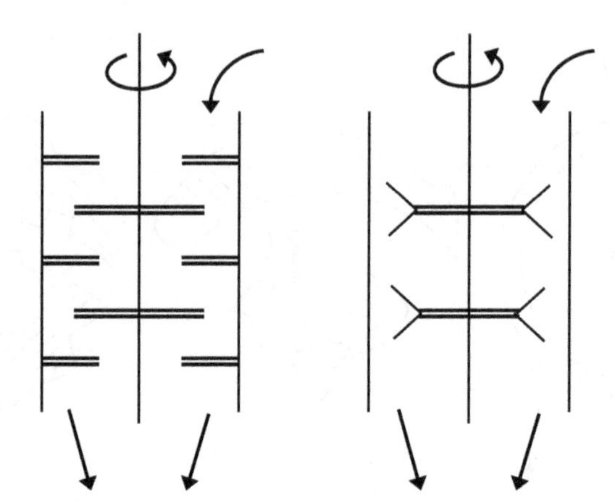

FIGURE 7.9 Disintegrators.

The product flows along the vertical axis of the rotating tools and is discharged from the bottom. The size distribution of particles may be influenced by the speed of the rotor and the type and number of rotating tools.

Disintegrators are often used for breaking up packaging or other waste materials. For disintegrating lumps, food by-products, or products containing high amount of fatty substances down to 100 μm, there are machines rotating at 4000 RPM.

7.3.3.3.7 Colloid Mills

Colloid mills are a variation of the flat grinding discs and may be used in disintegration of viscous products and in fine grinding of various particles. They are used in the homogenization of fluids, in manufacturing of mustard, mayonnaise, and salad dressings, and in fine grinding of animal or plant tissues for manufacturing baby foods and soups. The gap between the discs can be adjusted automatically down to less than 25 μm. When using discs for fluids, their speed is in the range 1500–3000 RPM. The forces exercised on products are shear forces. For reducing wear, plates are made of toughened steel or corundum.

Compared to common homogenizers, colloid mills are simpler in construction and they require no high pressure for their operation.

7.3.3.3.8 Jet Mills

Jet mills are based on the collision of particles, blown against each other by two high velocity air streams. There are three variations: (a) the counter jet mill, (b) the spiral jet mill, and (c) the oval jet mixer. Jet mills are used for very fine grinding (down to 1–3 μm) of hard or medium hard, temperature sensitive, particles (Figure 7.10).

In the counter jet mills, particles collide and break up as two air jets, containing the particles, impact when they are blown from exactly the opposite directions. Both jet streams have the same velocity, about 400 m/s (near to the velocity of sound). Particle size can be reduced to 1–3 μm.

In the spiral jet mill the product is "injected," by means of compressed air, into another high-speed air stream, rotating in a circular channel. Size reduction

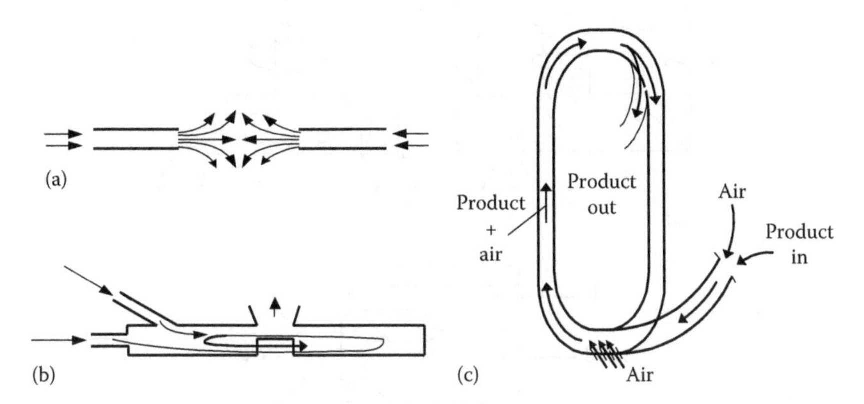

FIGURE 7.10 Jet mills: (a) counter jet, (b) spiral jet, (c) oval jet.

to 1–20 μm can be achieved through intensive attrition among particles and between particles and channel walls.

The oval jet mixer consists of a vertical oval channel in which air and particles circulate. New product is continuously fed into the lower part of the channel where it meets high-pressure air or steam at 1.5–20 bar. The incoming particles are size-reduced by the pressurized air/steam at the lower part of the channel by colliding with each other as they circulate in the oval channel. In the upper part, the larger particles continue their way through the channel downward, while the smaller particles are removed. This method can combine drying with grinding operations.

Jet mills have the advantages of no moving parts, grinding of hard products, low operating temperature, and reduced equipment wear.

7.3.4 Droplet Formation

Spray formation in liquids, known also as spraying or atomization, is concerned with the formation of liquid droplets of various sizes, which are used in some food processing operations, such as cooling, spray drying, and cleaning of equipment. Liquid droplets have increased external surface, which facilitates heat and mass transfer, and increases the rate of chemical reactions (Perry and Green, 1997).

Droplets can be produced from various liquids by three main devices: (a) pressure nozzles, (b) rotary nozzles, and (c) two-fluid (pneumatic) nozzles (Figure 7.11).

7.3.4.1 Pressure Nozzles

In the pressure nozzles, the liquid is compressed and passed through a small swirl (centrifugal) chamber where a liquid film is formed by centrifugal action. The liquid film exits the chamber at high velocity and breaks into numerous droplets in the size range of 1–300 μm.

The capacity of the pressure nozzle (m^3/h) is proportional to the square root of the pressure to liquid density ratio $(P/\rho)^{0.5}$. The operating pressure depends on the desired size range and the properties of the liquid, especially surface tension $\sigma^{1/3}$ and viscosity $\eta^{1/4}$.

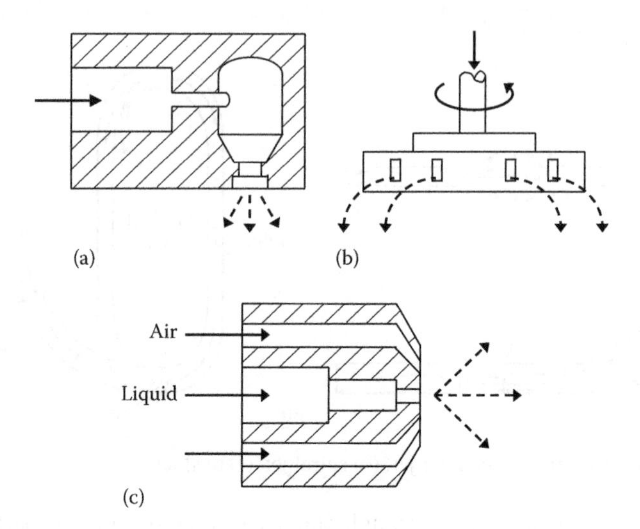

FIGURE 7.11 Spray nozzles: (a) pressure, (b) rotary, and (c) two fluid.

7.3.4.2 Rotary Nozzles

Rotary nozzles consist of a rotating horizontal disc that disperses droplets of various sizes. The liquid is fed through the center of the disc and exits at the periphery. The disc rotates at very high speed (up to 10,000 RPM) and the liquid exits the nozzle at high velocity (above 100 m/s).

Rotary discs have a diameter of up to 1.00 m and they have high capacity in spraying various solutions and suspensions, for example, in spray drying.

7.3.4.3 Two-Fluid Nozzles

Two-fluid or pneumatic nozzles utilize air or steam to disperse the liquid. Air is blown at a high velocity against a liquid layer, which is broken to droplets of small size.

Pneumatic nozzles consume more energy than the other two types of nozzles, but they can produce droplets of small size.

7.4 SIZE ENLARGEMENT

7.4.1 Introduction

Size enlargement or agglomeration is used in food processing to increase the size of food particles, resulting in various products of desirable properties, such as starch, instant drinks, milk powder, spices, and chocolate (Schubert, 1987).

In agglomeration, solid particles attach to each other to form new products, mainly through physical forces. Agglomerated products are aggregates of particles of appropriate size and porosity, which are strong enough to withstand handling and can be dispersed readily, when used in liquids. The desired size of aggregates is usually about 100–250 μm. Agglomeration (enlargement) depends on adhesion forces between similar or different materials, created by compression or extrusion. These forces contribute to the formation of large granules or pieces for direct consumption (e.g. instant products, candies, pastas, and flakes). Undesirable forces form clumps, deteriorate the homogeneity of products or have negative influence on the flowability of powders and grains.

Agglomeration can be achieved through free structuring and compression. In free structuring, the adhesion forces between solid particles are due to material association, nonmaterial association, and form-related association. Compression agglomeration may be carried out in different types of presses and extrusion devices.

In food agglomeration, binders are used, which increase the strength and the homogeneity of granules. Agglomerates may also contain flow aids, wetting agents, surfactants, and substances determining the final characteristics of the products, such as flavor modifiers and colors. Properties that determine the quality of agglomerates are the size and the porosity of powders or granules, their wetting ability, and their strength.

For foods, water is the most common binder. Water forms bridges between particles when crystallization of dissolved solids occurs. If the moisture of agglomerated material is high and agglomeration lasts relatively long, porous granulates are produced. If moisture is low and agglomeration lasts relatively long, high density granules of insignificant size variation are produced.

Food agglomerates, in addition to water, may contain other binders, such as carbohydrates, proteins (e.g. powder of milk and cocoa, soy flour), and starches or starch derivatives (amylopectin). In some cases, combinations of binders and other additives are used, such as emulsifiers and antioxidants, and substances that produce CO_2 when dissolved in water, accelerating the dispersability of the agglomerates.

7.4.2 Free Structuring Agglomeration

Free structuring agglomeration can be achieved by roll processes, mixing processes, and drying-related processes. The agglomerated products have a low strength and a nearly spherical shape. Binders (e.g. water) are much more important in producing granules by these methods than in compressed agglomerates.

Agglomerates produced by structuring methods are nearly spherical with a diameter of 0.3–2.0 mm. They have large external surface of granules; they can be transported mechanically, and they can exchange heat fast. The instantized granules are stable in storage and they are used easily.

In roll processes, the initial agglomeration particles are enlarged during rolling in rotating pans or drums. Enlargement of particles through mixing can be done through intensive mechanical agitation or in fluidized beds. Combination of mechanical agitation and fluidized bed is also possible. In drying-related processes, such as spray drying and fluidized bed drying, fine particles build agglomerates with granules before granules get completely dry.

Instantized products are very important in food processing of soups, drinks, and in several manufacturing processes, in which dried food in the form of powders or granules is used, such as in chocolate manufacturing. Instantization of a powder decreases the time required for the powder to be dissolved, but it does not change its solubility at equilibrium.

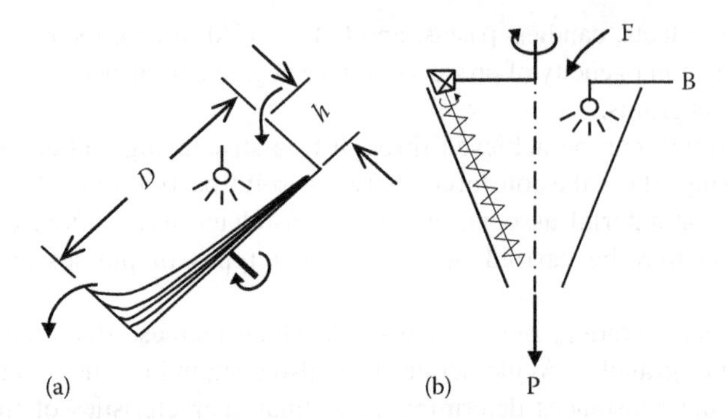

FIGURE 7.12 Rolling agglomeration: (a) rotating drum, (b) granule elevation–agitation. F, feed; B, binder; P, product.

Characteristic properties of instantized products are dispersability (quick dissolving), wettability (quick liquid penetration), sinkability (agglomerates should not float or sink), and flowability (product transport and use).

Rolling agglomeration (Figure 7.12) is achieved by means of (a) rotating walls that roll up the material, which subsequently falls down, before being taken upward again by the moving wall and (b) agitators that move up granules and then allow them to roll or fall back. This is done in vat equipment, in which helical coils or paddles rotate or swing slowly.

In mixing agglomeration, the particles are enlarged by bringing them into contact by means of high-speed rotating agitators, which mix the content of vats, or fluidization of small granules in a gas stream (Figure 7.13). Particles come in contact with rotating small granules, forming agglomerates.

In drying agglomeration, drying is required when the moisture content, due to wetting with binders, is high. Agglomeration is achieved by bringing into contact already dried-out small granulates with the key substance, which is wet powder of the same material (Figure 7.14).

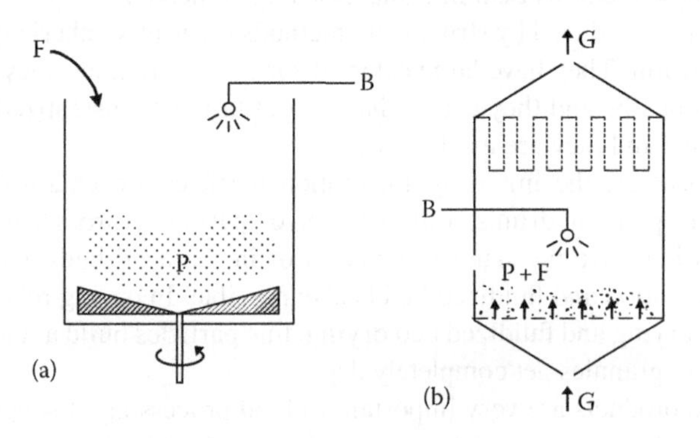

FIGURE 7.13 Mixing agglomeration: (a) rotating agitator; (b) fluidized bed. F, feed; P, product; B, binder; G, gas.

7.4.3 Pressure Agglomeration

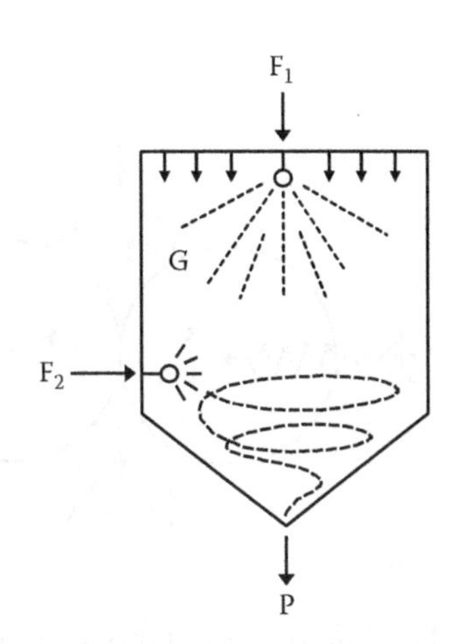

FIGURE 7.14 Drying agglomeration. F_1, feed; P, product; F_2, binder; G, gas.

Pressure agglomeration of food powders is mainly used in confectionery (chocolate processing, sugar further processing, production of candies, etc.) and in extrusion (production of flakes, pellets, pasta). Pressure agglomeration may be classified into tableting, roll pressing, and pelletizing systems. It requires more expensive equipment and more energy (two to four times) than structuring agglomeration.

Pressure agglomeration is affected by the following factors: raw material used, power applied, duration of compression, temperature of product during compression, and binders. A variety of shapes and sizes can be produced with an exact dosage of components. Pressure agglomeration is achieved by tableting, rolling, or briqueting operations.

In tableting, the agglomerates are produced by pressing of powders or granules in special molds, positioned under a filling funnel. The filled molds are compressed by pistons to the desired pressure and the tablets are removed in a continuous operation.

Powder is filled in molds and compressed by reciprocating pistons. Often, two pistons are used, one forming the bottom of the mold, in which the material is compressed and the second exerts the pressure required. The molds may move continuously under the filling funnel on a rotating table.

Large tableting machines are fully automated and can produce up to 1 million pieces per hour of tablet diameter 11 mm, and height 8–9 mm. The depth of the cylinder that is filled with powder and compressed between the two pistons is 18 mm. The compression force depends on the capacity of the machines and is 50–100 kN.

Roll presses consist of two identical rolls running reversely, with adjustable clearance. The surface of the rolls may be smooth or structured (toothed). Powders or granules are fed at the upper side between the rolls, and they are carried along and compressed between the two rolls. Roll diameters 0.8–1.5 m, low rotation speeds, and pressures 0.1–10 kbar are used.

Figure 7.15 shows the two basic types of roll pressing agglomeration equipment. The structured rolls may have tooth or gear-wheel shape, depending on the compressed product. They are used in agglomerating industrial by-products, transforming them economically to easy-to-handle useful products. Roll press machines can produce large quantities of agglomerated products. However, these products are less uniform than those produced by tableting machines. In calculations of roll systems, processing pressure, torque, gap opening, and angle of nip are important.

In pelletizing agglomeration, the materials are pressed through perforated surfaces or dies, which determine the final shape of the pellets. Pelletizing equipment includes screen pelletizers, hollow rolls, and extruders. Extruders are discussed in Section 7.5 of this chapter.

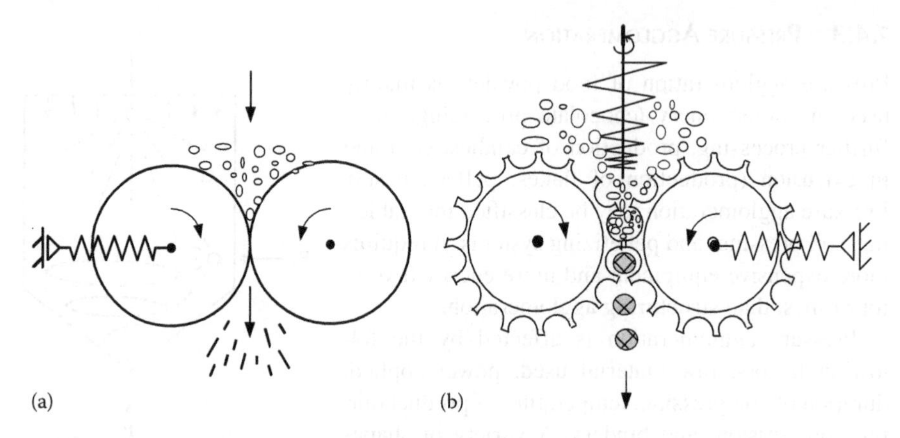

FIGURE 7.15 Roll pressing agglomeration: (a) smooth and (b) structured rolls.

Screen pelletizers consist of stationary screens with rolls or blades rotating over the material pressing it through screens underneath, forming pellets, which are cut to the desired length through scrapers. The capacity of such equipment depends on the kind of compressed material and the type of pellets produced. Pellets are usually cylindrical and their diameter may vary over a wide range (0.32–5.0 cm). The blades or rollers rotate at about 2 m/s and the energy required for pressing is about 10–20 kWh/ton.

In hollow rolls, the material is pressed through a perforated cylinder by a counter rotating nonperforated cylinder. The diameter of such pellets is usually up to 5 mm and the capacity of hollow rolls may be up to 3 ton/h.

7.5 MIXING AND FORMING

Mechanical mixing operations are used widely in the food processing industry to give new physical, rheological, and organoleptic properties to food products, to disperse components in multiphase mixtures, to improve heat and mass transfer, and to develop new food structures. Different mixing systems are required for gas/liquid, liquid/liquid, solids/liquid, and solid/solid mixing. The theory of mixing is more developed in fluid (liquid/liquid) systems, while solids mixing is treated empirically.

7.5.1 Fluid Mixing

The primary engineering characteristics of fluid mixers are the power (energy) requirements and the efficiency (uniformity) of mixing. Mixing of gases or liquids in a liquid is based on the mechanical agitation to disperse a component or a phase into another phase (Uhl, 1985; Walas, 1988).

7.5.1.1 Agitated Tanks

The basic mixing unit is the agitated tank, that is, a vertical cylindrical vessel equipped with one or more impellers and baffles with specified dimensions for a given application. Various types of impellers are used depending on the volume of

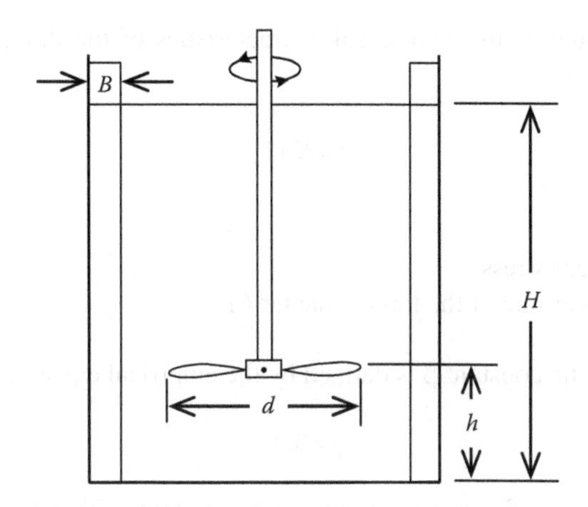

FIGURE 7.16 Agitated tank: *B*, baffle; *d*, propeller.

the vessel and the viscosity of the liquid. High-speed agitators include propellers, turbines, and hydrofoils, while anchors, paddles, ribbons, and screws are used for low-speed applications.

Mixing of low-viscosity liquids is improved by baffles on the walls of the vessel, which are ineffective in mixing very viscous or non-Newtonian fluids. Figure 7.16 shows a typical agitated tank, equipped with a propeller agitator and four-wall baffles. Typical geometries of agitated tanks for fluid mixing are:

$$\left(\frac{H}{D}\right) = 1, \quad \left(\frac{d}{D}\right) = \frac{1}{3}, \quad \left(\frac{h}{D}\right) = \frac{1}{3}, \quad \text{and} \quad \left(\frac{B}{D}\right) = \frac{1}{10} \tag{7.12}$$

The mixing of Newtonian fluids in agitated tanks is expressed by the Reynolds (*Re*) number, which is defined by the equation

$$Re = \frac{N d \rho}{\eta} \tag{7.13}$$

where

N (1/s) is the impeller rotational speed
d (m) is the impeller diameter
ρ and η are the density (kg/m^3) and viscosity (Pa s) of the liquid, respectively

In agitated tanks, the following flow regimes are distinguished: (a) laminar flow ($Re < 10$), (b) intermediate flow ($10 < Re < 10{,}000$), and turbulent flow ($Re > 10{,}000$). In liquids of low-viscosity, turbulent flow ($Re > 10{,}000$) can be obtained at high speeds. For non-Newtonian fluids, the *Re* number is estimated from the equation

$$Re = \frac{\rho d^2}{K \beta^{n-1} N^{n-2}} \tag{7.14}$$

where K (Pa s^n) and N are rheological characteristics of the fluid, defined by the power law equation

$$\tau = K\gamma^{n} \tag{7.15}$$

where
τ (Pa) is the shear stress
γ (1/s) is the shear rate of the fluid (Chapter 6)

The characteristic constant β is defined by the empirical equation

$$\gamma = \beta N \tag{7.16}$$

The empirical constant β depends on the agitated system, with typical values 10–13.

The apparent viscosity of pseudoplastic (non-Newtonian) fluids decreases ($n < 1$) as the speed N is increased. Thus, the mixing will be faster close to the agitator blades (propeller or turbine) than away from it, creating a well-mixed volume of liquid, within a surrounding volume of unmixed liquid. For such systems, paddle of anchor agitators are more efficient, since they can mix the entire volume of the liquid.

The power of mixing in agitated tanks is given by the empirical equation

$$\frac{Po}{Fr} = cRe^{m} \tag{7.17}$$

where

$$\text{Power number } Po = \frac{P_A}{\rho N^3 d^5} \tag{7.18}$$

$$\text{Froude number } Fr = \frac{N^2 d}{g} \tag{7.19}$$

P_A (W) is the agitator power
$g = 9.81$ m/s^2
c is a characteristic constant of the agitated system and the flow regime

The Froude Fr number is a measure of the vortex, formed in the center of unbaffled agitated tanks. For baffled tanks in laminar flow $Fr = 1$.

In laminar flow, Equation 7.17 yields

$$P_A = c_L N^2 d\eta \tag{7.20}$$

Thus, the power is proportional to the viscosity but independent of the liquid density.

In turbulent flow, the power is proportional to the density but independent of the liquid viscosity. In the intermediate flow range, the agitation power is estimated from empirical equations or diagrams of log(*Po*/*Fr*) versus log(*Re*) for both Newtonian and non-Newtonian fluids (Appendix Figure A.4).

7.5.1.2 Industrial Mixers

Selection of industrial mixers depends primarily on the viscosity of the liquid and the volume of the vessel. Propellers and turbines are used for low-viscosity liquids, $\eta < 10\,\text{mPa s}$.

For vessel volumes up to $1\,m^3$ high impeller speeds of about 1800 RPM are used, while lower speeds (500–1200 RPM) are used for larger volumes. Large paddles are used for fluids of viscosity 10–100 mPa s, while anchors or ribbons are used for higher viscosities (0.1–5 Pa s). Very high viscosities (>5 Pa s) require special screw agitators.

The agitators are top- or side-entering, depending on the volume of the vessel. The superficial fluid velocity near the rotating impeller depends on the viscosity (0.1–0.3 m/s). The power requirements of industrial mixers vary from 2 to 100 kW.

7.5.2 Food Mixing

In food processing, mixing is carried out, in addition to process engineering, for improving food quality, such as texture and color development. Food mixtures involve many ingredients, including liquids, powders, and gases. Some important ingredients are contained only in minor quantities, which should be dispersed evenly and efficiently in the final mixture (Levine and Behmer, 1997).

High viscosity and non-Newtonian fluids require special mixing equipment. Mixing patterns and product characteristics are related in a complex manner. Scale-up of food mixing is based more on constant food properties than on constant power/volume ratios. High shear stresses, induced by agitators, are required for making fine dispersions and emulsions, while low shears are used for mixing solid particles/pieces in solid or liquid phases. Segregation of particles from mixed products should be prevented.

The mixing of food pastes and doughs is accomplished in specialized equipment, developed empirically by equipment manufacturers and industrial users. Dough mixing and processing are important operations in the baking, pasta, and cereal process industries. Doughs are more solid-like and viscoelastic than usual pastes.

Mixing of protein doughs increases their viscoelasticity and gas-holding capacity, essential requirements for bread making. Gluten development during dough mixing (kneading) is based on the alignment, uncoiling, extension, and folding of the protein molecules. Dough mixers are usually batch units, and they can be vertical (most common) or horizontal (Matz, 1988).

Kneaders or Z-blade mixers are used for doughs and pastes, which cannot be handled by anchors and helical ribbons. They are mounted horizontally and have two counter-rotating blades, which have very close clearances with the walls of the trough for preventing sticking of the material on the wall. The kneaders achieve nixing by a combination of bulk movement and intense shearing as the material passes between the two blades or between the wall and one blade (Figure 7.17).

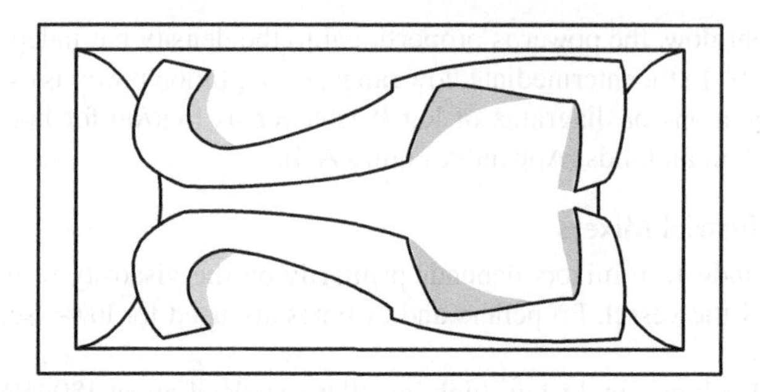

FIGURE 7.17 Diagram of a Z-kneader.

7.5.3 Homogenization

Homogenization is a size reduction technology, based on the application of mechanical forces, used in the emulsification of liquids, dispersion of solid particles in a product, and disruption of cell membranes.

Homogenization of large dispersed particles in highly viscous liquids (e.g. fibrous particles), can be achieved by agitation. However, homogenization is more important in low viscosity liquids, where fluid droplets and solid particles are broken down into smaller units mainly by pressure, rotor–stator movement, and ultrasound.

Homogenization in food processing is applied, among others, to pulps, fruit and vegetable juices, vegetable oil, baby food, salad sauces and creams, milk and milk products, cocoa in candy manufacturing, etc.

The homogenizer used depends on the final size of particles and the output required. Pressures may vary from about 120 to 600 bar, and the higher the pressure, the smaller the particles, which may be lower than 1 μm. If low pressure is applied, two-stage homogenization may be required.

Pressure homogenization is accomplished in special valves in which the fluid is forced through small gaps, breaking down the particles into smaller sizes. One- or two-stage valves are normally used. Figure 7.18 shows a two-stage pressure homogenization valve. The size of homogenized particles is reduced sharply as the pressure is increased. Particle sizes below 1 μm can be obtained at pressures higher than 50 bar.

The homogenization valve consists of a plunger and a valve seat, with a ring gap formed between them. The compressed material flows radially through the narrow ring gap and is then impacted on the surrounding wall ring. The valve efficiency increases if a flat-seated valve has grooved surface, which can compress and expand the particles as they flow along the peaks of each groove. The product velocity in the ring is very high (200–300 m/s), causing a rapid pressure reduction and cavitation, which weakens the coherence and disrupts the particles (droplets) into smaller sizes.

7.5.4 Solids Mixing

Solids mixing is used for the mixing and blending of solid particles and pieces. Uniform blending of solid particles is very important for the quality of the food

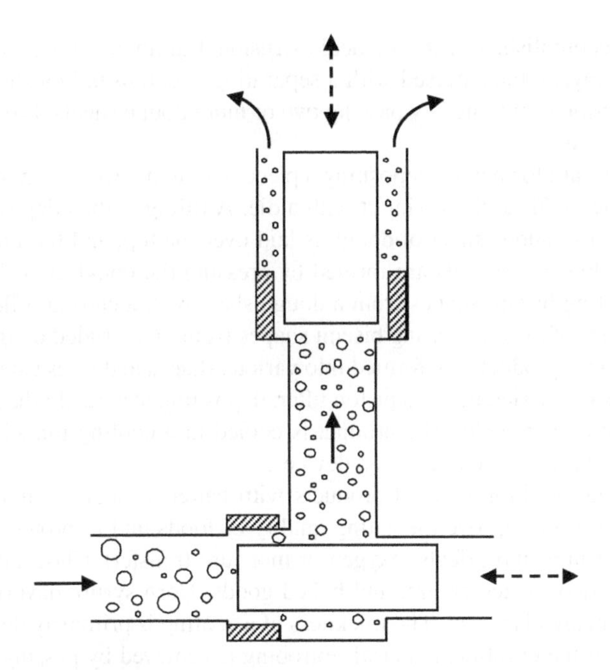

FIGURE 7.18 Two-stage pressure homogenization.

products. The blended product should be stable and de-mixing or agglomeration should be prevented during storage and use.

Industrial solids mixers/blenders were developed by equipment manufactures and end users; they are specific for various applications:

Ribbon mixers consist of helical blades, rotating horizontally, which can mix and convey particles in a horizontal U-shaped trough. Dual helical ribbons, rotating slowly at 15–60 RPM, can mix particles of bulk density about 500 kg/m^3 and capacities up to 50 m^3.

Tumbling mixers, consist of drum, double-cone, V-cone, or Y-cone blenders rotating at 20–100 RPM. The particulate material splits and refolds in the legs of the blender as it rotates. The capacity of these mixers varies from 0.1 to 7 m^3.

Conical screw blenders are large inverted cone vessels of capacity about 30 m^3, filled with particulate material, which is mixed with a vertical screw, orbiting around the periphery (epicyclic path) at about 3 RPM.

7.5.5 Forming Operations

Forming food products into various shapes and sizes is accomplished using various types of equipment, developed mostly empirically. Forming is important in the processing of bread and other baked products, biscuits, pies, and confectionary products.

Forming of a product may involve more than one operation; for example, bread rolls are prepared in three stages, that is, sheeting, curling, and sealing. Sheeting equipment (sheeters) is based on the reduction of dough thickness by passing a slab through two, three, or more sets of rolls of varying distance between them. Sheeting can also be accomplished by mechanical extrusion. Laminated dough products consist of dough layers, interspersed with a separating agent, usually a shortening (fat) layer. Lamination is obtained by passing two or more dough sheets through sheeting rolls several times.

Pie coatings are formed by depositing a piece of dough into aluminum containers or reusable pie molds and pressing it with a die. A filling is then deposited into the casing and a continuous sheet of dough is laid over the top, and the lids are cut by reciprocating blades. Biscuits are formed by pressing the dough in a shaped molding roller, cutting biscuit shapes from a dough sheet with a cutting roller, extrusion through a series of dies, or cutting biscuit shapes from an extruded dough sheet.

Confectionery products are formed into various shapes and sizes using individual molds, which are carried below a piston filler, depositing accurately the required hot sugar mass into each mold. The product is cooled in a cooling tunnel and ejected from the molds using special ejection devices.

Coating and enrobing of food products with batter, chocolate, and other components are used to improve the eating quality of foods and to protect the product from the environmental effects (oxygen or moisture transfer). Chocolate is used to enrobe confectionery, ice cream, and baked goods. Corn syrup, flavorings, colors, and emulsifiers are also used. The thickness of a coating is primarily determined by the viscosity of the enrobing material. Enrobing is achieved by passing the product on a stainless steel conveyor beneath a curtain of hot liquid coating. The coating is applied by passing the product through a slit in the base of a vessel or by coating rollers. A single layer of viscous batter is applied by passing the product through a bath of batter between two submerged mesh conveyors. Seasonings are applied from a hopper over a conveyor followed by passing the product through a rotating drum fitted with internal flights. Fluidized beds are also used for coating flavors on food particles. Coating of fruits and vegetables for protection against moisture loss and microbial contamination is practiced by dipping, spraying, or brushing of paraffin and beeswax dispersed in an organic solvent.

7.5.6 Extrusion

Low-temperature (cold) extrusion is a forming (shaping) operation used in the processing of pasta, cereal, and special food products, for example, coextruded foods and confectionery products. Extruders operate continuously and efficiently and they can replace some other shaping equipment, provided that equipment cost can be justified for a given application. Pasta extruders operate at about 50% moisture, at 50°C.

Extrusion cooking is operated at higher temperatures and pressures, producing several starch-based food products. Extrusion cooking, operated at high temperatures, can be considered, in addition to forming, as a continuous high temperature short time (HTST) thermal process, reducing microbial contamination, and inactivating spoilage enzymes (Kokini and Karwe, 1992).

Extrusion technology is based on the application of food chemistry and food process engineering to physical, chemical, and mechanical changes of food materials subjected to the flow, pressure, shearing, and temperature conditions of the continuous food extruder (Chang and Wang, 1999; Guy, 2001).

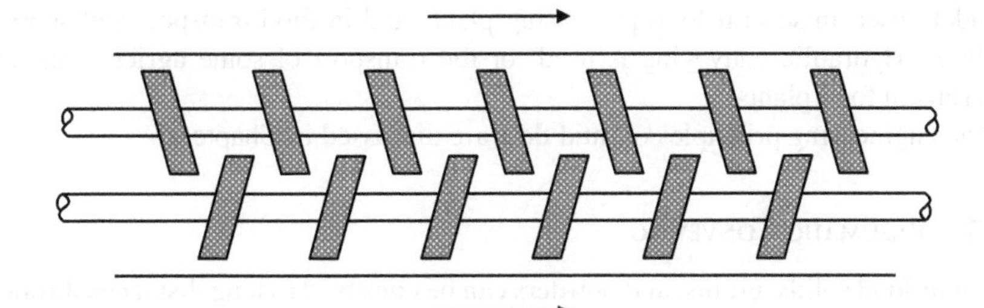

FIGURE 7.19 Diagram of a twin-screw extruder.

The physical properties (density, texture) of the extruded products can be designed by controlling the operating conditions of a given extruder.

Twin-screw extruders are preferred over the single-screw units because of their advantages. The material is usually a cereal powder (corn, wheat, etc.) at a moisture content of 15%–20%, which is compressed and heated above the gelatinization temperature and then expanded through a die to a puffed (porous) product of desired shape. Figure 7.19 shows a diagram of a twin-screw extruder.

The extruder can be divided into three sections: (1) conveying, (2) kneading, and (3) pressurizing. The extruder screws are designed with decreasing pitch so that the product is compressed as it is conveyed from the entrance to the exit of the barrel.

Most extruders operate with no external heating, utilizing the heat produced by mechanical dissipation of the viscous forces in the pressurized particulate food material.

The relative effect of external heat transfer to heat developed by heat dissipation of mechanical energy is characterized by the Brinkman number *Br*, defined by the equation

$$Br = \frac{\text{Mechanical energy}}{\text{Heat energy at wall}} = \frac{\eta u^2}{\lambda \Delta T} \tag{7.21}$$

where

η (Pa s) is the viscosity of the material
u (m/s) is the velocity of the material in the extruder
λ [W/(m K)] is the thermal conductivity
ΔT (K) is the temperature difference between the barrel wall and the product in the extruder

In most food extrusion processes, using low moisture particulate materials, the mechanical heat dissipation is higher than the heat transferred though the wall, that is, $Br > 1$.

7.6 CONVEYING OF SOLIDS

Solid food materials in the form of pieces, granules, and powder can be transported by pneumatic or hydraulic conveying, that is, suspending in an inert fluid medium (air or water), and by using fluid transport equipment and piping. Pneumatic conveying is widely used in several food processing plants and in food transport and storage facilities. Hydraulic conveying is used for the transport of some agricultural raw materials in food plants.

The engineering principles of fluid flow are discussed in Chapter 6.

7.6.1 Pneumatic Conveying

Granular food solids, grains, and powders can be conveyed to long distances through ducts with high velocity air streams. Typical applications include the unloading of granular foods from railroad cars and ships and the transport of such materials within the food processing plant. Food materials handled include wheat, corn, flour, beans, coffee, and granular sugar. Although requiring more energy than mechanical conveyors, pneumatic conveying is preferred for its important advantages, particularly in large, continuous food processing plants (Stoes, 1983; Mills, 1990).

Engineering properties of food materials, needed in the design of pneumatic conveying systems, include particle density, bulk density, particle size distribution, moisture content and hygroscopicity, coefficient of sliding friction (angle of repose), and flowability. Particular care should be taken to prevent fires and explosions of some sensitive food powders. Two basic systems of pneumatic conveying are applied, the air pressure and the vacuum system (Figure 7.20). Mixed system pneumatic conveyors are also used.

The basic equipment of pneumatic conveyors consists of the following units: (a) air blower, usually of the positive displacement rotary type, providing pressure drops up to 0.8 bar; (b) solids feeder, usually of the rotating valve type; (c) transfer line (duct), diameter 200–300 mm, length up to 100 m; (d) smooth bends of long radius; and dust collection equipment at the receiving bin (filter bags).

The pressure conveying system receives particulate material from one source and delivers it to more than one bin, while the vacuum system can receive material from several sources and deliver it to one receiving bin. Vacuum conveying is more expensive than pressure conveying, but it is preferred in transporting dusty products, since it reduces air pollution and explosion hazards.

The design of pneumatic conveying systems is based on empirical equations and data. The calculations lead to the estimation of the pressure drop and the required power of the conveying system. Required data include bulk density, air velocity, and length of the pipeline. Table 7.11 shows typical technical data for pneumatic conveying systems, used in food processing.

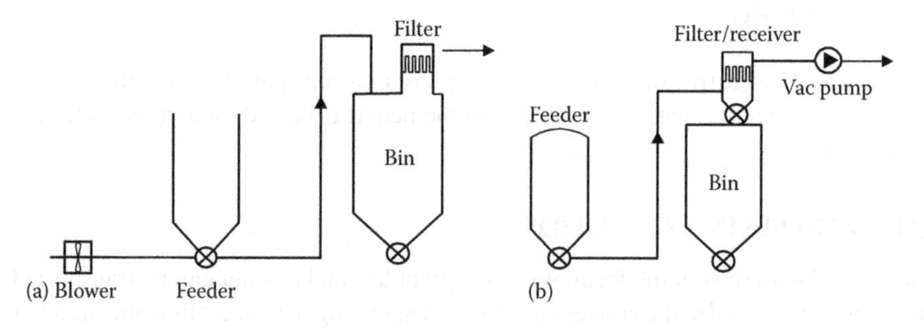

FIGURE 7.20 Pneumatic conveying systems: (a) air pressure and (b) vacuum.

TABLE 7.11
Typical Data for Pneumatic Conveying Systems

Food Material	Bulk Density, kg/m³	Air Pressure System Saturation, m³/kg	Velocity, m/s	Power, kWh/t	Vacuum System Saturation, m³/kg	Velocity, m/s	Power, kWh/ton
Sugar, granulated	800	0.093	18	2.7	0.186	33	4.6
Wheat	770	0.074	17	1.9	0.161	33	3.2
Corn	720	0.074	14	2.0	0.161	32	3.2
Coffee beans	673	0.068	14	1.9	0.136	23	3.2
Wheat flour	641	0.062	11	2.0	0.136	27	3.3

The saturation in the conveying air stream is expressed in standard m³ of air per kg of solids being transported. The power requirement of the system refers to kWh/ton.

Table 7.11 shows the significant effect of bulk density on the saturation (carrying air) and power requirements of pneumatic conveying. The power requirement for vacuum conveying is almost double the requirements of air pressure conveying, evidently due to the higher air velocities in the low pressure operation.

Explosion hazards exist in air-conveying of particles and powders of size lower than 200 μm. Explosion of air/particle mixtures requires solids concentration within the explosiveness range, for example, from 50 to 2000 g/m³. Solids concentrations higher than 5 kg/m³ (i.e. air saturation lower than 0.2 m³/kg), used in pressure and vacuum conveying, are resistant to explosion.

Hot surfaces inducing explosion include sparks caused by mechanical, electrical, or electrostatic forces. The minimum ignition temperatures for sugar, coffee, and cocoa are 350°C, 410°C, and 420°C, respectively. Explosion in closed conveying systems or tanks can create a rapid rise of pressure up to 7 bar. The collection tanks of the air conveying systems are the most vulnerable units to explosion damage. Relief valves are required.

Although air is used as the normal conveying medium, an inert gas, such as nitrogen, may be required for very explosive materials.

7.6.2 Hydraulic Conveying

Agricultural raw materials are usually transported and stored in bulk in the yard of the food processing plant. They are conveyed into the processing area by open water channels and flumes. Floating food materials, such as tomatoes, citrus fruit, and sugar beets, are transported to the juice extractors by mechanical elevators, which may act also as washing equipment (Saravacos and Kostaropoulos, 2001).

Transportation to longer distances against a pressure drop (e.g. to higher elevation) requires pumping and piping systems, specially designed for the particular application. Volute type centrifugal pumps, with special impellers, which can handle large food pieces, are used. Food materials, such as whole fruit, beets, carrots, potatoes, and fish can be pumped without damage to the quality of the product. The pump may be constructed of stainless steel or less expensive cast iron, when allowed.

Empirical equations and data, obtained from the minerals industry, can be utilized in the design of food conveying systems. The velocity of a suspension in the pipeline should be such that the particles will not settle due to gravity, but at the same time it should not be too high, which could damage the product quality and require excessive power. The critical velocity u_c for this purpose will be

$$u_c = 34.6 C_v d u_t \left[\frac{g(S-1)}{d_p}\right] \tag{7.22}$$

where

u_t (m/s) is the terminal (Stokes) velocity of the largest particle present
d (m) is the pipe diameter, d_p (m) is the particle diameter
S is the ratio of suspension to liquid densities
C_v is the solids volumetric fraction
$g = 9.81\ \text{m/s}^2$ is the gravity constant

An important parameter in the design of hydraulic conveying is the ratio of the pressure drops of the suspension and the liquid $\Delta P_s/\Delta P_L$, which can be calculated from the empirical equation

$$\left(\frac{\Delta P_s}{\Delta P_L}\right) = 1 + 69 C_v \frac{\left[gd(S-1)\right]}{\left[u^2 C_d^{1/2}\right]} \tag{7.23}$$

The drag coefficient C_d is calculated from the equation

$$C_d = \frac{1.333 g\, d_p (S-1)}{u_t^2} \tag{7.24}$$

For approximate calculations, the design velocity (u) can be taken as 30% higher than the terminal (Stokes) velocity u_t, that is, $u = 1.3u_t$.

From empirical data, the ratio $\Delta P_s/\Delta P_L$ is approximately equal to the ratio of the densities of suspension and liquid, that is, $(\Delta P_s/\Delta P_L) = \rho_s/\rho_L$.

7.6.3 Mechanical Conveying

Several types of mechanical conveyors are used in the food processing plants. Their selection is based on the type and properties of the product, the capacity, the conveying distance and direction, the hygienic and safety requirements, etc.

7.6.3.1 Belt Conveyors

Belt conveyors are used to convey, horizontally or inclined, granular products or larger packed or non-packed pieces. They consist of an endless belt driven by a drum (shaft) at one end of the belt, while a second drum lies at the other end. The drum is driven either directly or indirectly by a geared electrical motor.

The belt is stretched by a spring mechanism to offset thermal or overweight belt stresses, belt wear, or material accumulation on the shafts (drums). Belts are 0.4–2.5 m wide and up to 100 m long, and their speed is up to 5 m/s.

Belts are usually made of synthetic rubber, steel, or canvass combined with steel wire and plastic materials. Coated solid woven polyester is resistant to stretch and to bacterial contamination and it is cleaned and sanitized easily. The polyester is usually coated with polyurethane or polyvinyl chloride (PVC).

In long belts, supporting rolls are necessary every 0.8–2.0 m. The angle of inclined belts should not exceed 22°. Special support devices are used to distribute the food product evenly on the belt or remove the product from the belt.

Segmented belts consisting of adjoined segments of, for example, steel plates, placed tightly parallel to each other are used when too heavy products are to be conveyed or when the temperature during conveying is higher than 100°C–120°C, as for example, in belt ovens or belt can pasteurizes. The segmented belt can move on slide bars or on rolls at a maximum speed of 1.5 m/s. When a high capacity is required, (e.g. biscuit ovens), wider belts are used.

7.6.3.2 Roll and Skate Wheel Conveyors

Roll conveyors are often used in conveying heavy products. They consist of cylindrical steel rolls, which may be powered or idle. The powered roll conveyors are chain-driven.

In some cases, conveying occurs by the rotation of the powered single rolls. Free rotating rolls are used in conveying, using gravity.

The skate wheel conveyors consist of coaxial wheels, which are usually idle (conveying by gravity) or chain driven. The skate wheel conveyors are more flexible than the rolls. They are mainly used in interconnecting belt conveyors and in conveying products in curved paths. The speed of belt driven roll conveyors is 13–20 m/s.

7.6.3.3 Chain Conveyors

Chains conveyors transport products horizontally, vertically or inclined, in straight or curved paths. Chains may be used in direct conveying of products or in conveying with auxiliary devices attached to them. Direct conveying is "open conveying," that is, parallel chains transporting large pieces, such as barrels, milk containers, or pallets.

The auxiliary devices of the chains are flights or anchors, used at a maximum inclination of 30° at speeds up to 0.60 m/s. They need more specific energy than the plain belt but less than the screw conveyors.

In the overhead chain conveyors, hooks, horizontal plates, or buckets are used, which usually swing or are pivoted to the conveyor chain. Overhead hooks are often used in the meat and the poultry processing industries and in conveying packaging materials in the fruit and vegetable processing plants. The horizontal plate conveyors are used to transport products in trays up and down, maintaining the horizontal position. The maximum speed of chain conveyors is 1.3 m/s, when chains are used as the driving device, or 2 m/s, when belts are used.

7.6.3.4 Bucket Elevators

Bucket elevators are used to transport bulk materials vertically from a lower to a higher level (Figure 7.21). The material is emptied from the buckets by either centrifugal force or as a continuous discharge.

7.6.3.5 Screw Conveyors

Screw conveyors are used in transporting high-consistency, nonfree flowing products in all directions and in emptying silos. They consist of a helical device (shaft) rotating along the axis of a cylindrical vessel (Figure 7.22). The helical shaft may be a screw type spiral, a row of individual blades, or a continuous ribbon.

The length and diameter of common types of screw conveyors used in the food industry are 100–120 and 40–60 cm, respectively. The speed of rotation of the screw varies from 16 to 140 RPM. The diameter of a screw shaft d_h depends on the diameter of the conveyed largest grain or particle d_k, for example, $d_h = 12d_k$ for sorted large grain. The pitch of the screw is $l_p = (0.5–1.0)d_h$.

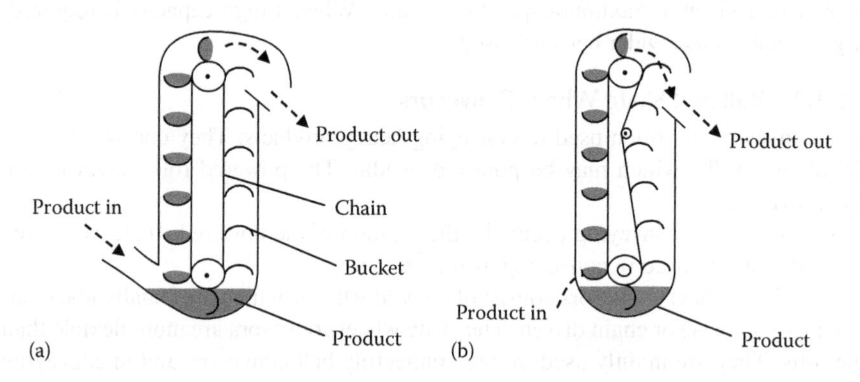

FIGURE 7.21 Bucket elevator: (a) centrifugal and (b) continuous discharge.

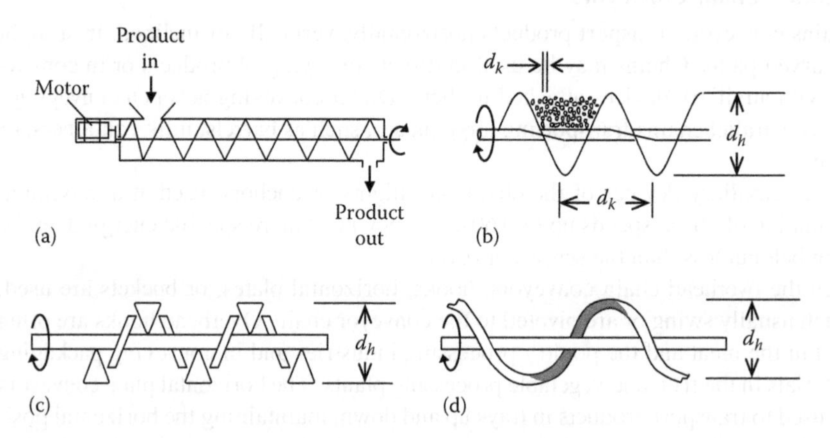

FIGURE 7.22 Screw conveyors: (a), (b) screw spiral; (c) individual blades; (d) continuous ribbon.

Long screw conveying is avoided because of the relatively high energy required due to friction and the problems that may arise, especially if dry, dust-containing products pack between the screw and the wall of the vessel. In such cases, there is a danger of explosion if the temperature of the product is high or if some sparking (friction or electrical) occurs.

In screw conveying of dry food, only 25%–50% of the vessel is filled. The capacity of food screw conveyors depends on the material, the usual range being 25–40 m^3/h.

7.6.4 Hygienic Considerations

Sensitive foods should be transported fast, avoiding microbial, or chemical/biochemical deterioration. Fresh foods, such as fruits and vegetables, should not be bruised or damaged mechanically by the conveying equipment.

Belt conveyors are more hygienic than other mechanical transport equipment, for example, screw conveyors. Smooth belts can be kept more hygienic than segmented, indented, or screen (perforated) belts.

Belts are cleaned and washed with detergents and water using mechanical scrubbing devices installed at the return loop of the belt. Antimicrobial chemicals can be used to sanitize the belts provided they are approved by public health authorities.

Hygienic operation and cleaning of screw conveyors is difficult in wet-food application, because the interior screw does not sweep the interior surface thoroughly leaving “dead” spaces for microbial growth. Well-designed removable covers can protect sensitive food products during operation.

Bucket elevators are mainly used in transporting dry food materials. They are easily contaminated in wet-food applications, especially at the lowest point, where food material may accumulate during operation.

7.7 MECHANICAL SEPARATIONS

7.7.1 Introduction

Mechanical separations are used extensively in food processing, either in preparatory operations or in the main manufacturing and preservation processes. Mechanical processing operations such as size reduction, size enlargement, agglomeration, homogenization, and mixing, discussed earlier in this chapter, may involve some forms of mechanical separation.

The materials involved in mechanical separations are discrete particles and solids, which behave differently than solutions or suspensions of dispersed materials. The mechanical equipment used in handling and processing particles has been developed empirically from practical experience.

Most mechanical separations are based on differences in density and size/shape of the food particles/pieces. They include solid/solid, solid/liquid, liquid/liquid, and solid/air operations (Purchas and Wakeman, 1986).

Solid/solid separations involve screening (sieving) of particulate foods (e.g. flour), cleaning and sorting of food pieces/particles (e.g. fruits, vegetables), and peeling and pitting of various plant foods (e.g. fruits/vegetables and cereals).

Solid/liquid separations are based on the size/shape of the food particles/pieces and the rheological properties of the liquid suspension. They include sedimentation (clearing of wastewater effluents), filtration, and centrifugation (juices, oils, milk).

Mechanical liquid/liquid separations are based on differences in liquid densities (e.g. oil–water separation). Solid/air separations are based on differences in densities of the materials (e.g. particles in cyclones) and on the difficulty of solids to pass through fine meshes and openings.

Physical properties of solid particles of importance to separation processes are particle shape, size, and size distribution (part A of this chapter); particle density, bulk density, and porosity; elastic, plastic, and viscoelastic properties; wetting and flow properties; and electric, dielectric, and optical properties (Chapter 5, Engineering Properties of Foods).

The particle (solids) density of dry food materials is about 1500 kg/m^3. The density of "wet" foods varies in the range of 560 kg/m^3 (frozen vegetables) to 1070 kg/m^3 (fresh fruit). The bulk density of food particles varies from 330 kg/m^3 (instant coffee) to 800 kg/m^3 (granulated sugar). The porosity of solid foods varies from about 0.1 to near 0.95 (freeze-dried and extruded products).

The flowability of food powders is characterized as very cohesive, nonflowing, cohesive, easy flowing, and free flowing. The hydrodynamic properties of food particles are characterized by the Stokes diameter (part A of this chapter).

7.7.2 Classification Operations

Classification of raw food materials is important before any further processing. Classification includes grading and sorting. Grading is the separation of the products according to their quality, while in sorting, acceptable products are classified according to predetermined physical or other characteristics.

7.7.2.1 Grading

Grading classifies food materials on the basis of commercial value, end usage (product quality), and official standards. Grading is necessary for example, for avoiding the further processing of spoiled food materials or products not meeting the quality requirements. Grading is done mostly by hand (e.g. inspection of fruits after washing) or through machinery. Thus, in rice, the white kernels are separated from the spoiled or from foreign matter, optically, and the unripe tomatoes can be separated from the ripe according to their density.

The effectiveness of hand grading depends on the (1) quality of the product, (2) quantity per inspector and min, (3) experience and physical condition of the inspector, (4) kind of inspection, and (5) the speed at which products move in front of the inspector.

The products should be properly illuminated and they should move relatively slowly in front of the inspectors. The products must not exceed a distance of more than 80 cm from the worker and the speed they are moving in front of the workers inspecting them is product specific, for example, 15 m/min (berries)–45 m/min (apples).

Grading of larger quantities of food or products, such as grains, is based on testing of smaller quantities. The samples are taken out randomly and they are subsequently evaluated in the laboratory.

Modern physical methods are recently used for continuous grading, such as color measurement, use of x-rays, lasers and infrared (IR) rays, and microwaves. Most of the machines used in grading can be also used in sorting of food. X-rays are also used in detecting foreign matter such as glass splits and stones in food products.

7.7.2.2 Sorting

Sorting facilitates subsequent processing operations, such as peeling, pitting, blanching, slicing, and filling of containers. It is desirable in heat and mass transfer operations, where processing time is a function of the size of the product (e.g. heat conduction, mass diffusion). Specific equipment is used for each product or product group.

Most of the mechanical sorters are based on the size of the materials, but some equipment utilizes differences in shape, density, and surface properties of the food pieces/particles.

Screens (flat or drum type) are used widely in sorting various grains, seeds, crystals, and other food pieces/particles of relatively small size. Inclined screens, one on top of the other with horizontal and vertical oscillations are effective in grain and seed sorting.

Sorting of fruits and vegetables is related to quality classification, for example, small-size peas and okra are considered more tender and desirable than larger sizes. Large potatoes are desirable for long French fries and corn cobs should not be too long. The shape and size of fruits and vegetables should be suitable for mechanical harvesting, handling, and processing.

Several fruits can be sorted in (a) diverging belt/cable sorters or (b) roller sorters (Figure 7.23), which separate the fruits into various sizes by allowing them to pass through an increasing slot (opening) sizes.

Cylindrical and disc separators are used to separate nearly round grains (e.g. wheat) from long grains or particles. The cylindrical unit consists of a horizontal cylinder with hemispherical indents on the inside surface. The mixed grains are picked up by the indents and they are separated on the basis of their length, as they

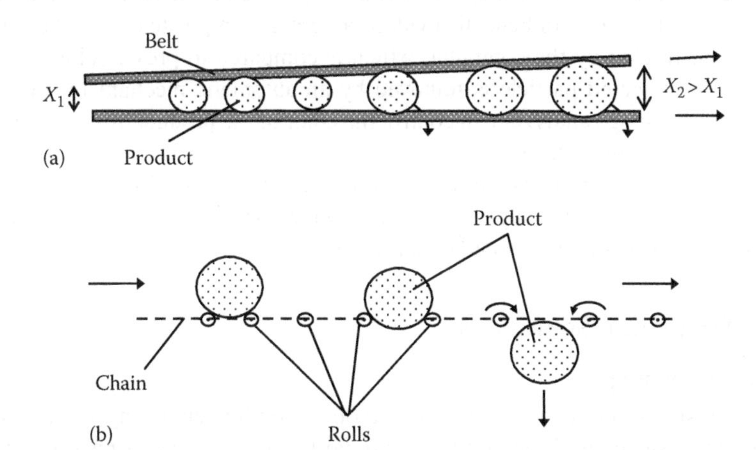

FIGURE 7.23 Size sorting equipment: (a) diverging belts, cables; (b) roller sorter.

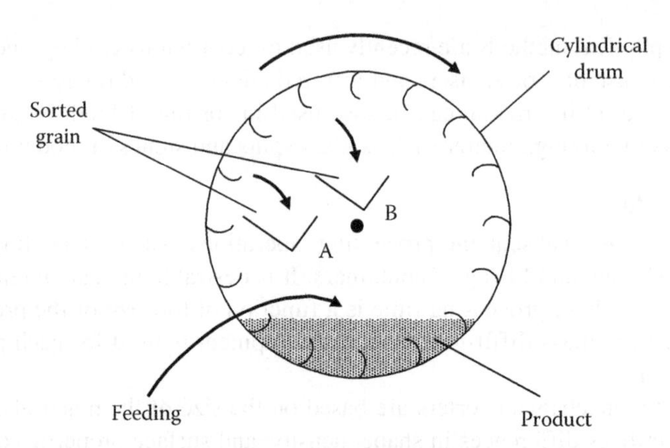

FIGURE 7.24 Sorting of grains in a cylindrical separator.

move up the cylinder. The longer grains fall down first, leaving the smaller grains, which fall in a different compartment (Figure 7.24). The fineness of the separation is controlled by the speed of rotation and the adjusted position of the separation edge in the cylinder.

The disc separator operates on the same principle with the cylindrical unit. The disc contains slightly undercut pockets, which can pick up and retain short grains, but long grains fall out. Thus, wheat grains can be separated from rye, oats, barley, etc.

Weight sorters are used for valuable foods, such as eggs, cut meats, and sensitive fruits, where accuracy in size separation and caution in handling are needed. Eggs are first inspected visually over tungsten lights (candling) to remove unfit eggs.

Color sorting is widely used in the food processing industry. The operation of the color sorters is based on the reflection of incident light on a food piece/particle, which is measured by a photodetector. The reflected light is compared with preset color standards and the rejected particle is removed from the product mixture by a short blast of compressed air.

Image processing can be used as a color sorting system. The food pieces/particles are fed on a roller conveyor beneath a video camera. A composite image of each food piece is constructed by the computer, which is compared to preset color specifications and the rejected particle is removed by an automatic mechanism. In bakery operations, the image analyzer can control the color of the product by controlling the gas/electricity supply to the oven.

Sorting of liquid products can be based on differences of viscosity. Automated refractometers indicate the sugar content (°Brix), while microwaves can be used for nondestructive water and fat content measurements.

7.7.3 Solid/Solid Separations

7.7.3.1 Screening

Screening (sieving) is a simple mechanical operation for separating solid particles in a series of sieves with openings of standard size. The screening surface may consist of perforated, woven wire, silk, or plastic cloth. The screens are either flat or cylindrical, and a relative motion such as vibration, shaking, or rotation, between product and screen is applied.

The openings of the sieves are usually squares, the dimension of which determines the size of the particles that can pass through (undersize) or remain on the screen (oversize). The industrial sieves are characterized by standard dimensions, which may be different in the various countries. Typical sizes of the standard sieve series used in the United States, Canada, Britain, Germany, and France are given in Table 7.4.

The screening process is facilitated by some kind of movement of the screening surface (vibration or shaking), which prevents the blocking of the screen openings with particles (blinding) and decreases the product flow rate and the separation efficiency of the screen.

Screens used in food processing are mainly flat/vibrating screens and rotating sifters.

Efficient operation of screens is obtained when the particle bed is stratified, that is, when the particles form layers of different sizes. Stratification is facilitated by vibration or shaking of the screens, which moves the small particles down to the screen surface so that they can pass easily though the screen (undersize). At the same time, the larger particles concentrate on the surface of the bed, being removed as the oversized product. Moisture can cause agglomeration of the fine particles, which will not pass though the screen.

7.7.3.1.1 Trommels

Trommels or revolving screens consist of perforated cylinders (1–3 m diameter, 3 m long) rotating below the critical velocity at about 15–20 RPM. They are inclined at 10°–20° and they are used to separate particles in the range of 10–60 mm (Figure 7.22).

The critical velocity N_{cr} (RPM) of the trommel at which the particles will not fall down because of the centrifugal force is estimated from the equation

$$N_{cr} = \frac{42.3}{d^{0.5}} \tag{7.25}$$

This simple relation is derived by equating the centrifugal and gravity forces

$$\frac{2mu^2}{d} = mg \tag{7.26}$$

where

d (m) is the diameter of the cylinder
$u = 3.149\ (N\ d/60)$ is the peripheral velocity (m/s)
m (kg) is the mass of the particle
$g = 9.81\ \text{m/s}^2$

As shown in Figure 7.24, actual sieving takes place only in a small part of the trammel, which is about 15%–20% of its circumference.

Drum (trommel) screens are often used in various combinations (series, parallel, or concentric) in the separation of grains and seeds.

7.7.3.1.2 Flat Screens

Flat screens consist of flat screening surfaces of several sieve sizes, arranged vertically or in line and usually inclined, which can separate and classify various solid particles. Flat screens are usually vibrated at 600–7000 strokes/min and they can separate particles of sizes down to 400 mesh (38 μm).

Vibration can be vertical or horizontal (shaking or reciprocating screens). Shaking and reciprocating screens are inclined slightly and vibrate at 30–1000 strokes/min, separating particles in the size range of 0.25–25 mm.

The width of the screens defines the capacity (kg/h) of the system, while the length affects strongly the screening (separating) efficiency.

7.7.3.1.3 Rotating Sifters

Rotating sifters consist of a series of square or round sieves (0.6–1.0 m) attached on springs and placed atop one another, which rotate in a gyratory motion. Sifting systems may consist of more than 24 rotating sieves, which are grouped so that flour is classified in various grades. Some sifters use bouncing balls on the sifting surface for auxiliary vibration and efficient separation.

7.7.3.1.4 Screening Cloths

Most of the wire cloths are square-mesh, but in some cases an oblong weave may be used, which provides greater open area and higher capacity. Screens with relatively large length-to-width ratio are preferred when moist and sticky materials tend to blind the square of short rectangular openings.

Synthetic woven materials made from monofilaments (e.g. nylon) and Swiss silk are used in light, standard, and heavy weights. The finer the wire of the cloth, the higher the screening capacity, although the operating life of the screen is shorter. Worn or damaged screens should be replaced because they let oversized particles pass through and reduce separation efficiency.

7.7.3.1.5 Magnetic Separators

Pieces and particles of iron and other ferrous metals (nickel and cobalt) are easily removed from food materials by magnetic separators. Electromagnets are preferred over permanent magnets because they can be cleaned more easily.

Two simple magnetic separators that can be used in food processing are the magnetic drum and the magnetic pulley systems. Belt magnetic detectors can detect iron pieces as small as 0.5 mm at belt speeds of 6–60 m/min.

7.7.3.1.6 Electrostatic Separators

Electrostatic separation is based on the differential attraction or repulsion of charged particles in an electrical field. Electrical charging is accomplished by contact, induction, and ion bombardment. Some particles in a mixture are charged and they can be removed electrically, while the rest are separated by gravity. Particle sizes up to 1.5 mm (granular) can be separated. An application to food materials is the electrostatic separation of nuts from the shells.

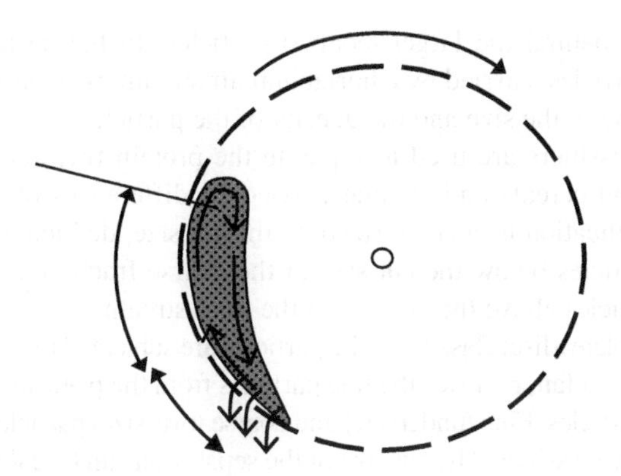

FIGURE 7.25 Trommel (drum) screen.

7.7.3.1.7 Sieving of Flour

One of the major applications of sieving in food processing is the separation of the various fractions of flour during the milling of wheat and other cereal grains. Scalping is the removal of the large particles from the flour, while dedusting is the removal of the very fine powder.

Grading of the flour is the classification of the flour into fractions of desired particle size, such as semolina and middlings. Narrow size distribution within a fraction can be obtained by closed cycle milling, in which the oversize stream from a sieve is returned to the mill for further size reduction (Section 7.3).

7.7.3.2 Fluid Classification

Separation and classification of solid particles by fluids (air or liquids) is based on differences of density, shape, and hydrodynamic, surface, electrical, and magnetic properties of the materials in the mixture. Air classification is used to separate various fractions of food components, while wet sieving and hydrocyclones are used in some separations of fractions of food materials. Subsieve-size particles in the range of 2–40 μm can be separated effectively using various fluid classifiers.

7.7.3.2.1 Air Classifiers

Air classification of solid particles is a dry separation process used in various food processing operations, such as cleaning of raw food materials and fractionation of particulate food components (Grandison and Lewis, 1996).

Simple air classifiers are based on drag forces acting on particles by the air stream, which counteract gravity. Aspiration classifiers are used to separate chaff (skins) from peas and grain in harvesting machines.

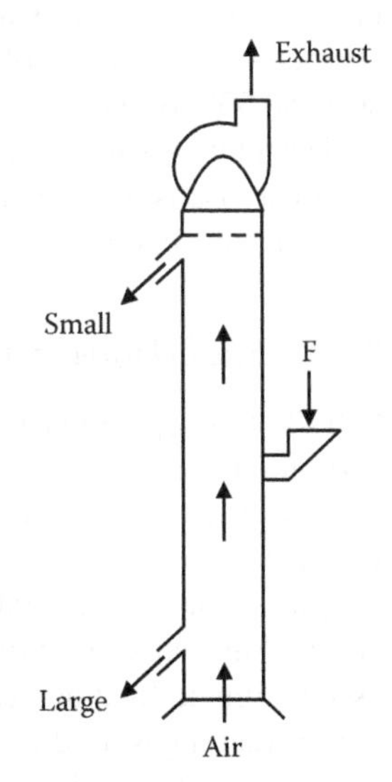

FIGURE 7.26 Diagram of a vertical air classifier.

Figure 7.26 shows the diagram of a simple vertical classifier in which the air stream carries away the fine particles, leaving behind the larger (coarse) particles. In the horizontal classifier, the mixture of particles, carried by a horizontal air stream, is separated into various fractions according to the size and the density of the particles.

Modern air classifiers are used to separate the protein fraction from the starch granules of ground cereals and legumes, based on differences of size, shape, and density. Air classification is characterized by the cut size, defined as the size where the weight of particles below the cut size in the coarse fraction is the same as the weight of the particles above the cut size in the fines stream.

In the rotating plate (disc) classifiers, the particles are subjected to a centrifugal force, while air, sucked by a fan, removes the fine particles from the plate, and separates them from the coarse particles. Fine (undersize) and coarse (oversize) particles are discharged separately from the classifier. The cut size of the separation can be calculated by equating the centrifugal, gravity, and drag (Stokes) forces acting on the particle.

7.7.3.2.2 Wet Classifiers

Wet classification finds some applications in food processing, such as wet sieving, hydrocyclones, and separation of tomatoes in water tanks. Wet sieving is used to separate small-size particles that are difficult to go through the standard sieves because they are sticky or they form agglomerates at high humidities, for example, wet sieving of starch products suspended in water. The suspending medium should be a liquid other than water, such as ethanol, when the particles are water soluble, for example, sugar crystals.

Hydrocyclones are similar to the normal solids/air cyclone separators, which are discussed later in this chapter. They are small inexpensive units, which can separate particles in the range of 5–300 μm, which are suspended usually in water, for example, protein/starch particles. The separation is based on differences in density of the particles and it is also affected by the viscosity of the fluid suspension. The characteristic diameter of the particles, d_{50}, separates the mixture into 50% undersize and 50% oversize.

The high shear rates, developed within the cyclone, reduce significantly the apparent viscosity of the non-Newtonian (pseudoplastic) suspension, improving the efficiency of separation.

7.7.4 Solid/Liquid Separations

Solid/liquid separation is used in food processing for cleaning food liquid from undesirable particles and for recovering useful food particles from water suspensions. It is also applied in the expression (expulsion) of juices and oils from fruits/vegetables and oilseeds.

The mechanical separation methods of solids are based on the particle size, shape, density, and concentration in the water suspension. Four separation methods are used mainly in food processing: screening, sedimentation, filtration, and centrifugation. Screening is used for particles larger than 200 μm and concentration 5%–30%. Sedimentation is applied to particles in the range of 1–300 μm and concentrations up to 25%. Filtration and centrifugation are applied over a wider range of particle size and concentration (Cheremisinoff, 1995).

7.7.4.1 Screening

Large pieces of food and waste materials are removed easily from water by grate screens consisting of curved parallel bars. Smaller pieces and particles are separated by screens of various sizes and shapes (rectangular or circular). Vibration of the screens is required in most cases to facilitate filtration and removal of solids (see Section 7.3).

7.7.4.2 Sedimentation

Gravity sedimentation is applied to the clarification of food liquids from suspended particles (clarifiers) or to the concentration of solid particles (thickeners). Gravity settling of suspended particles in water solutions/suspensions can take place in three mechanisms: (1) Particulate settling (Stokes equation), with the settling velocity a function of the particle diameter and density, the viscosity of the liquid, and the gravitational force. (2) Zone or hindered settling in which the particles fall together as a zone creating a distinct clear water layer. (3) Compression regime, in which the particles are compressed by gravity to form a compressed bed.

Gravity sedimentation is used in the treatment of drinking, industrial, and wastewater. Due to the large volumes of water involved, large sedimentation tanks are required. The sedimentation tanks are fed with the water suspension at the center, while the clear water overflows from the sides and the particles (sludge) are removed from the bottom.

The settling velocity of water suspensions is estimated from laboratory tests in long or short tubes. The depth of the sedimentation tank should be sufficient for settling the smallest solid particles, while the diameter should be such that the upward velocity of the clear water should be lower than the settling (Stokes) velocity of the particles.

Depending on the particle concentration, the tank charge (load) of the settling tanks varies in the range (0.3–$3.0\,m^3/m^2\,h$) with residence times of about 10 h. Diameters of 10–50 m are used, while the length of rectangular tanks may be as high as 80 m and the width 5–10 m. The depth of the sedimentation tanks varies from 3 to 5 m. Long agitating arms with scraping rakes installed on a bridge and rotating slowly are used to move and collect the settled particles (sludge) to the center of the tank from where they are removed with special pumps.

Small metallic settling tanks are used for the clarification or thickening of water or other liquid suspensions in the food processing plants. They consist of a cylindrical tank with a cone bottom, which are fed with liquid suspension in the center, while the clear liquid (water) overflows from the top, and the sludge is removed from the bottom.

Sedimentation of colloidal particles in water is facilitated by the use of flocculating agents, such as alum (aluminum sulfate) and some polymeric materials, which form large agglomerates that settle faster into more compact sludge.

7.7.4.3 Cake Filtration

In cake filtration, the particles form a layer of particles on the surface of the filter medium, which acts as a screen of the particles during operation.

The pressure drop Δp in cake filtration is given by the equation

$$\Delta p = \Delta p_m + \Delta p_c \tag{7.27}$$

where Δp_m and Δp_c are the pressure drops through the filter medium and filter cake, respectively, given by the following empirical equations

$$\Delta p_m = (\eta R_m)\left[\frac{dV}{A\,dt}\right] \tag{7.28}$$

$$\Delta p_c = \left(\frac{\eta R C V}{A}\right)\left[\frac{dV}{A\,dt}\right] \tag{7.29}$$

where

R_m and R are the resistances to flow of the filter medium and filter cake, respectively
V (m^3) is the volume of filtrate
C (kg/m^3) is the mass of particles deposited as cake per unit volume of filtrate
A (m^2) is the surface area of filtration
η (Pa s) is the viscosity of the liquid

The units of R_m are (1/m) and of R (m/kg). In most industrial filters, the resistance of the filter medium R_m is negligible compared to the resistance of the filter cake R and the pressure drop Δp_m can be neglected. For constant pressure drop Δp, the filtration time t to obtain a filtrate volume of V is found by integrating equation (7.29)

$$t = \left(\frac{\eta R C}{2\Delta p}\right)\left(\frac{V}{A}\right)^2 \tag{7.30}$$

The total mass of cake deposited is calculated from the relation $m = VC$.

The resistance R of compressible cakes of colloid and gelatinous particles increases almost linearly with the pressure drop Δp during filtration. The filter media used in industrial filtration are woven fabrics of various materials.

7.7.4.3.1 Plate-and-Frame Filters

Plate-and-frame filters (filter presses) are batch-operated units, in which the suspension is filtered through the surface of plates, forming a cake within the supporting frames. The filtrate is collected though special piping and the cake is discharged when the operation is stopped for cleaning (Figure 7.27).

The square filtering plates are made of stainless steel, dimensions 15–150 cm and thickness 1–5 cm. The filtering surface is made of a strong metallic screen covered with a woven fabric.

When the filter cake is a useful product, for example crystals, it may be washed with water on the filter at the end of the operating cycle.

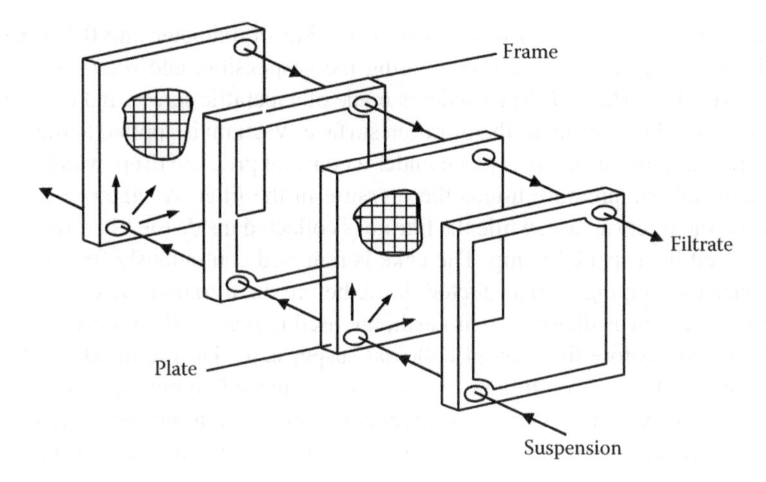

FIGURE 7.27 Diagram of plate-and-frame filter.

Filtration of colloidal particles is facilitated by filter aids, that is, inert powders, which increase the porosity and permeability of the cakes. Typical filter aids are diatomaceous earth, perlite, and cellulose (paper). The filter aid is mixed with the water suspension before filtration.

7.7.4.3.2 Rotary Vacuum Filters

The rotary vacuum filters are used extensively in the process industries because of their advantages over the plate-and-frame filters. They are continuous, faster, and they require less labor, but they are more expensive than the batch filter presses. Figure 7.28 shows diagrammatically the principle of operation of a rotary vacuum filter.

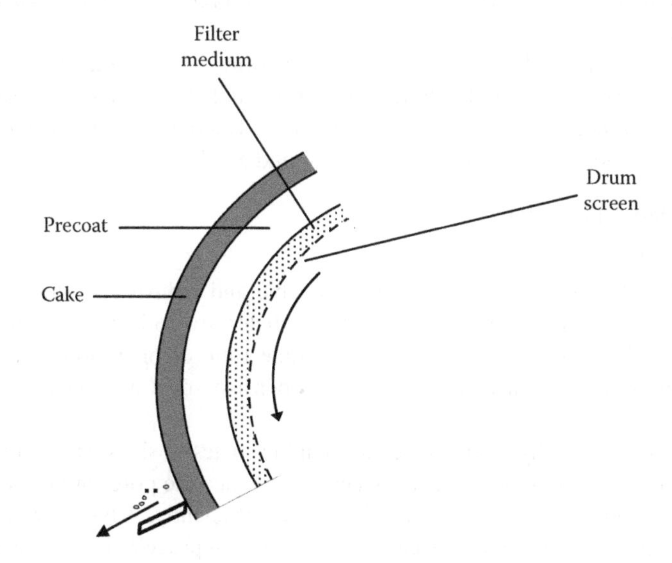

FIGURE 7.28 Principle of rotary vacuum filtration with precoat.

The filter consists of a horizontal drum 0.3–3 m in diameter and 0.3–4 m long, partially submerged in a trough containing the suspension, and rotating slowly at 0.1–2 RPM. The surface of the cylinder is made of a metallic screen and it is covered with a filter medium acting as the filtration surface. Vacuum is applied to the interior of the drum and filtration takes place under a constant pressure drop, which is equal to the atmospheric pressure minus the pressure in the filter. A cake of particles is formed on the filter surface, while the filtrate is collected inside the filter, from where it is removed by a special pump. The cake is removed continuously from the filtering surface by scraping with a "doctor" knife before the filtration cycle is repeated.

Filter aids, such as diatomaceous earth, are used to precoat the filtering area with a 5–16 cm layer before filtration of colloidal suspensions. During filtration, the particles form a cake layer on the precoat, which is removed continuously by a slowly advancing "doctor" knife. Part of the precoat is removed continuously together with the product cake; it becomes necessary to precoat again the filter after some period of operation.

7.7.4.3.3 Pressure Coat Filters

The pressure coat filters are used when pressure drops more than those obtained in rotary vacuum filters are required. They include pressure leaf filters and cartridge filters. The pressure leaf filters consist of a horizontal pressure vessel, containing a series of parallel leaf filters, made of perforated metal or metallic screens, which act as coated filters. The liquid suspension is forced by air pressure though the leaves and the clean product is collected in a manifold.

Cartridge filters are small and inexpensive units that are used for the clarification of relatively small volumes of industrial liquids containing low concentrations of solid particles. Filter cartridges consist of tubes 6–8 cm in diameter and 10–120 cm in length, with a cylindrical filtration surface. The cartridges are placed in pressure housings and the liquid to be filtered is forced through the cartridge by air pressure.

Two types of cartridges are used: (a) throwaway (expendable) filters made of woven fibers, like cotton and synthetic materials and (b) cleanable (reusable) cartridges made of porous ceramics or stainless steel. Liquids, cleaned with cartridges include boiler and cooling water, mineral oils, and alcohols.

7.7.4.4 Depth Filtration

7.7.4.4.1 Sand Filters

Depth or bed filters are used in cleaning potable and industrial water from small concentrations of small solid particles. The particles are collected within the mass of the bed, which should be cleaned after some time of operation. Larger particles are removed previously by some less expensive separation process, such as sedimentation.

Filter beds are usually made of cleaned sand particles of sizes 0.6–1.2 mm. Close size distribution of the sand particles is necessary since the pores are more uniform and they can collect the suspended particles more efficiently. A bed of wide particle size distribution would be blocked early in the filtration process, increasing the pressure drop and reducing the filtration rate.

Gravity bed filtration is commonly used in water filtration with sand beds of 70–80 cm depth on a layer of gravel 15–25 cm deep. Filtration takes place usually at a constant pressure drop and the filtration rate decreases gradually with time.

The filtration rate in sand filters is in the range of 4–10 m^3/m^2 h and the maximum operating pressure drop is 2 m of water (0.2 bar). When the maximum pressure drop is reached, the filtration is stopped and the sand filter is cleaned by backwashing with water. Clean water is forced from the bottom through the bed at a high flow rate (30 m^3/m^2 h) dislodging the collected particles and carrying them out of the system. Bed washing is facilitated by simultaneous blowing of compressed air.

7.7.4.4.2 Dual Media Filters

Dual media filters, consisting of a coal bed on top of a sand bed, are used in the filtration of potable and industrial water. The filters are enclosed in pressure vessels and they are operated in a similar manner with the gravity sand filters. The carbon filter removes by adsorption the undesirable odors and dissolved chlorine.

The filter bed consists of a carbon layer 0.25–0.50 m on a sand layer of 0.15–0.30 m, supported on a layer of gravel and an underdrain plate. High filtration rates are obtained (10–40 m^3/m^2 h) with maximum pressure drops in the range of 0.8–1.7 bar.

7.7.4.5 Sterile Filtration

Sterile filters are used in the laboratory and the plant for the removal of microorganisms from various liquids, which are too sensitive for thermal sterilization. Microorganisms of sizes 0.5–10 μm (bacteria to fungi) can be removed with membrane filters of known openings and porosity. Typical sterile filters (Millipore) are made of cellulose membranes, 130 μm thick, with openings 0.22 μm and porosity 0.75. They are operated at pressure drops of 1–4 bar and temperatures up to 120°C. For viruses (<0.1 μm), or for effective sterilization, fine filters or two-stage sterile filtration may be required.

7.7.4.6 Filtration of Clear Food Liquids

Filtration of clarified fruit juices (e.g. apple and grape), wine, and beer is applied widely to remove various small particles and colloids, which may precipitate during storage and affect product quality. Normal cake filtration is difficult because the filter cake formed is compressible, resulting in reduced filtration rate and increased pressure drop. Precoating the filter surface with a filter aid, which forms a porous layer, reduces the flow resistance.

A precoat layer of about 0.5–1.0 kg/m^2 is formed on the filtration surfaces of plate, leaf, or disc filters, by filtering a water slurry of 0.3%–1.0% of filter aid (diatomaceous earth, perlite, or cellulose-paper), at about 20 L/m^2 for about 30 min. Filtration is improved by adding continuously a small amount of filter aid to the juice during filtration. Filtration is stopped when the pressure drop and the filtration rate reach preset limits.

7.7.4.7 Centrifugation

Centrifuges are expensive equipment for efficient mechanical separations used in both sedimentation (separation) and filtration applications of solid/liquids and liquid/liquids.

7.7.4.7.1 Centrifugal Separators

Centrifugal sedimentation is based on the application of the centrifugal force to separate particles and liquids of different size and density.

The Stokes equation for settling velocity u (m/s) of particles of size d (m) in a centrifugal field of rotational speed ω (1/s) at a distance r (m) from the center of rotation is written as

$$u = \frac{[\omega^2 r(\rho_s - \rho)d^2]}{(18\eta)} \tag{7.31}$$

where ρ_s and ρ (kg/m³) are the densities of the particles and the liquid, respectively. The liquid flow through a centrifuge is considered as a plug flow with a residence time $t = V/Q$, where V (m³) is the holdup volume and Q (m³/s) is the flow rate. The time t_c (s) required to remove 50% of the particles (of "cut diameter" d_c) will be $t_c = z/(2u)$, where z (m) is the thickness of the liquid in the centrifuge. The flow rate to remove 50% of the particles will be $Q_c = 2u\ V/z$. Substituting the settling velocity u from Equation 7.31, the last relation becomes

$$Q_c = \left[\frac{\omega^2 r(\rho_s - \rho)\, d_c^{\,2}}{(9\eta)}\right]\left(\frac{V}{z}\right) \tag{7.32}$$

Equation 7.32 is equivalent to

$$Q_c = 2u_g\,\Sigma \tag{7.33}$$

where

u_g (m/s) is the gravity settling velocity of the particles (diameter d_c)

Σ (m²) is a characteristic parameter of the system, equivalent to the cross sectional area of a gravity settling tank, which has the same settling capacity with the specific centrifuge

Combining Equation 7.33 with the Stokes (gravity) equation (7.1), the "Sigma" parameter Σ is calculated from the relation

$$\Sigma = \frac{V\omega^2 r}{gz} \tag{7.34}$$

Equation 7.34 for cylindrical centrifuges becomes

$$\Sigma = \frac{\pi b\omega^2\left(3r_2^2 + r_1^2\right)}{2g} \tag{7.35}$$

where

r_1 and r_1 (m) are the distances of the internal and external surfaces of the liquid from the center of rotation

b (m) is the length of the active cylinder

For disc centrifuges the following equation is used:

$$\Sigma = \frac{\left[2\pi\omega^2(N-1)\left(r_2^3 - r_1^3\right)\right]}{(3g\tan\theta)} \quad (7.36)$$

where
N is the number of discs
r_1 and r_2 are the internal and external radii
2θ is the cone angle of the discs

Estimated values of the Σ parameter of industrial centrifuges vary from 300 to 1000 m². It is evident that the separating capacity of industrial centrifuges is quite high, for example, equivalent to a gravity settling tanks of cross section 300–1000 m² (equivalent tank diameters 20–35 m).

Cylindrical and cone and disc centrifuges are used for separation of particles and liquids of different densities. They are combined in efficient continuous centrifuges, such as the screw settler of Figure 7.29. Cylindrical or tubular (bowl) centrifuges are used in edible oil processing and in clarification of fruit and vegetable juices, and sugar syrups.

Disc centrifuges (Figure 7.30) are used in dairy processing (separation of cream from milk) and in the clarification of various liquids, such as fruit juices. They consist of a centrifugal bowl 20–50 cm diameter with a series of cone discs. Perforated

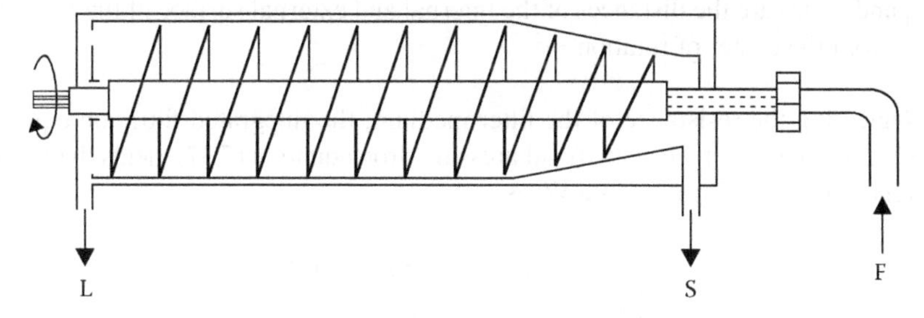

FIGURE 7.29 Cylindrical and cone centrifuge. F, feed; L, liquid; S, solids.

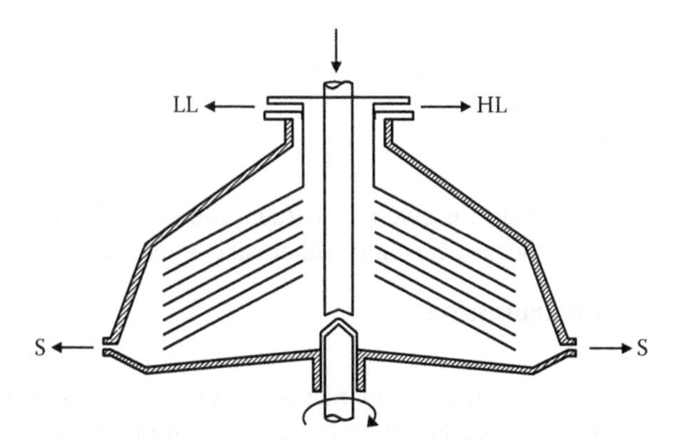

FIGURE 7.30 Disc centrifugal separator. LL, light liquid; HL, heavy liquid; S, solids.

discs are used to facilitate the centrifugal separation of liquids of different density. The liquid feed enters at the center of the bowl and it is separated by centrifugal force into a light and a heavy stream, which are removed separately with special piping. Nozzle-discharge centrifuges are used when significant amounts of solid particles settle in the centrifugal field. They have small openings at the bottom sides of the bowl through which the settled particles are removed continuously.

7.7.4.7.2 Filtering Centrifuges

Filtering centrifuges are used in food processing to separate solid particles from water suspensions, for example, in recovering sugar crystals from a crystallizer. The filter consists of a horizontal or vertical basket with perforated or wire mess wall, which is rotated at high speed. The suspension is fed in the center of the basket and the solid particles are forced to the walls, forming a cake, through which filtration takes place, like in the normal pressure or vacuum filters.

The pressure drop through the filter cake in a rotating basket is given by the equation

$$\Delta p = \frac{\rho\omega^2\left(r_2^2 - r_1^2\right)}{2} \tag{7.37}$$

where

- ρ (kg/m^3) is the density of the particles
- ω (1/s) is the rotation speed
- r_1 and r_2 (m) are the distances of the internal and external surface of the cake ring from the center of rotation

Neglecting the resistance of the filter medium, the integrated filtration equation (7.30), combined with the centrifugal pressure drop equation (7.37) yields the following relation for the filtration rate V/t as

$$\left(\frac{V}{t}\right) = \frac{\left[\rho\omega^2\left(r_2^2 - r_1^2\right)\right](A_a A_L)}{(2\eta R_m)} \tag{7.38}$$

where

- $A_a = 2\pi\, b(r_1 + r_2)$ and $A_L = [2\pi\, b\,(r_2 - r_1)]/[\ln(r_2/r_1)]$
- b (m) is the length of the basket
- $m = CV$ (kg) is the cake mass

Equation 7.38 shows that, for given centrifugal filtration, the filtration rate will decrease as the mass of the particles in the filter cake m increases with time.

7.7.4.8 Mechanical Expression

7.7.4.8.1 Introduction

Mechanical expression is used in the extraction of juices and oils from fruits, vegetables, and oilseeds. They are based on the application of pressure to disrupt the plant cells and release the contained juice or oil constituents. The by-products of mechanical expression are solid residues such as pomace or peels, which are either processed into animal feeds or they are disposed in the land.

Solvent extraction (Chapter 13) may be required to recover significant amounts of residual solutes from mechanically expressed materials, such as oilseeds.

The expression process depends on the (a) applied mechanical pressure, (b) yield stress of the food material, (c) porosity of the cake formed, and (d) the viscosity of the expressed liquid.

Mechanical expression is a complex physical process, which is treated by the empirical equation

$$\log(p) = \frac{k + k'}{V_c} \tag{7.39}$$

where

p (bar) is the applied pressure
V_c (m^3) is the volume of solids
k, k' are empirical constants

Mechanical expression of a liquid from a solid matrix is possible only if the solid is compressible. In incompressible solids, mechanical pressure cannot express the liquid, and other separation methods must be used.

7.7.4.8.2 Batch Presses

The material to be expressed is wrapped in a canvas (cotton) cloth and is placed and compressed in a series of steel boxes, fitting the fixed and moving heads of a vertical hydraulic press. The material is enclosed in a cylindrical pot, with fiber pads or screens in the bottom and on the top, and it is compressed with a hydraulic ram entering from above. The pot press can handle fluid materials. It is used, for example, for oil expression from olives and separation of cocoa butter from chocolate. Pot load per cycle is about 250 kg and the final expression pressure can reach 400 bar.

7.7.4.8.3 Continuous Presses

Continuous presses are used widely for the expression of fruit juices and oils from various oilseeds. They are preferred over the batch presses because they require less labor and they are more efficient in processing large volumes of material.

7.7.4.8.3.1 Screw Presses The screw press is the most popular equipment because of its many advantages. It consists of a horizontal or vertical rotating screw, fitting closely inside a slotted or perforated curb (frame). Both screw and curb are tapered toward the discharge to increase the pressure on the material. The pressure can also be increased by varying the pitch of the screw. As the material is pressed by the screw, the liquid escapes through the openings of the curb. Shaft speeds of 5–500 RPM are used with very high pressures, up to 2500 bar. The capacity of the screw presses can reach 200 ton/24 h and the residual oil in the press cake can be as low as 2%.

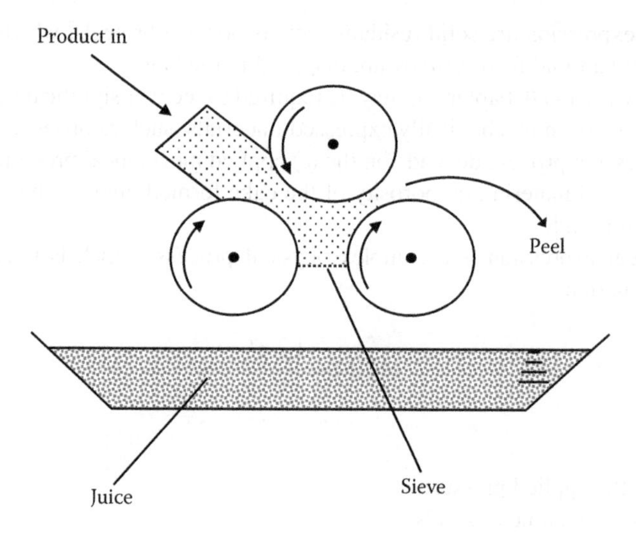

FIGURE 7.31 Roller press.

7.7.4.8.3.2 Roller Presses Continuous roller presses are used principally for expressing juice from fibrous food materials such as sugar cane and some fruits. They consist normally of three rollers, which squeeze the material as it is forced to pass between them successively. The rolls are made of cast iron and they are corrugated or grooved in various patterns (Figure 7.31).

7.7.4.8.3.3 Belt Presses The belt press combines the filtering and expression actions in one continuous operation. The solids are enclosed between two serpentine belts and they are pressed gradually by a series of rolls, forcing the liquid out. The pressure developed is relatively low and expression is confined to easily removed solutes such as fruit juices. A typical press of 60cm wide belt can process 3.5–5 ton/h of apples into juice.

7.7.4.8.4 Fruit Juice Expression

Pressing or expression is an important operation in fruit juice processing since it is related immediately to both economics (yield) and quality (composition) of the product. Expression equipment for three different types of juices, apple, citrus, and grapes, has particular commercial importance (Nagy et al, 1993).

7.7.4.8.4.1 Citrus Juice Extractors The simple home reamers are used industrially in expressing orange and other juices. The fruit is sliced with a sharp knife and the rotating reamers extract the juice and pulp. The reamers are the basic elements of the industrial Brown citrus extractor.

In the FMC extractor the juice is extracted from the whole fruit, without halving and reaming, as shown diagrammatically in Figure 7.32. The fruit is placed into the lower extraction cup and the upper cup descends pressing the fruit, while a circular cutter below cuts a bore, which is removed from the bottom. The fruit is squeezed and the expressed juice is separated from the fruit residue (seeds, rag, and peel fragments)

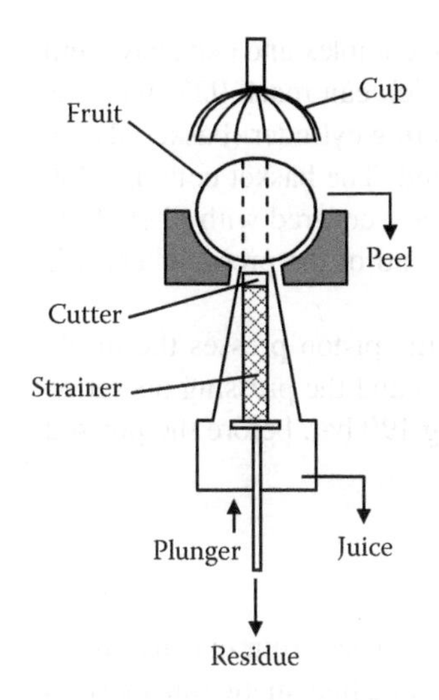

FIGURE 7.32 FMC citrus juice extractor.

through the small-diameter strainer. The juice is collected in the cup below and the residue (pulp) inside the strainer is discharged from the bottom through the orifice tube (plunger) (Kimball, 1999).

7.7.4.8.4.2 Expression of Apple and Grape Juices Expression of juice from apple fruits can be accomplished by quick pressing, so that the juice is removed from the slowly moving bulk material, allowing an exit path to remove the expressed juice. The pressing operation is affected by the fruit quality, ripe fruits yielding the best quality and quantity of juice. The fruit pieces and particles should be of the proper size (not too small or too large). Special grinding mills with knives are used (McLellan, 1993).

Pectolytic and cellulose or starch splitting enzymes are added to the fruit mash, facilitating juice expression. Enzyme pretreatment should be done at the optimum pH, temperature, and time.

Press aids help juice expression by increasing the permeability of the fruit mash (cake). Common press aids are mixtures of wood fibers, paper fibers, and rice hulls. The proportion of each of these press aids added to the fruit mass is about 3%–4%.

Water is usually added to the press cake to dissolve the residual solids and pressing is repeated, increasing the overall yield of juice.

Grapes are prepared for pressing in a stemmer/crusher, which is a rotating drum with perforations of 2.5 cm. The grapes are removed from the stems and are crushed by passing through the holes, while the stems are discharged from the center of the drum.

The crushed grapes are pressed between two rotating cylinders to express the juice. Grape pigments, for example, from red grapes, can be extracted into the juice by heating the crushed grapes to about 60°C.

Expression equipment, used in apple and grape juice processing, includes screw presses, belt presses, and screening centrifuges, discussed earlier in this section. In addition, the following special presses are used.

7.7.4.8.4.3 Rack and Frame Press "Cheese" cloths, containing fruit mash 5–8 cm thick, are stacked and pressed under a hydraulic ram, forcing the juice out. At the end of pressing, the cake, about 1 cm thick, is removed from the cloths and the operation is repeated.

7.7.4.8.4.4 Willmes Press The Willmes (bladder) press is used mostly in grape juice and wine processing. It is a pneumatic system, consisting of a perforated, rotating, horizontal cylinder with an inflatable rubber tube in the center. The cylinder is filled with grape mass and the air bag compresses the material forcing the juice out. The bag is then collapsed and the cylinder retracted. The rotation and pneumatic compression of the mash is repeated many times with increasing pressure.

7.7.4.8.4.5 Bucher Press The Bucher press is a large complex and expensive unit processing 5–7 ton/h of fruit with yield about 85%, which can reach 92% with the use of enzymes and leaching. The unit consists of a rotating cylinder (basket) 2 m in diameter and 2 m long with a hydraulic piston at one end. The basket contains 280 small filter elements, which are flexible, grooved openings, covered with filter cloth. Juice flows through the cloth, down the grooves to the end of the press, where it is collected.

The fruit mash (e.g. apple) is added to the basket, the piston presses the mash, and it is retracted. Then a new amount of mash is added and the pressing operation is repeated, until a high pressure is developed, reaching 190 bar, before the pressed mash is washed and dumped from the basket.

7.7.5 Solid/Air Separations

Solid/air separators are used in the recovery of solid particles from exhaust air in various food processing operations, such as spray drying and pneumatic transport, in reducing air pollution from industrial air effluents, and in cleaning the atmospheric air in food processing plants. The industrial air exhaust streams may contain high particle concentrations (up to 45 g/m^3), while the concentration of particles in the atmospheric air is less than 1 mg/m^3. The size of particles ranges from 1 to 1000 μm.

Solids/air separators are similar to the air classifiers (Section 7.2), that is, the hydrodynamic, centrifugal, and gravity forces. In addition, the electrostatic forces play a significant role in the separation of the small-sized particles.

7.7.5.1 Cyclone Separators

Cyclones are simple and inexpensive units, which can remove effectively solid particles (and liquid droplets) larger than 10 μm from industrial gases and air. Figure 7.33 shows the flow pattern and the dimensions of a standard industrial cyclone separator.

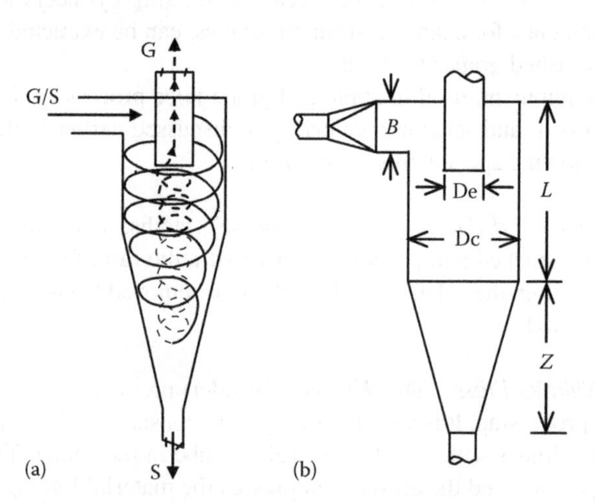

FIGURE 7.33 (a) Flow pattern in cyclone separator. (b) Standard cyclone dimensions. S, solids; G, air; Dc, cyclone diameter; $L = Z = 2\text{Dc}$; $B = \text{De} = \text{Dc}/2$.

The particle/air mixture enters the cyclone tangentially following a spiral flow pattern from top to bottom, a vortex flow from bottom to top, and an exit from the collector. The solid particles are subjected to centrifugal force and thrown to the cyclone walls from which they fall and are collected at the bottom. The air exits from the top of the cyclone and it may contain significant amounts of small-sized particles, which are collected by bag filters or wet scrubbers, installed after the cyclone.

In general, small diameter cyclones (about 25 cm diameter) are used in practice, because they are more efficient in removing the relatively small-sized particles.

The centrifugal force, developed by the cyclone spiral flow, is very large, due to the small cyclone diameter, reaching up to 1000 times the gravitational force. The separation efficiency of the cyclone is characterized by the cut diameter (d_c) of the particles, which is defined by the equation

$$d_c^2 = \frac{9\eta B}{2\pi N\rho u} \tag{7.40}$$

where

B (m) is the entrance width
N is the number of spiral "turns" of the cyclone
ρ (kg/m^3) is the particle density
u and η are the air velocity (m/s) and viscosity (Pa s), respectively

For normal cyclone collectors, $N = 5$ and entrance air velocity $u = 15$ m/s. The cut diameter corresponds to a collection efficiency of 50%.

The cut diameter is reduced and, therefore, the efficiency of the cyclone collector is increased by increasing the air velocity and/or the particle density or by reducing the width of the cyclone entrance.

The efficiency of a cyclone collector is estimated from the Lapple diagram (Appendix, Figure A.5) as a function of the particle size ratio d/d_c. The efficiency drops sharply for small particles to lower than 10% at $d/d_c = 0.3$ and exceeds 90% at $d/d_c = 3$.

Installation of another more efficient collector after the cyclone, such as a bag filter, may be necessary in order to collect most of the escaping particles and discharge a clean air stream into the environment.

A number of small diameter collectors, operated in parallel, will be required in order to handle large volumes of industrial gas (air) streams. The collectors are usually installed in parallel in compact structures (multiclones).

The air fan is installed either before (pressure) or after (suction) the collector. The suction installation is preferred, because the exhaust air from the cyclone will be free of the large particles, which might damage the fan rotor.

7.7.5.2 Bag Filters

Bag filters are usually made of woven cloth or felt, which act as surface filters. Depending on their size (d), the particles are collected mainly by inertia ($d > 1\,\mu$m) or electrostatic forces ($d < 0.5\,\mu$m). In bag filters, the particles form a mat on the surface, which acts as a filter medium, increasing the efficiency of filtration and the

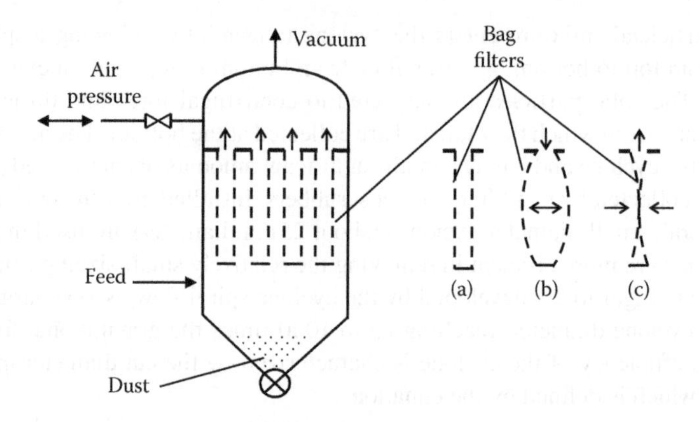

FIGURE 7.34 Bag filter: (a) normal position; (b) and (c) filter shaking for cleaning.

pressure drop. Filtration is interrupted when the pressure drop exceeds a preset limit, and the filter is cleaned (Figure 7.34).

The pressure drop through a bag filter is given by the empirical equation

$$\Delta p = K_c \eta u + K_d \eta w u \tag{7.41}$$

where

K_c (1/m) is the fabric resistance coefficient
K_d (m/kg) is the particle layer resistance coefficient
u (m/s) is the superficial air velocity
w (kg/m^2) is the particle loading of the filter
η (Pa s) is the air viscosity

The particle layer resistance coefficient K_d is related to the particle diameter d, the particle shape factor, φ and the porosity ε of the fabric by the Carman–Kozeny equation

$$K_d = \frac{160\,(1-\varepsilon)}{(\varphi^2 d^2 \varepsilon^3)} \tag{7.42}$$

The fabric resistance coefficient K_c is related to the pressure drop through the fabric Δp_c, according to the equation

$$K_c = \frac{\Delta p_c}{\eta u} \tag{7.43}$$

The pressure drop through the particle layer Δp_d is given by the following equation, which is derived from Equation 7.41:

$$\Delta p_d = K_d \eta C u^2 t \tag{7.44}$$

where

C is the particle concentration in the air (kg/m^3)

t is the filtration time (s)

The fabrics used in bag filters are made of cotton, wool, nylon, dacron, and teflon. The bag filters have diameter 12–20 cm and length 2.5–5.0 m, and they are often assembled in compartments, called "bag houses," of 100–200 m^2 cloth surface.

The bag filters are cleaned either by shaking or by reverse flow of air. Shaking may be periodic or after a preset pressure is built up determined by using a differential pressure instrument. Cleaning may be necessary when the pressure drop reaches 5–15 mbar. Superficial air velocities (air filtration rates) for woven filters are 0.5–2 m/min and for felt filters 25 m/min.

7.7.5.3 Air Filters

Air filters are used to clean atmospheric air from small particles ($d < 0.5\,\mu m$) and produce very clean air mainly for the pharmaceutical, biotechnological, semiconductor, and nuclear industries. They are also used in clean room technology (hospitals and some advanced food processing industries).

Concentration of particles in atmospheric air is normally lower than 12 mg/m^3, which is much lower than the particle concentration in industrial gas (air) streams. Air filters are essentially deep-bed filters, made of porous cellulose materials. The mechanism of deep air filtration includes mechanical sieving and electrostatic forces. Two types of fibrous filters are normally used: (1) viscous filters, in which the filter medium is coated with a viscous mineral oil, which retains the dust. The used filters are cleaned periodically and returned to service and (2) dry filters, which are cheaper, are made up of cellulose pulp, cotton, or felt, and are discarded after use.

Size of normal filters is 0.5 m × 0.5 m, which can handle up to 4500 m^3/h at superficial air velocities of 1.5–3.5 m/s. They have a collection efficiency higher than 90%, which can be increased at higher air velocities. The operating cycles of air filters are about 1 week for the dry and 2 weeks for the viscous.

7.7.5.4 Electrical Filters

Electrical filters are used to remove small particles from industrial gases and atmospheric air based on electrical charging of the particles, followed by collection on charged electrodes. Two types of filters are used: (1) electrical precipitators and (2) positively charged filters.

The electrical (or Cottrell) precipitators are large industrial installations used mainly in the chemical process industries and in power generating stations to remove various particles and fly ash from gaseous effluents, reducing air pollution. The particles, charged negatively from ionized gases, are collected in large positive plate electrodes, operated at about 50 kV, with efficiencies about 90%. The gas velocity in the electrodes is about 2–3 m/s and the precipitation (migration) velocity of the particles ranges from 5 to 15 cm/s. Electrical precipitators are complex installations and they have a high investment and maintenance cost.

Positively charged or two-stage precipitators are relatively small units, used mainly to clean atmospheric air from dust, smoke, and other particles, often as part of air conditioning systems. The particles are charged positively by dc electrodes at about 13 kV and then collected on negative (grounded) electrodes operated at 6 kV. The collection efficiency is about 85%–90% and the filters should be cleaned, depending on particle loading, every 2–6 weeks.

7.7.5.5 Wet Scrubbers

Wet scrubbers, or wet particle collectors, are used to clean industrial gases and air from small solid particles that escape simpler separators, such as cyclones. The main collection mechanism is inertial deposition of the particles on the liquid (water) droplets. Wet scrubbers are also used to absorb various gases from air streams in connection with air pollution control.

The absorption of gases and vapors in liquids is basically a mass transfer operation, such as distillation and solvent extraction, which is analyzed in Chapter 13. The particulate scrubbers consist of two parts: (a) the contactor stage and (b) the entrainment stage. The entrained sprays and deposited particles are removed from the cleaned gas (air) by cyclone or impingement separators.

The wet scrubbers are divided into two classes: (a) low energy equipment, which includes spray towers, packed towers, and cyclone scrubbers, which can remove particles larger than 1 μm and (b) the high energy units, which include the Venturi and jet scrubbers, which can remove particles smaller than 1 μm.

Wet scrubbers are used extensively in the chemical process industry and in some food processing plants in connection with anti-pollution systems for cleaning exhaust gases and air from undesirable gases, for example, odorous compounds in the refining of edible oils. An important disadvantage of wet scrubbers is the production of a stream of polluted wastewater, which must be treated with some wet separation method before it is discharged into the environment.

7.7.6 Removal of Food Parts

7.7.6.1 Introduction

Food related separations include the removal of product-own parts (e.g. cherry stems or cherry stones) and the removal of foreign parts (e.g. dust, insects). Separation operations of product-own parts include the recovery of separated materials or substances (e.g. juice from fruit, sugar from beets), or the removal of undesired material or substances (e.g. fruit peels and filtrate residues).

In the removal of undesired external parts, dry, wet, or mixed methods are used. Dry methods include knife peeling of onions and apples, brushing of oranges, cutting and de-boning of meat, and burning of chicken hair after plucking. Wet methods include washing of vegetables and peeling of peaches in lye solutions. Mixed methods include the removal of potato peels by abrasion, just after steaming, the removal of chicken feathers by beating with rubber strap-wheels, just after scalding, the suction of blood after slaughtering, and the removal of corn seeds after soaking. Mechanical wet and dry separations in foods include cleaning operations, which remove undesired foreign parts from foods (Luh and Woodroof, 1988).

7.7.6.2 Removal of Undesired Food Parts

7.7.6.2.1 Animal Products

Special mechanical operations are used for the separation of external and internal parts of animal products (meat and seafood). They are designed to remove bones (deboning), meat strings, fish fins, hair from pigs and chicken, feathers from chicken, skins from animal and fish, shells from mussels, and blood from animals.

Typical operations, used in food processing are: skinning of animals, skinning of fish, dehairing of pigs and poultry, cutting of hoofs, etc., screening of meats from bones and fibers, removal of mussel shells, and removal of blood from slaughtered animals.

7.7.6.2.2 Plant Products

Removal of undesired own parts of plants includes external parts (shells, hulls, stems, and peels) and internal parts, such as coring the seeds of fruits, for example, of apples or pears.

Typical examples are the breaking of nuts; the dehulling of onions, grains, and rice; the brushing and polishing of fruits; the destemming of fruits; the pitting of stone fruits, coring and scooping, cutting and slicing; and peeling (mechanical or chemical).

7.7.6.3 Removal of Desired Food Parts

The goal of separation is the recovery of certain food parts. Internal food parts are the main goal of separation operations, such as juices (fruit and vegetable), oil (seeds, olives), fat (animal, milk), starch (cereals), and nuts. Several animal by-products, such as collagen (jellies), bones (feedstuffs), and intestines (sausages), can be also utilized.

7.7.6.4 Food Cleaning Operations

The cleaning processes refer to the separation and removal of external undesired material that either adheres on food or on food equipment. In cleaning of equipment, lye and acid solutions, and several sophisticated chemicals are used. Automatic cleaning, such as the clean-in-place (CIP) system is preferred. Cleaning is usually followed by rinsing and washing with detergents.

For raw food, cleaning is an important preprocessing operation, which removes foreign materials and contaminants. The cleaning operation should neither waste a large portion of the product nor pollute the environment. Cleaning from heavy foreign materials such as stones and metal pieces is necessary for protecting size reduction and milling equipment used in processing operations.

7.7.6.4.1 Wet Cleaning

Most of the cleaning methods use water as a cleaning medium. Wet cleaning is effective in removing firmly adhering soils from raw fruits and vegetables allowing the use of detergents and sanitizers. Water conservation methods should be used (e.g. recirculation) to reduce the large amounts of water needed in some processing operations, such as in the canning of fruits and vegetables, where up to $15\,m^3$/ton of product may be used.

In wet cleaning operations, a relative motion between the cleaning fluid and the product is applied, which is achieved by movement of the cleaning fluid, movement of the product, or by movement of both. In soaking, the cleaning medium diffuses and separates the undesired dirt.

The main operations of wet cleaning are soaking, spraying, and mixed systems, in which two or more cleaning systems are involved. Soaking of raw materials in long tanks removes heavy contaminants, such as stones and adhering soil. Mechanical paddles or air currents at the bottom of the tank are used for effective mixing. Detergents may be used to remove the spray residues of agrochemicals from the surfaces of the fruits and vegetables, and chlorination of water may be required to prevent the growth spoilage microorganisms.

Spray washers use less water and they are more efficient, due to mechanical action, than water soakers. Effective spraying nozzles are used, which are replaced when worn out. In belt-type spray washers, the raw material is transported slowly on rollers or vibratory conveyors under water spays. Brushes and special rubber discs can remove the adhering dirt and contaminants from the sound product. Figure 7.35 shows a mixed wet cleaner. Simple flotation in water may be used to separate bruised or rotten fruits or vegetables from sound products. Froth flotation can be used to separate foreign materials of the same size and density with the main product, for example, green peas. The raw material is immersed in an emulsion of mineral oil/detergent, through which air is blown. The contaminants float at the surface and the product is separated from the bottom.

7.7.6.4.2 Dry Cleaning

Dry cleaning by an air stream is based on the same aerodynamic principles discussed in Air Classifiers (Section 7.7.3.2.1). The most common operation of aspiration (winnowing) removes light materials (skins, leaves, etc.) from heavier food pieces, for example, onions, peas, and beans.

Screening, discussed under solid/solid separations, is used widely in dry cleaning of various food pieces and particles, such as grains and seeds. In a three-scene setup, the top screen (scalper) removes the largest pieces/particles, the second screen collects the main product, while the bottom pan collects the undesirable products, the soil and the debris. Depending on the shape of the grains, round, triangular, or

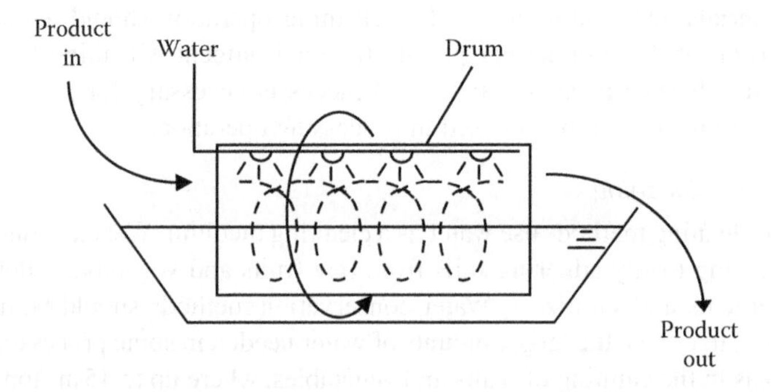

FIGURE 7.35 Mixed soak and spray washing equipment for fruits and vegetables.

slotted holes can be used. Pneumatic separators or fanning mills consist of a set of screens and a fan for moving air though the grain, which removes chaff, dirt, and lightweight weed. A blowing or suction fan, that is, an aspiration system, is used.

Combined cleaning methods are used for thorough separation of some raw materials, for example, wheat before milling: The wheat goes through a series of separations to remove various contaminants and separate the oversize and undersize fractions, for example, magnetic separation, screening, disc separation, washing, centrifugation, and drying. Brushing is applied in the dry cleaning of some fruits and vegetables.

APPLICATION EXAMPLES

Example 7.1

Particle size analysis of a sample of wheat flour gave the following results:

x, particle size (μm) 2, 3, 5, 10, 20, 30, 50, 100, 200
R, larger than x (%) 95, 92, 88, 75, 56, 47, 28, 12, 3

It is assumed that the particle size follows both the log-normal and the Rosin–Rammler distributions.

(a) Plot the data on log-normal distribution coordinates of R (probability scale) versus log(x) and determine the empirical constants.
(b) Plot the data as log[log(1/R)] versus log(x) and estimate the Rosin–Rammler constants.

Solution

The data of Table 7.11 are plotted in Figures 7.36 (log-normal distribution) and 7.37 (Rosin–Rammler distribution).

The characteristic constants of the log-normal distribution are estimated as shown in Figure 7.36. The mean size of particles is estimated at $\%R = 50$, corresponding to point (A), at which $\log x = 1.31$ and $x = 20\,\mu m$. The standard deviation (s) is estimated from the relation, $s = [\%R(16)]/[\%R(50)]$. The $\%R(16)$ corresponds to point (B) on the diagram, at which $\log x = 1.87$ and $x = 74$. Thus, $s = 74/20 = 3.7$. It should be noted that in the log-normal distribution, the standard deviation s is a dimensionless quantity.

In the Rosin–Rammler distribution, the uniformity index (n) of Equation 7.7 is estimated as the slope of the straight line of Figure 7.37, $n = [\log \Delta y]/[\log \Delta x] = 1$.

The characteristic constant (x') of the Rosin–Rammler equation is estimated at $\%R = 37$ or $R = 0.37$, on the diagram $\log[\log(1/0.37)] = -0.37$, $\log x' = 1.63$ and $x' = 43\,\mu m$ (Table 7.12).

Example 7.2

An aqueous suspension of starch containing 0.1 kg solids/kg water is filtered through a rotary vacuum filter at a filtrate rate of 1 tons/h. The filter has a diameter of 1 m and length 1 m, it is immersed in the liquid trough by 20% and it rotates at 10 turns/h.

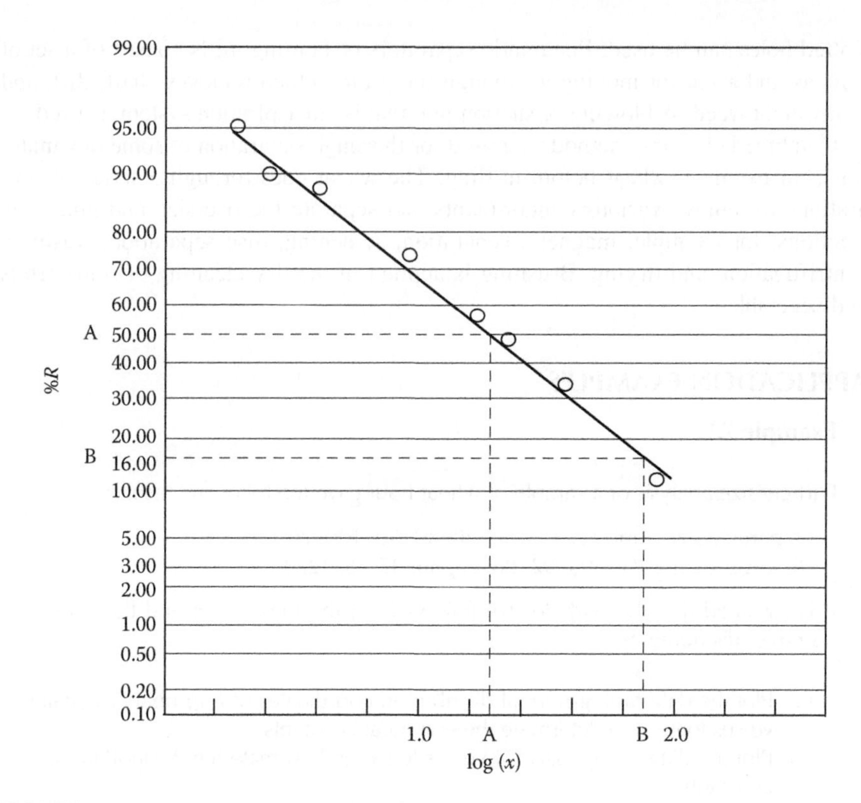

FIGURE 7.36 Log-normal distribution of wheat flour particles.

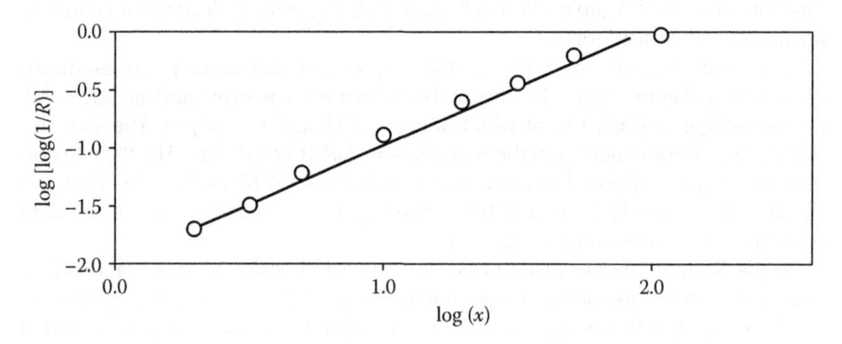

FIGURE 7.37 Rosin–Rammler distribution of wheat flour particles.

The vacuum filter operates at an absolute pressure of 0.5 bar, the atmospheric pressure is 1 bar, and the resistance of the filter medium is negligible. Viscosity of the liquid suspension is 10 mPa s.

Calculate: (a) The mean thickness of the cake formed, if it has a bulk porosity of 0.5 and particle density 1200 kg/m^3. (b) The specific resistance of the filter cake.

TABLE 7.12
Calculated Values Obtained from the Given Data

x	2	3	5	10	20	30	50	100	200
log(x)	0.30	0.48	0.70	1.00	1.30	1.48	1.70	2.00	2.30
R	0.95	0.92	0.88	0.75	0.56	0.47	0.28	0.12	0.03
1/R	1.05	1.07	1.14	1.33	1.78	2.13	3.57	8.33	33.33
log[log(1/R)]	−1.7	−1.5	−1.2	−0.91	−0.60	−0.48	−0.23	−0.036	−0.18

Solution

(a) The surface area of the rotary filter will be $A = 3.14 \times 1.0 \times 1.0 = 3.14\,\text{m}^2$. The filtration area will be $A_F = 0.2 \times 3.14 = 0.628\,\text{m}^2$. Mean pressure drop in the filter cake $\Delta p = 1.0 - 0.5 = 0.5$ bar.
The mass of filter cake will be $m = 0.1 \times 1000 = 100\,\text{kg/h}$.
The total volume of the filter cake formed will be $V_c = 100/(1200 \times 0.5) = 0.167\,\text{m}^3\text{/h}$. The cake volume formed per revolution will be $V_c = 0.167/10 = 0.0167\,\text{m}^3$/revolution. Therefore, the mean thickness of the cake on the filter will be $x = V_c/A = 0.0167/3.14 = 0.0053\,\text{m} = 5.3\,\text{mm}$.

(b) The filtration time (t) to obtain 1000 kg or 1 m³ of filtrate (water) is given by the equation

$$t = (\eta\, R\, C/2\Delta p)\,(V/A)^2, \text{ from which } R = (2t\,\Delta p\, A^2)/(\eta C V^2)$$

where
$t = 3600\,\text{s}$
$\eta = 1 \times 10^{-2}$ Pa s
R = cake resistance (m/kg)
C = 100 kg solids per m³ of filtrate
$\Delta p = 0.5\ \text{bar} = 5 \times 10^4$ Pa pressure drop through the filter
$V = 1\,\text{m}^3$ filtrate volume
$A = 0.2 \times 3.14 = 0.628\,\text{m}^2$ filtration area (20% of the total filter area)

Thus, $R = [2 \times 3600 \times 5 \times 10^4 \times (0.628)^2]/(1 \times 10^{-2} \times 100 \times 1) = 1.42 \times 10^8$ (m/kg).

Example 7.3

A non-Newtonian fluid is mixed in an agitated tank of 1 m diameter using a paddle agitator $d_a = 0.75\,\text{m}$ diameter, operated at $N = 100$ RPM. Density of the fluid 1100 kg/m³, rheological constants, $K = 10\,\text{Pa s}^n$ and $n = 0.5$. The tank is not baffled.

Calculate the required power for agitation, if the mean shear rate at the paddle surface is $\gamma = 10\,N$.

Solution

Assuming that the power model is applicable, the apparent viscosity of the fluid at the paddle surface will be $\gamma = 10 \times 100/60 = 16.7$ 1/s and the apparent viscosity of the fluid at the paddle surface $\eta_a = K\gamma^{1-n}$ or $\eta_a = 10 \times (16.7)^{-0.5} = 2.44$ Pa s.

Reynolds number $Re = (\rho N d_a^2)/\eta_a = 1100 \times (100/60) \times (0.75)^2/2.44 = 423$, intermediate flow.

From a power/*Re* diagram (Appendix, Figure A.4) of nonbaffled agitator for $Re = 423$ we get $(Po/Fr) = 3$, where *Po* is the power number and *Fr* the Froude number, defined as, $Po = P/(N^3 d_a^5\,\rho)$, $Fr = (N^2\,d_a)/g$, $Re = (N^2\,d_a\,\rho)/\eta$.

Thus, $Fr = (100/60)^2 \times (0.75)/9.81 = 0.212$. Since $(Po/Fr) = 3$, $Po = 3 \times 0.212 = 0.636$.

$P/(N^3\,d_a^5\,\rho) = 0.636$ and $P = 0.636 \times (100/60)^3 \times (0.75)^5 \times 1100 = 1031$ W = 1.031 kW.

If the efficiency of the system is 75%, the required power will be $P - 1.031/0.75 =$ 1.37 kW.

Example 7.4

The exhaust air/gases of a spray dryer are cleaned from most of the milk particles in a multiple cyclone system, consisting of several cyclones operated in parallel. A standard cyclone, diameter $D_c = 20$ cm, is used.

Calculate (a) the number of cyclones needed for a gas flow rate of 2500 m³/h, if the entrance velocity of the gases into the cyclones is 15 m/s, and (b) the efficiency of the cyclone collectors, if the mean particle size is 10 μm.

Density of the gases $\rho = 1$ kg/m³, viscosity $\eta = 0.02$ Pa s. Density of particles $\rho = 1200$ kg/m³.

Solution

The dimensions of the standard cyclone collector (Figure 7.33), based on its basic diameter $D_c = 0.2$ m, will be $L = Z = 2D_c = 0.4$ m, $D_e = D_c/2 = 0.1$ m. The rectangular entrance of the cyclone will have dimensions $(D_c/2)\,(D_c/4) = 0.1 \times 0.05 = 0.005$ m².

(a) The volumetric flow through each cyclone will be $V_1 = 15 \times 0.005 = 0.075$ m³/s or $V_1 = 0.075 \times 3600 = 270$ m³/h. The number of cyclones (n), operated in parallel, will be

$$n = \frac{2500}{270} = 9.26, \quad \text{or} \quad n = 10.$$

(b) Equation 7.40 can be used to calculate the cut diameter of the particles d_c, which corresponds to 50% collection efficiency.

$$d_c^2 = (9\eta B)/(2\pi N\,\rho u)$$

where

$B = D_c/4 = 0.2/4 = 0.05$ m is the entrance width

$N = 5$ is an empirical constant, corresponding to the number of spiral "turns" in the collector

$u = 15$ m/s is the entrance velocity

Thus, $d_c^2 = (9 \times 0.02 \times 10^{-3} \times 0.05)/(2 \times 3.14 \times 5 \times 1200 \times 15) = 15.92 \times 10^{-12}$ and $d_c = 3.95 \times 10^{-6}$ m, or $d_c = 3.95$ μm.

The separation efficiency (η) of the cyclone collectors is estimated from the Lapple diagram (Appendix) as a function of the ratio (d/d_c). Thus, for a mean particle size $d = 10$ μm, we have $(d/d_c) = 10/3.95 = 2.53$, and from the Lapple diagram $\eta = 0.9$ or 90% efficiency.

Note

The overall efficiency of a cyclone collector will be lower than the efficiency based on a relatively large particle size. A large number of smaller particles will pass through the collector. The overall efficiency can be determined by integrating the collection efficiencies of all particles.

PROBLEMS

7.1 A roll mill is used to grind 1 ton/h of dry soybeans from mean size 6 mm to 300 μm. The mill is operated by an electrical motor supplying a power of 30 kW.
Estimate the empirical constants (K) of the Rittinger, Kick, and Bond equations. What will be the Bond work index W_b?

7.2 A paddle agitator, diameter 50 cm, is used to mix a pseudoplastic pulp, density 1100 kg/m^3 and rheological constants $K = 50$ Pa s^n, $n = 0.5$. The agitator is rotated at 20 RPM and the shear rate $\gamma(1/s)$ is given by the empirical relation $\gamma = 10\text{N}$.
Estimate the Reynolds number and the agitator power P_A (W) if the empirical constant is $c_L = 300$ in appropriate units. Calculate also the Power and Froude numbers.

7.3 A mixture of dry starch and protein particles of sizes 0.5–1.0 mm is separated in a vertical air separator, in which air is blown from the bottom and removed from the top. The densities of starch and protein are 1500 and 1200 kg/m^3, respectively.
Calculate the minimum air velocity to remove all starch particles from the top of the unit.

7.4 A rotary vacuum filter of diameter 1.5 m and length 1.0 m is used to clarify completely 1 ton/h of a liquid suspension, containing 0.2 kg solids/kg liquid. The filter is operated at a pressure difference of 500 mbar and at 20 turns/h. The filter drum has 20% of its filtering surface immersed in the liquid suspension.
Calculate the maximum thickness of the cake on the filter surface if its porosity is 50%.

7.5 Differentiate cake and depth filtration and cite typical examples. Explain the operation of dual media filtration and cite one typical application.

7.6 Propose two different filtration systems for clarification a cloudy fruit juice and explain the underlying mechanisms.

LIST OF SYMBOLS

A	m^2	Area
a	m	Distance
Br	—	Brinkman number ($\eta u^2/\lambda\Delta T$)
C	kg/m^3	Concentration
C_D	—	Drag coefficient
D, d	m	Diameter
E	kWh	Energy

f	—	Shape factor
Fr	—	Froude number ($N^2\ d/g$)
G	m/s^2	Acceleration of gravity (9.81)
H, h	m	Height
m	kg	Mass
N	—	Number
N	1/s	Rotational speed
P	W	Power
Po	—	Power number $[P/(\rho N^3 d^5)]$
R	—	Residue cumulative distribution
Re	—	Reynolds number ($du\rho/\eta$)
r	m	Radius
s	—	Standard deviation
T	s	Time
U	m/s	Velocity
V	m^3	Volume
x	m	Size
z	m	Distance
γ	1/s	Shear rate
δ	m	Thickness
ε	—	Porosity
Δp	Pa, bar	Pressure difference
ΔT	°C, K	Temperature difference
η	Pa s	Viscosity
λ	W/(m K)	Thermal conductivity
π	—	3.14
ρ	kg/m^3	Density
Σ	—	Summation
Ψ	—	Sphericity

REFERENCES

Allen, T. 1990. *Particle Size Measurement*, 4th edn. Chapman & Hall, London, U.K.

Chang, Y.K. and Wang, S.S. eds. 1999. *Advances in Extrusion Technology*. Technomic Publications, Lancaster, PA.

Cheremisinoff, P.N. 1995. *Solids/Liquids Separation*. Technomic Publications, Lancaster, PA.

Grandison, A.S. and Lewis, M.J. 1996. *Separation Processes in Food and Biotechnology Industries*. Technomic Publications, Lancaster, PA.

Guy, R. 2001. *Extrusion Technology and Applications*. CRC Press, New York.

Kimball, D.A. 1999. *Citrus Processing*, 2nd edn. Aspen Publications, Gaithersburg, MD.

Kokini, J.L., Ho, C.T., and Karwe, M.V. eds. 1992. *Food Extrusion Science and Technology*. Marcel Dekker, New York.

Levine, L. and Behmer, E. 1997. Dough processing systems. In: *Handbook of Food Engineering Practice*. K.J. Valentas, E. Rotstein, and R.P. Singh, eds. CRC Press, New York.

Luh, B.S. and Woodroof, J.G. 1988. *Commercial Vegetable Processing*, 2nd edn. Van Nostrand Reinhold, New York.

Matz, S.A. 1989. *Equipment for Bakers*. Elsevier Science Publications, London.

McLellan, M.R. 1993. An overview of juice filtration technology. In: *Juice Technology Workshop*. Special Report No. 67, D.L. Downing, ed. N.Y. Agricultural Experiment Station, Cornell University, Geneva, NY.

Mills, D. 1990. *Pneumatic Conveying Guide*. Butterworth-Heinemann, London, U.K.

Nagy, S., Chen, C.S., and Shaw, P.E. eds. 1993. *Fruit Juice Processing Technology*. AgScience Inc., Auburndale, FL.

Perry, R.H. and Green, D. 1997. *Perry's Chemical Engineer's Handbook*, 7th edn. McGraw-Hill, New York.

Purchas, D.B. and Wakeman, R.J. 1986. *Solid/Liquid Equipment Scale-Up*. Uplands Press, London, U.K.

Rumpf, H.1975. Mechanisch Verfahrenstechnik. C. Hanser Verlag, Munich, Germany.

Saravacos, G.D. and Kostaropoulos, A.E. 2002. *Handbook of Food Processing Equipment*. Springer, New York.

Schubert. H. 1987. Food particle technology. I. Properties of particles and particulate food systems. *J. Food Eng*. 6: 1–32.

Stiess, M. 1992. *Mechanische Verfahrenstechnik*, vol. 1. Springer Verlag, Berlin, Germany.

Stiess, M. 1994. *Mechanische Verfahrenstechnik*, vol. 2. Springer Verlag, Berlin, Germany.

Stoes, H.A. 1982. *Pneumatic Conveyors*. Wiley, New York.

Swientek, R.J. 1985. Expression-type belt press delivers high juice yields. *Food Process*. pp. 110–111.

Uhl, Y. and Gray, J.B. 1966. *Mixing Theory and Practice*. Academic Press, New York.

Walas, S.M. 1988. *Chemical Process Equipment*. Butterworth-Heinemann, London, U.K.

8 Heat Transfer Operations

8.1 INTRODUCTION

Heat transfer operations (heating and cooling) are used widely in food processing, usually as an important part of various food processes, such as pasteurization, sterilization, evaporation, drying, refrigeration/freezing, cooking, roasting, and frying. In these operations, the aim is to affect a desirable microbiological, biochemical, chemical, or sensory quality change to the food product being processed, which is generally temperature sensitive (Saravacos and Kostaropoulos, 2002).

The principles of heat transfer mechanism are reviewed in Chapter 2, while kinetics is introduced in Chapter 3.

The major heat transfer operations, used in food processing, pasteurization/sterilization, evaporation, drying, and refrigeration/freezing are treated in detail in Chapters 9 through 12, respectively (Maroulis and Saravacos, 2003).

In this chapter, the principles of heat transfer operations (heating/cooling) are discussed, together with the required processing equipment.

8.2 HEAT TRANSFER COEFFICIENTS

Heat transfer is one of the basic transport phenomena operations of process engineering, the other two being mass transfer and momentum transfer (fluid flow). There are basic similarities of the transport mechanism of these phenomena, which are reviewed in Chapter 2 (Saravacos and Maroulis, 2001).

Most heat transfer operations in food processing involve heat transfer between heating or cooling media and fluid or solid food materials in bulk or packaged form. Heat may be transferred by one or more of the basic mechanisms of heat conduction, heat convection, or heat radiation.

For practical reasons, one-dimensional (x-direction) heat transfer is considered mainly in this book. In some special cases three-dimensional analysis may be necessary.

Heat transfer, in general, is expressed by semi-empirical equations of the general form:

$$q = h\Delta T \tag{8.1}$$

where

- q (W/m^2) is the heat (energy) flux per unit surface area of the exposed material
- ΔT (K or °C) is the temperature difference between the heating or cooling medium and the product
- h [W/(m^2 K)] is the heat transfer coefficient

Note that the temperature in the SI system is degrees Kelvin (K) and it is related to the common temperature in degrees Celsius (°C) by the equation K = °C + 273 and that ΔK = Δ°C.

In the simplest case, heat is transferred by convection through a surface (A) by a temperature difference (ΔT), and the convection heat transfer coefficient (h) is defined by Equation 8.1. However, in most cases, heat is transported through a more complex system, involving convection, conduction, and sometimes radiation. An overall heat transport equation, similar to Equation 8.1 may be written to describe such cases.

The values of heat transfer coefficients (h) depend on several parameters, such as the physical properties of the fluid (mainly viscosity and density), the fluid velocity near the heat transfer surface, the temperature, and the geometry of the transport system (Geankoplis, 1993).

Table 8.1 shows some typical values of (h) of interest to food process engineering. In the English system of units, (h) are expressed in 1 Btu/(h ft^2 °F) = 5.678 W/(m^2 K).

8.2.1 Plane Walls

A common heat transfer system of plane (flat) walls is shown in Figure 8.1. Heat is transported through a wall from an outside heating medium (o) to an inside heated medium (i), due to a temperature difference $\Delta T = (T_o - T_i)$. The heat energy has to flow through an outside film, the wall, and an inside film.

Under steady-state conditions (constant temperature profile with time), the heat flux q/A (W/m^2) from the heating medium (o) to the wall surface through the exterior film will be

$$\frac{q}{A} = h_o \, \Delta T_1 \tag{8.2}$$

where

$\Delta T_1 = T_o - T_1$ (K) is the temperature difference between the outside and wall temperature

h_o [W/(m^2 K)] is the surface (convection) heat transfer coefficient

TABLE 8.1
Ranges of Heat Transfer Coefficients (h)

System	h, W/(m^2 K)
Atmospheric air, heating/cooling	2–50
Superheated steam	20–120
Vegetable oils, heating/cooling	50–1,000
Aqueous food fluid	200–2,000
Water, heating/cooling	200–10,000
Water, boiling	1,500–20,000
Water vapors, condensing	5,000–40,000

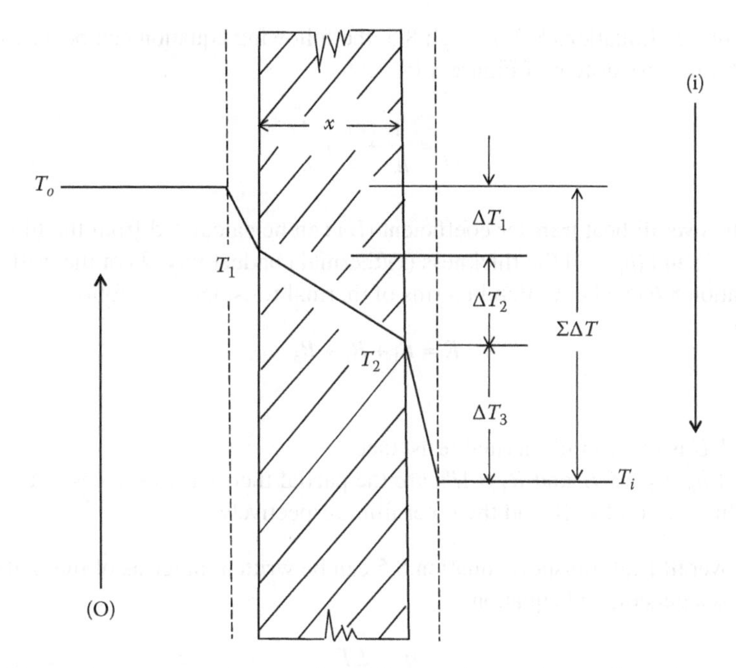

FIGURE 8.1 Heat transport through a flat wall separating heating and heated streams.

Assuming no accumulation or loss of heat energy, the heat flux (q/A) through the outer film of the system of Figure 8.1 will be the same as the flux of heat through the solid wall and the inner film (McCabe et al, 2004):

$$\frac{q}{A} = \lambda\left(\frac{\Delta T_2}{x}\right) \tag{8.3}$$

$$\frac{q}{A} = h_i \Delta T_3 \tag{8.4}$$

Equation 8.3 describes the heat conduction flux through a wall of thickness x (m) with a constant thermal conductivity λ [W/(m K)] and $\Delta T_2 = (T_1 - T_2)$ (K).

Equation 8.4 expresses the heat transfer from the inner wall to the bulk of stream (i), where $\Delta T_3 = T_2 - T_i$ (K) is the temperature difference between the wall and inner bulk temperature, and h_i [W/(m^2 K)] is the inner surface (convection) heat transfer coefficient.

An overall heat flux equation can be written, using the overall temperature difference (ΔT) as the driving force:

$$\frac{q}{A} = U\Delta T \tag{8.5}$$

where

$\Delta T = \Delta T_1 + \Delta T_2 + \Delta T_3 = T_o - T_i$
U [W/(m^2 K)] is the overall heat transfer coefficient

Combining Equations 8.2 through 8.5, the following equation can be derived for the heat transport system of Figure 8.1:

$$\frac{1}{U} = \frac{1}{h_o} + \frac{x}{\lambda} + \frac{1}{h_i} \quad (8.6)$$

Thus, the overall heat transfer coefficient (U) can be calculated from the film coefficients (h_o) and (h_i) and the thickness (x)/thermal conductivity (λ) of the wall.

Equation 8.6 can be written in terms of thermal resistances, as follows:

$$R = R_1 + R_2 + R_3 \quad (8.7)$$

where

$R = 1/U$ is the overall thermal resistance

$R_1 = 1/h_o$, $R_2 = x/\lambda$, and $R_3 = 1/h_i$ are the partial thermal resistances of the outer film, the solid wall, and the inner film, respectively

The overall heat transport equation 8.5 can be written, in terms of thermal resistances, as a generalized equation:

$$\frac{q}{A} = \frac{\Delta T}{R} \quad (8.8)$$

Equation 8.7 can be applied to heat transport through several parallel plane walls, involving heat conduction and heat convection, provided that the wall thicknesses and thermal conductivities, the film heat transfer coefficients, and the overall temperature difference are known.

8.2.2 Cylindrical Walls

Several industrial applications involve tubes and pipes in which heat is transported through cylindrical walls. Equations similar to the transport through plane walls are used, taking into consideration that the transfer area of the walls changes as the diameter is changed.

Heat conduction through cylindrical walls is treated in Chapter 2.

In a cylindrical system, similar to Figure 8.1, the heat flux from the outside heating medium (temperature, T_o) through an outer film (heat transfer coefficient, h_o), a wall (thickness/thermal conductivity, x/λ), and an inner film (heat transfer coefficient, h_i) to the inner heated product (temperature, T_i) will be given by the equations

$$\frac{q}{A_o} = U_o \Delta T \quad \text{or} \quad \frac{q}{A_i} = U_i \Delta T \quad (8.9)$$

where

$(\Delta T) = (T_o - T_i)$, (A_o, A_i) are the outside and inside surface areas

(U_o, U_i) are the overall heat transfer coefficients, based on the outside and inside surface areas, respectively

These coefficients are calculated from the equations

$$\frac{1}{U_o} = \frac{1}{h_o} + \left(\frac{d_o}{d_L}\right)\left(\frac{x}{\lambda}\right) + \left(\frac{d_o}{d_i}\right)\frac{1}{h_i}$$

or (8.10)

$$\frac{1}{U_i} = \frac{1}{h_i} + \left(\frac{d_i}{d_L}\right)\left(\frac{x}{\lambda}\right) + \left(\frac{d_i}{d_o}\right)\left(\frac{1}{h_o}\right)$$

where $d_L = (d_o - d_i)/\ln(d_o/d_i)$ is the log mean diameter of tube diameters (d_o, d_i).

In Equation 8.10 the diameters d_o, d_i, and d_L can be replaced by the corresponding surface areas A_o, A_i, and A_L, since the ratios of the surface areas are equal to the ratios of the diameters. For thin walls (close values of d_o, d_i), $d_L = (d_o + d_i)/2$.

It should be noted that in all the above equations, the heat transported (q) is the same, independently of (U) or (A) used. Since the overall temperature difference (ΔT) is the same, it follows that $U_o A_o = U_i A_i$.

Equation 8.10 can be applied to cylinders of multiple walls, provided that the thicknesses/thermal conductivities of the walls are known.

For large diameter tubes/pipes ($d_o > 25$ mm (1 in.)) of thin walls, the diameters (d_o, d_i, and d_L) can be considered as approximately equal and, therefore, $U_o = U_i = U$. In this case, Equation 8.10 can be replaced by Equation 8.6 for plane (flat) walls.

In most industrial applications, the heating or cooling medium is separated from the product side by thin metal walls. Thus, the thermal resistance (x/λ) of the wall is very small, compared to the thermal resistances of the two films ($1/h_o$ and $1/h_i$), and the overall heat transfer coefficient can be calculated from the approximate equation: $1/U = 1/h_o + 1/h_i$.

In some cases the main thermal resistance may be in the film of one side, for example, a very low heat transfer coefficient (h_o), when the overall heat transfer coefficient will be $U = h_o$. A typical such case is the heating of water in tubes from hot flue gases in a steam boiler.

8.2.3 Natural Convection

Heat transfer by natural convection has limited industrial applications because of the low heat fluxes attained. It is of importance to the loss of thermal energy from hot surfaces to the surrounding air.

In natural convection, heat is transported by natural flow of a gas (air) or a liquid near a heated or cooled surface. Natural flow is caused by density differences in the gas or liquid, created by temperature differences in the fluid.

Figure 8.2 shows a simplified diagram of heating air in contact with a heated surface.

Due to differences in density, the air moves upward. The air velocity increases gradually and it reaches a maximum, as the distance from the heated wall is increased. The air temperature decreases sharply near the heated surface, leveling off to the bulk temperature.

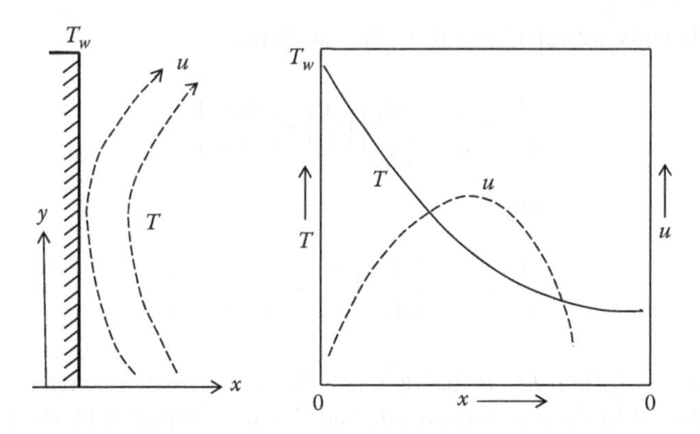

FIGURE 8.2 Natural convection near heated surface.

If the temperature of the contact surface is lower than the bulk air (or liquid) temperature, the flow pattern is reversed (flow downward), the velocity profile is the same, and the fluid temperature increases as the distance increases.

Natural convection in more complex systems, such as horizontal tubes, is characterized by empirical heat transfer equations, derived by regression analysis of experimental data.

The heat transfer coefficient (h) in such a system is a function of the following parameters: outside diameter (d_o) and length (L) of the tube, density (ρ), viscosity (η), thermal conductivity (λ), specific heat (C_p), temperature difference (ΔT), coefficient of thermal expansion (β), and gravity acceleration ($g = 9.81\ \mathrm{m/s^2}$).

Dimensional analysis of the parameters involved in natural convection of single horizontal tubes yields the following generalized dimensionless equation:

$$Nu = f(Gr, Pr) \tag{8.11}$$

where

Nusselt number $Nu = hd_o/\lambda$
Grashof number $Gr = (d_o^2\rho^2\beta g\Delta T)/\eta^2$
Prandtl number $Pr = (C_p\eta)/\lambda$

The Nusselt number (Nu) is the ratio of convective to conductive heat transfer. The Prandtl number (Pr) is the ratio of momentum to thermal conductivity. The Grashof number (Gr) is the ratio of buoyancy to viscous forces of a fluid.

The dimensionless numbers Nu, Pr, Gr, and Re are used in most heat transfer dimensionless equations. The Gr number is applied only in the natural convection. It contains a characteristic parameter of natural flow, the volumetric change coefficient (β), defined as $\beta = (\Delta V/\Delta T)/V$, which for perfect gases reduces to $\beta = 1/T$.

In estimating the Gr number, the physical properties of the fluid are taken at the mean film temperature $T_f = (T_s + T)/2$, where (T_s, T) are the surface and bulk temperatures, respectively.

Regression analysis of experimental natural convection data yielded the following generalized dimensionless equation:

$$Nu = \alpha(Gr\ Pr)^m \quad (8.12)$$

The constants (α, m) have the following values:

1. $Gr\ Pr > 10^9$ horizontal cylinders, vertical plates $\alpha = 0.13$, $m = 1/3$
2. $Gr\ Pr < 10^9$ horizontal cylinders, vertical plates $\alpha = 0.54$, $m = 1/4$
3. $Gr\ Pr < 10^3$ use literature diagrams of log(Nu) versus log ($Gr\ Pr$)

For atmospheric air the Pr number is in the range of 0.6–1.0 and the Gr number is between 10^9 and 10^3; therefore, $\alpha = 0.54$, $m = 1/4$. In this range the flow of air is considered to be turbulent.

In the previous equations, the diameter (d_o) in the horizontal tubes is replaced by the height (L) in the vertical plates.

The following simplified dimensional equations, derived from the generalized Equation 8.12 can be used for the natural convection of atmospheric air in certain heat transfer systems:

1. Horizontal tubes

$$h = 1.42\left(\frac{\Delta T}{d_o}\right)^{1/4} \quad (8.13)$$

2. Vertical planes

$$h = 1.42\left(\frac{\Delta T}{L}\right)^{1/4} \quad (8.14)$$

3. Horizontal heated plates
   a. Facing up

$$h = 2.48(\Delta T)^{1/4} \quad (8.15)$$

   b. Facing down

$$h = 1.3(\Delta T)^{1/4} \quad (8.16)$$

where

h is the heat transfer coefficient, W/(m^2 K)
ΔT is the temperature difference between hot surface and air, K
d_o is the outside tube diameter, m
L is the plate height, m

8.2.4 Forced Circulation

In most industrial operations, forced circulation (use of pump or fan) of heating or cooling fluids is used, increasing substantially the heat transfer rate. Empirical dimensionless equations of various forms are applied, depending on the type of flow (laminar or turbulent).

For well-developed turbulent flow in pipes ($Re > 8000$), the dimensionless Sieder–Tate equation is used:

$$Nu = 0.023 Re^{0.8} Pr^{1/3} \left(\frac{\eta}{\eta_w}\right)^{0.14} \tag{8.17}$$

where (η, η_w) is the viscosity of the fluid at the bulk and film temperature, respectively. The factor $(\eta/\eta_w)^{0.14}$ is important in viscous fluids, the viscosity of which decreases sharply as the temperature is increased (e.g., concentrated sugar solutions, oils). It can be neglected when the viscosity does not change much with temperature (e.g., water).

Equation 8.17 is applied to long tubes, i.e., tubes of $L/d_i > 60$, where (L, d_i) are the tube length and inside diameter, respectively. For shorter tubes, the right-hand side of Equation 8.17 should be multiplied by the correction factor $[1 + (d_i/L)^{0.7}]$.

For the laminar flow ($Re < 2100$), the following Sieder–Tate equation is used:

$$Nu = 1.86 \left[Re\ Pr \left(\frac{d_i}{L}\right)\right]^{1/3} \left(\frac{\eta}{\eta_w}\right)^{0.14} \tag{8.18}$$

Equation 8.18 can be written in the form

$$Nu = 2.0 Gz^{1/3} \left(\frac{\eta}{\eta_w}\right)^{0.14} \tag{8.19}$$

where $Gz = (GAC_p/\lambda L)$, Graetz number (dimensionless).

G (kg/m^2 s) is the mass flux, and A (m^2) is the cross-sectional area of the flow. Note that $GA = m$ (kg/s), mass flow rate.

The Graetz number $Gz = (d/L)\ Re\ Pr$ characterizes laminar flow in a conduit.

For the intermediate flow ($2100 < Re < 6000$) there are no accurate equations available, and the heat transfer coefficient can be estimated from empirical diagrams, such as Figure 8.3. In this diagram, the heat transfer factor (j_H) is defined by the equation

$$j_H = \left(\frac{h}{C_p G}\right) Pr^{2/3} \left(\frac{\eta}{\eta_w}\right)^{0.14} \tag{8.20}$$

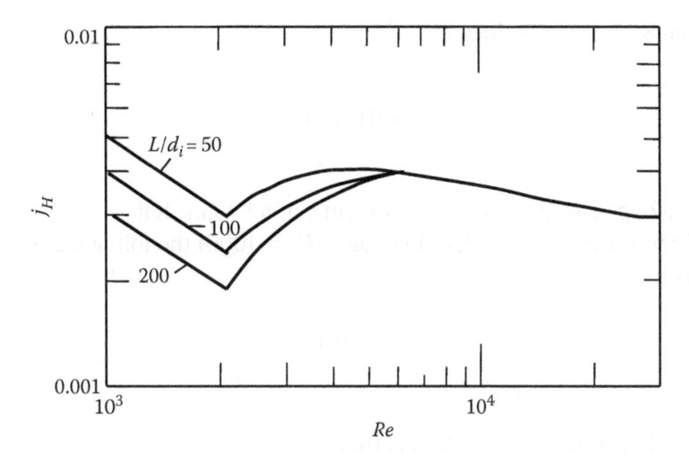

FIGURE 8.3 Heat transfer factor (j_H) in laminar, intermediate, and turbulent flow.

For laminar flow of power-law non-Newtonian fluid, the following equation is applied:

$$Nu = 2\left[\frac{(3n+1)}{4n}\right]^{1/3} Gz^{1/3}\left(\frac{m}{m_w}\right)^{0.14} \tag{8.21}$$

where

$m = 8^{n-1}K$ and K (Pa s^n) is the rheological constant
n (–) the flow behavior index
The constants m, m_w are taken at the bulk and wall temperature, respectively

8.2.5 Falling Films

Falling films in vertical tubes (inside and outside), and in vertical plates are used in the heating/cooling and in the evaporation of various liquids during food processing. High heat transfer coefficients are obtained with minimum expense of pumping energy. Film flow is reviewed in Chapter 6.

Heat transfer in a vertical falling film in the laminar region is expressed by the empirical equation

$$Nu = 0.008Re^{2/3}Pr^{1/3} \tag{8.22}$$

where

$Nu = (h\delta)/\lambda$, $Re = 4\Gamma/\eta$, $Pr = C_p\eta/\lambda$, $\delta = (3\Gamma\eta/\rho^2 g)^{1/3}$ (m) is the film thickness
$\Gamma = m/b$ [kg/(m s)] is the flow rate per unit of wetted perimeter
m (kg/s) is the mass flow rate
b (m) is the wetted perimeter

For a tube of inside diameter d_i(m), wetted perimeter $b = 3.14d_i$. The properties of the fluid are estimated at the temperature of mean film thickness T_f.

Equation 8.22 is also written as

$$\frac{h}{\varphi} = 0.01(Re\ Pr)^{1/3} \tag{8.23}$$

where $\varphi = [(\lambda^3 \rho^2 g)/\eta^2]^{1/3}$ is a factor with units of h [W/(m² K)].

For falling water films at high flow rates ($Re > 1000$) the following dimensional equation is used:

$$h = 9150\Gamma^{1/3} \tag{8.24}$$

where

h [W/(m² K)] is the heat transfer coefficient

Γ [kg/(m s)] is the flow rate

8.2.6 Condensing Vapors

The condensation of vapors on cooled surfaces can take place by two different mechanisms, (1) film condensation, which is a slow process, and (2) drop-wise condensation, which is much faster. In heating operations, condensation of saturated steam or water vapor is applied for rapid heat transfer, and drop-wise condensation is desirable. Vapor condensation is also important in refrigeration systems.

In the design of condensers, film condensation is considered for practical reasons, because drop-wise condensation, although much faster, is not a predictable process.

In short vertical tubes, the flow of the condensed film is laminar, but turbulent flow may develop in longer tubes, which enhances considerably the heat transfer coefficient. Empirical dimensional equations have been developed for the estimation of heat transfer coefficients of condensing water vapor (or saturated steam) at atmospheric pressure. Similar equations have been developed for vapor condensation in horizontal tubes.

For vertical tubes or plates of height L (m):

$$h = \frac{13.9}{(L^{1/4} \Delta T^{1/3})} \text{ kW/(m}^2\text{ K)} \tag{8.25}$$

For horizontal tubes of outside diameter d_o (m):

$$h = \frac{10.8}{(Nd_o)^{1/4} \Delta T^{1/3}} \text{ kW/(m}^2\text{ K)} \tag{8.26}$$

where

ΔT is the temperature difference between the saturated vapor and the cold surface of condensation

N is the number of horizontal tubes in a vertical plane

Superheated vapors (or steam) must be cooled to the saturation temperature before condensation. Heat transfer coefficients in cooling superheated vapors are very low, similar to air (gas) cooling systems.

8.2.7 Boiling Liquids

Heat transfer in (thermally) saturated liquids takes place by nucleate boiling, i.e., by the formation of numerous bubbles near the heated surface, which are removed continually from the fluid phase. Heat transfer coefficients (h) during boiling can be determined by measuring the heat flux (q/A) from a heated surface (e.g., an electric wire) at a given temperature difference (ΔT).

Figure 8.4 shows a diagram of heat flux versus temperature difference in logarithmic coordinates. In the first section (ABC) of the boiling curve, the heat flux increases continuously with the applied (ΔT), and heat is transported mainly by nucleate boiling. After a critical point (C), the heat flux decreases as the (ΔT) is increased until point (D), due to the formation of a vapor layer near the heated surface. Beyond points (D) and (E), the heat flux increases again continuously with increasing (ΔT), due to the combined effect of convection, conduction, and radiation from the heated surface. Radiation heat transfer becomes very important at higher temperatures.

In industrial applications, heat transfer by nucleate boiling at relatively low temperature differences is preferred. Thus, experimental measurements of heat transfer in water boiling at (ΔT) = 20°C–30°C have yielded (h) values up to 40 kW/(m^2 K).

The boiling heat transfer coefficient (h) of water increases substantially with increasing boiling temperatures. Thus, evaporation of aqueous solutions (e.g., fruit juices) at low temperature results in significantly lower (h) values, which should be taken into consideration in the design of vacuum and multiple-effect evaporation systems.

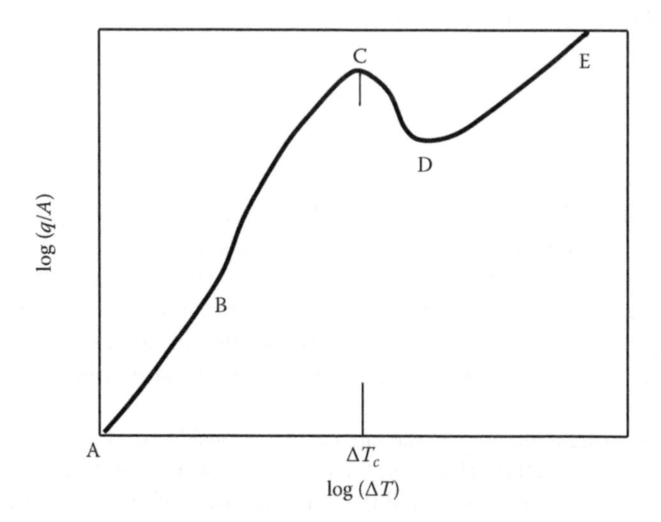

FIGURE 8.4 Heat flux (q/A) in a boiling liquid at various temperature differences (ΔT).

8.2.8 Radiation Heat Transfer

The principles of radiation heat transfer are reviewed in Chapter 2. Thermal radiation becomes the controlling heat transfer mechanism at high temperatures, while heat convection and conduction are more important at lower temperatures. In food engineering applications, radiation heat transfer is considered mainly between two surfaces, separated by a nonabsorbing medium, such as air.

The radiation incident to a surface is either absorbed, reflected, or transmitted. The fractions of absorbtivity (α), reflectivity (ρ), and transmissivity (τ) add up to unity:

$$\alpha + \rho + \tau = 1 \tag{8.27}$$

Surfaces that absorb all incident radiation ($\alpha = 1$) are called a black body.

The absorption and reflection of radiant energy takes place on the surface of the material, from which it may be transmitted inside by conduction.

Most engineering materials are opaque, which absorb and reflect the radiant energy without any transmission ($\tau = 0$). Some materials, such as glass and some plastics are transparent, i.e., they transmit a significant portion of the incident radiation.

Kirchhoff's law of radiation states that at thermal equilibrium, the ratio of radiation energy to the absorptivity at a given temperature is constant:

$$\frac{w_1}{\alpha_1} = \frac{w_2}{\alpha_2} = \cdots = w_\alpha \tag{8.28}$$

where

$w = q/A$ (W/m^2) is the radiation energy
w_α is the radiation energy of black body

The emissivity (ε) is the ratio of energy emitted by a material to the energy radiated by a black body at the same temperature:

$$\varepsilon = \frac{w}{w_\alpha} \tag{8.29}$$

For a black body $\varepsilon = 1$, and for all other materials $\varepsilon < 1$.

At thermal equilibrium, combination of Equations 8.28 and 8.29 yields $\alpha = \varepsilon$, i.e., the absorptivity is equal to the emissivity of the material.

In engineering applications, the surface of the material is assumed to be a gray body, which has a constant emissivity, not affected by temperature or radiation wavelength.

The emissivity (ε) depends on the condition of the radiation surface. It is higher in rough and oxidized surface, but it becomes very low at polished metal surfaces (Table 8.2).

For engineering applications, the emissivity is considered as approximately constant with the temperature. However, the emissivity of the metals increases slightly as the temperature is increased, and the opposite effect of temperature is observed in nonmetallic materials.

TABLE 8.2
Emissivities (ε) of Engineering Materials

Material	Temperature, °C	ε
Polished copper, silver	100	0.02
Polished aluminum, nickel	100	0.05
Tin plate	100	0.07
Polished stainless steel	100	0.08
Mercury	100	0.12
Polished cast iron	200	0.20
Stainless steel	200	0.44
Oxidized steel	200	0.78
Aluminum paint	100	0.52
Wood	25	0.90
Building brick, gypsum	25	0.92
Glass	25	0.94
Water	0–100	0.95–0.96

The radiation rate of a black body is given by the Stefan–Boltzmann equation:

$$w_B = \sigma T^4 = 5.67\left(\frac{T}{100}\right)^4 \tag{8.30}$$

where
$w_B = q/A$ (W/m²) is the radiation rate of black body
T (K) is the temperature
$\sigma = 5.67 \times 10^{-8}$ W/(m² K⁴) is the Stefan–Boltzmann constant

The rate of radiation heat exchange $q_{1,2}$ (W) between surfaces A_1 and A_2, which are at equilibrium temperatures T_1 and T_2, respectively, will be

$$q_{1,2} = 5.67 A_1 F_{12}\left[\left(\frac{T_1}{100}\right)^4 - \left(\frac{T_2}{100}\right)^4\right] \tag{8.31}$$

where $F_{1,2}$ (–) is the view factor of surface A_1 to surface A_2.

The view factor F_{12} is a fractional number ($0 < F_{12} < 1$) representing the fraction of thermal radiation emitted from surface A_1 that is intercepted by surface A_2. It expresses, quantitatively, how surface A_1 "sees" surface A_2.

The view factors F_{11} and F_{22} refer to the fraction of radiation intercepted by the surfaces A_1 and A_2 themselves, respectively. For parallel planes, the surfaces do not "see" themselves and $F_{11} = F_{22} = 0$. For the inside surface of hollow spheres $F_{1,1} = F_{2,2} = 1$.

It is obvious that

$$A_1 F_{12} = A_2 F_{21} \tag{8.32}$$

For a given surface A_1 the sum of all view factors equals unity:

$$F_{11} + F_{12} + F_{13} + \cdots = 1 \tag{8.33}$$

Refractory materials reflect all incident radiation without transmitting it to the interior. They are used in heating systems, such as furnaces to confine the radiation energy in the system, increasing their thermal efficiency.

The view factors (F) of black body in various geometric shapes, encountered in radiation heat transfer, can be calculated by mathematical analysis. Radiation heat transfer in engineering applications takes place from a hot to a colder surface, separated by a nonabsorbing medium (air).

The simplest geometric system is radiation between two parallel infinite planes, in which all energy radiated from surface A_1 is intercepted by the opposite surface A_2, i.e., $F_{12} = 1$. It should be noted that in this system, $F_{21} = F_{12} = 1$.

The view factors of surfaces of finite size are determined for each geometry by taking differential surface elements and integrating over the entire surfaces. In engineering applications, the view factors of some common geometries are estimated from the Hottel diagrams of the literature, such as Figures 8.5 and 8.6.

Reradiating (refractory) walls, connecting parallel black body surfaces reradiate all incident radiation, increasing the energy efficiency of industrial equipment, such as furnaces. The refractory walls do not absorb or conduct the incident radiation.

The view factor ($\bar{F}_{12}$) of two parallel surfaces (A_1 and A_2) connected with refractory walls is given as a function of the view factors F_{12} and F_{21} (black bodies) by the equation

$$\bar{F}_{12} = \frac{(A_1 - A_1F_{12})}{(A_1 + A_2 - 2A_1F_{12})} - \frac{\left[1-(A_1/A_2)F_{12}^2\right]}{\left[(A_1/A_2 + 1 - 2(A_1/A_2)F_{12})\right]} \tag{8.34}$$

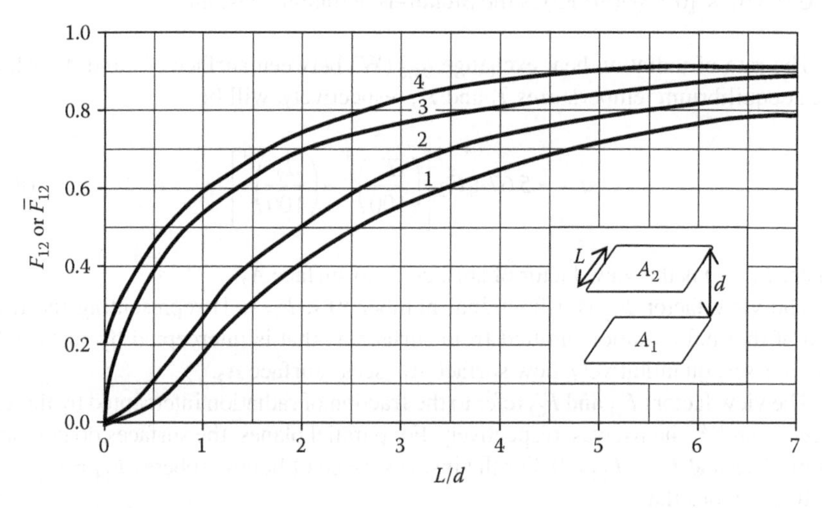

FIGURE 8.5 View factors between parallel black planes. F_{12} single planes: (1) disks, (2) rectangular planes 2/1. $\bar{F}_{12}$ refractory walls, (3) disks, and (4) rectangular planes 2/1. L, smaller dimension, d, distance between planes.

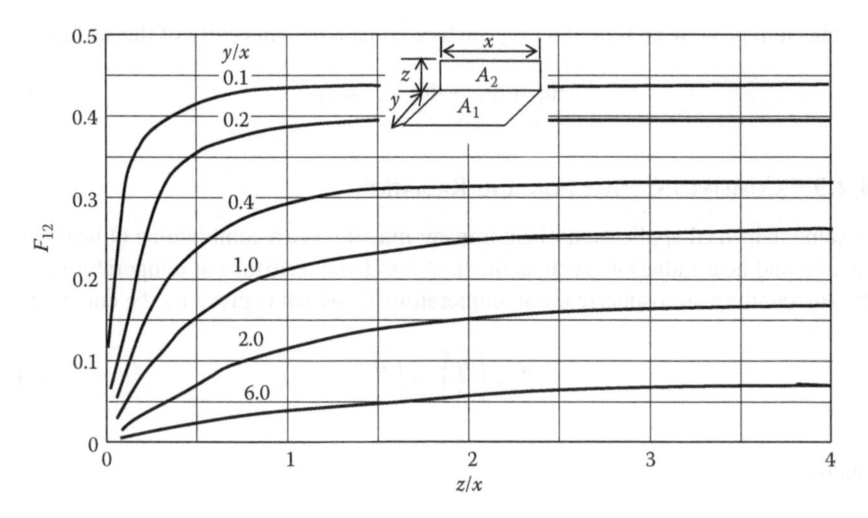

FIGURE 8.6 View factors (F_{12}) between adjacent perpendicular black planes.

Values of the view factor (F_{12}) for refractory systems of parallel planes and disks are given in Figure 8.4. The sum of view factors (F) of a surface equals unity, similar to Equation 8.33 ($F_{11} + F_{12} + F_{13} + \cdots = 1$).

Heat transfer by radiation in gray surfaces is given by the equation

$$q_{1,2} = 5.67 A_1 \Phi_{12} \left[\left(\frac{T_1}{100} \right)^4 - \left(\frac{T_2}{100} \right)^4 \right] \tag{8.35}$$

where $\Phi_{1,2}$ is the view factor of gray surface A_1 to surface A_2, which is a function of the view factor for black body (F_{12}) and the emissivities of the two surfaces:

$$\frac{1}{\Phi_{12}} = \frac{1}{F_{12}} + \left(\frac{1}{\varepsilon_1} - 1 \right) + \left(\frac{A_1}{A_2} \right) \left(\frac{1}{\varepsilon_2} - 1 \right) \tag{8.36}$$

Equation 8.36 is simplified for some special geometric systems, such as gray parallel planes of infinite dimensions, $F_{12} = 1$, $(A_1/A_2) = 1$:

$$\frac{1}{\Phi_{12}} = \frac{1}{\varepsilon_1} + \frac{1}{\varepsilon_2} - 1 \tag{8.37}$$

Gray surface A_1 surrounded entirely by another gray surface A_2

$$\frac{1}{\Phi_{1,2}} = \frac{1}{\varepsilon_1} + \left(\frac{A_1}{A_2} \right) \left(\frac{1}{\varepsilon_2} - 1 \right) \tag{8.38}$$

Gray surface A_1 surrounded entirely by a black body ($\varepsilon_2 = 1$)

$$\Phi_{12} = \varepsilon_1 \tag{8.39}$$

The last case is similar to the radiation of a gray body surrounded by air.

The sum of view factors of a gray surface equals the emissivity of the surface:

$$\Phi_{11} + \Phi_{12} + \Phi_{13} + \cdots = \varepsilon_1 \tag{8.40}$$

8.2.9 Combined Convection and Radiation

In some industrial applications, heat transfer may involve a combination of heat convection and heat radiation, such as the heat loss from a surface at temperature T_1 to the surrounding atmospheric air at temperature T_2, which is given by the equation

$$\frac{q}{A} = \left(\frac{q}{A}\right)_c + \left(\frac{q}{A}\right)_r \tag{8.41}$$

where

$$\left(\frac{q}{A}\right)_c = h_c \Delta T \quad \text{heat loss by convection} \tag{8.42}$$

and

$$\left(\frac{q}{A}\right)_r = 5.65 A_1 \Phi_{12} \left[\left(\frac{T_1}{100}\right)^4 - \left(\frac{T_2}{100}\right)^4\right] = h_r \Delta T \tag{8.43}$$

heat transfer by radiation, and $\Delta T = T_1 - T_2$.

Combination of the last three equations yields

$$\frac{q}{A} = (h_c + h_r)\Delta T \tag{8.44}$$

The convection heat transfer coefficient (h_c) is calculated from specific empirical equations, while the radiation heat transfer coefficient (h_r) is calculated from Equation 8.43.

In practical applications, the sum of heat transfer coefficients ($h_c + h_r$) is estimated from empirical relations, such as the diagram of Figure 8.7 for heat loss from horizontal pipes, surrounded by atmospheric air at 20°C.

From the last diagram it follows that the heat transfer rate from horizontal pipes to the surrounding air is inversely proportional to the external diameter of the pipes.

8.2.10 Heat Transfer Factor

Regression analysis of experimental heat transfer coefficients has resulted in empirical equations useful in food engineering applications. The analogy of heat, mass, and momentum transfer is discussed in Chapter 2.

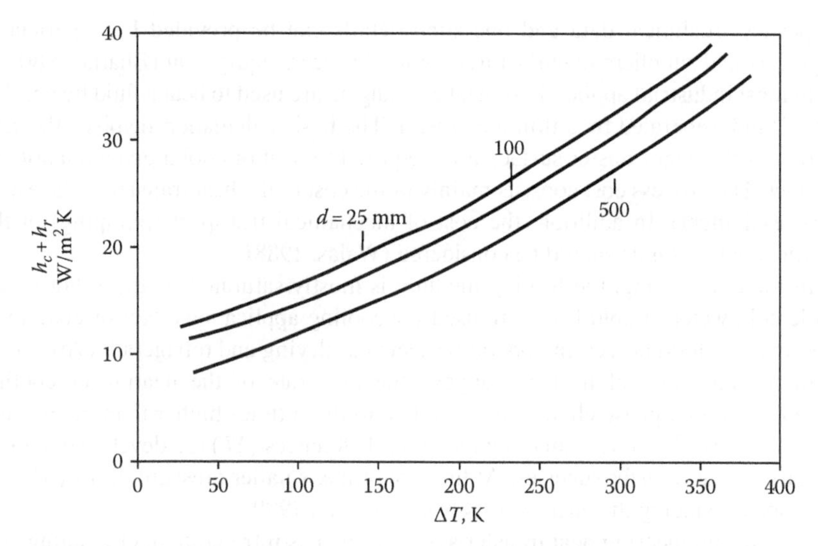

FIGURE 8.7 Heat loss by combined convection and radiation to air at 20°C.

Correlations of heat transfer coefficients (h) have yielded a generalized equation for the heat transfer factor (j_H) as a function of the Reynolds number (Re):

$$j_H = 0.344Re^{-0.433} \tag{8.45}$$

where $j_H = St_H\, Pr^{2/3}$, $St_H = h/(GC_p)$ is the Stanton number for heat transfer.

The St_H is the ratio of heat transferred into a fluid to the heat capacity of the fluid. It is related to the Reynolds (Re), Prandtl (Pr), and Nusselt (Nu) numbers by the equation

$$St_H = \frac{Nu}{(Re\ Pr)} \tag{8.46}$$

The Prandtl number is $Pr = (C_p\eta)/\lambda$, Nusselt $Nu = (hL)/\lambda$, and Reynolds $Re = (Lu\rho)/\eta$.

$G = u\rho$ (kg/m²) is the mass flux, u (m/s) is the velocity, ρ (kg/m³) is the density, η [kg/(m s)] is the viscosity, λ [W/(m K)] is the thermal conductivity, and C_p [J/(kg K)] is the specific heat of the fluid. The flow conduit dimension L (m) is the inner diameter of pipes (tubes) or a characteristic length dimension for other flow geometries.

Literature data on heat transfer coefficients of various food process operations, such as heating, cooling, freezing, drying, and sterilization have been published in diagrams of log(h) versus log(Re).

8.3 HEAT EXCHANGERS

Heat exchangers are used in several food processing operations to heat or cool various food materials and products. The design and operation of heat exchangers is based on the principles and data of heat transfer, fluid flow, and food engineering properties. Technical data and operating details can be provided by experienced engineers and suppliers/manufacturers of food process equipment (Bhatia, 1980).

In most industrial applications, heat exchangers are used to heat a fluid by another hotter fluid, separated by a thin metal wall. The basic calculation involves the estimation of the heat transfer surface area required to heat or cool a given quantity of product. The process cost consists mainly of the cost of the heat transfer surface area (heat exchanger). In addition, the cost of mechanical transport (pumping) of the heating/cooling fluids should be considered (Walas, 1988).

In food processing, the heating medium is mostly saturated steam or hot water, while cold water or cold brine are used for cooling applications. Hot or cold air is used in some food processing operations, such as drying and refrigeration/freezing.

In the operation of heat exchangers, the flow rate of the heating or cooling medium, without phase change, is about two to three times higher than the process stream. At low flow rates, small temperature differences (ΔT) are developed, requiring larger heat transfer surfaces. At the same time, smaller pressure drops (ΔP) are developed, reducing the energy cost (Knudsen et al, 1999).

In some engineering heat transfer systems, such as refrigeration, circulating systems are used, involving two heat exchangers, one between the heating or cooling medium and the intermediate heat transfer fluid and the other between the intermediate and the process fluid. Recirculating systems are safer and they can be controlled more easily than the primary heating/cooling media.

A brief discussion of the plant utilities, used in food processing, is presented in the Appendix of this book.

8.3.1 Heat Transfer Rate

In most heat exchangers, the rate of heating of the cold stream is equal to the rate of cooling of the hot stream, i.e., the heat losses to the environment and the heat of friction are neglected. The heating rate of a fluid q (W) from temperature $T_1 - T_2$ (°C), without phase change, will be

$$q = mC_p\Delta T \tag{8.47a}$$

where

m (kg/s) is the product flow rate
C_p [J/(kg K)] is the specific heat, assumed to be constant
ΔT (K) is the temperature difference (note that ΔT (K) = ΔT (°C))

If there is a phase change in the temperature range ($T_1 - T_2$), the heating rate becomes

$$q = m[C_p\,\Delta T + \Delta H) \tag{8.47b}$$

where ΔH (J/kg) is the heat of phase change, which for water evaporation (or condensation) is, for example, at 100°C, $\Delta H = 2.26\,\text{MJ/kg}$.

8.3.2 Double-Pipe Heat Exchangers

The simplest two-fluid heat exchanger is the double-pipe system, in which the heating medium usually flows in the outer pipe (annulus) and the cold fluid in the inner pipe.

Double-pipe heat exchangers are operated in either co-current (parallel) or countercurrent flow (Figure 8.8).

Figure 8.9a shows a four-pass double-pipe heat exchanger in countercurrent operation. Multi-pass pipe heat exchangers, consisting of straight pipes connected with U-bends, are used for compact heat exchangers, which avoid the long straight pipes, needed to obtain large heat transfer areas. However, they develop high pressure drops (ΔP), increasing the energy cost. Heating or cooling coils have similar advantages and shortcomings.

Figure 8.9b shows a coil steam heater with the cold product (c) in the spiral tube and the heating medium (s) in the jacket. A similar system can be used for cooling a fluid product in the tube with the cooling medium in the jacket.

Coils reduce the length of straight pipes and increase the heat transfer coefficient by the turbulence, which is developed at lower Reynolds (Re) numbers. The turbulence in coils is expressed by the Dean number, defined as $Dn = Re\ (d/D)^{1/2}$, where (d, D) are the tube and coil diameters, respectively.

Triple-pass pipe heat exchangers, shown in Figure 8.10 are sometimes used in special applications, such as thermal processing (pasteurization or sterilization) of liquid foods. In the diagram of Figure 8.10, the process product (c) flows in the annulus between the pipes, while the heating fluid (h) flows in the inner and outer pipes. Thus, the product is heated faster in the annulus, utilizing the double contact surface of both pipes. Fast heating up is essential in the thermal processing of heat-sensitive food products.

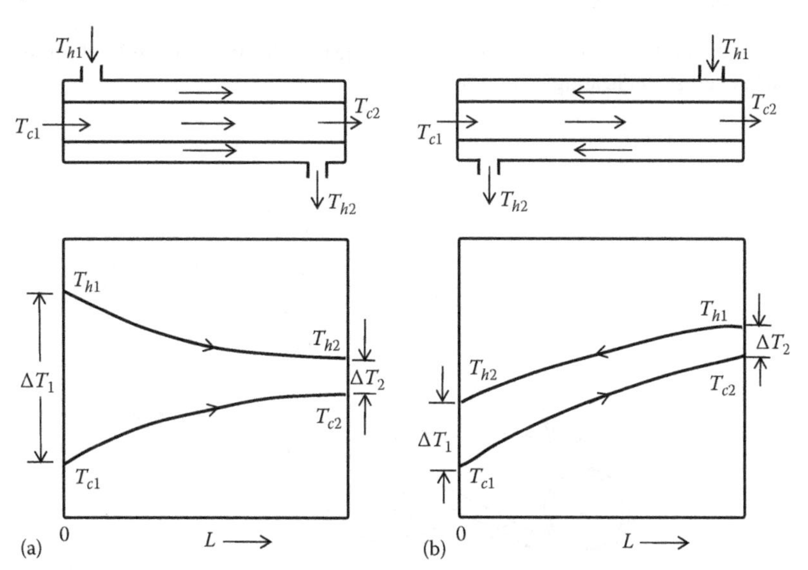

FIGURE 8.8 Double-pipe heat exchangers: (a) co-current and (b) countercurrent.

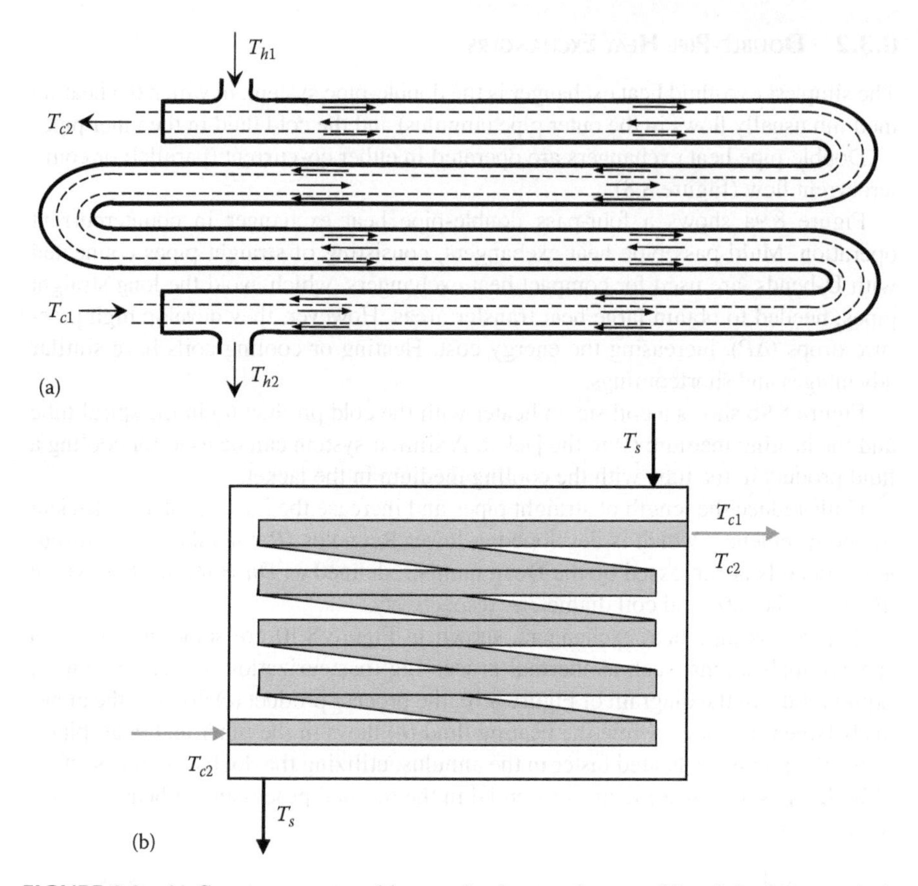

FIGURE 8.9 (a) Countercurrent multi-pass pipe heat exchanger; (b) coil (multi-pass pipe) heat exchanger: steam heating (*s*) of product (*c*).

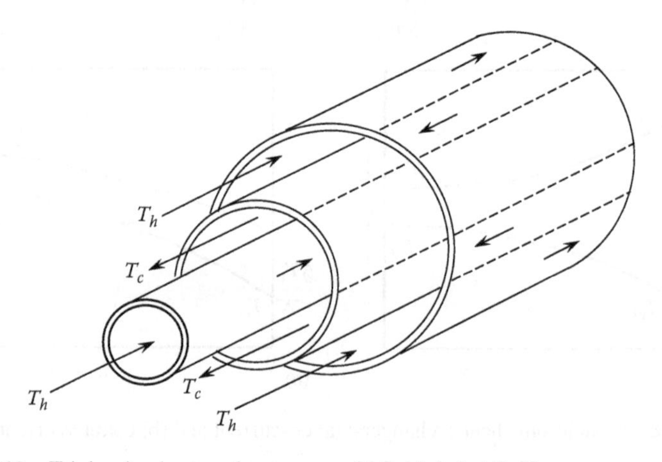

FIGURE 8.10 Triple-pipe heat exchanger: *c*, cold fluid; *h*, hot fluid.

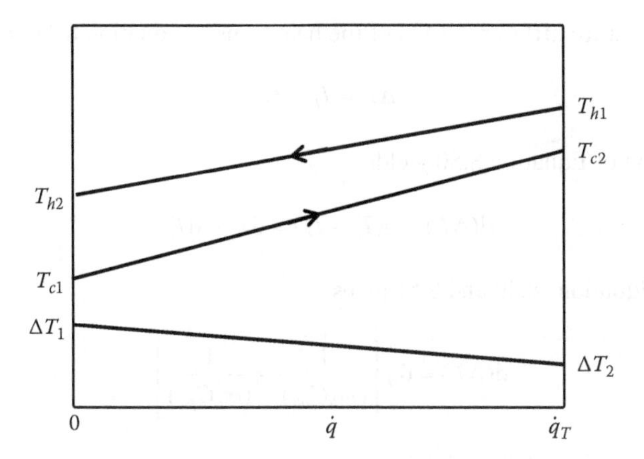

FIGURE 8.11 Fluid temperature versus heat flow rate in a countercurrent double-pipe heat exchanger.

In Figure 8.8, the heating medium (h) enters the heat exchanger at a temperature T_{h1} and exits at T_{h2}, while the corresponding temperatures of the cold fluid are T_{c1} and T_{c2}, respectively. The temperature differences between the two streams at the ends of the heat exchanger are $\Delta T_1 = (T_{h1} - T_{c1})$ and $\Delta T_2 = (T_{h2} - T_{c2})$ for co-current flow, and $\Delta T_1 = (T_{h2} - T_{c1})$ and $\Delta T_2 = (T_{h1} - T_{c2})$ for counterflow. It should be noted that although the temperatures (T) are usually expressed in degrees Celsius (°C), the temperature difference (ΔT) is numerically equal to degrees Kelvin (K), which are compatible with the temperature units of specific heat (C_p) and heat transfer coefficients (h, U).

Assuming that the specific heats C_{ph} and C_{pc} of the two fluids are constant during heat exchange, the temperature–heat flow rate relationship in a countercurrent heat exchanger is linear, as shown in Figure 8.11.

The rate of heat transfer dq (W) at a differential surface area dA (m^2) is given by the differential equation

$$\frac{\mathrm{d}q}{\mathrm{d}A} = U\Delta T \tag{8.48}$$

where

U [W/(m^2 K)] is the overall heat transfer coefficient
ΔT (K) is the temperature difference between the hot and cold streams

The integration of Equation 8.48 requires some substitutions, since (ΔT) does not refer to one fluid, but it is the temperature difference between two different fluids.

Assuming no heat losses, the differential heat transfer rate (dq) between the two fluids will be equal to the heat given up by the heating medium (h), or the heat absorbed by the cold fluid (c):

$$\mathrm{d}q = m_h C_{ph} \mathrm{d}T_h = m_c C_{pc} \mathrm{d}T_c \tag{8.49}$$

where

m_h and m_c (kg/s) are the mass flow rates
C_{ph} and C_{pc} [J/(kg K)] are the specific heats of the hot and cold fluids, respectively

The temperature difference (ΔT) of the heat transfer equation 8.48 is defined as

$$\Delta T = T_h - T_c \tag{8.50}$$

Differentiation of Equation 8.50 yields

$$\mathrm{d}(\Delta T) = \mathrm{d}(T_h - T_c) = \mathrm{d}T_h - \mathrm{d}T_c \tag{8.51}$$

Combining Equations 8.49 and 8.51 gives

$$\mathrm{d}(\Delta T) = \mathrm{d}q\left[\frac{1}{(m_h C_{ph})} - \frac{1}{(m_c C_{pc})}\right] \tag{8.52}$$

Combining Equations 8.48 and 8.52 gives

$$\frac{\mathrm{d}(\Delta T)}{\Delta T} = U\left[\frac{1}{(m_h C_{ph})} - \frac{1}{(m_h C_{pc})}\right]\mathrm{d}A \tag{8.53}$$

Assuming a constant overall heat transfer coefficient (U), Equation 8.53 can be integrated between (ΔT_1, ΔT_2) and (0, A) to obtain

$$\ln\left(\frac{\Delta T_2}{\Delta T_1}\right) = UA\left[\frac{1}{(m_h C_{ph})} - \frac{1}{(m_c C_{pc})}\right] \tag{8.54}$$

Integrating Equation 8.49 we obtain

$$q = m_h C_{ph}(T_{h1} - T_{h2}) = m_c C_{pc}(T_{c2} - T_{c1}) \tag{8.55}$$

Equation 8.55 yields, for countercurrent flow,

$$\left[\frac{1}{(m_h C_{ph})} - \frac{1}{(m_c C_{pc})}\right] = \frac{1}{q}[(T_{h1} - T_{c2}) - (T_{h2} - T_{c1})] \tag{8.56}$$

where $\Delta T_2 = (T_{h1} - T_{c2})$ and $\Delta T_1 = (T_{h2} - T_{c1})$.

Equation 8.54 yields the general equation for heat transfer in heat exchangers:

$$q = UA\frac{(\Delta T_2 - \Delta T_1)}{\ln(\Delta T_2 - \Delta T_1)} = UA\frac{(\Delta T_1 - \Delta T_2)}{\ln(\Delta T_1 - \Delta T_2)} \tag{8.57}$$

which is usually written as

$$\frac{q}{A} = U(\Delta T)_{LM} \tag{8.58}$$

where $(\Delta T)_{LM} = (\Delta T_1 - \Delta T_2)/\ln(\Delta T_1/\Delta T_2)$ is the log mean temperature difference.

The general heat transfer equation 8.58 is applied also to co-current or parallel flow heat exchangers.

If the overall heat transfer coefficient (U) changes linearly with the temperature, but the specific heats (C_{ph}, C_{pc}) are constant, the heat flux is given by the equation

$$\frac{q}{A} = \frac{(U_2\Delta T_1 - U_1\Delta T_2)}{\ln\left[(U_2\Delta T_1)/(U_1\Delta T_2)\right]} \quad (8.59)$$

where U_1, U_2 are the heat transfer coefficients at the two ends (1, 2) of the heat exchanger, respectively.

Equation 8.59 applies to both co-current and countercurrent heat exchangers.

If the heat transfer coefficient (U) and the specific heats (C_{ph}, C_{pc}) change nonlinearly with the temperature, the required heat transfer area (A) can be estimated by integrating the differential quantity $\mathrm{d}q/U\Delta T$) from $q = 0$ to $q = q$:

$$A = \sum_{0}^{q}\left(\frac{\mathrm{d}q}{U\Delta T}\right) \quad (8.60)$$

If the temperature differences at the two ends of the heat exchanger (ΔT_1 and ΔT_2) are close numerically (e.g., less than 10% difference), their arithmetic mean value $\Delta T_M = (\Delta T_1 + \Delta T_2)/2$ can be used, instead of the log mean temperature difference ΔT_{LM}.

The ratio of the two mean temperature values can be estimated from the equation

$$\frac{\Delta T_M}{\Delta T_{LM}} = \frac{[(\Delta T_1 + \Delta T_2)/2]}{[(\Delta T_1 - \Delta T_2)/\ln(\Delta T_1/\Delta T_2)]} = \frac{[(\Delta T_1/\Delta T_2) + 1]\ln(\Delta T_1/\Delta T_2)}{[2(\Delta T_1/\Delta T_2 - 1)]} \quad (8.61)$$

The ratio ($\Delta T_M/\Delta T_{LM}$) approaches the unity (1) when ($\Delta T_1/\Delta T_2$) = 1.

The countercurrent heat exchangers have a higher thermal efficiency than the co-current units, and they are generally preferred in industrial applications. This advantage can be proven by engineering analysis of the two heat exchanger systems, which takes into consideration the specific heats of the two streams (C_{ph} and C_{pc}).

8.3.3 Heat Transfer Coefficients

Design and operation of heat exchangers in food processing operations requires reliable heat transfer coefficients, which are difficult to predict by theoretical calculations or obtain from the literature, due to the complex structure of food materials. Experimental data and empirical correlations are limited to specific food products and specified applications.

Heat transfer coefficients calculated from theoretical equations refer to clean heat transfer surfaces. However, in industrial practice, the heat transfer surfaces are fouled by various deposits or scales, which reduce considerably the heat transfer coefficients.

The food products are heated or cooled flowing inside tubes or other conduits, and the inner heat transfer coefficient (h_i) is usually the controlling factor in the overall heat transfer operation, due to the high viscosity and the fouling tendency of the food fluid.

The heating or cooling medium in the outside has a higher heat transfer coefficient (h_o), due to the lower viscosity and the absence of any significant fouling.

The heat transfer coefficients (h_i, h_o) in a heat exchanger are normally based on the mean temperature of each fluid. They can be determined experimentally by two methods:

1. Using a system in which the main resistance to heat transfer is on one surface, usually the product being heated or cooled, e.g., heating a high viscosity fluid or air inside a tube (h_i) with condensing steam in the outside (h_o). In this case, the inside heat transfer coefficient (h_i) is approximately equal to the overall heat transfer coefficient (U), determined experimentally using Equation 8.58, since the resistances to heat transfer of the metal wall (x/λ) and the outside surface ($1/h_o$) are very small and can be neglected. Here, (x) is the wall thickness and (λ) is the metal thermal conductivity. The thermal resistances involved in heat transfer are expressed by Equations 8.6 through 8.8.
2. The graphical (Wilson) method, in which only one thermal resistance is changed, while the other is kept constant. This method is applied to fluids in turbulent flow inside tubes, heated with condensing steam in the outside (high heat transfer coefficient).

In turbulent flow inside tubes, the Sieder–Tate equation 8.17 is applicable. Assuming constant physical properties of the fluid at a mean temperature, this equation is simplified to the equation,

$$h_i = \alpha G^{0.8} \tag{8.62}$$

where

$G = u\,\rho$ = mass flux [kg/(m^2 s)]

(α) is a constant (SI units)

The overall heat transfer coefficient, based on the inside surface of the heat exchanger (U_i), is given by Equation 8.10, which, when combined with Equation 8.62, gives

$$\frac{1}{U_i} = \frac{1}{(\alpha G^{0.8})} + b \tag{8.63}$$

where $b = (d_i/d_L)\,(x/\lambda) + (d_i/d_o)\,(1/h_o)$, and $d_L = (d_o - d_i)/\ln(d_o/d_i)$ is the log mean diameter of tube diameters (d_o, d_i).

For thin walls $d_L = (d_o + d_i)/2$.

Experimental data are obtained from the given heat exchanger for the variation of the overall heat transfer coefficient (U_i), calculated from Equation 8.58, as a function of the mass flow rate (G), which are plotted in a diagram of ($1/U_i$) versus ($1/G$). If the Sieder–Tate equation is applicable, then this plot is approximated by a straight line with a slope (α) and an intercept (b), as shown in Figure 8.12.

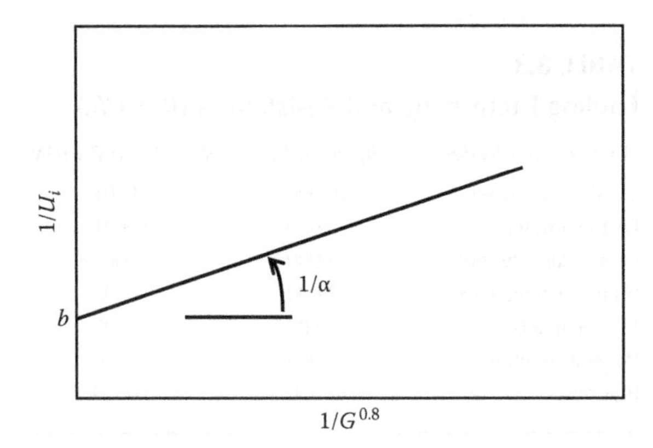

FIGURE 8.12 Plot of experimental heat transfer data (Equation 8.63).

From the experimentally determined values of (α) and (b), the heat transfer coefficients h_i and h_o are determined, using Equations 8.62 and 8.63.

8.3.4 Fouling of Heat Exchangers

In most industrial applications, the heat transfer surfaces are covered with various deposits, which reduce considerably the heat transfer coefficients and, therefore, the thermal efficiency of the heat exchangers. The deposits are formed by various physicochemical processes, caused by heating of complex foods, such as polymerization, denaturation of proteins, agglomeration of colloid particles, and crystallization of salts. Large deposit quantities are formed on the heat surfaces of food evaporators.

The process of fouling or scaling of heat transfer surfaces can be reduced by increasing the flow rate or by agitation of the fluid near the heat transfer surface, or by physicochemical treatments of the fluid. Periodic cleaning of fouled heat transfer surfaces is necessary in several food processing operations.

Fouling is taken into consideration in the design of heat exchangers. Empirical fouling factors (h_f) or fouling resistances ($R_f = 1/h_f$) are added to the heat transfer equations to account for anticipated fouling. For plane heat transfer surfaces or for thin tubes of large diameters (d_i close to d_o), the overall heat transfer coefficient (U) is given by the equation

$$\frac{1}{U} = \frac{1}{h_\iota} + \frac{x}{\lambda} + \frac{1}{h_o} + \frac{1}{h_f} \tag{8.64}$$

For normal size tubes ($d_i < d_o$):

$$\frac{1}{U_i} = \frac{1}{h_\iota} + \frac{1}{h_{f\iota}} + \left(\frac{d_i}{d_o}\right)\frac{x}{\lambda} + \left(\frac{d_i}{d_L}\right)\left(\frac{1}{h_o} + \frac{1}{h_{fo}}\right) \tag{8.65}$$

where

$h_{f\iota}$ and h_{fo} are the fouling factors of the inner and outer surface, respectively
$d_L = (d_o - d_i)/\ln(d_o/d_i)$ is the log mean diameter difference

TABLE 8.3
Fouling Factors (h_f) and Resistances ($R = 1/h_f$)

Heat Transfer System	h_f, W/(m² K)	$R = 1/h_f$, (m² K)/W
Condensing steam	10,000	0.0001
Distilled water	10,000	0.0001
Brine (salt solution)	5,000	0.0002
Fruit/vegetable juices	2,000	0.0005
Vegetable oils	1,500	0.0007
Polymer solutions	1,000	0.0010
Flue gases	500	0.0020

Table 8.3 gives some empirical fouling factors (h_f) and fouling resistances ($R = 1/h_f$) of industrial heating systems.

In the preliminary design of industrial heat exchangers, empirical values of the heat transfer coefficient (U) are frequently used, determined either experimentally or estimated from operating industrial systems. Typical U values of interest to food engineering operations are given in Table 8.4.

8.3.5 Shell and Tube Heat Exchangers

Shell and tube heat exchangers are an extension of the simple double-pipe heat exchanger and they are used widely in the process industries, because they offer large heat transfer surfaces in a small volume (compact units) and they are economical.

A shell and tube heat exchanger consists of a bunch of parallel tubes through which flows the fluid to be heated (or cooled), while the heating (or cooling) medium flows in the shell (jacket). Most units are operated in counterflow, as shown in the diagram of Figure 8.13. The liquid in the tubes flows at relatively high velocity, which increases the heat transfer coefficient (h_i) and reduces the fouling (formation of deposits). The shell contains a number of baffles, i.e., metal deflectors, which increase the turbulence of the heating medium and improve the heat transfer coefficient (h_o).

TABLE 8.4
Empirical Overall Heat Transfer Coefficients (U)

Heat Transfer System	U, W/(m² K)
Condensing steam/water	2,000–10,000
Water/water	1,000–2,500
Steam or water/liquid food	1,000–1,500
Steam or water/viscous food	500–1,000
Steam or water/vegetable oils	500–800
Steam or water/air	50–100
Air(gas)/air(gas)	20–50

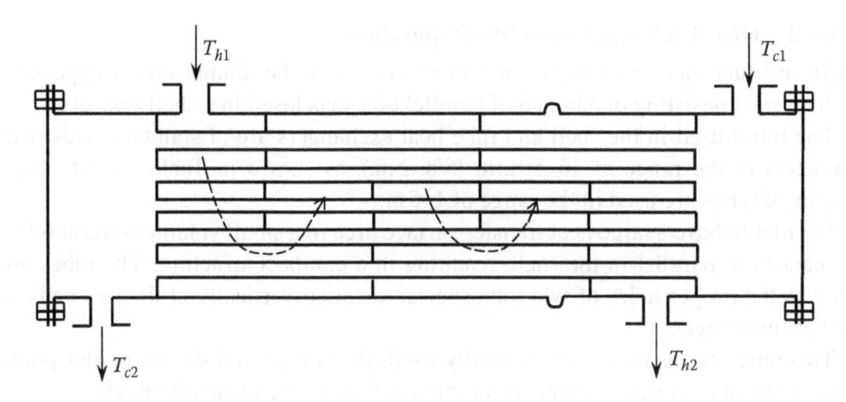

FIGURE 8.13 Countercurrent flow (1–1) shell and tube heat exchanger.

The shell and tube heat exchanger of Figure 8.13 is a countercurrent flow (1–1) unit, meaning that there is only one pass of the cold and one pass of the hot streams.

In industrial applications more cold and hot streams may be included, e.g., shell and tube heat exchangers (1–2), (2–2), (2–4).

The cold fluid enters at one end of the heat exchanger at temperature T_{c1} and leaves the other end at T_{c2} (Figure 8.13). At the same time, the hot fluid enters the opposite end at temperature T_{h1} and leaves at T_{h2}. The end temperature differences between the two streams, which are needed for heat transfer calculations are, $\Delta T_1 = (T_{h1} - T_{c2})$ and $\Delta T_2 = (T_{h2} - T_{c1})$ for countercurrent flow. The end temperature differences in a co-current (parallel) flow (1–1) heat exchanger are $\Delta T = (T_{h1} - T_{c1})$ and $\Delta T_2 = (T_{h2} - T_{c2})$.

The two end temperature differences are the same as in the double-pipe heat exchangers, shown in Figure 8.8.

8.3.5.1 Heat Transfer Surface

The heat transfer area (A) of a one-pass (1–1) shell and tube heat exchanger, required for a heating or cooling load (rate of heat transfer) q (W), is calculated from the basic heat transfer equation

$$\frac{q}{A} = U(\Delta T)_{LM} \tag{8.58}$$

where $(\Delta T)_{LM} = (\Delta T_1 - \Delta T_2)/\ln(\Delta T_1/\Delta T_2)$ is the log mean temperature difference.

The thermal efficiency of multi-pass heat exchangers (1–2, 2–4, etc.) is lower than the efficiency of one-pass (1–1) units. The reduced efficiency is caused by the presence of co-current heat exchange sections in the complex system. It is expressed by a correction factor $Y < 1$ of the log mean temperature difference $Y(\Delta T)_{LM}$ of Equation 8.58. The correction factors (Y) are given in diagrams of the literature as functions of the temperature differences ΔT_1 and ΔT_2. The dimensionless factor Y varies in the range of 0.5–1.0.

8.3.5.2 Heat Exchanger Structure/Operation

Shell and tube heat exchangers are an extension of the simple double-pipe heat exchangers, consisting of a bunch of parallel tubes enclosed in a shell (jacket).

The tubes used in the shell and tube heat exchangers are of standard sizes with diameters in the range of 16–50 mm (5/8–2 in.), as shown in Table 6.4. Standard lengths of tubes are used in the range of 1–5 m.

In order to have a large heat transfer surface area in a small volume, several tubes are packed in parallel in the shell, resulting in a compact structure. The tubes are fitted to the proper holes of two tube sheets, installed vertically at the two ends of the heat exchanger.

Two tube arrangements are normally used, the square and the triangular pitch. A larger number of tubes can be accommodated using the triangular pitch.

In heat transfer calculations, the inner surface of the tubes (A_i) along with the corresponding overall heat transfer coefficient (U_i) are normally used in Equation 8.58, since the heat transfer resistance of the inner film ($1/h_i$) is generally much higher than the resistances of the outer film ($1/h_o$) and the metal wall (x/λ).

The metal wall resistance (x/λ) may control the heat transfer rate in some cases, particularly when stainless steel (SS) tubes are used, since the thermal conductivity of SS is low [18 W/(m K)] compared to the thermal conductivity of other metals, e.g., copper, 380 W/(m K), and aluminum, 320 W/(m K). However, stainless steel is used widely in food heat exchangers because of its advantages (noncorrosive, nontoxic, cleanable, sturdy).

The pressure drop of the fluids through heat exchangers should be considered for an economic operation. The pressure drop (ΔP) of the fluid in the tubes is generally higher than outside, since high velocities and small diameter tubes are used in order to increase the heat transfer coefficient. An optimum fluid velocity may be calculated in heat exchange system for a minimum of heat transfer surface area and a maximum (ΔP).

8.3.6 Plate Heat Exchangers

The plate heat exchangers (PHE) are used widely in the heating and cooling of fluid foods because they are efficient compact units, packing large heat transfer areas in a small volume. High heat transfer coefficients are obtained with small pressure drops, and the plates can be cleaned thoroughly by disassembling the unit. The PHE are used widely in the thermal processing of heat-sensitive fluid foods of low-to-moderate viscosity, such as the pasteurization of milk and fruit juices.

Figure 8.14 shows a diagram of a typical PHE.

The PHE consist of several corrugated plates of effective heat transfer area 0.02–4.75 m^2 and plate thickness 0.6–0.9 mm. They are clamped together in a frame with a channel opening (gap) of 2–6 mm. Special gaskets made of rubber or Teflon are used to prevent fluid leakage from the plate joints. The temperature differences between the hot (h) and cold (c) fluids at the two ends of the exchanger are defined the same way as in the tubular heat exchangers. Thus, for the countercurrent flow system of Figure 8.14, $\Delta T_1 = (T_{h1} - T_{c2})$ and $\Delta T_2 = (T_{h2} - T_{c1})$.

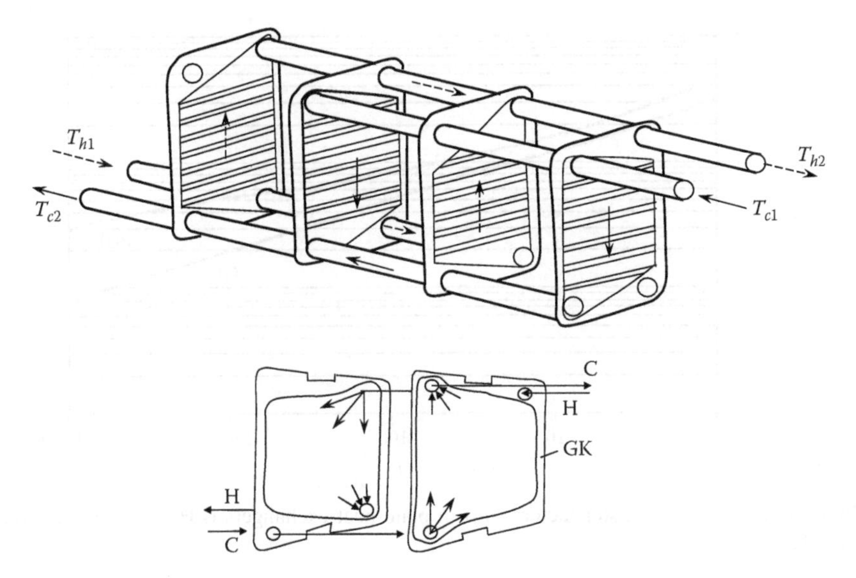

FIGURE 8.14 Diagram of a plate heat exchanger.

The plates are connected with tubes of 25–400 mm diameter, which supply the two fluids in countercurrent or co-current operation. PHE are used in a wide range of sizes, up to 1000 plates per unit, with flow rates of 0.5–5000 m³/h.

The structure of heat transfer plates induces turbulent flow at much lower Reynolds numbers than in normal pipe (tube) flow ($Re > 5$). The heat transfer coefficient (h) in a plate channel can be estimated from empirical diagrams or equations, such as

$$Nu = 0.352 Re^{0.54} Pr^{1/3} \tag{8.66}$$

The Re number is based on the equivalent diameter of the flow channel, which is four times the hydraulic diameter r_h, defined in Equation 6.7.

Heat transfer in PHE is, in general, higher than in tubular units. Figure 8.15 compares the heat transfer factors (j_H) in PHE with a generalized correlation (GL) of published heat transfer data.

The sizing of the PHE is based on the heat load q (W) and the overall (total) heat transfer coefficient (U), which may vary in the range of 0.8 kW/(m² K) (vegetable oils) to 3.0 kW/(m² K) (aqueous liquid foods). In the PHE, the log mean temperature difference is approximately equal to the arithmetic temperature difference.

The total number of plates (N) will be

$$N = \frac{A}{A_o} \tag{8.67}$$

where

A (m²) is the total heat transfer surface
A_o (m²) is the surface of one plate

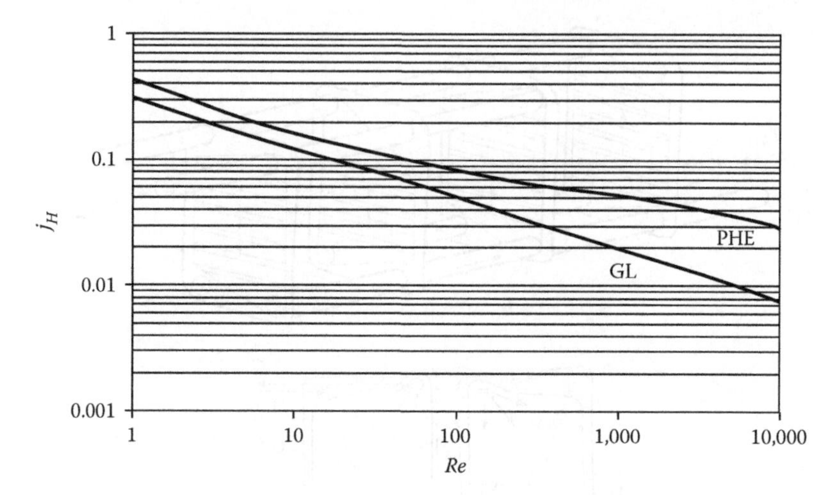

FIGURE 8.15 Heat transfer factors (j_H) in plate heat exchangers (PHE) and generalized (GL) data.

The pressure drop Δp (Pa) in a PHE is estimated from an empirical equation analogous to the Fanning relation:

$$\Delta p = 8j_p\left(\frac{L}{d_e}\right)\left(\frac{\rho u^2}{2}\right) \tag{8.68}$$

where

L (m) is the total length of flow path
d_e (m) is the equivalent diameter
j_p is a pressure drop factor (dimensionless) which is related to the *Re* number

$$j_p = 1.25Re^{-0.3} \tag{8.69}$$

The design pressure drop (Δp) decreases from 25 to 5 bar, as the operating temperature is increased from about 0°C to 200°C. There is an upper limit in the operating temperature, due to the heat damage of rubber gasket materials. Typical industrial PHE, used in food processing, have heat transfer surface areas up to 100 m^2 (about 250 plates).

8.3.7 Agitated Heat Exchangers

Agitated heat exchangers are applied to the heating or cooling operations of viscous foods, suspensions or pulps, which are difficult to handle in tubular or plate heat transfer units. They are more expensive to construct and operate, and their capacity is lower than the conventional units.

Two major types of agitated heat exchangers are used in food processing: (a) the scraped-surface and (b) the agitated-vessel, as shown in Figure 8.16.

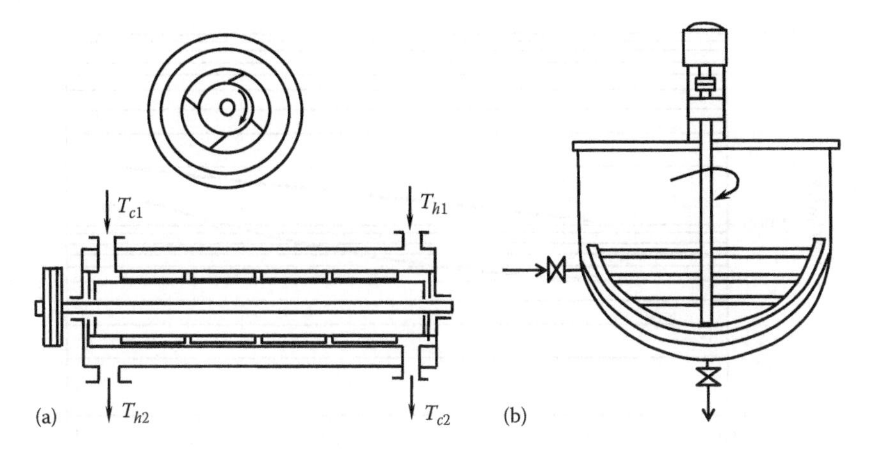

FIGURE 8.16 Agitated heat exchangers: (a) scraped-surface and (b) agitated-vessel.

8.3.7.1 Scraped-Surface Heat Exchangers

Scraped-surface heat exchangers (SSHE) are used for viscous, fouling or crystallizing food liquids, such as ice cream and margarine. They consist of an inner pipe, usually 15 cm diameter, carrying the product, and an outer pipe, 20 cm diameter, with the heating or cooling medium. The wall thickness is higher than in the conventional tube or PHE, for withstanding the high operating pressure. The inner pipe contains a shaft with several knives which scrape and transport the product by a screw action at rotational speeds of 500–700 RPM. The product is transported through the SSHE with a positive displacement pump at pressure drops (Δp) up to 10 bar. The SSHE can be operated in the horizontal or vertical position in systems of two or more connected units.

Heat transfer in the SSHE is by a complex mechanism of scraping the heated (or cooled) product from the surface and mixing it into the bulk product. Theoretical analysis is difficult and, for practical purposes, empirical overall heat transfer coefficients (U) are used with typical values of 0.5–1.0 kW/(m^2 K).

Because of the thicker walls used in the SSHE, temperature differences between the heating (or cooling) medium and the product are much higher than in the tubular or PHE, e.g., 25°C versus 5°C.

8.3.7.2 Agitated-Vessel Heat Exchangers

Figure 8.16 shows a scraping anchor-type agitator in a kettle used for batch heating of food products with saturated steam in the jacket of the kettle. Types of agitators, used in agitated-vessels, include paddles, blades, and propellers.

The heat transfer coefficient (h) in agitated-vessels is calculated from the equation

$$Nu = \alpha Re^{0.67} Pr^{1/3} \left(\frac{\eta}{\eta_w} \right)^{0.14} \tag{8.70}$$

where the parameter α (dimensionless) depends on the type of agitator and product. For scraping anchor and Newtonian fluids $\alpha = 0.55$; for non-Newtonian fluids $\alpha = 1.47$.

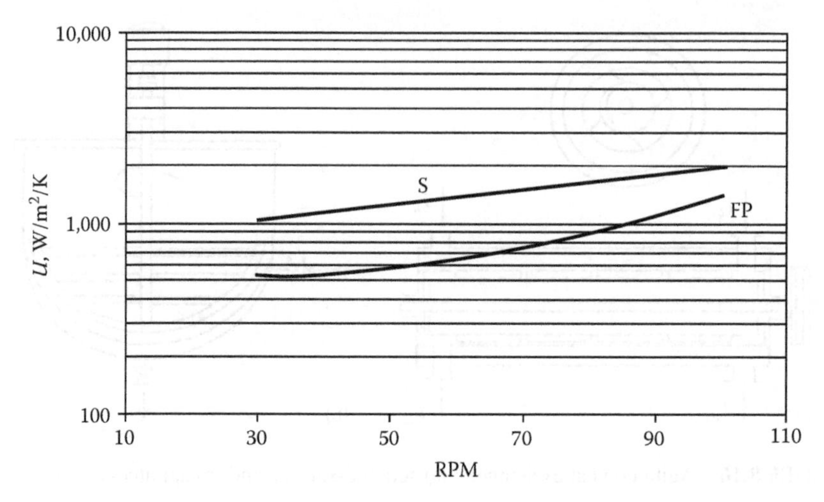

FIGURE 8.17 Heat transfer in an agitated kettle: (S) 40% sucrose solution and (FP) fruit puree.

For other types of agitators the literature values of parameter (α) are in the range of 0.36–0.74. The parameter (α) is affected by the geometry of the tank, the location of the agitator, and the presence of baffles in the vessel.

The *Re* number in Equation 8.70 is defined as $Re = (d_i^2 N\rho)/\eta$, where d_i (m) is the diameter of the impeller (anchor), N (1/s) is the rotational speed (N = RPM/60) and η, η_w (Pa s) are (the apparent) viscosities of the fluid product at the bulk and wall temperature, respectively.

The Nusselt number is defined as $Nu = hd_T/\lambda$, where d_T (m) is the tank (kettle) diameter and λ [W/(m K)] is the thermal conductivity of the product.

Heating of food products in agitated-vessels is normally performed using saturated steam in the jacket of the vessel. The steam condensate should be removed continuously from the jacket through a steam trap, keeping free the heating surface for steam condensation and thus increasing the heat transfer coefficient.

Figure 8.17 shows a diagram of the overall heat transfer coefficient (U) as a function of the agitation speed in an agitated kettle. The heat transfer coefficient increases exponentially with increasing the speed of agitation. Agitation has a stronger effect on the heating of pseudoplastic fruit puree than on Newtonian sucrose solutions.

8.3.7.3 Batch Heating

Agitated-vessels heat or cool food products in a batch operation, contrary to all other heat exchangers discussed previously, where continuous operation is the rule. The temperature change in a batch operation can be estimated from engineering data of the system, assuming that the product is thoroughly mixed throughout the operation. The same engineering analysis can be applied to closed containers (e.g., cans) containing low viscosity products, which can be mixed thoroughly by mechanical agitation.

The heat (enthalpy) change of the product by a temperature change (dT) at time (dt) will be ($V\rho C_p$) (dT/dt), which will be equal to the heat transferred in the system:

$$(V\rho C_p)\left(\frac{dT}{dt}\right) = (UA\Delta T) \tag{8.71}$$

where

V, ρ, C_p are the volume (m^3), density (kg/m^3), and specific heat [kJ/(kg K)] of the product, respectively
U [kW/(m^2 K)] is the overall heat transfer coefficient
A (m^2) is the heat transfer surface area

Assuming constant thermophysical properties of the product, and constant overall heat transfer coefficient, Equation 8.71 can be integrated for product temperature $T_1 - T_2$, yielding the following equation:

$$\ln\left[\frac{(T_{h/c} - T_1)}{(T_{h/c} - T_2)}\right] = \left[\frac{UA}{(V\rho C_p)}\right]t \tag{8.72}$$

where

($T_{h/c}$) is the temperature of the heating (or cooling) medium, assumed to be constant during the operation
(T_1, T_2) are the product temperatures at the beginning and end
t (s) is the time required for the operation

8.3.8 Fin Heat Exchangers

Fin or extended surface heat exchangers are used for the heating or cooling of air or other gases, where low external heat transfer coefficients at the metal/air surface (h_o) are limiting the heat transfer rate. In these exchangers, the heat resistance at the metal/air surface ($1/h_o$) is much higher than the thermal resistance of the metal wall (x/λ) and the inner resistance of the fluid ($1/h_i$) and the overall heat transfer coefficient (U) is very close to the external heat transfer coefficient (h_o), i.e., $U_o = h_o$, based on the external heat transfer area A_o. Thus, large (extended) heat transfer areas (A_o) are needed for high heating or cooling loads (q), according to the basic equation $q = UA\Delta T$.

As an example, for heating of air by forced convection, about $h_o = 0.10$ kW/(m^2 K), saturated steam can be used in a tube, where approximately $h_i = 10$ kW/(m^2 K). The metal wall thermal resistance is relatively negligible and, therefore, the overall heat transfer coefficient, based on the outer surface, will be about $U_o = 0.10$ kW/(m^2 K).

The extended heat exchangers consist of a series of tubes with the heating or cooling medium inside and several cross or longitudinal fins outside, as shown in Figure 8.18.

Large extended surface heat exchangers are constructed in compact form using several tubes connected in parallel with several fins. The air is forced by fans through the exchanger, and the required pressure drop should be taken into consideration.

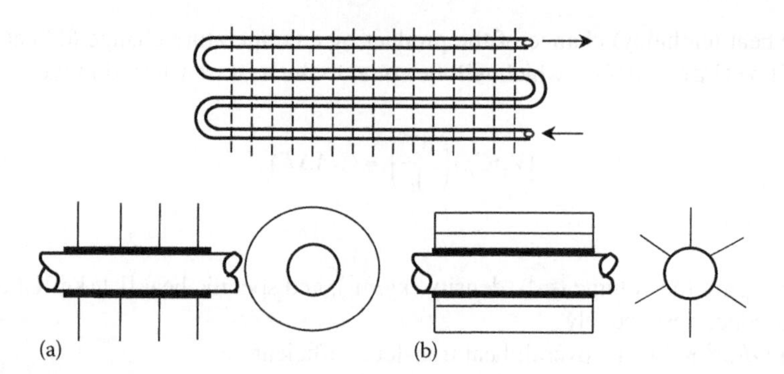

FIGURE 8.18 Fin heat exchangers: (a) cross-fins and (b) longitudinal fins.

Fin-fan coolers may be used for the cooling of large amounts of industrial water in closed circuits, instead of the cooling towers, which require large space and they may pollute the environment.

8.3.9 Direct Heat Transfer

Direct heat transfer consists of mixing steam with a liquid food product, resulting in very fast heating. The saturated steam used for direct heating of, for example, milk, must be clean, tasteless, and free of any dissolved gases and toxic components, which might come from a conventional steam boiler (culinary steam). The feed water to the boiler should be of potable quality and contain no chemical additives.

Mixing of steam with the liquid food can be accomplished by two methods: (a) steam injection, i.e., injecting steam into the fluid food through a special nozzle, and (b) steam infusion, i.e., mixing the steam with a film and droplets of the liquid food, sprayed in a special infusion vessel (Figure 8.19).

Direct heating of foods minimizes fouling, by eliminating the heat transfer surface. The condensed water from the heating steam is usually removed by vacuum

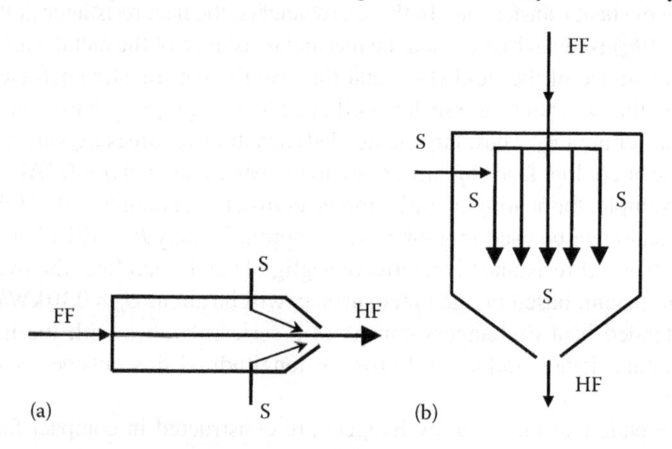

FIGURE 8.19 Direct steam heating of liquid food: (a) steam injection and (b) steam infusion. FF, fluid food; HF, heated food; S, steam.

evaporation (flashing), which reduces also the temperature and removes off-flavors and odors produced in the liquid product, as for example, in the UHT sterilization (Chapter 9).

8.4 ELECTRICAL/RADIATION HEATING

The need for faster heating of heat-sensitive food products has led to the development of the heat generation processes, which are based on the conversion of electrical energy into heat within the food material. Electrical heating can be achieved either as a result of electrical resistance of the material to the passage of electric current (ohmic heating) or by friction during molecular rotation of water and other molecules (microwave [MW] or dielectric heating).

8.4.1 Ohmic Heating

Ohmic or electric resistance heating generates heat within the food material by electrical current. The heat rate produced q (W) is a function of the electrical resistance R (ohm), the current I (ampere), and the applied electrical potential V (volts) in the food product. According to the Joule's law, the heat generation rate is given by the equation:

$$q = VI = I^2 R \tag{8.73}$$

This process is suitable for heating particulate foods suspended in liquid, which cannot be heated evenly by conventional heating. The food suspension is heated by passing through special electrodes. Ohmic heating is under development and it could be applied to the sterilization (inactivation of spoilage microorganisms) or in blanching (enzyme inactivation) of food products, such as particulate foods, which are difficult to process in conventional thermal processing (Chapter 9).

Alternating current (50–60 Hz) is supplied to an ohmic heating column, which may contain four or more electrodes (Figure 8.20). The electrodes are connected using stainless steel spacer tubes, lined with insulating plastics, e.g., polyvinylidene.

The heating column is operated in a vertical position with upward flow of the product in the upward direction. Commercial-scale ohmic heating systems have been designed for power outputs of 75 and 750 kW, corresponding to product capacities of approximately 750 and 3000 kg/h, respectively, for a temperature rise of water of 75°C.

8.4.2 Microwave and Dielectric Heating

The most common radiations used in electrical heating of foods are the MW at frequencies 915 or 2450 MHz, and dielectric or radio frequency (RF) at frequencies 3–30 MHz (1 Hz = 1/s). The wavelengths (λ) in the air, corresponding to these frequencies are 32.8 cm (915 MHz), 12.3 cm (2450 MHz), and 10–1 m (3–30 MHz).

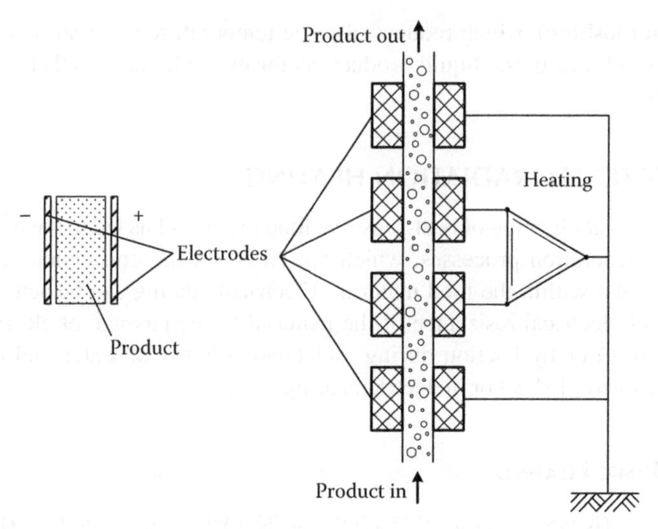

FIGURE 8.20 Diagram of an ohmic heating system.

The heat generation rate q (W) per unit volume of a material is given by the equation

$$q = 0.56 \times 10^{-10} E^2 \varepsilon' \omega \tan \delta \tag{8.74}$$

where

- E (V) is the electric field strength
- ω (1/s) is the frequency
- ε' is the dielectric constant
- $\tan \delta$ is the loss tangent of the material, defined as $\tan \delta = \varepsilon''/\varepsilon'$, where ($\varepsilon''$) is the dielectric loss

The dielectric constant (ε') is a measure of the MW rate of penetration in the food. It indicates the ability of the material (food) to store electrical energy. The dielectric loss (ε'') is a measure of how easily is this energy dissipated, and in general, ($\varepsilon'/\varepsilon''$) > 1.

The dielectric properties are difficult to predict and they are normally measured experimentally. They depend on the composition of the food material, and they usually decrease with increasing temperature, with liquid water having the highest values. Typical values of dielectric constants and dielectric losses (ε', ε'') are: water (80, 20), fruit (55, 15), meat (40, 10), vegetable oils (2.5, 1.5), and ice (3.2, 0.003).

The very low dielectric constants of ice are of importance in the MW thawing of frozen foods, since the liquid water tends to be heated much faster than the melting of ice, damaging the food quality. Thus, MW thawing should be controlled carefully.

In addition to the dielectric constants, the penetration depth of MW and RF radiation in the materials is of fundamental importance in heating applications. The penetration depth (d), is defined as the depth where the radiation intensity decays by 37% (1/e) of its surface value:

$$d = \frac{\lambda_o(\varepsilon')^{0.5}}{2\pi\varepsilon''} = \frac{\lambda_o}{2\pi(\varepsilon')^{0.5}\tan\delta} \tag{8.75}$$

where (λ_o) is the wavelength of the radiation in the air.

According to Equation 8.75, the penetration of radiation in a material increases with increasing wavelength. Thus, MW radiation at 915 MHz penetrates more than the 2450 MHz radiation, $d = 30$ mm versus 10 mm.

Application of MW and RF to the heating of food products requires the knowledge of dielectric properties (ε', ε'') and the thermal and transport properties (C_p, λ, α, D) of the food materials.

Microwave equipment, such as magnetrons, waveguides, and applicators are used in various arrangements to produce and direct the radiation to the food product, which is usually moved on a conveyor belt. For pasteurization and cooking at temperatures below 100°C, open conveying systems are used with a hood to keep the water vapors and prevent surface drying-out of the product. For high-temperature sterilization (110°C–130°C), closed systems are used. When food in plastic trays of pouches is sterilized, overpressure (air or steam) is required to prevent bursting of the packages. The presence of steam during MW–RF treatment reduces also the corner and edge effects and the cold spots of the food products.

Food applications of MW–RF, in addition to drying (Chapter 11), include tempering of frozen foods, precooking of meat and other foods, and pasteurization and sterilization.

In tempering of frozen foods, the frozen product is heated to a temperature just below the freezing point and then allowed to fully thaw at low temperature, reducing sharply the normal thawing time. The process is applied to the packaged frozen food, without taking the packages apart, and it reduces significantly drip losses (more economical than conventional thawing). Thawing of frozen foods is discussed in Chapter 12.

Precooking of meat, poultry, and other foods by MW–RF is faster than conventional heating. The products are packaged in plastic or tray containers.

8.4.3 Infrared Heating

The temperature of the infrared (IR) radiators determines the spectral distribution and the maximum emitted radiation flux. The penetration of IR radiation is limited. Short wavelength radiation ($\lambda < 1.25\,\mu$m) is preferred in food processing, because it has higher penetrating power. Radiation heating should avoid the overheating and burning of the food surface.

Two types of IR radiators are used in food processing: (a) gas heated, which give large wavelengths, and (b) electrically heated, which include the tubular and ceramic heaters (long λ) and the quartz and halogen heaters (short λ). Some high-intensity radiators require water or air-cooling to avoid overheating.

Water vapor and carbon dioxide absorb part of the IR radiation, reducing the efficiency of IR heating of foods, e.g., during the end of the baking process, when the product may be sprayed with water.

IR radiation is used for the drying of vegetables, in baking/roasting, and in frying of foods. The main advantages are fast heating, high efficiency, and easy process control.

8.4.4 Solar Heating

The solar radiation consists of a wide range of wavelengths, some of which, such as IR radiation, can be used for heating operations. Thermal solar energy is used mainly for heating water for domestic or for drying agricultural and food products. From the engineering standpoint, thermal solar energy is a low-intensity energy source, which has a limited industrial potential. Large surfaces of solar collectors are needed for industrial applications.

Usually, flat-plate collectors are used with either natural or forced circulation of the air. The collectors are usually installed on the roofs of houses or industrial buildings, facing the sun. For very large surfaces, the collectors can be installed in open areas. Solar energy and, in general, renewable energy sources are important and economical, particularly during energy crises, when the cost of fuel energy increases sharply; they have the added advantage of not polluting the environment, e.g., by carbon dioxide (fuel combustion).

The common flat-plate collector consists of a black plate, which absorbs the incident solar radiation, a transparent cover, and insulation material. The incident solar energy (insolation) varies with the geographical location and the season of the year. A typical insolation for a hot climate would be 0.6 kW/m^2 with an average sunshine time of 7 h/day. This energy corresponds to about $0.6 \times 3600 = 2.16\,MJ/m^2\,h$ or 15 MJ/m^2 day.

Two brief examples of solar heating are given here: (1) For water heating to be used in industrial or home applications. The energy 2.16 MJ/m^2 h corresponds to the heating of 8.6 kg/h water from 20°C to 80°C (for intermittent 7 h operation). (2) For a drying application, the evaporation of water at 40°C requires theoretically 2.4 MJ/kg and practically about 3 MJ/kg. Therefore, the mean evaporation rate of water in a solar dryer will be about $2.16/3 = 0.72\,kg/m^2\,h$ (intermittent operation 7 h/day). It is evident that large surface areas of solar collectors are needed in order to heat or dry industrial quantities of water or product.

A serious limitation of solar heating is intermittent or periodic operation of the collectors due to partial availability of solar radiation during the day (night, clouded sky, etc.). Solar systems are used mostly as an auxiliary or supplementary heat source to conventional fuel (gas or oil) heating systems. Excess solar energy collected during high insolation can be stored as hot water in large (insulated) water tanks and used during the night or during periods of low insolation.

APPLICATION EXAMPLES

Example 8.1

Air at atmospheric pressure and temperature 20°C is transported at the rate of 100 kg/h through a pipe of inner diameter 3 cm and length 10 m.

Calculate

1. The convection heat transfer coefficient (h)
2. The pressure drop (Δp)
3. The values of (h) and (Δp) when the airflow rate is doubled

Solution

The flow of air is considered noncompressible and the equations described in Chapters 6 and 8 of this book are applicable.

1. The mass flow rate of air in the pipe will be $m = 100/3600 = 0.028$ kg/s. The Reynolds number will be $Re = (du\rho)/\eta = (dG)/\eta$, where $d = 0.03$ m (pipe inner diameter), $\eta = 0.018$ mPa s $= 1.8 \times 10^{-5}$ Pa s $= 1.8 \times 10^{-5}$ kg/(m s) is the air viscosity (Chapter 5).

   $G = (u\rho) = 4m/(3.14\, d^2) = 4 \times 0.028/(3.14 \times 0.03^2)$ is the mass flux [kg/(m^2 s)] of the air.

   Therefore, $Re = (0.03 \times 4 \times 0.028)/[(3.14 \times 0.03^2)(1.8 \times 10^{-5})] = 66{,}053$ and the flow is well-developed turbulent.

   The heat transfer coefficient (h) is calculated from the Sieder–Tate equation 8.17:

   $Nu = 0.023\ Re^{0.8}\ Pr^{1/3}(\eta/\eta_w)^{0.14}$. For atmospheric air, approximately $(\eta/\eta_w)^{0.14} = 1$.

   The Prandtl number is calculated as $Pr = (C_p\eta)/\lambda$. From Chapter 5, for atmospheric air at 20°C, $C_p = 1{,}000$ J/(kg K), $\eta = 1.8 \times 10^{-5}$ kg/(m s), $\lambda = 0.02$ W/(m K), and $Pr = (1{,}000 \times 1.8 \times 10^{-5})/0.02 = 0.90$.

   $Nu = (hd)/\lambda = 0.03h/0.02 = 0.023 \times (66{,}053)^{0.8} \times 0.90^{1/3} = 0.023 \times 7{,}176 \times 0.966 = 159.4$ and $h = (159.4 \times 0.02)/0.03 = 106.3$ W/(m^2 K).
2. The pressure drop in the 3 cm × 10 m pipe is calculated from the Fanning equation $\Delta P/\rho = 4f(L/d)(u^2/2)$, where the friction factor (f) is calculated from the Blasius equation 6.16:

   $$f = \frac{0.079}{Re^{0.25}} \quad \text{or} \quad f = \frac{0.079}{(66{,}053)^{0.25}} = \frac{0.079}{16} = 0.005$$

   The air velocity (u) is calculated from the mass flow rate $m = 0.028$ kg/s, $u = 4m/(3.14d^2\, \rho)$, where the air density is $\rho = 1.2$ kg/m^3 (Chapter 5). Therefore, $u = (4 \times 0.028)/(3.14 \times 0.03^2 \times 1.2) = 33.0$ m/s. The pressure drop in the pipe will be $\Delta p/\rho = 4 \times 0.005 \times (10/0.03)(33.0^2/2) = 3630$ and $\Delta p = 3630 \times 1.2 = 4356$ Pa $= 0.0436$ bar.
3. Doubling the airflow rate will increase the heat transfer coefficient, according to the Sieder–Tate equation by $Re^{0.8}$, or $h(2) = 2^{0.8}\ h = 1.74 \times 106.3 = 185$ W/(m^2 K). The thermophysical properties of the air and, therefore, the

Prandtl number (*Pr*), are assumed constant. The pressure drop will increase by u^2 or fourfold, according to the Fanning equation. Therefore, $\Delta p(2) = 2^2 \Delta p = 4 \times 0.0436 = 0.1774$ bar.

Example 8.2

Saturated steam at 150°C is transported in a horizontal pipe of 150 mm outer diameter, surrounded with atmospheric air at 20°C. The pipe is made of oxidized steel, which has an emissivity $\varepsilon = 0.75$. The thermal resistance of the pipe metal wall is negligible for the calculations of this example.

Calculate

1. The convection and radiation heat transfer coefficients, comparing their sum $(h_c + h_r)$ to the literature data, cited in this book
2. The steam condensate formed in kg/h per 10 m of the pipe, due to the heat losses to the surrounding atmosphere—heat of steam condensation at 150°C, $\Delta H = 2.11$ MJ/kg
3. The % reduction of steam condensate when the pipe is covered with an aluminum foil of low emissivity $\varepsilon = 0.05$

Solution

1. The natural convection heat transfer coefficient of air in a horizontal tube is given by the simplified equation 8.13, $h_c = 1.42(\Delta T/d)^{1/4}$ or $h_c = 1.42(130/0.15)^{1/4} = 1.42 \times 867^{1/4} = 1.42 \times 5.6 = 7.96$ W/(m² K).

   The radiation heat transfer coefficient from the pipe will be given by the equation

   $$h_r = 5.67\varepsilon\ [(T_1/100)^4 - (T_2/100)^4]/\Delta T$$
   $$= 5.67 \times 0.75(4.23^4 - 2.93^4)/130 = 8.1\ \text{W/(m}^2\ \text{K)}.$$

   The sum of the two heat transfer coefficients $(h_c + h_r) = 16.06$ W/(m² K) compares well with the literature data of Figure 8.7.
2. The temperature of the external surface of the steam pipe will be approximately equal to 150°C, since the pipe wall has a negligible thermal resistance.

   The outer surface of a 10 m long pipe will be $A = 3.14 \times 0.15 \times 10 = 4.71$ m².

   The heat loss per 10 m of pipe will be $q = (h_c + h_r)\, A\, \Delta T = 16.06 \times 4.71 \times 130 = 9896.7$ W or $q = 9.9$ kW.

   The heat (energy) loss per 10 m of pipe will be $E = 9.9$ kWh $= 9.9 \times 3.6 = 35.64$ MJ. Since the heat of condensation of steam at 150°C is 2.11 MJ/kg, the amount of steam condensate formed will be $m = 35.64/2.11 = 16.9$ kg/h.

   It should be noted that steam condensates are removed continuously by steam traps, installed in the steam pipeline system.
3. Covering the steam pipe with an aluminum foil of low emissivity $\varepsilon = 0.05$ will reduce significantly the radiation heat transfer coefficient. Assuming that the temperature of the covered external surface of the pipe will not change $(T_1 = 150°\text{C})$, the temperature difference (ΔT) will be the same, and from Equation 8.43 it follows that

$h_{r2} = (\varepsilon_2/\varepsilon_1)\, h_{r1}$ or $h_{r2} = (0.05/0.75) \times 8.1 = 0.54\,W/(m^2\,K)$. The convection heat transfer coefficient will not change (same ΔT), $h_c = 7.96\,W/(m^2\,K)$. The sum of the two heat transfer coefficients will be $(h_c + h_r) = 7.96 + 0.54 = 8.5\,W/(m^2\,K)$. Therefore, the amount of steam condensate formed will be $16.9 \times (8.5/16.06) = 8.94$ kg/h. The percent reduction of steam condensate will be $100 \times (16.90 - 8.94)/16.90 = 47.1\%$.

Example 8.3

Design a shell and tube heat exchanger to heat 8 ton/h of "hot break" tomato juice concentrate (TJC) of 30% TS from 60°C to 100°C. Stainless steel tubes of 5/8 in. (15.87 mm) outer diameter, 10.33 mm inner diameter, and 2.77 mm wall thickness (BWG 12) are used. The heated TJC is held at 100°C for 15 s (sterilization), cooled to 30°C and packed aseptically in plastic-lined drums.

The following engineering data are given:

Rheological properties of TJC 30% TS, $K = 50\,Pa\,s^n$, $n = 0.33$ (20°C), density $\rho = 1100\,kg/m^3$, specific heat $C_p = 3600\,J/kg\,K$, thermal conductivity $\lambda = 0.60\,W/(m\,K)$.

The rheological constant (K) is related to the temperature with the Arrhenius equation with energy of activation $E_a = 20$ kJ/mol. The flow behavior index (n) is independent of temperature. The inner tube diameter will be $d_i = 15.87 - 2 \times 2.77 = 10.33$ mm. The mean velocity of the concentrate in the tubes is $u = 1$ m/s. Saturated steam at 120°C is the heating medium, and the steam condensate leaves the HE as saturated liquid (120°C).

Solution

Mass flow rate of TJC in the heat exchanger system, $m = 8000/3600 = 2.23$ kg/s.
Shear rate of the paste in the tube $\gamma = 8u/d = 8 \times 1.0/0.0103 = 777$ 1/s.
Rheological constant (K) at the mean temperature $(60 + 100)/2 = 80$°C.
$K(80) = K(20)\exp\{E/R(1/353 - 1/293)\}$, $(E/R) = 20{,}000/8.31 = 2407$.
$\ln[K(80)/K(20)] = -1.4$ and $K(80) = 0.246\, K(20) = 12.3\,Pa\,s^n$.
Apparent viscosity of the TJC at $\gamma = 777$ 1/s and 80°C, $\eta = 12.3 \times (777)^{-0.67} = 0.152$ Pa s.

Heat Transfer Coefficient

Reynolds number in the tubes $Re = (1.0 \times 0.0103 \times 1100)/0.152 = 74.5$ (laminar flow).

For heat transfer of non-Newtonian fluids in tubes in laminar flow, Equation 8.19 gives:

$Nu = 2.0\, Gz^{1/3}(K/K_w)^{0.14}$, where the Graetz number is $Gz = (mC_p)/(\lambda L) = (2.22 \times 3600)/(0.60 \times 10) = 1332$. The length of the pipe is assumed $L = 10$ m (for a more detailed analysis, a trial and error calculation can be used). $Gz^{1/3} = 1332^{1/3} = 11$. The ratio of the rheological constants at the bulk and wall temperatures (assumed constant) is $(K/K_w) = 50/12.3 = 4.1$, and $(K/K_w)^{0.14} = 1.2$, $Nu = 2 \times 11 \times 1.2 = 26.4$, $Nu = (h\,d)/\lambda = 0.0103h/0.60 = 0.0172h = 26.4$ and $h = 1535\,W/(m^2\,K)$.

The heat transfer coefficient (h) can also be estimated from the generalized correlation of the heat transfer factor (j_H) for food materials (Equation 8.45):

$j_H = 0.344 \times (74.5)^{-0.423} = 0.055$. The Prandtl number will be $Pr = (C_p\eta)/\lambda = (3600 \times 0.152)/0.60 = 912$ and $Pr^{2/3} = 90$. From Equation 8.20 it follows that $h = (j_H u\rho C_p)/Pr^{2/3} = (0.055 \times 1.0 \times 1100 \times 3600)/90 = 2420\,W/(m^2\,K)$. There is a significant difference in the estimated values of (h) by the two methods. For the purposes of this example, we will accept the lower value of first calculation, $h = 1535\,W/(m^2\,K)$.

The overall heat transfer coefficient (U) in this heat exchanger includes the resistances of the heating (steam) side and the tube wall, which are considered small, compared to the thermal resistance of the product side. The wall thickness of the stainless steel tube is $x = 2.77\,mm$ and its thermal conductivity $\lambda = 18\,W/(m\,K)$. Assuming a steam-side heat transfer coefficient $h_o = 10\,kW$, the overall heat transfer coefficient (clean surface) is given by Equation 8.10:

$1/U_i = 1/1535 + (0.0103/d_L)(0.0028/18) + (0.0103/0.0159) \times (1/10{,}000)$, where ($d_L$) is the log mean tube diameter, $d_L = (d_o - d_i)/\ln(d_o/d_i) = (0.0159 - 0.0103)/\ln(0.0159/0.0103) = 0.013\,m = 13\,mm$. Thus, $1/U_i = 6.5 \times 10^{-4} + 1.2 \times 10^{-4} + 0.65 \times 10^{-4} = 8.35 \times 10^{-4}$ and $U_i = 1198\,W/(m^2\,K)$, estimated overall heat transfer coefficient of clean surface.

In this example, the wall and steam-side thermal resistances are, respectively, (1.2/8.35) = 14% and (0.65/8.35) = 6.6% of the overall resistance. It should be noted that, in this example, the tube wall thickness is relatively high (BWG 12) and the thermal conductivity of stainless steel is low, resulting in a significant thermal resistance. In most heat exchangers thinner tube walls (e.g., BWG 16) and metals of higher thermal conductivity (e.g., copper) may be used in order to reduce drastically the wall thermal resistance. Thick tube walls are necessary for high operating pressures.

For design and operating applications, the fouling of the heat transfer surface should be taken into consideration. For this example of heating TJC, a fouling factor $h_f = 2000\,W/(m^2\,K)$ is assumed (Table 8.3). The fouling resistance of the steam side can be neglected. The overall heat transfer coefficient of the fouled heat exchanger is estimated from the simplified Equation 8.65:

$1/U_i = 1/1535 + 1/2000 + (0.01033/0.013)(0.0028/18) + (0.0103/0.0159) \times (1/10{,}000)$, from which $U_i = 787\,W/(m^2\,K)$.

Heat Transfer Surface

The heat flux in the HE will be $q = mC_p\Delta T = 2.23 \times 3.6 \times 40 = 321.1\,kW$. The log mean temperature difference in the heat exchanger will be, $\Delta T_M = (\Delta T_1 - \Delta T_2)/\ln(\Delta T_1/\Delta T_2)$, where $\Delta T_1 = (120 - 60) = 60°C$ and $\Delta T_2 = (120 - 100) = 20°C$. Therefore, $\Delta T_M = (60 - 20)/\ln(60/20) = 36°C$. The required heat transfer area will be $A = q/U\,\Delta T_{LM} = 321.1/(0.787 \times 36) = 11.33\,m^2$.

The product velocity in the tubes is assumed to be 1 m/s. If the total cross-sectional area of the tubes is A, the mass flow rate will be $m = u\rho A = 1 \times 1100\,A = 2.23\,kg/s$ and, therefore, $A = 2.23/1100 = 2.03 \times 10^{-3}\,m^2$. The cross-sectional area of one tube 10.3 mm inner diameter is $0.084 \times 10^{-3}\,m^2$ (Table 6.4), and $N = (2.03/0.084) = 24$ tubes. The inner surface area of each tube will be $A = 11.33/24 = 0.47\,m^2$ and its length $L = (0.47)/(3.14 \times 0.0103) = 14.5\,m$.

A heat exchanger with 24 straight tubes 14.5 m long and 10.3 mm diameter is not practical for industrial application. Instead, a four-pass compact heat exchanger

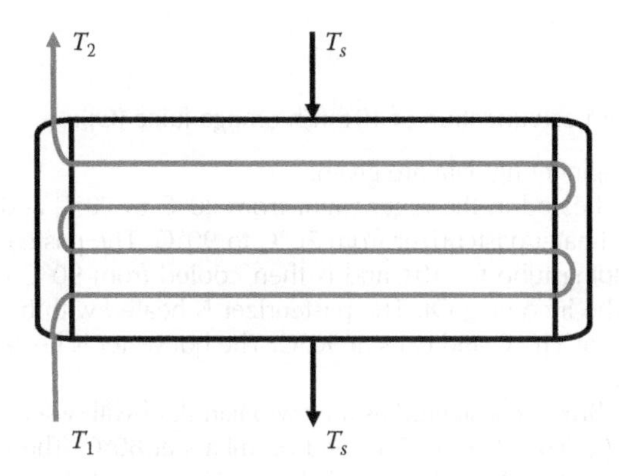

FIGURE 8.21 Four-pass shell and tube heat exchanger: T_1, T_2 product; T_s, steam.

4 m long could be used, in which the bended tubes are fitted in the steam jacket (Figure 8.21).

The pressure drop Δp (Pa) in the product side of the heat exchanger can be calculated from the Poiseuille equation for laminar flow, $d\ \Delta p/4L = \eta\ (8u/d)$, where $d = 10.3$ mm is the inner tube diameter, $u = 1$ m/s is the fluid velocity, $L = 21$ m is the tube length, and $\eta = 0.152$ Pa s is the apparent viscosity of the product at the given flow rate.

The pressure drop is considered that it is caused by flow of the product in the total length L of straight tube, neglecting any effects of the flow in the connecting bends.

$$\Delta p = 4L\eta(8u/d^2) = 4 \times 14.5 \times 0.152 \times 8 \times 1/(0.0103)^2 = 664{,}794 \text{ Pa} = 6.6 \text{ bar.}$$

The pressure drop developed is relatively high for a heat exchanger, but the selected thick-walled stainless steel tube (5/8 in. BWG 12) can withstand it.

As an alternative, a compact 4–1 pass heat exchanger can be used, with four bunches of tubes 4 m long, fitted in the heating (steam) jacket (Figure 8.22). The pressure drop in this heat exchanger would be similar to the previous exchanger (about 6.6 bar).

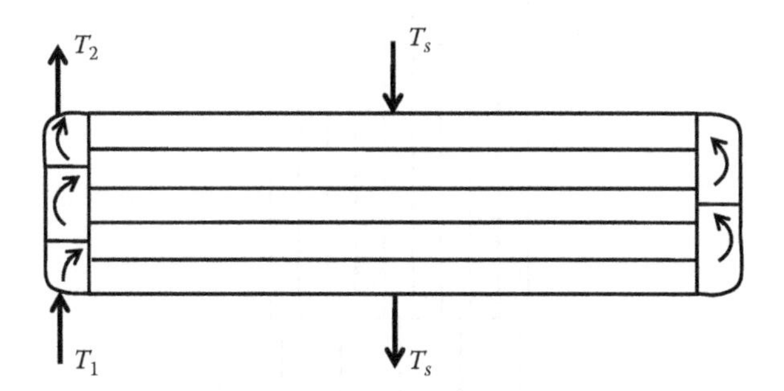

FIGURE 8.22 Shell and tube heat exchanger 4–1: T_1, T_2 product; T_s, steam.

Example 8.4

Design a PHE for pasteurization of 10 ton/h orange juice (OJ).

The following engineering data are given:

The OJ is preheated in the regenerator from 20°C to 70°C and subsequently is heated in the heater/pasteurizer from 70°C to 90°C. The pasteurized OJ flows through the holding tube for 10 s and is then cooled from 90°C to 40°C in the regenerator by the incoming OJ. The pasteurizer is heated with hot water, which enters the heater at 110°C and exits at 90°C. The hot water is heated to 110°C by steam injection.

The OJ of 12°Brix is considered as a Newtonian fluid with viscosity (η) equal to 1.5 mPa s at 20°C, 0.6 mPa s at 55°C, and 0.4 mPa s at 80°C. The viscosity of hot water at 100°C is 0.25 mPa s. The specific heats of water and OJ are assumed to be constant with temperature, C_p(water) = 4.18 kJ/kg and C_p(OJ) = 3.86 kJ/kg (Chapter 5).

The thermal conductivity of water and OJ is taken approximately as λ = 0.65 W/(m K).

The density of OJ is taken as 1000 kg/m^3 and of water 958 kg/m^3 at 100°C.

Assume a stainless steel PHE for both regeneration and heater sections. Plate dimensions 1.0 m × 0.6 m with an effective heat transfer area 0.6 m^2/plate. The plate thickness is 0.6 mm and channel spacing 3 mm.

Figure 8.23 shows a diagram of the flow pattern in a PHE section. Figure 8.24 shows the principal components of the PHE pasteurizer. The fresh OJ is preheated in the regenerator (R) in countercurrent flow with the hot pasteurized product (HP). The preheated product (PH) is heated in the heater (H) in countercurrent flow with hot water (HW). The heated OJ is held at the pasteurization temperature in the holding tube (HT).

Solution

The following flow and temperature data are considered:

Flow rate of feed OJ, m = 10 ton/h = 10,000/3,600 = 2.78 kg/s. Mean velocity of OJ and water in the heat exchanger channel, u = (2.78)/(1,000 × 0.6 × 0.003) = 1.5 m/s.

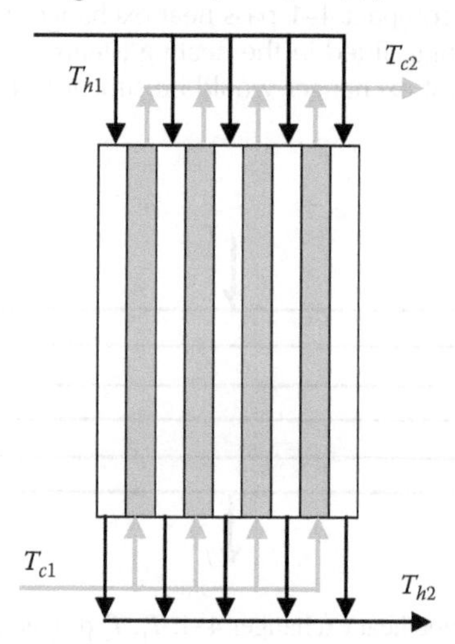

FIGURE 8.23 Section of a countercurrent PHE: c, cold stream; h, hot stream.

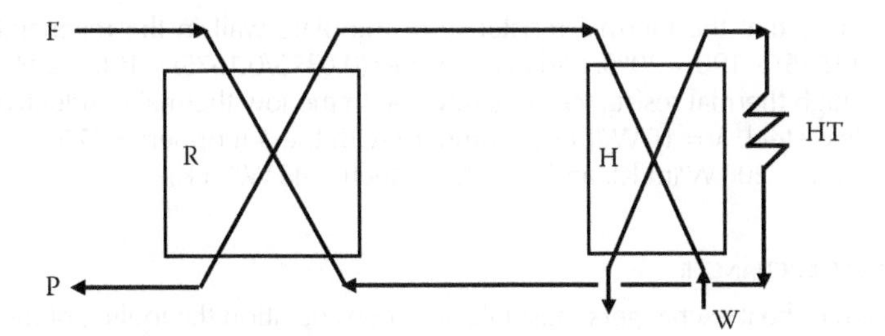

FIGURE 8.24 Pasteurization PHE, consisting of preheating the feed (F) in the regenerator (R), heating the preheated product (PH) in the heater (H) by hot water (HW), holding the product in the holding tube (HT), and cooling the hot product (P) in the regenerator (R).

The effective flow diameter (L) of the channel is defined as $L = 4 \times$ (cross-sectional flow area)/(wetted perimeter), or $L = 4 \times (0.6 \times 0.003)/[2(0.6 + 0.003)] = 0.006\,\text{m}$.

The mean temperature of cold OJ in the regenerator will be $T = (20 + 70)/2 = 45°\text{C}$; the hot OJ in the regenerator $T = (90 + 40)/2 = 65°\text{C}$; the hot OJ in the heater $T = (70 + 90)/2 = 80°\text{C}$; and the hot water in the heater $T = (110 + 90)/2 = 100°\text{C}$.

The Reynolds number of cold and hot OJ in the regenerator will be

$Re = L\,u\,\rho/\eta = (0.006 \times 1.5 \times 1{,}000)/0.0006 = 15{,}000$, turbulent flow. For the hot OJ in the heater, $Re = (0.006 \times 1.5 \times 1{,}000)/0.0004 = 22{,}500$, turbulent flow. For the hot water in the heater, $Re = (0.006 \times 1.5 \times 958)/0.00025 = 34{,}488$, turbulent flow.

Heat Transfer Coefficients

The generalized correlation for the heat transfer factor j_H (Equation 8.45) is used in this example, which gives lower (conservative) heat transfer coefficients than the empirical correlation for PHE of Figure 8.15. For the OJ in the regenerator $j_H = 0.344 \times (15{,}000)^{-0.423} = 0.0054$. For the OJ in the heater, $j_H = 0.344 \times (23{,}500)^{-0.423} = 0.0045$. For the hot water in the heater, $j_H = 0.344 \times (34{,}480)^{-0.433} - 0.0037$.

The Prandtl numbers are, $Pr = (C_p\eta/\lambda)$: For the OJ in the regenerator, $Pr = (3860 \times 0.0006)/0.65 = 3.56$ and $Pr^{2/3} = 2.33$, for the OJ in the heater, $Pr = (3860 \times 0.0004)/0.65 = 2.37$ and $Pr^{2/3} = 1.78$, and for the hot water at 100°C, $Pr = (4180 \times 0.00025)/0.65 = 1.61$ and $Pr^{2/3} = 1.37$.

From the heat transfer factors (j_H), $h = (j_H u \rho C_p)/Pr^{2/3}$:

For the OJ in the regenerator, $h = (0.0054 \times 1.5 \times 1000 \times 3860)/2.33 = 13.76$ kW/(m^2 K), for the OJ in the heater, $h = (0.0045 \times 1.5 \times 1000 \times 3860)/1.78 = 14.64$ kW/(m^2 K), and for the hot water in the heater, $h = (0.0037 \times 1.5 \times 958 \times 4180)/1.37 = 16.22$ kW/(m^2 K).

The calculated overall heat transfer coefficients (U), assuming no fouling (clean heat transfer surfaces), are

For the regenerator, $1/U = 1/13.76 + 0.6/16 + 1/13.76$.
$1/U = 0.1835$ and $U = 5.45$ kW/(m^2 K).
For the heater, $1/U = 1/14.64 + 0.6/16 + 1/16.22 = 0.1675$ and $U = 5.97$ kW/(m^2 K).

It is noted that the thermal resistance of the plate wall in the regenerator is (0.0375/0.1835) × 100 = 20%, and in the heater (0.0375/0.1675) × 100 = 22%. The relatively high thermal resistance is mainly due to the low thermal conductivity of the stainless steel, λ = 16 W/(m K). Compare with the λ (copper) = 375 W/(m K), λ (aluminum) = 206 W/(m K), and λ (carbon steel) = 45 W/(m K).

Plate Heat Exchanger

The design of heat exchangers must take into consideration the fouling of the heat transfer surfaces. In this example, the cloudy OJ contains suspended particles, colloids, and dissolved biopolymers, which may cause fouling during operation. If reliable heat transfer data from similar operating systems are not available, the overall heat transfer coefficients can be corrected with empirical fouling factors (Table 8.3). As an approximation, we accept a fouling effect in the PHE similar to the shell and tube heat exchanger (Example 8.3), that is a 35% reduction of (U). Thus, the overall heat transfer coefficients in the fouled PHE will be, approximately, U (regenerator) = 3.6 kW/(m^2 K) and U (heater) = 4 kW/(m^2 K). Lower (U) values should be taken with more viscous and severely fouling fluids.

The heat transfer duty of the regenerator will be q = 2.78 × 3.86 × (70 − 20) = 536 kW and that of the heater q = 2.78 × 3.86 × (90 − 70) = 215 kW. The temperature difference in the regenerator will be ΔT = 65 − 45 = 20°C, and in the heater ΔT = 100 − 80 = 20°C. In this case the arithmetic mean temperature difference (ΔT = 20°C) should be used instead of the log-mean temperature difference, which gives an indeterminable value for $\Delta T_1 = \Delta T_2$. The heat transfer area of the regenerator will be A = 536/(3.6 × 20) = 7.4 m^2, and in the heater A = 215/(4 × 20) = 2.7 m^2.

Each plate has an effective surface area of (1.00 × 0.60) = 0.60 m^2. The number of plates in the regenerator will be N = 7.4/0.6 = 12.3 or 13 plates, and in the heater, N = 2.7/0.6 = 4.5 or 5 plates. The PHE will have a total of 18 plates of (1.00 × 0.60) m.

Example 8.5

The auxiliary requirements for hot water in a milk processing plant (cleaning of equipment, personnel use) can be covered by solar heating. The solar energy is used during the sunny hours to heat water, which is stored in a 5 ton tank and used throughout the day.

Calculate

1. The total surface of the flat solar collectors
2. The heat losses from a stainless steel storage tank

   The following engineering data are given:

   The mean solar insolation (incident radiation) is 0.5 kW/m^2 for an average of 4 h/day. The thermal efficiency of the solar collector is 80%.

   The water is heated in the solar collector from 15°C to 55°C. Mean temperature of the hot water in the storage tank, 50°C. The stainless steel tank has a wall thickness of 3 mm and thermal conductivity of 16 W/(m K). The tank has a diameter of 1.5 m and a height of 3 m. It is installed in the vertical position on an insulated support at the plant floor, and it is closed at the top.

The mean room temperature is 20°C, and the specific heat of water C_p = 4.18 kJ/(kg K).

Solution

1. The heat required per day (24 h) will be 5000 × 4.18 × (55 − 20) = 731.5 MJ = 731.5/3.6 = 203.2 kWh. Since the mean solar heating time is 4 h/day, the solar heating rate will be 203.2/4 = 50.8 kW. The required incident solar radiation rate will be 50.8/0.80 = 63.5 kW, and the required surface of solar collectors A = 63.5/0.5 = 130 m^2. If the surface of a flat solar collector is 2 m^2 (2 m × 1 m), the required number of collectors is 130/2 = 65. The collectors are normally installed on the roof of the building. A piping system and a pump are needed to supply the collectors with cold water and transport the hot water to the storage tank in the plant.

   It should be noted that the insolation is a function of the geographical latitude, the time of the year, and the time of the day. It is higher in the summer and low latitudes (near the equator) and lower in the winter and high latitudes (away of the equator).
2. The heat losses from the storage tank are mainly due to natural convection of air, while radiation losses can be neglected, since the temperature is relatively low. The tank loses heat from the side wall and the top, while the bottom is considered as insulated. Since the thermal resistance of the tank metal wall is very small, compared to the resistance of the outer wall air film, the outer wall temperature will be approximately the same as the water temperature in the tank (50°C).

   The side wall will have a surface of A_w = 3.14 × 1.5 × 3.00 = 14.13 m^2. The surface of the tank top will be A_t = 3.14 × (1.5)^2/4 = 1.77 m^2.

   The natural convection heat transfer coefficients from the tank wall (h_w) and top (h_p) can be calculated from the simplified Equations 8.14 and 8.15, respectively, which are applicable to the atmospheric air. Due to the large diameter of the tank (1.5 m), the vertical tank wall can be considered as a vertical plane of height L = 3.00 m. The temperature difference is ΔT = (50 − 20) = 30°C.

$$h_w = 1.42(\Delta T/L)^{1/4} = 1.42 \times (30/3)^{1/4} = 2.52\ \text{W/(m}^2\ \text{K)}$$

and

$$h_t = 2.48(\Delta T)^{1/4} = 2.48 \times (30)^{1/4} = 5.8\ \text{W/(m}^2\ \text{K)}.$$

The total heat loss from the tank will be q = 2.52 × 14.3 + 5.8 × 1.77 = 45.2 W.

The heat loss is relatively low (0.045 × 24 = 1.08 kWh/day) to consider tank insulation.

PROBLEMS

8.1 The inside of a storage room for frozen food is maintained at −40°C. The room has dimensions (5 m) × (6 m) × (3.5 m). The walls and ceiling are made of stainless steel 2 cm thick (λ = 18 W/(m K), covered with 10-cm foam insulation (λ = 0.035 W/(m K) and 2-cm wood siding (λ = 0.12 W/(m K). The ambient air outside the freezer is 30°C and its dew point is 25°C.

Calculate

1. The thickness of a layer of corkboard insulation ($\lambda = 0.12$) mm to cover the foam insulation, so that the temperature outside the wood siding is 26°C
2. The rate of heat transfer by convection from the storage room to the surrounding air, neglecting the heat transfer by conduction through the floor and by radiation from the whole storage room

8.2 The convection heat transfer coefficient from a horizontal plane facing upward is significantly higher than from a plane facing downward. Explain physically the difference between the two coefficients.

8.3 A viscous Newtonian fluid is heated in the tubes of a shell and tube heat exchanger by condensing steam in the jacket. The heat transfer coefficient increases with the velocity of the fluid, resulting in a smaller heat transfer area and less expensive heat exchanger. Discuss the limitation of this beneficial effect of high velocity, considering the total operating cost of the heat exchanger system.

8.4 Saturated steam flows in a horizontal steel pipe at 150°C, while the surrounding atmospheric air remains at 20°C. In order to reduce heat losses to the environment, two approaches are considered: (a) cover the pipe with a thin aluminum foil and (b) cover the pipe with a layer of insulation which will reduce the external surface temperature to about 40°C.

Discuss, qualitatively, the effect of the two approaches to the combined convection and radiation heat transfer coefficients ($h_c + h_r$) and to the heat losses from the pipe. The emissivity of the aluminum foil is very low, compared to the emissivities of steel and insulation.

8.5 Compare the advantages and disadvantages of a coil heater with a shell and tube heat exchanger. In both systems a viscous Newtonian fluid is heated with saturated steam in the jacket.

8.6 Design a shell and tube heat exchanger (surface area and tube arrangement) to be used as condenser of 2 ton/h saturated water vapors at 100°C. The vapors are condensed in the shell of the condenser without subcooling, while cooling water flows in the tubes at a velocity of 2 m/s, entering at 20°C and exiting at 50°C. Horizontal tubes of 1 in. outer diameter 16 BWG made of a copper alloy of thermal conductivity 100 W/(m K) are used. Number of tubes at a vertical plane $N = 5$. Density of water $\rho = 1000\,kg/m^3$, viscosity (35°C) $\eta = 0.72$ mPa s, thermal conductivity $\lambda = 0.60$ W/(m K), specific heat $C_p = 4.18$ kJ/kg, heat of condensation $\Delta H = 2.26$ MJ/kg.

Calculate

1. The heat transfer coefficients of the two sides of the tube
2. The overall heat transfer coefficient, assuming a clean surface of vapor condensation, and a fouling factor of water side $h_f = 8000\,W/(m^2\,K)$
3. The required surface area of the condenser
4. The number of water tubes in the heat exchanger
5. The number of passes of cooling tubes for a reasonable tube length in the heat exchanger, e.g., $L = 3$ m

8.7 An agitated jacketed kettle is used to heat a batch of 200 kg of tomato sauce from 15°C to 100°C with saturated steam at 120°C. The hemispherical kettle has an internal diameter 1.00 m and it is filled at 80% of its volume. The kettle is agitated with a scraping anchor agitator, rotating at 120 RPM. The sauce is a power-law non-Newtonian fluid with an apparent viscosity of 100 mPa s at the operating conditions.

Properties of tomato sauce: ρ = 1050 kg/m³, C_p = 3.8 kJ/(kg K), λ = 0.50 W/(m K), apparent viscosity at product heating surface η_w = 50 mPa s, bulk η = 250 mPa s. Volume of sphere $V = (3.14/6)\, d^3$, surface $A = 3.14\, d^2$, where d (m) is the sphere diameter.

Calculate

(a) The product-side heat transfer coefficient
(b) The overall heat transfer coefficient, based on the inner surface, assuming that the steam-side heat transfer coefficient is 10 kW/(m² K) and the thermal resistances of kettle wall and inner fouling are 0.0002 and 0.0003 (m² K)/W, respectively
(c) The time required to complete this batch process (heat the product from 20°C to 100°C), assuming that the product is well mixed to an average temperature at any given time

8.8 A can containing a fluid food at an initial temperature of 30°C is heated by saturated steam at 120°C on the outside surface. The cylindrical can has dimensions (diameter) × (height) (99 mm) × (120 mm) and it contains 0.9 kg of food product (similar to the US No. 21/2; can). Product properties: ρ = 1040 kg/m³, C_p = 3.7 kJ/(kg K), λ = 0.55 W/(m K), apparent viscosity of pseudoplastic product η = 100 mPa s.

Calculate the time required to heat the product to 90°C, assuming perfect mixing at any time during the process. The overall heat transfer coefficient is 100 W/(m² K).

8.9 A swept surface heat exchanger (SSHE) is used to cool 2 ton/h TJC from 90°C to 35°C, using cooling water which enters the SSHE at 20°C and leaves at 28°C. The overall heat transfer coefficient, based on the inside heat transfer surface, is 1 kW/(m² K). Inside diameter of the SSHE is 25 cm.

Calculate the length of the cylindrical heat exchanger, (a) for countercurrent flow, and (b) for co-current flow operation.

LIST OF SYMBOLS

A	m²	Surface area
C_p	kJ/(kg K)	Specific heat
d, D	m	Diameter
Dn	—	Dean number [$Re(d/D)$]
f	—	Fanning friction factor
g	m/s²	Acceleration of gravity (9.81)
G	kg/(m² s)	Mass flux ($u\rho$)

Gr	—	Grashof number $[(d_o^3\rho^2\beta g\Delta T_o)/\eta^2]$
Gz	—	Graetz number $[(mC_p)/\lambda L]$
h	W/(m² K)	Heat transfer coefficient
j_H	—	Heat transfer factor $[St\ Pr^{2/3}]$
K	Pa s^n	Flow consistency coefficient
L	m	Length
m	kg/s	Mass flow rate (GA)
Nu	—	Nusselt number $[(hd)/\lambda]$
q	W	Heat flow rate
p	Pa	Pressure
r	m	Radius
Pr	—	Prandtl number $[(C_p\eta)/\lambda]$
Re	—	Reynolds number $[(du\rho)/\eta]$
St	—	Stanton number $[h/(GC_p)]$
t	s	Time
T	°C or K	Temperature
u	m/s	Velocity
U	W/(m² K)	Overall heat transfer coefficient
n	—	Flow behavior index
x	m	Wall thickness
β	—	Expansion coefficient
γ	1/s	Shear rate
Δp	Pa	Pressure difference
ΔT	K or °C	Temperature difference
e	—	Emissivity
η	Pa s	Viscosity
λ	W/(m K)	Thermal conductivity
ρ	kg/m³	Density
τ	Pa	Shear stress

REFERENCES

Bhatia, M.V. ed. 1980. *Heat Transfer Equipment*, Process Equipment Series, vol. 2. Technomic Publ., Westport, CT.

Geakoplis, C. 1993. *Transport Processes and Unit Operations*, 3rd edn. Prentice Hall, Upper Saddle River, NJ.

Knudsen, J.G., Hottel, H.C., Sarofim, A.F., Wankat, P.C., and Knaebel, K.S. 1999, Heat and mass transfer. In: *Perry's Chemical Engineers' Handbook*, 7th edn. McGraw-Hill, New York, pp. 5–1 to 5–41.

Maroulis, Z.B. and Saravacos, G.D. 2003. *Food Process Design*. CRC Press, Boca Raton, FL.

McCabe, W., Smith, J., and Harriott, P. 2004. *Unit Operations of Chemical Engineering*, 7th edn. McGraw-Hill, New York.

Saravacos, G.D. and Maroulis, Z.B. 2001. *Transport Properties of Foods*. Marcel Dekker, New York.

Saravacos, G.D. and Kostaropoulos, A.E. 2002. *Handbook of Food Processing Equipment*. Springer, New York.

Walas, S.M. 1988. *Chemical Process Equipment*. Butterworths, London.

9 Thermal Processing Operations

9.1 INTRODUCTION

Thermal processing operations are used extensively in food processing in various technologies of food preservation and thermal treatment of foods. Heat can inactivate food pathogens and food spoilage agents, such as microorganisms (m.o.), enzymes, and pathogenic molecules. At the same time, thermal preservation may cause undesirable changes in the sensory and nutritive quality of the foods, such as softening, color changes, loss of vitamins, and production of off flavors.

Thermal treatment processes, such as baking, toasting, and frying, are used to affect desirable quality and sensory changes in foods, such as flavor enhancement, digestibility, and nutrient availability.

Heat transfer is also involved in some other important food processing operations, such as evaporation, dehydration, and refrigeration/freezing, which are discussed in separate chapters of this book.

9.2 THERMAL PRESERVATION PROCESSES

9.2.1 INTRODUCTION

Thermal preservation is primarily concerned with the application of heat to destroy (inactivate) m.o. and enzymes, which can cause spoilage of foods and health hazards to the consumers. Thermal preservation involves heating of foods at various time–temperature combinations, which define the three main thermal processes, i.e., sterilization, pasteurization, and blanching (Kessler, 1981).

The objective of sterilization, traditionally called "thermal processing," is the long-term and safe preservation of sensitive foods, preferably at ambient (room) temperatures. Traditionally, thermal processing has been applied to the canning of foods, which are packaged in metallic or plastic containers, and preserved for long time.

Thermal preservation includes, in addition to canning, the following food processing operations: (a) blanching (heat inactivation of spoilage enzymes, normally in vegetables, prior to further processing), (b) pasteurization (inactivation of pathogenic and some spoilage m.o. and enzymes), and (c) aseptic processing (long-term preservation of foods with minimum heat damage).

High-temperature processing may have, in addition to preservation, some other desirable effects on the food product, such as improvement of eating quality (cooking), softening of some hard foods, and destruction of some undesirable components, such as the trypsin inhibitor in legumes.

Thermal preservation is also used in the bulk storage of fluid foods, usually combined with refrigeration. Gentle thermal processing is used recently in the minimal processing and preservation of foods. The design of thermal processing or thermal preservation is based on the heat inactivation of undesirable enzymes and m.o., and the heat transfer from the heating medium to the food product.

A major problem of thermal processing (sterilization) is the significant damage to the nutritional (vitamins, proteins) and organoleptic (sensory) quality (taste, color, and texture) of foods, particularly when exposed to high temperature for a relatively long time. This problem is resolved by using high temperature—short-time processing methods, aseptic packaging, and special food containers.

9.2.2 Kinetics of Thermal Inactivation

The kinetics of thermal inactivation deals with the destruction of enzymes and m.o. involved in the spoilage of foods. The same kinetic principles are applicable to the effect of thermal processing on the nutritional and sensory quality of processed foods.

9.2.2.1 Inactivation of Microorganisms and Enzymes

The kinetics of thermal inactivation usually follows a first-order chemical reaction (Chapter 3), although the mechanism may be more complex. At a given temperature T, the rate of inactivation of a population (N) of m.o. is given by the equation:

$$\frac{\mathrm{d}N}{\mathrm{d}t} = -kN \tag{9.1}$$

where

k (1/s) is the rate constant
t (s) is the time

Equation 9.1 upon integration yields:

$$\ln\left(\frac{N}{N_0}\right) = -kt \tag{9.2}$$

where (N_0 and N) are the initial and final numbers of m.o. after time of heating (t) at the given temperature T.

In thermal processing, Equation 9.2 is usually written as

$$\log\left(\frac{N}{N_0}\right) = -\left(\frac{t}{D}\right) \tag{9.3}$$

where $D = 2.3/k$ is the decimal reduction time, expressed usually in minutes.

The D value (min) is the time required for 90% reduction of a population. It is determined from the slope ($-1/D$) of a semi-log survivor plot of $\log(N/N_0)$ versus time (t).

The heat resistance of spore-forming bacteria is much higher than the vegetative cells. Equation 9.3 is applicable to enzyme inactivation. Some enzymes can be as heat resistant as spore-forming bacteria, e.g., catalase and lipoxidase.

The thermal death time curve is obtained by plotting $\log(D)$ versus the corresponding temperature (T), according to the equation:

$$\log\left(\frac{F}{F_0}\right) = \log\left(\frac{D}{D_0}\right) = -\frac{(T - T_0)}{z} \tag{9.4}$$

where the decimal reduction times (D, D_0) correspond to temperatures (T, T_0).

The z value, calculated from the slope of the thermal death time curve, is the temperature rise required to reduce the D value by one log cycle (90%). For most m.o. of interest to thermal processing, the z value is about 10 K or 10°C (18°F).

The thermal death time (F) required to obtain the specified inactivation (reduction of the m.o. population) is a multiple of the decimal reduction time (D), $F = m\,D$, where m (–) is the reduction exponent. In the thermal processing of low-acid foods (pH > 4.5), a reference temperature of $T_0 = 121$°C (250°F) is normally used. The (F) value (usually in minutes) at a given temperature (T) is converted to the equivalent (F_0) at the reference temperature (T_0) by the equation:

$$\log\left(\frac{F_0}{F}\right) = (T - T_0)z \tag{9.5}$$

where z-value (°C or K) is the thermal resistance parameter, i.e., the rise in temperature needed to increase the inactivation of m.o. by 90%.

A reduction (sterilization) exponent $m = 12$ is used in process calculations of low-acid foods, in which toxin-producing *Clostridium botulinum* has a thermal resistance $D_0 = 0.30$ min. Thus, the thermal death time at 121°C in this case will be $F_0 = 12 \times 0.30 = 3.60$ min.

The decimal reduction time (D_0) is too high for very resistant spoilage m.o., such as *Bacillus stearothermophilus*, which has a $D_0 = 3.00$ min. In such cases, a lower reduction (sterilization) exponent is used, e.g., $m = 5$, which yields a high thermal death time of $F_0 = 15$ min. It is obvious that a process based on thermal resistant *B. stearothermophilus* will be more than adequate for inactivation of *C. botulinum*.

Table 9.1 shows some typical values of thermal inactivation data for m.o. and enzymes, used in thermal process calculations. Most enzymes are heat sensitive and they are inactivated easily. There are a few very resistant enzymes, such as peroxidase, which can survive a thermal process designed to inactivate common spoilage m.o.

TABLE 9.1
Decimal Reduction Time (D) of Microorganisms and Enzymes

m.o. or Enzyme	Temperature, °C	D, min	z, °C	m, –	Food Product
C. botulinum	121	0.30	11	12	Low acid pH > 4.5
C. sporogenes	121	1.50	11	5	Meat
B. stearothermophilus	121	3.00	10	5	Vegetables, meat
Bacillus subtilis	121	0.40	7	6	Dairy products
Bacillus coagulans	121	0.07	10	6	Tomato
Clostridium pasteurianum	100	0.50	8	5	Fruits
Lactobacilli, Yeasts, Molds	65	0.80	10	—	High acid pH < 4.5
Pectin enzymes	75	0.20	5	—	Tomato, citrus
Peroxidase	121	3.00	30	—	Vegetables

D, decimal reduction time; *z*, temperature parameter; *m*, reduction exponent.

The effect of temperature on the thermal death time (*F*) and the decimal reduction time (*D*) can be also expressed by the Arrhenius equation and the Q_{10} ratio. The Arrhenius equation for the (*D*) value is

$$\ln\left(\frac{D}{D_0}\right) = -\left(\frac{E_a}{R}\right)\left(\frac{1}{T} - \frac{1}{T_0}\right) \tag{9.6}$$

where

E_a (kJ/mol) is the energy of activation
R [8.31 J/(mol K)] is the gas constant
(T, T_0) are the temperatures (K) corresponding to (D, D_0).

By combining Equations 9.4 and 9.6, the following equation is obtained, which relates the energy of activation (E_a) to the *z*-value:

$$E_a = 2.3R\frac{(TT_0)}{z} \tag{9.7}$$

As an illustration, for $z = 10$ K and $T = 373$ K, $T_0 = 393$ K, and $R = 8.31$ J/(mol K), Equation 9.7 yields, $E_a = 2.3 \times 8.31 \times 373 \times 393/10 = 280.17$ kJ/mol.

Activation energies in the range of 210–476 kJ/mol are reported in the literature for the thermal inactivation of various m.o. These values are very high, compared to the activation energies of physical and chemical changes (reactions).

The Q_{10} ratio represents the increase of reaction rate by an increase of the temperature by 10°C, which in the case of (*D*) values is defined as

$$Q_{10} = \frac{D_T}{D_{T+10}} \tag{9.8}$$

By combining Equations 9.4 and 9.8, the following relation is obtained:

$$\log Q_{10} = \frac{10}{z} \tag{9.9}$$

Thus, for the usual value of $z = 10$, the ratio becomes $Q_{10} = 10$, i.e., the decimal reduction time (D) decreases by 10 times, when the temperature is increased by 10°C. It should be noted that for most chemical reactions the ratio Q_{10} is about 2, i.e., the effect of temperature on the inactivation of m.o. is about five times faster than on typical chemical reactions.

9.2.2.2 Heat Damage to Food Components

Although some thermal processes may cause desirable changes to foods, such as cooking and improvement of eating quality of foods, most heat-induced chemical and biochemical changes are undesirable, e.g., nonenzymatic browning, and vitamin, taste, texture, and color deterioration.

Most heat damage reactions are described by first-order kinetics, similar to the inactivation of m.o. and enzymes. The rate of thermal damage to food components is much slower than the thermal inactivation of the heat-resistant m.o. and enzymes, i.e., significantly higher D and z values, and lower Arrhenius activation energies. Table 9.2 shows examples of thermal damage indicators, important in thermal processing operations.

The cooking value (C) of potatoes at temperature (T) with respect to a reference temperature of 100°C, can be expressed by the relation $\log(C/C_0) = (T - 100)/z$, which is analogous to the effect of temperature on the (D) value of m.o. (Equation 9.4), but with a (z) value about three times higher (about 30°C).

The significant differences in (z) values indicate that high-temperature and short-time thermal processes can yield sterilized products with minimum heat damage to important food components.

Figure 9.1 shows typical thermal curves for inactivation of spoilage m.o. and heat damage to food components. It is evident that spoilage m.o. can be inactivated in short time at high temperatures ($T > 100$°C), while the quality of the food product is not damaged excessively (note that 1.E + 02 = 10^2).

Figure 9.2 shows a diagram of thermal preservation processes for milk, i.e., pasteurization and sterilization at low and high temperatures. For comparison, the thermal inactivation curves of some m.o. and food components are also shown.

TABLE 9.2
Thermal Damage Indicators of Food Components at 121°C

Food Component	D, min	z, °C	E_a, kJ/mol
Ascorbic acid	931	18	165
Nonenzymatic browning	384	35	85
Thiamine	264	25	120
Vitamin A	43	20	150
Chlorophyll	15	45	65

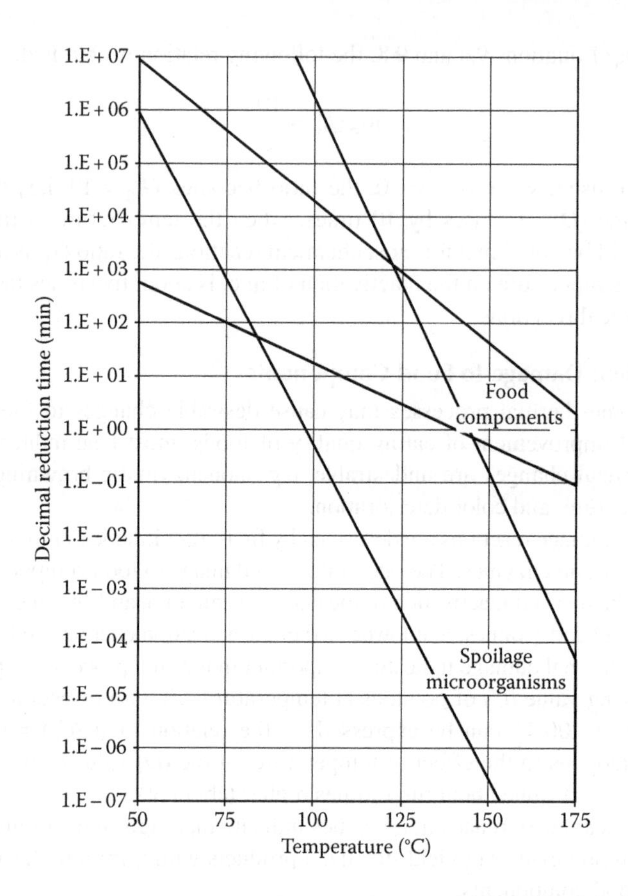

FIGURE 9.1 Thermal damage (*D*-values) of spoilage m.o. and food components. (Modified from Maroulis, Z.B. and Saravacos, G.D., *Food Process Design*, CRC Press, Boca Raton, FL, May 9, 2003, fig. 8.3.)

The activation energy (E_a, Arrhenius equation) of thermal damage to food components is generally lower than in microbial inactivation. In food materials undergoing phase change from glass to rubbery state, the Williams, Landel, and Ferry (WLF) equation is more accurate than the traditional Arrhenius equation near the glass transition temperature (Tg):

$$\log\left(\frac{k_g}{k}\right) = \frac{C_1(T - Tg)}{\left[C_2 + (T - Tg)\right]} \tag{9.10}$$

where

k_g, k are the reaction constants $k = 2.3/D$ at temperatures Tg, T, respectively
C_1, C_2 are empirical constants ($C_1 = 17.4$, $C_2 = 51.6$)

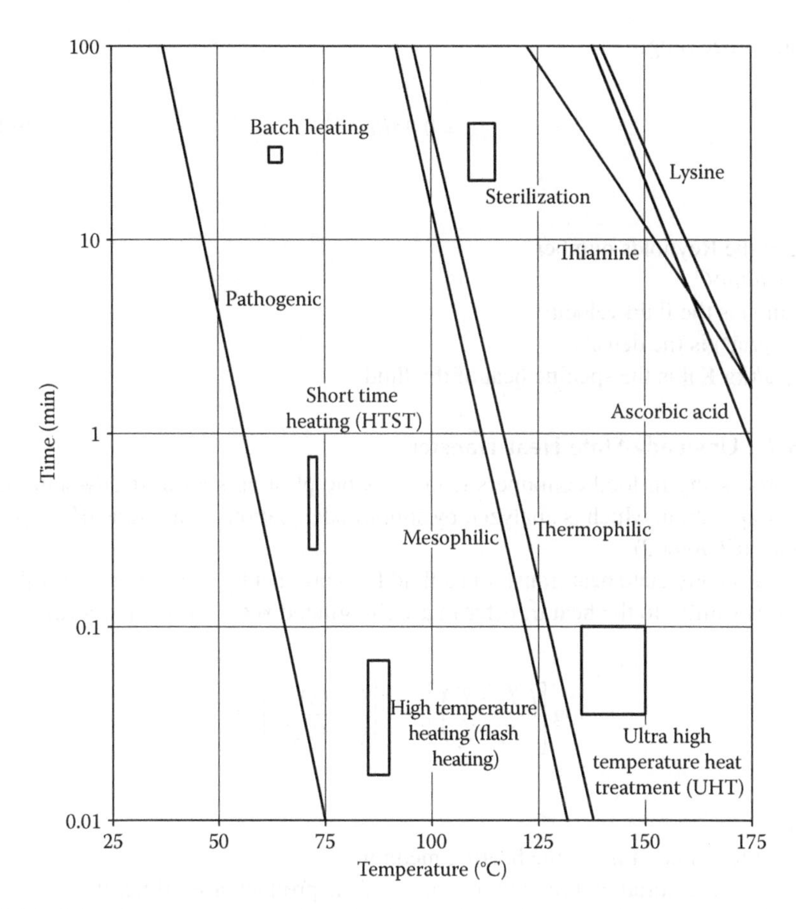

FIGURE 9.2 Temperature–time (*D*-values) data for milk pasteurization or sterilization. (Modified from Maroulis, Z.B. and Saravacos, G.D., *Food Process Design*, CRC Press, Boca Raton, FL, May 9, 2003, fig. 8.5.)

9.2.3 Heat Transfer Considerations

9.2.3.1 Heat Transfer Coefficients

The principles of heat transfer applied to thermal processing of foods are reviewed in Chapter 8. Heat transfer in fluid foods in continuous thermal processes, such as pasteurization and aseptic processing, is analyzed by classical steady-state heat transfer calculations.

Limited heat transfer coefficients (h, U) can be obtained from the literature or from experimental pilot plant or industrial plant operations. The following correlations of published data in the form of heat transfer factors (j_H), are useful:

Retort sterilization:

$$j_H = 1.03Re^{-0.50} \tag{9.11}$$

Aseptic processing:

$$j_H = 0.85Re^{-0.45} \quad (9.12)$$

where

Re is the Reynolds number
$j_H = h/(u\rho C_p)$
u (m/s) is the fluid velocity
ρ (kg/m^3) is the density
C_p [J/(kg K)] is the specific heat of the fluid

9.2.3.2 Unsteady-State Heat Transfer

Heat processing in food containers (cans, flexible plastic) is an unsteady-state heat transfer operation, which is analyzed by simplified equations or numerical/computer methods (Chapter 2).

The unsteady-state heat transfer in a fluid food product by convection (natural and forced) is similar to the heat transfer in a well-agitated vessel (Equation 8.72):

$$\log\left[\left(\frac{T_h - T}{T_h - T_0}\right)\right] = -\left[\left(\frac{UA}{2.3VC_p}\right)\right]t \quad (8.72)$$

where

T_h is the temperature of the heating medium
T_0, T are the initial and final temperature of the product (after time, t)
U is the overall heat transfer coefficient
A is the heat transfer area
V is the volume of the container
C_p is the specific heat of the product

Equation (8.72) can be simplified in the form:

$$\log\left[\left(\frac{T_h - T}{T_h - T_0}\right)\right] = \left(\frac{1}{f_h}\right)t \quad (9.13)$$

where $f_h = (2.3\ VC_p)/(UA)$.

The heating parameter (f_h) is the time (usually in minutes) required to reduce the temperature ratio by 90% (one log cycle), and it is analogous to the temperature parameter (z) of the thermal death curve.

Unsteady-state cooling of fluid foods in containers by convection can be expressed by an equation analogous to Equation 9.12, replacing T_h by T_c and f_h by f_c.

It is assumed that the product attains a uniform temperature T at any time, due to convection and complete mixing in the container.

Unsteady-state heating of solid and semi-solid foods in containers is expressed by the empirical equation:

$$\log\left[\frac{(T_h - T)}{j(T_h - T_0)}\right] = \left(\frac{1}{f_h}\right)t \tag{9.14}$$

Here, the temperature T refers to the slowest heating point (center) in the container.

The remaining symbols used in this equation are similar to those of Equation 9.13, except for the heating rate lag factor (j), which is defined by the equation:

$$j = \left(\frac{T_h - T_{po}}{T_h - T_0}\right) \tag{9.15}$$

The j and f_h parameters, defining the heat penetration curve (Equation 9.13), are estimated by plotting experimental heat penetration data of $\log(T_h - T)$ versus time t, Figure 9.3. The heating time parameter f_h is determined from the slope of the curve, and the lag factor j from the temperatures T_0 and T_{po}. The pseudo-initial temperature T_{po} is estimated from the extrapolation of the heating line to time zero ($t = 0$).

In the literature, the heat penetration diagram of Figure 9.3 may have the temperature scale of $\log(T_h - T)$ inverted, with the heat penetration line directed upward.

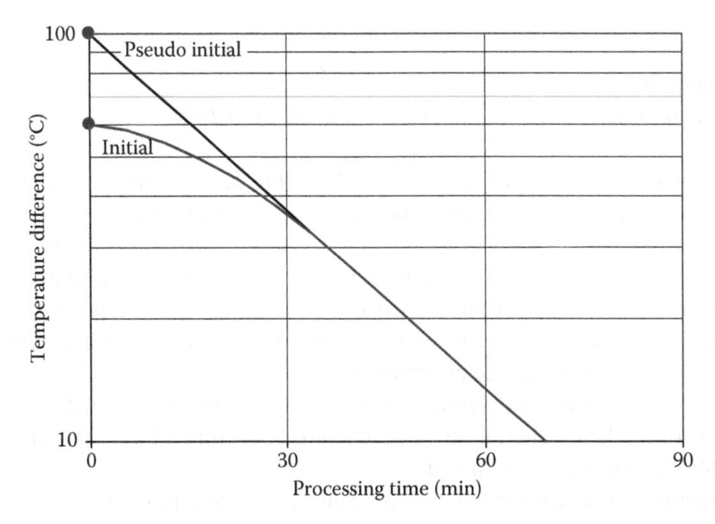

FIGURE 9.3 In-container heat penetration diagram. (Modified from Maroulis, Z.B. and Saravacos, G.D., *Food Process Design*, CRC Press, Boca Raton, FL, May 9, 2003, fig. 8.7.)

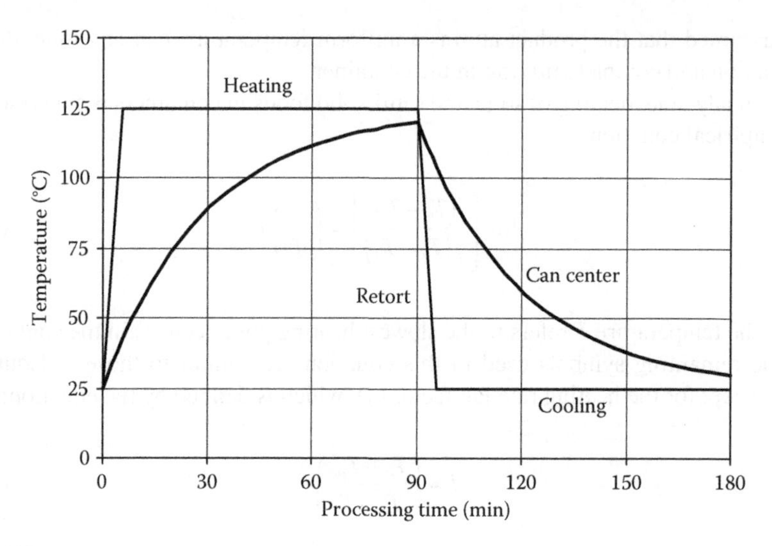

FIGURE 9.4 Heat penetration profile of in-container sterilization. (Modified from Maroulis, Z.B. and Saravacos, G.D., *Food Process Design*, CRC Press, Boca Raton, FL, May 9, 2003, fig. 8.6.)

The heating parameter f_h, in conduction-heated foods, is related to the thermal diffusivity α of the food product in a cylindrical container (can), according to the equation:

$$f_h = \frac{0.398}{\left[\left(\frac{1}{r_2} + \frac{0.427}{L^2}\right)\alpha\right]} \tag{9.16}$$

where

r is the radius

L is the half-height of the cylindrical container ($L = H/2$)

If SI units are used, the f_h parameter will be obtained in (seconds), which should be converted to the commonly used thermal process time (minutes).

The thermal diffusivity α of food materials at the thermal processing temperatures does not vary much ($\alpha = 1.0$ to 1.5×10^{-7} m²/s). Therefore, the value of f_h depends mainly on the dimensions (size) of the container, if the product is heated by conduction. However, in most food products, heat transfer by convection may be important and experimental determination of f_h is necessary.

Cooling curves for conduction-heating foods in cans are plotted in a similar manner with the heat penetration curve (Figure 9.3), from which the characteristic parameters (f_c and j_c) are estimated.

The heating (heat penetration) semi-log curves in some food products, may consist of two straight-line sections, from which two f_h parameters can be estimated. The "broken" heating lines are an indication of physico-chemical changes in the food product, induced by heat, which may have a significant effect on its thermal transport properties. Such a change has been observed in the heat-induced gelatinization of starch materials, which changes (increases) significantly both thermal conductivity and thermal diffusivity.

9.2.4 In-Container Sterilization

In-container sterilization is applied to solid, semisolid, and liquid foods, packaged in metallic or plastic containers. Calculation of thermal process time is based on the thermal destruction of spoilage and toxin-producing m.o. and enzymes, and the minimization of undesirable thermal damage to heat-sensitive food components. Thermal process design deals primarily with the sterilization of solid nonacid foods (pH > 4.5), packaged in cylindrical metallic containers (cans), heated by unsteady-state conduction (Downing, 1996; Teixeira, 1992).

Thermal process calculations are traditionally based on the attainment of sterilization at the slowest—heating point (normally the geometric center) of the can. An integrated approach, based on the sterilization of the entire container, yields approximately similar sterilization results with the traditional slowest—heating-point method. Figure 9.4 shows temperatures of in-container sterilization.

The thermal process time for in-container sterilization can be calculated by (i) the General method, (ii) the Formula (Ball) method, and (iii) a Computer method.

9.2.4.1 General Method

The General (or Bigelow) method is based on the integration of the lethality of the test m.o. with heat penetration data in the food container. Thermal processes are defined by the F_0 value delivered to the product, i.e., the equivalent time, or integrated lethality (minutes), at the reference temperature, which for low-acid foods is 121°C. In acid foods, the reference temperature is normally 100°C or lower. Essential data for process calculations are the inactivation parameters (D, z) and the heat penetration characteristics (f_h, j). In canning, the thermal process calculations are often based on the inactivation of the putrefactive anaerobe *Clostridium sporogenes* NFPA (National Food Processors Association) No. 3679 or PA 3670 (NFPA, 1982, 1984).

The lethality (L) of the test m.o. at a temperature (T) is calculated from the equation:

$$\log(L) = \frac{(T - 121)}{z} \tag{9.17}$$

By definition, the lethality at 121°C (250°F) is $L = 1$. The lethality drops sharply at temperatures below 121°C, and it becomes negligible at $T < 100$°C (when the reference temperature is 121°C). It increases sharply above the reference temperature (121°C).

For a correct sterilization process, the desired thermal death time F_0 should be equal to the integral or the summation of the lethality at all process temperatures:

$$F_0 = \sum (\Delta t) \tag{9.18}$$

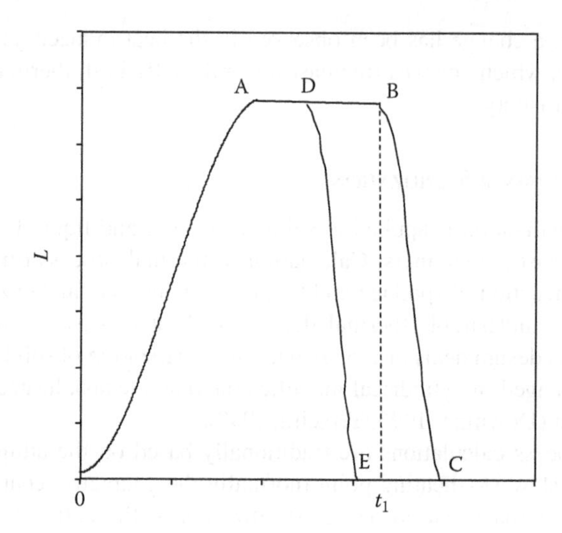

FIGURE 9.5 Thermal process calculation diagram (General method).

The summation of Equation 9.17 is estimated using time-temperature data from the heat penetration curve of the can center, and calculating the lethality L at each temperature. If the process lethality (summation of $L \times t$) is different than the fixed thermal death time F_0, a trial and error procedure may be used, with repeated calculations until the summation of lethality becomes equal to the desired F_0 value. The process calculations are facilitated by the application of computers.

The calculation of thermal process time by the general method, when the process lethality is higher than F_0, is shown in the lethality (L)—Time (t) diagram of Figure 9.5. In the process (OABC), heating is stopped at point (B), followed by cooling curve (BC), and the process (operator) time is (t_1). The total lethality is calculated from the area under the curve (OABC). If a smaller total lethality is desired in the same system, heating is stopped at point (D), followed by cooling curve (DE), parallel to the initial curve (BC). The new process time will be (t_2) and the new total lethality is calculated from the area under the curve (OADE). The trial and error procedure is repeated until the total lethality becomes equal to the desired F_0 value.

It is evident that the cooling period has a significant sterilizing effect.

9.2.4.2 Formula Method

The thermal process time (t_B) in a food container can be calculated from the following Ball formula:

$$t_B = f_h \log\left[\frac{j(T_h - T_0)}{g}\right] \tag{9.19}$$

where T_h, T_0 are the heating medium (retort) and initial temperatures, respectively, and the parameters f_h, j are obtained from the heat penetration curve (Figure 9.3). The thermal process parameter g is defined as the difference between the retort temperature and the (maximum) temperature of the center of the container (T_{max}) at the end of heating ($g = T_h - T_{max}$) (Ball, 1957).

The value of (g) is a function of the heating characteristics of the container (f_h, j) and the thermal inactivation of the test m.o. (D_{121}, z). The log(g) is estimated from tables and diagrams of the literature as a function of the ratio (f_h/U) at various (j) values (see Figure 9.18).

The parameter (U) is calculated from the thermal death time (F_0) according to the equation:

$$\log(U) = \frac{\log(F_0) + (121 - T_h)}{z} \tag{9.20}$$

(U) becomes equal to (F_0), when the retort and the reference temperatures coincide (T_h = 121°C).

In the absence of experimental heat penetration data, for approximate engineering calculations, the heating parameter (j_h) can be estimated from the heat conduction Equation 9.16a:

$$f_h = \frac{0.398}{\left[\left(\frac{1}{r^2} + \frac{0.427}{L^2}\right)\alpha\right]} \tag{9.16a}$$

where

r is the radius

L is the half-height of the cylindrical container

Conduction heat transfer is assumed and α is the product thermal diffusivity.

In the formula method, an approximate cooling rate factor of $j_c = 1.4$ is often used, which may be different from the experimental heating rate factor (j_h) of the specific food system. In detailed calculations, the experimental cooling factor should be used. The cooling factor is included in the calculations, since cooling has a significant effect on the lethality of the indicator spoilage m.o., as shown in the General method.

The thermal process time for food containers exhibiting two broken heat penetration lines is calculated from the equation:

$$t_B = f_{h1}\log[j(T_h - T_{po})] + (f_{h2} - f_{h1})\log(g_1) - f_{h2}\log(g_2) \tag{9.21}$$

where, f_{h1}, f_{h2} are the heating time parameters of Sections 9.2.1 and 9.2.2 of the broken penetration line, $g_1 = (T_h - T_1)$ and $g_2 = (T_h - T_2)$, where T_1 is the temperature at the break point, and T_2 is the temperature at the end of heating (T_{max}). The parameters log(g_1) and log(g_2) are determined from literature tables or diagrams as functions of (U/f_h). For simplification the heating and cooling time parameters are sometimes assumed to be equal ($f_c = f_h$).

The formula method (Equation 9.19) assumes that the retort (sterilizer) attains the processing temperature (T_h) instantaneously (at $t = 0$).

Improved mathematical procedures and shortcut methods have been suggested for estimating the sterilization time in food containers. Computerized data acquisition and evaluation of thermal processing (canning) of foods is provided by software packages.

In industrial thermal processing (canning), the retort (sterilizer) reaches the operating temperature (T_h) after the so-called come-up time (t_{cut}), after heating (steam) is turned on. In industrial practice, about 42% of the "come-up" time is considered as process time at (T_h). Thus, in thermal process calculations with the formula and other mathematical methods, the heat penetration curve (Figure 9.3) should start at a corrected ($t = 0$), after 58% of the "come-up" time. The heating-time parameter (f_h) does not change appreciably with the "come-up" time, but the lag factor (j) should be based on a pseudo-initial temperature, which is estimated from the extrapolation of the heating straight line to the corrected zero time ($t = 0$).

Thus, the total thermal process time (t_B) is calculated from the equation:

$$t_B = t_p + 0.42 t_{cut} \quad (9.22)$$

where

t_p is the thermal process time at the retort temperature T_0
t_{cut} is the "come-up" time

9.2.4.3 Commercial Sterility

Commercial sterility of the low-acid foods (pH > 4.6) requires the destruction of the toxin-producing m.o. *C. botulinum*, usually $F_0 = 12 \times 0.3 = 3.6$ min (at 121°C). According to the U.S. FDA, commercial sterility in canning is defined as the process in which all *C. botulinum* spores and all other pathogenic bacteria have been destroyed, as well as more heat-resistant organisms which, if present, could produce spoilage under normal conditions of non-refrigerated canned food storage and distribution. It is essential that strict sanitation principles be followed, while raw materials (foods and packaging materials) are prepared for processing (canning).

Inoculated packs, containing selected spoilage m.o., are required to validate (confirm) the thermal process times, calculated from microbial kinetics and heat penetration curves. These tests are particularly important for some food products, which exhibit irregular heating curves. Once the theoretical calculation is established and validated for a specific product, the procedure can be used to estimate the process time in a variety of container (can) sizes. Spoilage m.o., producing gases, are inoculated in test cans, which after sterilization develop high internal pressure and they are bulged (distorted).

9.2.4.4 In-Container Sterilizers

In-container sterilizers (retorts) are used in canning of several food products in various sealed containers, made of metallic cans (tinplate or aluminum), glass, and plastic materials (rigid or flexible-pouches). The heating medium may be saturated steam, steam/air mixtures, or hot water. Heating of the cans in still retorts by sprays of hot water provides faster heating and good temperature control. Relative motion of cans and heating (or cooling) medium increases considerably the heat transfer rate.

Modern retorts are automated and equipped with PLC (programmable logic controllers), which use product-specific software.

9.2.4.4.1 Can Sizes and Processes

The size of cylindrical cans is characterized by the outside diameter (D) and height (H), which include the wall thickness and the height of the double seam (nominal dimensions). In the United States, the can dimensions are traditionally given in inches/sixteenth of an inch, e.g., a can (401) × (411) = (4 1/16) × (4 11/16) in. or (103 × 119) mm i.e., (10.3) × (11.9) cm (Downing, 1996).

The volumetric capacity (mL) of a can is calculated from its nominal dimensions, assuming an average double seam 0.5 mm thick and 2.0 mm high. The wall thickness of the metal is neglected. Thus, the cylindrical dimensions of the can considered in volume calculation are $(D-0.1) \times (H-0.4)$ cm. As an example, the capacity of a can (401) × (411) or (10.3) × (11.9) cm will be $V = [3.14 \times (10.2)^2 \times 11.5]/4 = 940\,\text{cm}^3$ or 940 mL.

Table 9.3 shows the nominal dimensions and capacities of some common food cans.

In the EU (European Union), a typical can size is the (1/1) can, dimensions (99) × (122) mm and capacity 905 mL (close to the U.S. No. 2½ can).

A typical aluminum can, used in packaging beverages (beer and colas) has nominal dimensions (62) × (110) mm and capacity 330 mL.

Can preparation and sealing are discussed in Chapter 15.

Table 9.4 shows typical thermal process times which depend on the type of food product, the can size, the temperature, and the type of retort.

TABLE 9.3
Can Sizes and Capacities (United States)

Can Dimensions, in.	Can Dimensions, mm	Capacity (Volume), mL
(202) × (204)	54 × 57	117
(303) × (406)	81 × 111	538
(307) × (409), No. 2	87 × 116	584
(401) × (411), No. 2½	103 × 119	940
(603) × (700), No. 10	157 × 178	3324

TABLE 9.4
Thermal Process Times of Canned Food Products, Process Time, min

Food Product	Can Size	Temperature, °C	Still Retort	Agitated Retort
Whole tomatoes	No. 2	100	45	14
Whole tomatoes	No. 10	100	100	25
Peach halves	No. 2½	100	—	20
Peas	(303) × (406)	121	20	—
Cream-style corn	(303) × (406)	121	65	—
Boned chicken	(303) × (406)	121	80	—

9.2.4.4.2 Batch Sterilizers

Batch sterilizers (retorts) are used in many small and medium-size food processing plants, because of their low cost and simple operation. The batch sterilizers include the still retorts, the rotary batch retorts, the crateless retorts, and the retorts for glass and flexible containers. Batch retorts are convenient for thermal processing of several food products, particularly when the raw material is seasonal and relatively small volume, such as fruits and vegetables (Gould, 1996).

Two types of still retorts (autoclaves) are commonly used, the vertical and the horizontal units, which are shown diagrammatically in Figure 9.6. The vertical retorts consist of a steel cylindrical vessel of dimensions about (1.5 m diameter) × (2.5 m height), with a hinged large top cover, which can be closed hermetically during processing. They are equipped with all the necessary piping, valves, and instruments, specified by regulations and technical publications for the canning industry (NFPA). The retorts should be constructed following special mechanical and safety specifications, such as those of the ASME (American Society of Mechanical Engineers) code

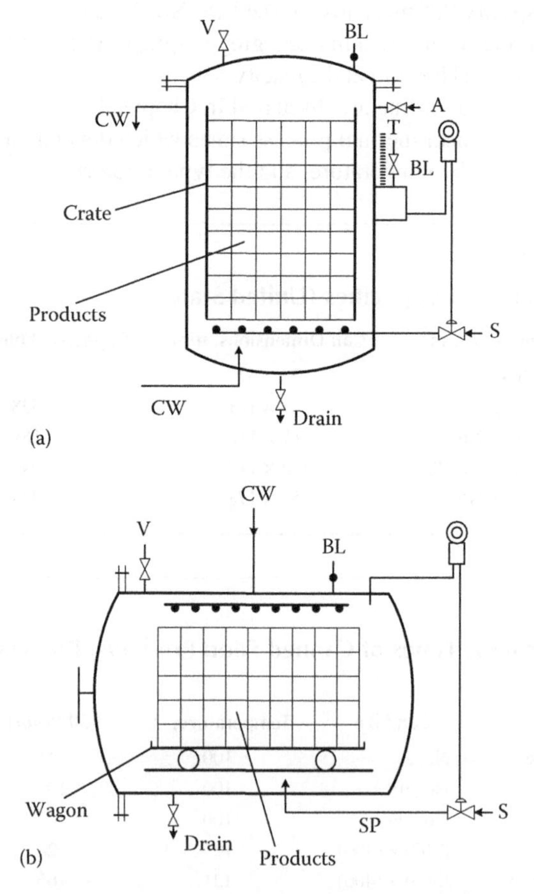

FIGURE 9.6 Still retorts: (a) vertical, (b) horizontal. S, steam; SP, steam spreader; T, thermometer; V, vent; BL, bleeder; CW, cold water; A, air.

for unfired pressure vessels. The sterilizers are located in a special area of the food processing plant (the "cook room"), which must comply with the regulations of the Public Health Authorities and the Good Manufacturing Practices.

Cooling of large (e.g., No. 10) and flat cans with water requires overriding air pressure to prevent bulging, i.e., mechanical distortion of the cans, due to excessive internal pressure, particularly during the initial stage of cooling. The high pressure developed within the cans is due to the increased pressure of water vapor at high temperatures and the pressure of the entrapped air or other gases.

Cooling of the cans with water, after retorting, should be fast, so that the inside can temperature should reach quickly a temperature of about 38°C, in order to prevent the growth of any surviving thermophilic bacteria. Lower temperatures should be avoided, since the metallic cans may be corroded (rusted).

The still retorts have the basic disadvantage of low heating rates of the cans, due to low heat transfer coefficients of natural convection between the heating medium (steam or water) and the cans. Improvement of the heat transfer rate is achieved by forced convection of the heating medium and/or agitation of the food containers, such as the batch rotary sterilizers, in which the cans are rotated, either axially or end-over-end.

Special batch retorts are used for sterilization of glass and flexible plastic containers, which are processed and cooled in water under overriding air pressure, which prevents the rejection of glass lids and the distortion of plastic packages due to internal pressure.

Crateless batch retorts reduce labor requirements for loading and unloading the cans.

In the "Flash 18" system, sterilization is accomplished in a pressurized chamber, maintained at an overpressure of 18 psig (1.24 bar). The food product is heated to the thermal processing temperature either directly by steam injection, or indirectly in a scraped surface heat exchanger, and the hot product is filled into cans, which are sealed and held at the processing temperature for the prescribed process time. The sterilized cans are subsequently discharged to atmospheric pressure and cooled with water.

Flame sterilizers are used for some solid food products, which are packed without a fluid medium, such as syrup or brine. Due to the high temperature differences developed, very high heat transfer rates are obtained, reducing significantly the processing time and, thus, improving the product quality. The cans are heated by direct exposure to the flames of gas burners at temperatures of 1200°C–1400°C. To prevent overheating and surface burning by the flames and combustion gases, the cans are rotated rapidly during heating Flame sterilization is applied to small food cans, which can withstand the high internal pressure, developed during thermal processing.

9.2.4.4.3 Continuous In-Container Sterilizers

The rotary cooker/coolers (FMC) consist of two horizontal pressure vessels equipped with rotating spiral reels, in which the cans move progressively through prescribed cycles of heating, holding, and cooling. Special pressure locks (PL) transport the cans into the pressure cooker, and from the cooker to the atmospheric water cooler (Figure 9.7). Steam (S) is applied at the bottom of the cooker, while the rotary cooler is cooled with counter-flow cold water (CW). The rotary system improves heat transfer to the canned product and reduces the use of CW in counter—flow with the cans.

Acid foods, such as tomato products, are processed in rotary cookers/coolers at atmospheric pressure, followed by continuous cooling with CW.

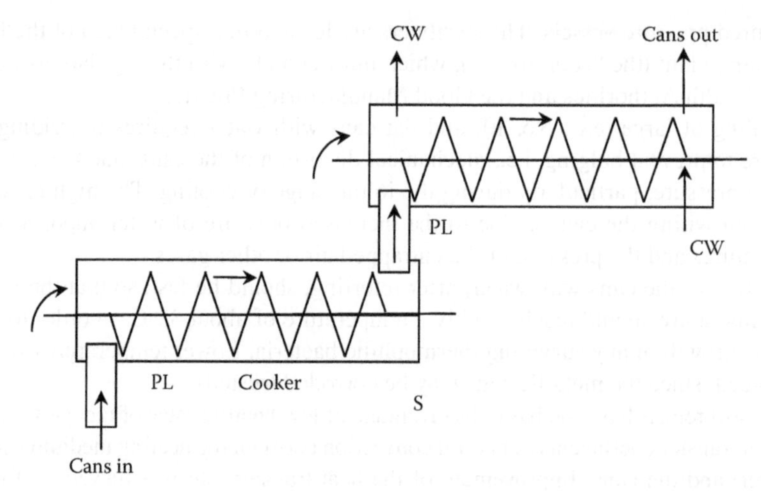

FIGURE 9.7 Diagram of a continuous cooker/cooler.

Hydrostatic sterilizers (Figure 9.8) are used in large canning operations. They can operate at pressures above atmospheric ($T > 100°C$), which are maintained by water columns (hydrostatic pressure), eliminating the need for closed vessels and PL. They are tall installations, e.g., a pressure of 1 bar (gauge) requires a water column of about 10 m.

The hydrostatic sterilizers consist of four chambers: (a) the hydrostatic feed leg, (b) the sterilization chamber, (c) the hydrostatic discharge leg, and (d) the cooling canal.

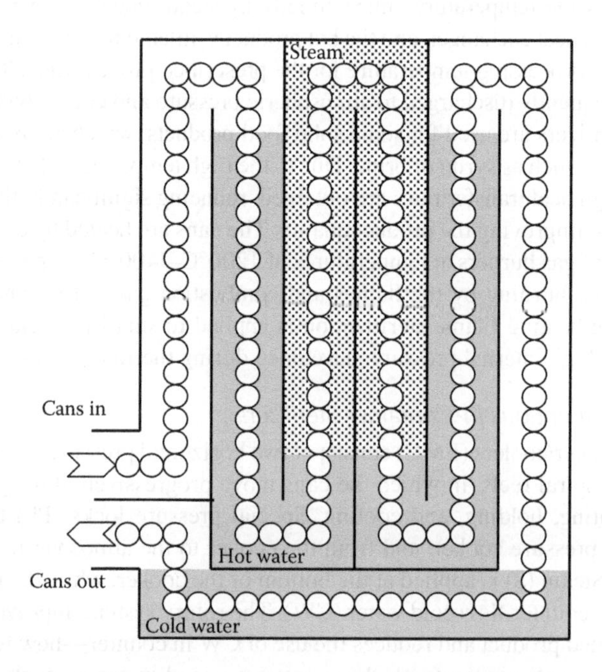

FIGURE 9.8 Diagram of a continuous hydrostatic sterilizer. (Modified from Maroulis, Z.B. and Saravacos, G.D., *Food Process Design*, CRC Press, Boca Raton, FL, May 9, 2003, fig. 8.9.)

The cans are transported through the sterilizer by chain conveyors. The residence time in the sterilization chamber is defined by the calculated thermal process time of the particular can size and product.

9.2.5 Continuous-Flow Sterilization

9.2.5.1 Sterilization Flowsheet

Homogeneous fluid foods and food suspensions can be sterilized in a continuous-flow system, similar to the traditional continuous-flow pasteurization of fluid foods (Figure 9.9). The system consists of four sections, i.e., heater, holding tube, regenerator, and cooler. The heating medium is steam, used either indirectly or by direct injection. A positive displacement (metering) pump is used to transport the fluid product, which is assumed to be well mixed and to flow at a uniform (mean) velocity in the heat exchangers and the holding tube. The food product is kept in the holding tube for the required process time, which is relatively short, because of the high process temperature. A back-pressure valve at the exit of the cooler maintains the pressure in the sterilization system (Lewis and Heppel, 2000).

Continuous flow or ultrahigh-temperature (UHT) sterilization, followed by aseptic packaging (Chapter 15), is used mainly to sterilize low-viscosity fluid foods, such as milk and fruit juices. It is also applied to viscous foods, while application to particulate (two-phase) foods is under development.

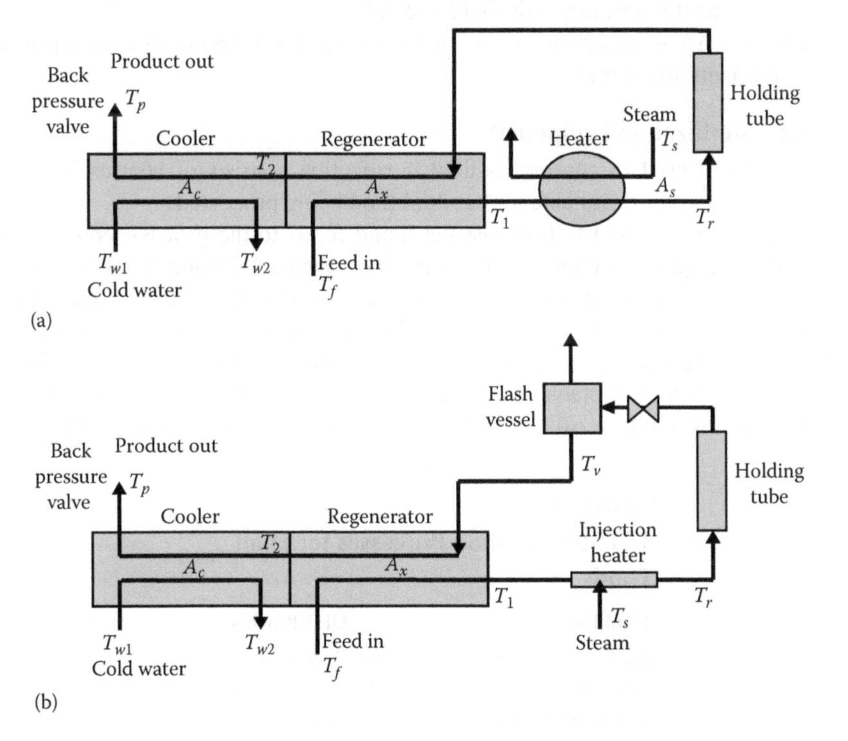

FIGURE 9.9 Continuous flow thermal processing of foods: (a) indirect steam heating and (b) injection steam heating.

UHT-sterilized food products are packaged in consumer containers (laminated cartons of various sizes) or in institutional and commercial—size packs of 55 U.S. gallon (208 L) or larger (fruit purees, tomato concentrates, etc.). The packaged products can be stored at ambient temperatures for several months, sometimes up to 2 years.

UHT processing results in better quality food products, due to the different kinetics of thermal inactivation of m.o. and food nutrients. High temperatures favor nutrient retention, while destroying more effectively the spoilage m.o.

The design of the UHT sterilizers is based on the assumption that all the required thermal lethality, or thermal death time (F_0) is delivered to the liquid food in the holding tube, neglecting the effects of the "come-up" and the early cooling periods. Such an assumption is valid for the direct (steam) heating of the liquid product, due to the very fast heating. However, in indirect heating systems, the contribution of preheating and cooling on the lethality is significant, amounting to significant overprocessing, which, in practice, is taken as a safety factor (Reuter, 1993).

The UHT sterilizer is divided into two sections, i.e., the sterile (downstream), and the non-sterile (upstream) sections. All parts of the downstream system should be sterilized by steam at 130°C for 30 min, immediately before processing.

The UHT-sterilization system operates at pressures above the atmospheric, and a back-pressure valve must be used to keep the fluid food at the high pressure before it exits the system at the packaging section.

The UHT-sterilization processes may range from (93°C/30 s) to (149°C/1 s) for acid and low-acid foods, respectively (Table 9.5).

Figure 9.10 shows diagrams of high temperature thermal processes (sterilization and pasteurization) for milk.

9.2.5.2 Sterilization Equipment

The equipment used in continuous-flow sterilization consists of heat exchangers, discussed in Chapter 8, suitable for the fluid food being processed.

Direct heating is used to heat fast the liquid foods to the process temperature, improving the quality of the sterilized product. Steam injection (introduction of steam into a stream of liquid), or steam infusion (introduction of the liquid into a steam chamber) can be used (Chapter 8). Heating rates of about 200°C/s can be obtained, i.e., the liquid product can reach the sterilization temperature (e.g., 150°C) in about 1 s. Culinary (potable) steam should be used, since part of the steam condensate remains in the sterilized product. The directly heated product is diluted by

TABLE 9.5
UHT-Sterilization Processes for Fluid Foods

Fluid Food	UHT Process
Acid foods pH < 4.5	93°C–96°C/30–15 s
Low-acid foods pH > 4.5	135°C–149°C/30–1 s
Milk (United States)	138°C/2 s
Milk (United Kingdom)	$T > 135$°C/$t > 1$ s

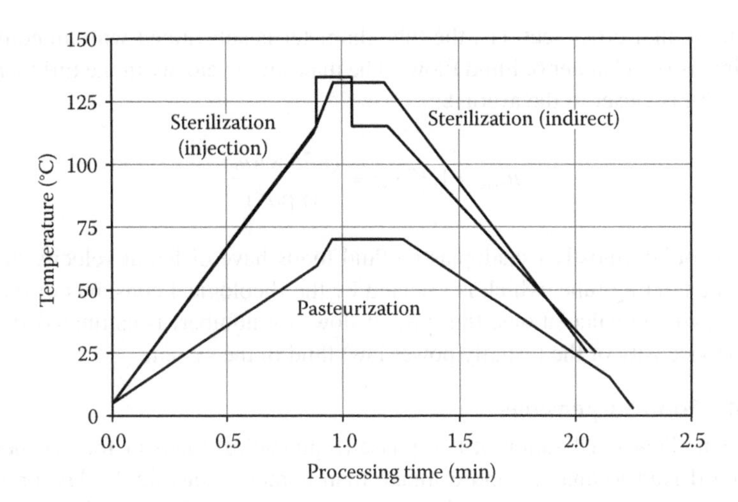

FIGURE 9.10 High-temperature thermal processes (sterilization and pasteurization) for milk.

the steam condensate, e.g., by 10%–15%, which is removed from the product, along with "cooked" milk flavors, by vacuum flashing.

Indirect heating in the UHT sterilizers utilizes plate heat exchangers, tubular heaters, or scraped surface heat exchangers, depending on the rheological properties of the product.

Ohmic heating is suitable for heating suspensions of food particles, resulting in uniform temperature distribution.

9.2.5.3 Holding Tube

The fluid food is heated to the process temperature and is held at this temperature for the specified sterilizing time (F_0). Thermal inactivation of the spoilage and health concern agents (bacteria or enzymes) is required to be accomplished entirely while the product is at the process temperature (T_0) in the holding tube of the sterilizer. The preheating and cooling effects are neglected, because of the very fast heat transfer rate.

The length of the holding tube (L) is calculated from the following equation:

$$L = (F_0)u_{max} \tag{9.23}$$

The maximum velocity (u_{max}) is the velocity of the fastest moving element of the fluid, i.e., the element with the shortest residence time in the holding tube. For a fluid in laminar motion ($Re < 2100$), the maximum fluid velocity in the center of a circular tube u_{max} is twice the average fluid velocity u_{avg} (m/s):

$$u_{max} = 2u_{avg} = \frac{(2 \times 4m)}{(\pi \rho d^2)} \tag{9.24}$$

where

m (kg/s) is the flow rate of the product
ρ (kg/m^3) is its density
d (m) is the tube diameter

For non-circular cross sections, the tube diameter is substituted with an equivalent flow dimension (Chapter 6, Fluid Flow). The maximum velocity in the turbulent flow ($Re > 2100$) is closer to the average:

$$u_{\max} = 1.25u_{\text{avg}} = \frac{(1.25 \times 4m)}{(\pi\rho d^2)} \tag{9.25}$$

Non-Newtonian (mostly pseudoplastic) fluid foods have different velocity distribution in the holding tube, which is affected by the rheological constants of the fluid. For approximate calculations, the type of flow (*Re* number) is estimated from the apparent viscosity of the (usually power-law) fluid in the system.

9.2.5.4 Food Suspensions

Continuous flow sterilization of two phase (liquid/solid) foods or food suspensions is more difficult to analyze and evaluate than homogeneous fluids. The process is under review and approval of health authorities, before wide industrial application. The main problem is to assure that sterilization is delivered to the slowest—heating point of the suspended food particle or piece (Sastry, 1994).

Heat transfer in food suspensions is controlled by heat conduction within the food particles, which have a relatively low thermal conductivity (λ), i.e., high Biot number ($Bi = h_p d_p/\lambda$).

The heat transfer coefficient (h_p) between a fluid and suspended particles is given by the empirical equation:

$$Nu = 2.0 + 1.51Re^{0.54}Pr^{0.24}\left(\frac{d_p}{d_i}\right)^{1.09} \tag{9.26}$$

where

$Nu = h_p d_p/\lambda$

$Re = (u\rho d_t)/\eta$

$Pr = \eta C_p/\lambda$

(d_p, d_i) are the particle and inside tube diameters, respectively

Sterilization of the slowest-heating point of food particles requires long process (heating) time for relative large particle sizes, e.g., spheres of diameter $d_p > 10\,\text{mm}$. Under such conditions, the quality of the sterilized food product may be damaged seriously.

Figure 9.11 shows a time–temperature diagram of continuous-flow processing of a food suspension of particles of diameter 15 mm, calculated from a simplified solution of the heat conduction equation, using typical thermo-physical properties.

9.2.6 Pasteurization

Pasteurization is a mild heat treatment, which is used to inactivate pathogenic and spoilage m.o. and enzymes with minimal changes of food quality. The pasteurized food products have a limited storage life, in contrast to sterilized packaged foods,

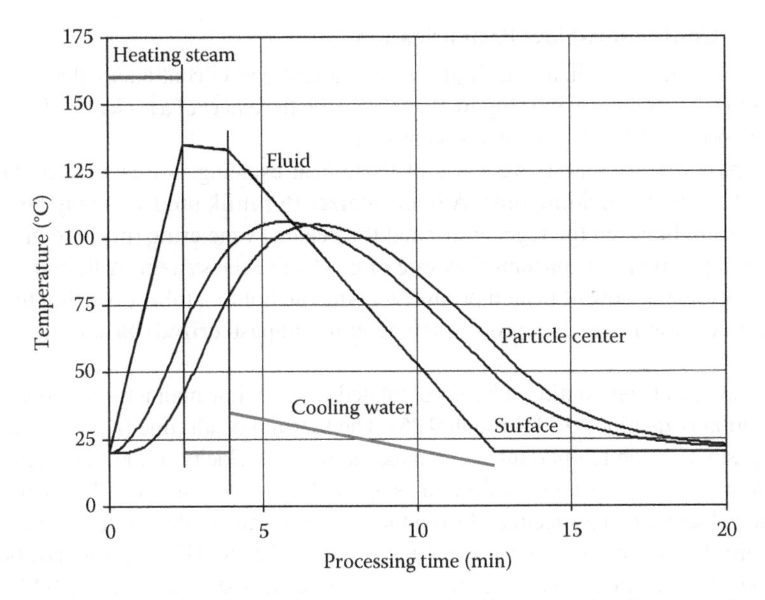

FIGURE 9.11 High-temperature thermal processing of food particles diameter 15 mm.

which can be stored at ambient temperatures for several months or even years. Originally developed for the elimination of pathogenic m.o., such as *Mycobacterium tuberculosis*, from milk and protect the public health, pasteurization is now applied to several other food products, e.g., fruit juices, beer, liquid eggs, and ice cream.

Public Health regulations require inactivation of some additional pathogenic and spoilage bacteria, such as *Salmonella*, *Listeria*, and *Escherichia coli* 0157.

The time–temperature combinations of the various pasteurization processes are based on the same principles with thermal sterilization, i.e., the kinetics of microbial/enzyme inactivation, the heat transfer rate, and the kinetics of food quality damage.

Table 9.6 shows some pasteurization processes for food products.

The pasteurization time–temperature combinations may have small variations in the various countries, particularly in the case of fluid milk, due to public health concerns.

TABLE 9.6
Pasteurization Processes for Liquid Food Products

Food Product	Temperature, °C/Time
Milk grade A (United States)	63/30 min, 77/10 s
Milk (United Kingdom)	63/30 min, 72/15 s
Fruit juices	85/15 s
Liquid eggs (United States)	60/3.5 min
Liquid eggs (United Kingdom)	64.4/2.5 min
Beer (containers)	65/20 min

9.2.6.1 Continuous-Flow Pasteurization

Fluid foods, such as milk and fruit juices, are pasteurized in continuous-flow systems, similar to the UHT sterilizers. Figure 9.12 shows the flowsheet of a typical high-temperature short time (HTST) pasteurization process.

The pasteurizer consists basically of three heat exchangers (regenerator, heater, and cooler) and the holding tube. A homogenizer (for milk products) may be added to the system between the regenerator and the heater. A metering (positive displacement) pump supplies a constant flow rate of product to the system. A flow-diversion valve controls the flow of heated product leaving the holding tube, diverting the flow of any product at lower than process temperature (unpasteurized) back to the supply tank.

The length of the holding tube is calculated in a similar manner with continuous sterilization (Equations 9.23 through 9.25). The thermal death time (F_0) is considered at the pasteurization temperature. The maximum fluid velocity in the flow channel is taken as $u_{max} = 2u_{avg}$ for laminar flow, or as $u_{max} = 1.25u_{avg}$ for turbulent flow, where the average velocity (u_{avg}) is calculated from the mass flow rate of the product in the tube.

Figure 9.13 shows a temperature–time diagram for the HTST pasteurization of milk. The inactivation of spoilage m.o. in this process is shown in Figure 9.14.

HTST pasteurizers should prevent post-pasteurization contamination from leaks of product in the heat exchangers and the holding tube. Special attention should be given to the cleaning and sanitizing of the equipment. Good hygienic conditions (sanitation) of raw milk eliminates the growth of heat-resistant bacteria, such as *Bacillus aureus*, which can grow at refrigeration temperatures and cause food poisoning.

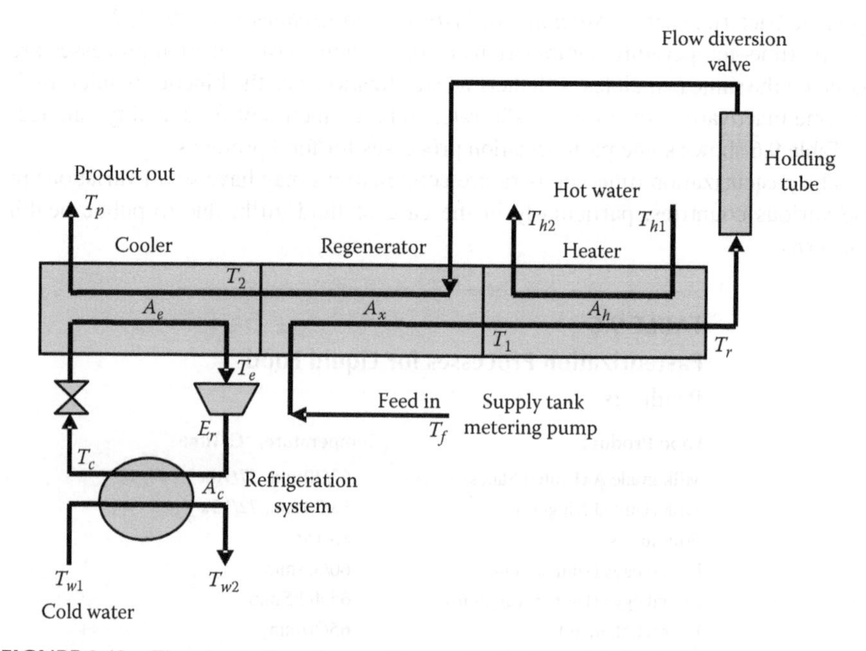

FIGURE 9.12 Flowsheet of continuous flow pasteurization of fluid foods.

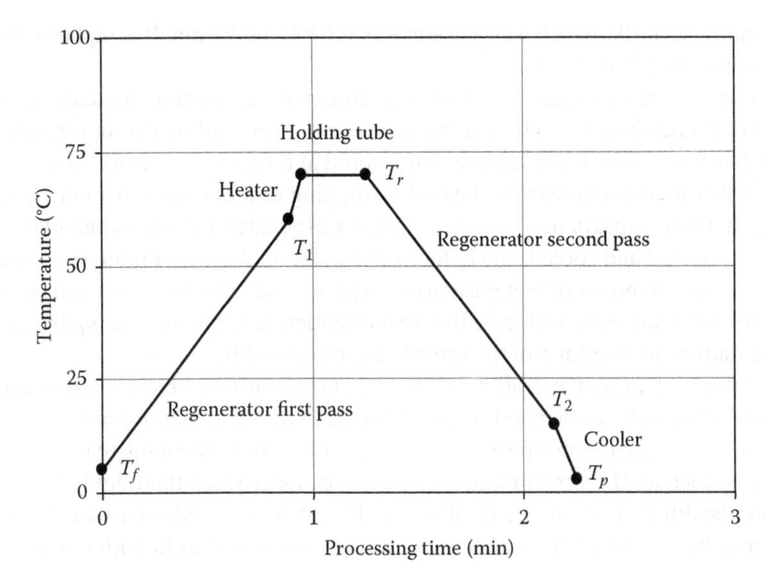

FIGURE 9.13 HTST milk temperature versus processing time.

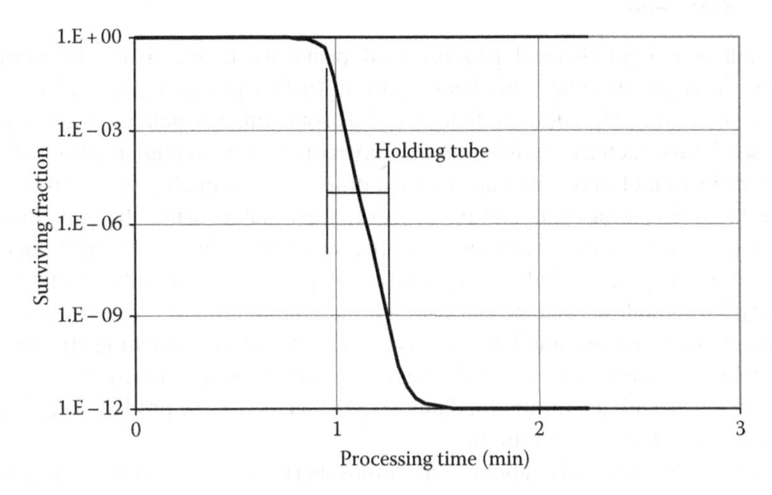

FIGURE 9.14 Inactivation of m.o. versus processing time in HTST milk.

Some heat-resistant enzymes, such as peroxidase and lipase may survive the pasteurization process. The enzyme lactoperoxidase in milk is inactivated at 80°C/15 s, which is a more severe process than normal milk pasteurization. Thus, detection of lactoperoxidase in pasteurized milk is an indication that the product has not been over-pasteurized.

9.2.6.2 Batch Pasteurization

The batch ("holder" or "vat") pasteurizers consist of a jacketed and agitated tank, which is filled with the fluid, heated to the desired temperature, and held for the specified time before cooling and filling the containers. The batch process was used traditionally for milk, which was pasteurized at 63°C for 30 min. It is replaced by the continuous flow HTST process.

In-container pasteurizers are normally used for the thermal pasteurization of canned or bottled juices, high-acid fruits in syrup, beer, carbonated beverages, and some other foods, which are subsequently stored at refrigeration temperatures.

The filled food containers are heated by dipping in a hot water bath or by recirculating hot water sprays in a conveyor tunnel. Carbonated beverages and glass containers are heated and cooled slowly for avoiding thermal shock. Figure 9.15 shows a diagram of the common tunnel pasteurizer with hot water bath. Steam can be used, instead of hot water bath, reducing the pasteurization time and increasing the capacity. The total residence time in the tunnel may be up to 1 h.

The efficiency of pasteurization is checked periodically by biochemical or microbiological tests, such as the alkaline phosphatase (ALP) test, used for milk and dairy products. The enzyme phosphatase is slightly more heat resistant than microbial pathogens, such as *M. tuberculosis*, which must be inactivated thoroughly according to public health regulations. A positive (ALP) test means under-pasteurized milk, which may have resulted from contamination of pasteurized milk with raw milk.

9.2.7 Blanching

Blanching is a light thermal process used primarily to inactivate deteriorative enzymes in vegetables and some fruits, prior to further processing (freezing, dehydration, or canning). In addition, blanching has some other beneficial effects on the processed foods, such as expulsion of air (oxygen) from intercellular spaces of vegetables, reduction of m.o., and improvement of texture and quality of the product.

The blanching time–temperature process is estimated using the principles of thermal processing, i.e., inactivation kinetics and heat transfer (D and z factors). Deteriorative enzymes include lipoxygenase, polyphenoloxidase, polygalacturonase, chloropyllase, catalase, and peroxidase. The last enzyme is the most resistant in fruit/vegetable materials, and it is used as an indicator of the blanching efficiency.

Blanching is optimized by considering the minimum time to inactivate the undesirable enzymes with the minimum damage to the quality of the product, keeping the cost of the operation at a minimum.

Blanching is normally applied at atmospheric pressure and temperatures 88°C–98°C, using steam, hot water, or microwave energy.

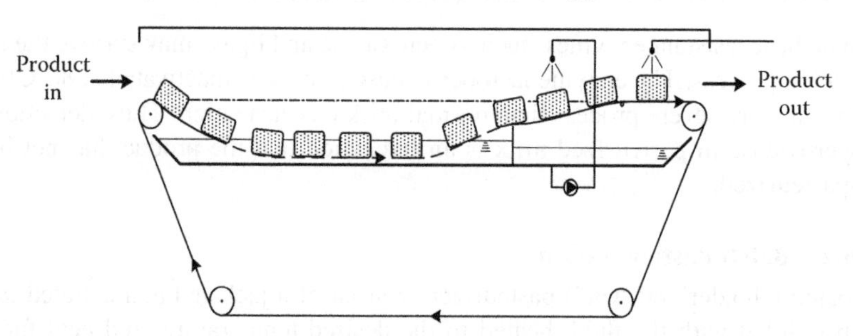

FIGURE 9.15 Tunnel pasteurizer with water bath for food containers.

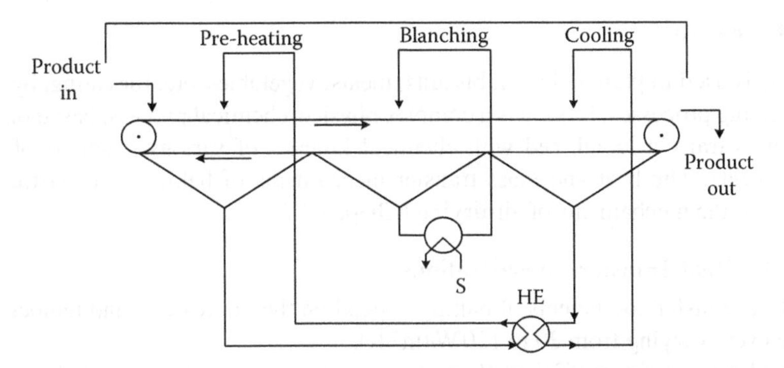

FIGURE 9.16 Diagram of a recirculating water blancher. S, steam; HE, heat exchanger.

Steam blanching has the advantages of smaller losses of water-soluble food components, and production of less waste, but it is less effective for some products, such as leafy vegetables, and it is more expensive than water blanching. Steam blanchers consist of a belt or chain conveyor, which transports slowly the vegetable product through the steam chamber, followed by water or air cooling. The residence time in the blancher depends mainly on the size of the vegetable pieces, varying from 2 to 10 min.

The rotary hot-water blanchers consist of cylindrical mesh drum, partially submerged in hot water, rotating slowly on a spiral reel, and moving forward the food pieces by a series of flights. The residence time is controlled by the rotational speed (RPM).

The immersion hot-water blancher transports the vegetable material through a hot-water tank by a conveyor belt, moving at the required speed.

The IQB (individual quick blanching) system is a three-stage blanching process, in which the vegetable pieces are heated rapidly in thin layers by steam, followed by holding in a deep bed, where temperature equilibration takes place, after which the material is cooled down. The method has the advantages of short residence time, improved yield, and reduced wastewater.

The recirculating water blancher consists of a conveyor belt, loaded with the product, which passes through three sections, i.e., preheating, blanching, and cooling. The hot water for blanching is heated by steam and is recirculated through the central section. The recirculating preheated water is heated in a heat exchanger by the recirculating cooling water (Figure 9.16).

9.3 THERMAL TREATMENT PROCESSES

Thermal treatment processes, including baking, toasting, and frying, are used to affect desirable quality and sensory changes in foods, such as texture, flavor enhancement, digestibility, and nutrient availability.

Cooking is an ancient empirical art of preparing foods for eating, based on physical and chemical changes of food materials induced by heating. Heat transfer to and within the product is involved.

9.3.1 Baking

Baking is used to prepare bread, biscuits, meats, vegetables, etc. for eating, by various heating processes. Baking is a complex physico-chemical process, based on heat and mass transfer, combined with chemical kinetics of various reactions of food components. The heat and mass transfer mechanisms of baking and roasting are similar to the mechanisms of air-drying (Chapter 11).

9.3.1.1 Heat Transfer Considerations

The heat transfer coefficients of baking depend on the air velocity and temperature in the oven, varying from 20 to 120 W/(m^2 K).

The heat transfer coefficient (h) in convection ovens can be estimated from the following empirical correlation of the heat transfer factor (j_H), which is based on published experimental data:

$$j_H = 0.801 Re^{-0.3} \tag{9.27}$$

where the (dimensionless) heat transfer factor is defined as $j_H = h/(u\rho C_p)$, h [W/(m^2 K)] is the heat transfer coefficient, ρ (kg/m^3) is the density, u (m/s) is the velocity, and C_p [J/(kg K)] is the specific heat of the air (gases).

Figure 9.17 is a graphical representation of the empirical equation (9.27).

9.3.1.2 Baking Ovens

Baking ovens consist of either a compartment of several shelves, or a tunnel through which the product is baked on a conveyor belt. The heating medium of the ovens is usually hot air, sometimes mixed with steam, moved by either natural convection or forced circulation. Radiation heating from the oven walls to the product may be also involved.

Ovens operate normally at atmospheric pressure, and the maximum temperature of the wet (high-moisture) product is 100°C. The wet product is heated by the hot air or hot combustion gases until the surface layer is dried, forming a crust at 100°C, while its interior remains at a lower temperature. Air temperatures of 150°C–250°C are used in baking bread and meat.

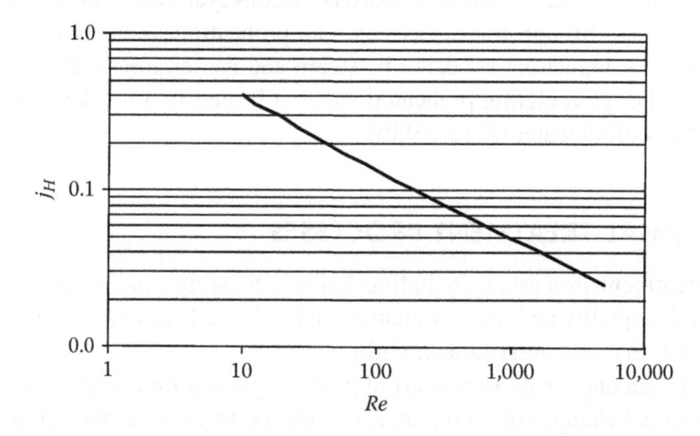

FIGURE 9.17 Heat transfer factor (j_H) in a baking oven. *Re*, Reynolds number.

Heat transfer to the baking ovens can be by (a) direct heating by combustion gases from a clean gas fuel, such as natural gas or LPG, or from a MW power source; (b) indirect heating from oven walls, heated by steam tubes, or from electrical resistances. Direct heating is preferred, because it is faster and more efficient.

Natural convection and forced circulation baking ovens are used in food processing and in catering operations. Bread and other dough products are placed on racks/trolleys, fixed/rotating shelves, or conveyor belts. Hot air (gas or oil burning) up to 250°C is circulated by a fan at velocities 1–10 m/s. Live steam or air impingement may be used, improving heat transfer and increasing product yield.

In large installations, tunnel ovens similar to belt dryers are used, which recirculate hot air, heated directly or indirectly in heat exchangers (metallic surfaces) with hot combustion gases. Conveyor belts about 0.5 m wide and up to 15 m long are used.

In the baking of biscuits, continuous ovens, with long metallic conveyor belts, are used. Such belts can be up to 45 m long, and they may be divided into three or more compartments, in which the temperature may rise progressively up to 210°C–270°C, with total residence time 4–6 min. Multi-deck traveling belts are used for reducing the length of continuous ovens, applied in baking various products.

9.3.2 Roasting

Roasting refers mainly to the thermal treatment of coffee, cocoa, and nuts for the development of characteristic color and aroma. Roasting brings about important changes in the microstructure of the food material (expansion, puffing), which facilitate mass transfer during processing, storage, or utilization of the product. It is accompanied by loss of moisture and carbon dioxide from the food.

Roasting is accomplished either at HTST (260°C, 3 min), or at low temperature long time (LTLT, 230°C, 12 min).

The roasting equipment normally consists of a metallic perforated drum, partially filled with the product. The drum rotates, while heated by hot air or flames.

9.3.3 Frying

Frying is used primarily to improve the eating quality of foods. At the same time, due to the high-temperature treatment, most of the spoilage m.o. are inactivated, and a surface crust is formed, improving the preservation and the quality of the food product.

Crust formation removes the free water from the food material, creating empty capillaries, which are filled with oil.

Frying can be accomplished either by shallow frying on a hot surface (pan) or by deep-fat frying in hot fat or vegetable oil. Film heat transfer coefficients in shallow (contact) frying are higher [250–400 W/(m^2 K)] than in deep-fat frying [200–300 W/(m^2 K)]. Oil temperatures 160°C–180°C are used in frying processes.

Batch fryers are used in small applications, while continuous units are applied in large installations. The oil is heated by electrical resistances, gas, fuel oil, or steam. Screw or belt conveyors are used to transfer the product through the hot oil in an inclined draining section. Continuous fryers, using baskets for transferring the food product through the heated oil may be used in large installations.

APPLICATION EXAMPLES

Example 9.1

Determine the spoilage probability of a thermal process time $F = 30$ min at 110°C, based on the inactivation of *C. botulinum* with thermal resistance parameters $D_{121} = 0.3$ min and $z = 10$°C.

Solution

The decimal reduction time of *C. botulinum* at 110°C (D_{110}) is calculated from the D_{121} by the Equation 9.4 $\log(D_{110}/D_{121}) = (121 - 110)/z = 1.1$, $(D_{110}/D_{121}) = 12.6$, and $D_{110} = 0.3 \times 12.5 = 3.78$ min.

The reduction exponent will be $m = 30/3.78 = 7.93$, approximately $m = 8$. Therefore, the probability of spoilage by *C. botulinum* will be about 1 in 10^8 or 10 to 10^9, i.e., 10 cans in one billion cans.

Example 9.2

The thermal inactivation parameters for a heat-resistant m.o. in a food product are $D_{121} = 1.5$ min and $z = 10$°C, while the heat damage parameters of a heat-resistant enzyme in the same product are $D_{121} = 3$ min and $z = 30$°C.

Determine

(a) The thermal death time at 121°C for a population reduction to one millionth of the heat resistant m.o.
(b) The retention of the enzyme in question in the above process (a).
(c) The thermal death time at 115°C for a reduction to one in million of the spoilage m.o., and the retention of the above enzyme in the same process.

Solution

(a) Thermal death time of the m.o. at 121°C, $F_0 = 6 \times 1.5 = 9$ min.
(b) Thermal damage exponent (m) of the enzyme in the above process, $m = F_0/D_{121} = 9/3$ or $m = 3$. Therefore, the survival rate of the heat-resistant enzyme will be 1–1000.
(c) The D_{115} of the spoilage m.o. is calculated from the D_{121} as follows:

$$\log\left(\frac{D_{115}}{D_{121}}\right) = \frac{(121-115)}{z} = 0.6, \quad \left(\frac{D_{115}}{D_{121}}\right) = 3.98, \quad D_{115} = 1.5 \times 3.98 = 5.97\,\text{min}$$

The D_{115} of the enzyme is calculated from the D_{121} as follows:

$$\log\left(\frac{D_{115}}{D_{121}}\right) = \frac{(121-115)}{z} = 0.2, \quad \left(\frac{D_{115}}{D_{121}}\right) = 1.6, \quad D_{115} = 3 \times 1.6 = 4.8\,\text{min}$$

For a survival of the spoilage m.o. of 1–1 million, the thermal death time becomes:

$$F_{115} = 6 \times 5.97 = 35.62 \text{ min}$$

For this F_{115} value, the (m) of the enzyme at 115°C will be

$$m = \frac{F_{115}}{D_{115}} \quad \text{or} \quad m = \frac{35.62}{4.8} = 7.4$$

Therefore, the survival rate of the heat resistant enzyme will be $1–10^{7.4}$ (1–25,118,864) or (40–1 billion).

Thus, because of the difference in the z-values, the m.o. is more easily inactivated at higher temperature than the enzyme, i.e., the enzyme survives more easily than the m.o. at higher than at lower temperatures.

Example 9.3

Calculate the energy of activation at $T_0 = 121°C$ and $T = 100°C$ for thermal inactivation of *C. botulinum* ($z = 10°C$), peroxidase ($z = 30°C$), thiamine ($z = 25°C$), and pectic enzymes ($z = 5°C$).

Solution

The energy of inactivation of m.o. and enzymes (E_a) is related to the temperature T (K) and the temperature parameter z (K) by the Arrhenius Equation 9.7:

$E_a = 2.3R(T\ T_0)/z$, where (T_0) is the reference temperature, usually $T_0 = 394\,K$, and $R = 8.31$ J/mol K. At (T_0), Equation 9.7 becomes

$$E_a = \frac{2.3R(T_0)^2}{z}$$

For *C. botulinum* $z = 10°C = 10\,K$ at $T = 373\,K$, $E_a = (2.3 \times 8.31 \times 373 \times 394)/10 =$ 280.9 kJ/mol, and at $T = 394\,K$, $E_a = [2.3 \times 8.31 \times (394)^2]/10 = 296.7$ kJ/mol.
For peroxidase $z = 30°C = 30\,K$ and $T = 373\,K$, $E_a = (2.3 \times 8.31 \times 373 \times 394)/30 =$ 93.6 kJ/mol, and at $T = 394\,K$, $E_a = [2.3 \times 8.31 \times (394)^2]/30 = 98.6$ kJ/mol.
For thiamine $z = 25\,K = 30\,K$ and $T = 373\,K$, $E_a = (2.3 \times 8.31 \times 373 \times 394)/25 =$ 112.3 kJ/mol, and at $T = 394\,K$, $E_a = [2.3 \times 8.31 \times (394)^2]/25 = 118.3$ kJ/mol.
For pectic enzymes $z = 5°C = 5\,K$ and $T = 373\,K$, $E_a = (2.3 \times 8.31 \times 373 \times 394)/5 = 561.8$ kJ/mol, and at $T = 394\,K$, $E_a = [2.3 \times 8.31 \times (394)^2]/5 =$ 593.4 kJ/mol.

It is concluded that the energy of activation (E_a) for the thermal inactivation of spoilage m.o., such as *C. botulinum*, is relatively high, meaning that heat sterilization is more effective at high temperatures. Very high E_a values (low z-values) characterize heat-sensitive spoilage and pathogenic m.o., such as *Lactobacilli*, *Salmonella*, and *E. coli*, and some spoilage enzymes (e.g., pectic enzymes). However, some very heat-resistant enzymes, such as peroxidase, and nutrients, such as vitamins, have relatively low activation energies (high z-values), meaning that they can survive better at high sterilization temperatures.

Example 9.4

(a) An acid fluid food is sterilized in a No. 2½ can in a boiling water retort at 100°C. Estimate the required heating time to reach a temperature of (i) 95°C and (ii) 98°C, assuming that the product is mixed thoroughly at any time during processing by can agitation (entirely convection heating), and that the initial temperature is 30°C. Overall heat transfer coefficient (boiling water/can contents) U = 0.3 kW/m² K, specific heat C_p = 3.5 kJ/kg K, and density ρ = 1100 kg/m³.

(b) Estimate an effective heating parameter (f_h), assuming that product heating follows the heat conduction equation.

(c) Estimate the holding time at (i) 95°C and (ii) 98°C required to sterilize the product, if the thermal death time is F_{100} = 10 min. The indicator m.o. has a z-value = 10°C. It is assumed that the sterilizing value of heating and cooling is equivalent to 25% of the heating time.

Solution

(ai) The heating time (t) to reach T = 95°C from an initial temperature T_0 = 30°C is calculated from Equation 8.72:

$$\log\left[\left(\frac{T_h - T}{T_h - T_0}\right)\right] = -\left[\left(\frac{UA}{2.3\rho V C_p}\right)\right] t$$

where T_h = 100°C, U = 0.3 kW/m² K, ρ = 1100 kg/m³, C_p = 3.5 kJ/kg K, capacity of can V = 0.94 L, and outside surface of can A = 532 cm². Note that the internal dimensions of the No. 2½ can are taken as (10.2) × (11.5) cm (Table 9.3).

$$t = \log\left[\left(\frac{100-30}{100-95}\right)\right] \times \frac{\left[2.3 \times 1100 \times 0.94 \times 10^{-3} \times 3.5\right]}{\left[0.3 \times 5.32 \times 10^{-2}\right]}$$

$$t = \frac{\left[\log(14) \times 2.3 \times 1.1 \times 0.94 \times 3.5 \times 100\right]}{1.6} = 597\,\text{s} \quad \text{or} \quad t = 10\,\text{min}$$

(aii) The heating time (t) to reach T = 98°C from an initial temperature T_0 = 30°C will be

$$t = \log\left[\left(\frac{100-30}{100-98}\right)\right] \times \frac{\left[2.3 \times 1.1 \times 0.94 \times 3.5 \times 100\right]}{1.6} = 803\,\text{s} \quad \text{or} \quad t = 13.4\,\text{min}$$

The heating time is affected strongly by the overall heat transfer coefficient (U) with the major thermal resistance in the product film inside the can. A conservative value of U = 0.3 kW/m² K is suggested for a medium viscosity product, agitated normally.

(b) The effective heating parameter (f_h) of the canned product, assumed to heat by conduction, is given by the Equation, $\log[(T_h - T)/(T_h - T_0)] = (1/f_h)t$.

Thus, for $T = 95°C$, $t = 10$ min, $f_h = t/\log[(100-30)/(100-95)] = 10/\log(14) = 8.7$ min.

The same value $f_h = 8.7$ min is obtained for heating to $T = 98°C$.

(c) The thermal death time at 95°C will be $F_{95} = F_{100} \times 10^{(100-95)/10} = 10 \times 3.16 = 31.6$ min, while the thermal death time at 98°C will be $F_{98} = F_{100} \times 10^{(100-98)/10} = 10 \times 1.6 = 16$ min.

The holding time at 95°C will be $t = 31.6 - 0.25 \times 10 = 29.1$ min, while the holding time at 98°C will be shorter $t = 16 - 0.25 \times 13 = 12.75$ min.

It is observed that the effect of heating/cooling on sterilization time is higher at high temperatures.

Example 9.5

(a) A solid food is sterilized in a No. 2½ can in a steam retort operated at 120°C. The product is heated by a mixed conduction/convection mechanism.

(i) Estimate the time required to reach a temperature of 115°C at the slowest heating point (geometric center) of the can, if the initial temperature is 30°C. The product is assumed to heat entirely by conduction and the thermal diffusivity is $\alpha = 1.5 \times 10^{-7}$ m²/s. Apply a simplified solution for the unsteady-state heat conduction equation.

(ii) Estimate the holding time at 115°C, required to sterilize the conduction-heating product, if the thermal death time is $F_{121} = 8$ min and the z-value is 10°C. It is assumed that the sterilizing value of heating and cooling is equivalent to 25% of the heating time.

(b) Calculate the thermal process time for the above product at 120°C by the Formula method. Assume mixed conduction/convection heating, and use an average heating parameter (f_h) of the values estimated in part (a) of this example and in Example 4(b). Assume a thermal death time $F_{121} = 8$ min and a heating factor $j = 1.4$.

Solution

(ai) The heating parameter of the conduction-heating product (f_h) can be estimated from the can dimensions $(D) \times (H)$ and the thermal diffusivity (α) of the product, Equation 9.16:

$$f_h = \frac{0.398}{\left[\left(\frac{1}{r^2} + \frac{0.427}{L^2}\right)\alpha\right]}$$

where

$r = D/2 = 10.2/2 = 5.1$ cm

$L = 11.5/2 = 5.7$ cm

$\alpha = 1.5 \times 10^{-7}$ m²/s, $f_h = (0.398 \times 10^3)/1.5 \times [(1/(5.1)^2 + 0.427/(5.7)^2] = 5142$ s, or $f_h = 85$ min

The time required to reach 115°C is calculated from the heat conduction equation

$$\log\left[\frac{(T_h - T)}{(T_h - T_0)}\right] = -\left(\frac{1}{f_h}\right)t \text{ and } t = f_h \log\left[\frac{(T_h - T_0)}{T_h - T}\right]$$

$$= 85 \times \log\left(\frac{90}{5}\right) = 85 \times \log(18) = 85 \times 1.25 = 106\,\text{min}$$

(aii) The thermal death time at 115°C will be $F_{115} = F_{121} \times 10^{(121-115)/10} = 8 \times 10^{0.6} =$ 32 min. The heating and cooling effect on lethality is assumed to be 25% of the heating time or 0.25 × 106 = 26.5 min. Therefore, the holding time of the can center at 115°C will be $t = 32 - 26.5 = 5.5$ min.

It is observed that the holding time at the process temperature is relatively low, compared to heating time. The long heating time results from the very long heating parameter f_h = 106 min of the conduction-heating product.

(b) The Formula method requires the (U) value at T = 121°C from the Equation 9.20, $U = F_{121} \times 10^{(121-T)/10} = 8 \times 10^0 = 8$ min.

The heating parameter of the food product is assumed to be equal to the mean value of the parameters calculated for conduction and convection heating, $f_h = (85 + 8.7)/2 = 47$ min. This is a normal value of heating parameter, determined experimentally, in canned foods for No. 2½ can size. The heating lag factor is assumed $j = 1.4$.

For a parameter $f_h/U = 47/8 = 5.8$, the log(g) of the process is 0.85 (Figure 9.18) and $g = 7.1$. The thermal process time will be,

$$t_B = f_h \log\left[j(T_h - T_0)/g\right], \quad \text{or} \quad t_B = 47 \times \log\left[(1.4 \times 90)/7.1\right]$$

$$= 47 \times 1.25 = 58.7\,\text{min}.$$

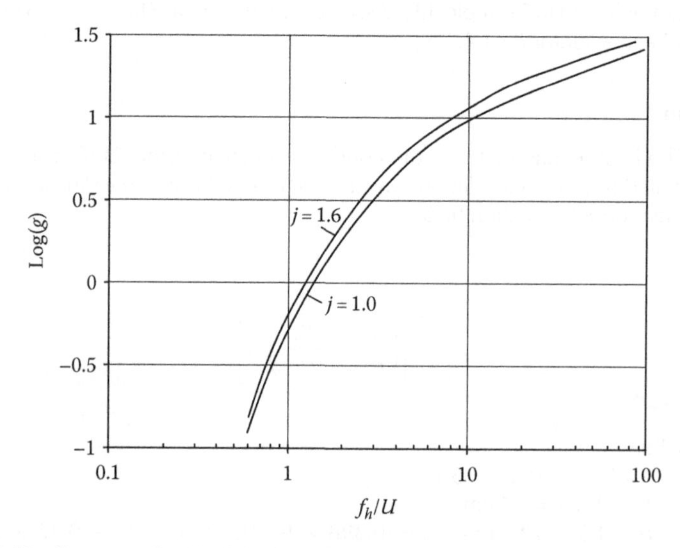

FIGURE 9.18 Parameter log(g) of the formula method. Upper line $j = 1.6$ and lower line $j = 0.8$.

Example 9.6

Using the Formula method, calculate the operator time for sterilizing a canned food product at a retort temperature of 121°C. The initial product temperature is 30°C and the heating factors are f_n = 40 min and j = 1.6. The indicator spoilage m.o. has D_{121} = 1 min, z = 10°C, and the reduction (sterilization) exponent is m = 9. The come-up time in the retort is 5 min.

Solution

The sterilization or operator time (t_p) is the time at the retort temperature. The thermal process time, calculated by the Formula (Ball) method (t_B) is related to the (t_p) by the approximate relation $t_B = t_p + 0.42\ t_{cut}$, where t_{cut} is the come up time of the retort.

The thermal death time at 121°C will be $F_0 = 1 \times 9 = 9$ min.

The Ball equation (9.19) yields:

$$t_B = f_h \log\left[\frac{j(T_h - T_0)}{g}\right] = 40 \times [\log(1.6 \times 91) - \log(g)]$$

The parameter log(g) is estimated from the literature Figure 9.18 as a function of $\log(f_h/U)$, where $\log(U)$ is given by Equation 9.20:

$\log(U) = \log(F_0) + (121 - T_h)/z$. In this case, $U = F_0 = 9$ min, and $(f_h/U) = 40/9 = 4.4$.

From Figure 9.18, for $(f_h/U) = 4.4$ and $j = 1.6$, the value $\log(g) = 0.75$ is obtained, and g = 5.6°C. Therefore, $t_B = 40 \times [\log(1.6 \times 91) - \log(5.6)] = 40 \times \log(145.6/5.6)$, or $40 \times 1.4 = 56.0$ min.

The sterilization or operator time at the retort temperature (t_p) will be

$$t_p = t_B - 0.42 t_{cut} = 56.0 - 0.42 \times 5 = 53.9 \text{ min.}$$

Example 9.7

Tomato concentrate is sterilized at the rate of 1 ton/h in a continuous flow system, followed by aseptic packing. The tomato product has a density of 1100 kg/m^3 and behaves as a non-Newtonian power-law fluid with rheological constants K = 20 Pa s^n and n = 0.5, which do not change appreciably with the temperature in the process system.

The product is preheated fast to 96°C and held at this temperature for the required thermal process time, followed by rapid cooling. The holding tube has an internal diameter of 50 mm. The thermal process is based on the reduction of the population m = 5 of a spoilage m.o. with D_{90} = 0.1 min and z = 10°C.

Calculate the length of the holding tube required for this process.

Solution

The average fluid velocity in the holding tube will be

$$u_{avg} = \frac{4m}{\gamma d^2} = \frac{4 \times (1000/3600)}{[3.14 \times 1100 \times (0.05)^2]} = 0.128 \text{m/s}$$

The shear rate at the tube wall will be $\gamma = 8u_{avg}/d = 8 \times 0.128/0.05 = 20.5$ (1/s). The apparent viscosity of tomato paste at the tube wall will be

$$\eta_a = K\gamma^{n-1} = 20 \times (20.5)^{-0.5} = \frac{20}{(20.5)^{0.5}} = \frac{20}{4.52} = 4.42\,\text{Pa s}$$

The Reynolds number of the tomato paste in the holding tube will be

$$Re = \frac{(du_{avg}\rho)}{\eta_a} = 0.05 \times 0.128 \times \frac{1100}{4.42} = 1.6$$

Therefore, the flow is definite laminar and the maximum fluid velocity in the tube will be, $u_{max} = 2u_{avg} = 2 \times 0.128 = 0.256$ m/s.

The reduction (sterilization) exponent of the thermal process will be $m = 5$, and the thermal death time at 90°C, $F_{90} = 5 \times 0.1 = 0.5$ min = 30 s. The thermal death time at 96°C will be

$$F_{96} = F_{90}10^{(90-96)/10} = 30 \times 10^{-0.6} = 7.5\,\text{s}$$

Therefore, the length of the holding tube will be $L = 7.5 \times 0.256 = 1.92$ m. The holding tube length can be taken as 2.0 m.

PROBLEMS

9.1 Calculate the lethal rate (L), relative to the L_{121}, at the following temperatures: 120°C, 115°C, 110°C, 105°C, and 100°C. Assume a temperature parameter z = 10°C.

9.2 Calculate the activation energy (E_a) and the thermal death time (F_0) at 121°C and sterilization exponent $m = 12$ for the spoilage m.o. in Table 9.7.

9.3 The temperature of the center of a food can during sterilization at steam temperature 120°C at various process times is shown in Table 9.8.
- (a) Plot the temperature versus time data.
- (b) Plot the lethal rate (L) versus time, assuming z = 10°C.
- (c) Calculate the total lethality of the process.
- (d) Using the same heating curve, estimate the process time (steam off) for thermal death time F_0 = 3 min.

TABLE 9.7
Data for Various Spoilage m.o.

m.o.	D_{121}, min	z, °C
C. botulinum	0.25	11
C. sporogenes	0.50	10
B. subtilis	0.75	10
B. stearothermophilus	2.00	10
PA 3814	3.00	7

9.4 Plot the heat penetration data of problem 3 as $\log(T_h - T_0)$ versus time and estimate the parameters (f_h) and (j) for the given product. Determine the thermal process time (t_B) by the Formula method. Sterilization of the product at 120°C will be based on the inactivation of a spoilage m.o. with $D_{121} = 1$ min and $z = 10$°C. The parameter $\log(g)$ can be estimated from Figure 9.18.

9.5 A sterilization process can be validated by inoculation tests of a known m.o. in a canned product. Usually gas-producing m.o. are used, which can be detected by the number of bulged spoiled cans after storage of the canned product.

A canned food product, containing 10 spoilage m.o./can of $D_0 = 1.0$ min is sterilized to obtain a probability of spoilage of $1\text{–}10^5$. An inoculation test is made in a pack of the same product, using 10^5 spores of *C. sporogenes* per can (gas-producing m.o.).

Calculate: (a) The F_0 value of the process; (b) What will be the probability of bulged cans, assuming that one surviving spore of *C. sporogenes* per can will spoil the product.

TABLE 9.8

Time, min	Temperature, °C
0 (steam on)	70
2	80
5	90
10	105
15	112
20	115
25	118
30 (steam off)	119
32	117
35	107
40	80
45	60
50	50
55	40
60	35
65	33
70	30

9.6 Orange juice (OJ) of 12°Brix is sterilized in a continuous flow UHT system at 95°C and 15 s, and packaged aseptically in cartons of 1 L capacity. The thermal process is based on the inactivation of a spoilage m.o. of $D_{90} = 0.1$ min and $z = 10$°C.

Calculate the probability of spoilage of packages, if the initial OJ contains 10 spoilage m.o. per mL.

9.7 Tomato paste is sterilized at the rate of 2 tons/h in a continuous flow system, followed by aseptic packing. The tomato product has a density of 1100 kg/m^3 and behaves as a non-Newtonian power-law fluid with rheological constants $K = 30$ Pa s^n and $n = 0.5$, which do not change with the temperature in the process system.

The product is preheated fast to 96°C and held at this temperature for the required thermal process time, followed by rapid cooling and aseptic packing in 55 gallon (208 L) drums. The holding tube has an internal diameter of 50 mm. The thermal process should have a probability of spoilage of $1\text{–}10^5$ drums, based on a spoilage m.o. with $D_{90} = 0.1$ min and $z = 10$°C, and an initial microbial load of 5 spores/mL product.

Calculate the length of the holding tube required for this process.

9.8 A canned food product of initial temperature 40°C is sterilized in a retort at 121°C.

From experimental heat penetration measurements it was obtained $f_h = 30$ min and $j = 1.4$. The thermal death time at 121°C is $F_0 = 4$ min, and $z = 10$°C.

Calculate the sterilization time (t_B) by the Formula (Ball) method, and the operator time in the retort (t_p), assuming a come-up time of 6 min.

9.9 The thermal sterilization time of a conduction-heating product in a No. 2 can (87) × (116) mm, calculated by the Formula method, is $t_B = 35$ min. Estimate the sterilization time of the same product in a larger No. 2½ can (103) × (119) mm, using the Formula method and assuming the same initial temperature, (*g*) value, and (*j*) parameter. The heating parameter (f_h) can be calculated from the heat conduction equation (9.16).

LIST OF SYMBOLS

A	m^2	Heat transfer area
Bi	—	Biot number (h L/λ)
C	—	Cooking value
C_p	kJ/(kg K)	Specific heat
d	m	Tube diameter
d	m	Particle diameter
D	m	Diameter (container)
D_T	min	Decimal reduction time at temperature T
E_a	kJ/mol	Activation energy
f_h	min	Heating parameter
f_c	min	Cooling parameter
F_T	min	Thermal death time at temperature T
F_0	min	Thermal death time at reference temperature (usually 121°C)
g	min	Temperature difference at end of heating ($g = T_h - T_{max}$)
H	m	Height (container)
$HTST$	—	High-temperature short time
h	W/(m K)	Heat transfer coefficient
j	—	Heating (j_h) or cooling (j_c) rate factor
j_H	—	Heat transfer factor
k	1/s	Rate constant
K	Pa s^n	Rheological constant
L	m	Tube length
L	m	Half height (container)
L	—	Lethality, $L = 1$ at reference temperature (usually 121°C)
m	—	Reduction (sterilization) exponent
m	kg/s	Mass flow rate
m.o.	—	Microorganism(s)
n	—	Flow behavior index
N	—	Population
Nu	—	Nusselt number (hL/λ)
Pr	—	Prandtl number ($C_p\eta/\lambda$)
Q_{10}	—	Rate increase for increasing temperature by 10°C
r	m	Can radius
R	J/(mol K)	Gas constant (8.31)
Re	—	Reynolds number ($du\rho/\eta$)
t	min	Time
t_B	min	Thermal process time, Ball formula

t_{cut}	min	Come-up time (retort)
t_p	min	Thermal process time at retort temperature (operator time)
T	°C	Temperature
T_g	°C	Glass transition temperature
T_h	°C	Heating (retort) temperature
T_0	°C	Initial temperature, reference temperature
T_{max}	°C	Maximum temperature
T_{po}	°C	Pseudo-initial temperature
u	m/s	Velocity
u_{avg}	m/s	Average velocity
u_{max}	m/s	Maximum velocity
U	min	Thermal death time parameter, $U = F_0$ at $T = 121°C$
U	W/(m² K)	Overall heat transfer coefficient
UHT	—	Ultrahigh temperature
V	m³	Volume
z	°C	Temperature parameter to reduce D or F by 90%
α	m²/s	Thermal diffusivity
η	Pa s	Viscosity
λ	W/(m K)	Thermal conductivity
ρ	kg/m³	Density
Σ	—	Summation

REFERENCES

Ball, C.O. and Olson, F.C.W. 1957. *Sterilization in Food Technology*. McGraw-Hill, New York.

Downing, D.L. ed. 1996. *A Complete Course in Canning*, 13th edn., Books I, II, and III. CTI Publications, Timonium, MD.

Fellows, P.J. 1990. *Food Processing Technology*. Ellis Horwood, London.

Gould, W.A. 1996. *Unit Operations for the Food Industries*. CTI Publications, Timonium, MD.

Kessler, H.G. 1981. *Food Engineering and Dairy Technology*. Verlag A. Kessler, Freising, Germany.

Lewis, M. and Heppell, N. 2000. *Continuous Thermal Processing of Foods*. Aspen Publications, Gaithersburg, MD.

Loncin, M. and Merson, R.L. 1979. *Food Engineering*. Academic Press, New York.

NFPA. 1982. *Thermal Processes for Low-Acid Foods in Metal Containers, Bulletin 26-L*, 12th edn. National Food Processors Association, Washington, DC.

NFPA. 1984. *Thermal Processes for Low-Acid Foods in Glass Containers. Bulletin 30-L*, 5th edn. National Food Processors Association, Washington, DC.

Rahman, M.S. ed. 2007. *Handbook of Food Preservation*, 2nd edn. CRC Press, New York.

Rahman, M.S. 2009. *Food Properties Handbook*, 2nd edn. CRC Press, New York.

Reuter, H. ed. 1993. *Aseptic Processing of Foods*. Technomic Publications, Lancaster, PA.

Ramaswamy, H.S. and Singh, R.P. 1997. Sterilization process engineering. In: *Handbook of Food Engineering Practice*. CRC Press, New York.

Saravacos, G.D. and Kostaropoulos, A.E. 2002. *Handbook of Food Processing Equipment*. Springer, New York.

Sastry, S.K. 1994. Continuous sterilization of particulate foods by ohmic heating. In: *Developments in Food Engineering: Part 2*. T. Yano, R. Matsuno, and K. Nakamura, eds. Blackie and Professional, London, U.K.

Sastry, S.K. 1997. Measuring residence time and modeling the aseptic processing system. *Food Technol.* 10(10): 44–46.

Stumbo C.R. 1973. *Thermobacteriology in Food Processing*, 2nd edn. Academic Press, New York.

Teixeira, A. 1992. Thermal process calculations. In: *Handbook of Food Engineering*. Heldman D.R. and D.B. Lund, eds. Marcel Dekker, New York.

Teixeira, A.A. and Shoemaker, C.F. 1989. *Computerized Food Processing Operations.* Van Nostrand Reinhold, New York.

Thijssen, H.A.C., Kerkhof, P.J.A.M., and Liefkens, A.A.A. 1978. Shortcut method for the calculation of sterilization yielding optimum quality retention for conduction heating of packed foods. *J. Food Sci.* 43: 1096–1101.

10 Evaporation Operations

10.1 INTRODUCTION

Evaporation is a physical separation process, which removes most of the water in a food liquid, resulting in a concentrated product, which may be used as such or processed further, e.g., by drying or freezing.

Evaporation is used in concentrating fruit and vegetable juices, milk, coffee extracts, and in refining sugar and salt. Evaporation is the major operation of concentrating liquid foods, although some new methods have been proposed as alternatives, such as freeze-concentration and reverse osmosis (Chapter 14) (APV, 1987).

The thermal efficiency of evaporators for removing water is much higher (e.g., 90%), compared to the efficiency of dryers (e.g., 60%). For this reason, whenever possible, evaporation is used to remove as much water as possible, prior to industrial drying (Kessler, 1981, 1986).

The engineering design of evaporators is based on the efficient transfer of heat from the heating medium (usually steam) to the liquid product, the effective vapor/liquid separation, and the utilization of energy. In food applications, evaporation should preserve the quality of heat sensitive products, and the evaporation equipment should conform to the hygienic requirements of cleaning and good manufacturing practices (Saravacos and Maroulis, 2001).

10.2 HEAT TRANSFER

Large amounts of heat energy must be transferred from the heating medium to the boiling liquid through the metallic walls of the evaporator. The heat requirements are determined by material and energy balances around each evaporator unit, and in the whole system. Heat transfer at the wall–liquid interface is the most important transfer operation in evaporation, since the thermal resistances of the wall and the heating medium (saturated system) are generally smaller. Heat transfer at the evaporation surface is directly related to the thermophysical and transport properties and the flow pattern of the liquid.

10.2.1 PHYSICAL PROPERTIES

The physical properties of the liquid, which are of direct importance to evaporation, are the viscosity (or rheological constants), the thermal conductivity, the density, the specific heat, the surface tension and the boiling point elevation. The transport properties of foods (viscosity and thermal conductivity) are very important.

The surface tension of water is 73 dyn/cm or 73 mJ/m^2 (25°C) and it decreases significantly when organic components are present in aqueous system. The surface tension of liquid food materials is lower (about 30 dyn/cm^2), due to the surface active components present.

Boiling point elevation (BPE) is caused by solute–water interaction and is undesirable in evaporation, since it requires a higher temperature of the heating medium to affect the same driving force (temperature difference). It is high in some chemicals, such as concentrated aqueous solutions of salts and alkalis (e.g., sodium hydroxide).

The BPE of liquid food is relatively low, and in most cases, it can be neglected in heat transfer calculations. It becomes important in concentrated solutions of sugars and other low molecular weight components. High molecular weight components dissolved or dispersed in water, such as starch, pectins, and proteins give negligible BPE.

For aqueous sugar solutions, similar to fruit juices, the BPE is a function of the molar concentration of the sugar. Monosaccharides (low-molecular-weight molecules) give higher BPE than polysaccharides and proteins. The following empirical equation can be used to estimate the BPE in food liquids (mainly fruit and vegetable juices):

$$\mathrm{BPE} = 0.33\exp(4x) \tag{10.1}$$

where x is the mass fraction of the sugar.

Thus, the BPE of a fruit juice will increase, during evaporation, from about 0.7°C (20°Brix) to 4.4°C (65°Brix).

10.2.2 Heat Transfer Coefficients

The heat transfer rate q (W) at the heating surface is given by the general equation,

$$Q = UA\Delta T \tag{10.2}$$

where

U [W/(m^2 K)] is the overall heat transfer coefficient
A (m^2) is the heating surface
ΔT (°C or K) is the temperature difference between the heating medium (steam, vapors) and the boiling medium

The overall heat transfer coefficient U is usually determined experimentally or is taken from operating similar industrial or pilot plant evaporators. Theoretical prediction of U is difficult because of the fouling resistance at the heating surface, which cannot be quantified accurately. However, heat transfer analysis is useful in evaluating the thermal resistances of evaporation systems.

The overall thermal resistance $1/U$ [(m^2 K)/W] of a heating system for evaporation is given by the equation

$$\frac{1}{U} = \frac{1}{h_s} + \frac{x}{\lambda} + \frac{1}{h_i} + \mathrm{FR} \tag{10.3}$$

where

$1/h_s$ and $1/h_i$ [(m^2 K/W)] are, respectively, the heat transfer resistances at the heating (steam) and evaporation sides

x/λ [(m^2 K)/W] is the thermal resistance of the evaporator wall

FR [(m^2 K)/W] is the fouling resistance, defined as the inverse of the fouling coefficient (FR = $1/h_f$)

Equation 10.3 refers to plane heat transfer surfaces, but it can be applied approximately to tubes of relatively large diameter, e.g., >50 mm. For smaller diameter tubes, the thermal resistances must be corrected by the ratio of outside to inside diameters, as discussed in Chapter 8.

The heating side is normally assumed to have negligible fouling resistance, since food evaporators use clean saturated steam and clean metallic surfaces. Thus, the resistance of the heating side is relatively low, since high h_s values are obtained with saturated steam or water vapors. The wall resistance x/λ is relatively low, since thin walls (low x) and high thermal conductivity λ characterize the evaporators.

The heat transfer coefficient at the evaporation surface h_i is a function of the physical properties (mainly the viscosity) and the flow conditions of the liquid. It increases at high flow rates and high temperatures and it can be estimated from empirical correlations, discussed in Chapter 8. Heat transfer coefficients in liquid films are of particular importance to falling film food evaporators.

10.2.3 Fouling in Evaporators

Fouling involves formation of undesirable deposits on the heat transfer surface, which reduce heat transfer and evaporation rates and may damage the quality of the product.

Fouling refers mainly to scaling, which is caused by precipitation of inorganic salts on the heating surface, and precipitation of suspended food particles, such as proteins, pectins, and polysaccharides. It may also involve corrosion fouling, biological fouling (attachment of microorganisms to the heat transfer surface), and solidification fouling (freezing or solidification of high melting components).

The mechanism of fouling involves initiation, mass transport and attachment to the heating surface, and removal into the fouling liquid. In food evaporators and other heat transfer equipment, fouling is caused mainly by the adsorption and denaturation of food biomolecules, such as proteins, pectins, and starch on the heated surface.

Empirical correlations of fouling resistance to the operating time of food evaporators for a specific application (e.g., sugar evaporators) are useful for determining the optimum cleaning cycle, i.e., how often will the evaporator be cleaned by interrupting its operation.

Use of fouling resistances (FR) or fouling factors in Equation 10.3 gives only approximate values for a specific evaporator and product, which are useful for preliminary process design. More accurate and reliable values of the overall heat transfer coefficient U may be obtained in operating pilot plant or industrial food evaporators. Typical values of U for food evaporators are shown in Table 10.1.

TABLE 10.1
Typical Values of Overall Heat Transfer Coefficients (*U*)

Type of Evaporator	Liquid Food	*U* (W/(m² K))
Falling film, tubular	Fruit juices 12°Brix–65°Brix	500–2000
Falling film plate	Milk 10%–30% TS	1000–2500
Rising film, tubular	Milk 10%–35% TS	1000–2000
Forced circulation	Sugar syrup 15°Brix–65°Brix	1500–2500
Agitated film	Fruit/vegetable pulp	800–2000

10.2.4 Evaporation of Fruit Juices

Falling films evaporators are used extensively for the concentration of fruit juices and experimental data of heat transfer coefficients from pilot plant or industrial installations are useful in the design and evaluation of the industrial units (Saravacos, 1970; Saravacos et al, 1970; Saravacos, 1974).

The overall heat transfer coefficients *U* decrease significantly, as the juice is concentrated (Figure 10.1). Thus, the *U* values of filtered grape juice decrease from about 2 kW/(m² K) to nearly 1.2 kW/(m² K), as the percentage soluble solids is increased from 20°Brix to 65°Brix. Lower *U* values are obtained in the evaporation of unfiltered grape juice (1.35–0.65 kW/m² K). This significant reduction in heat transfer rate is caused by fouling at the evaporator surface with particles and organic components, which precipitate on the heated surface at high concentrations. Similar heat transfer coefficients are obtained with filtered (depectinized) and unfiltered apple juice. Depectinization (removal of dissolved colloidal pectins by enzyme treatment and filtration) is normally practiced in the production of apple juice concentrates.

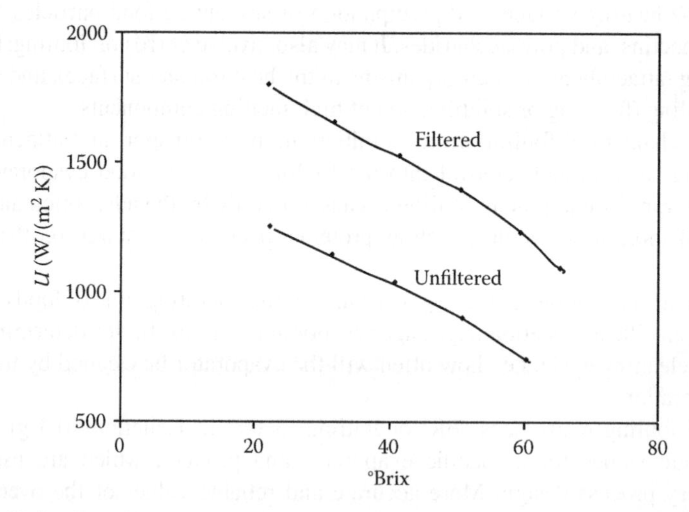

FIGURE 10.1 Overall heat transfer coefficients *U* of filtered and unfiltered grape juice at 55°C.

10.3 DESIGN OF EVAPORATORS

Several types of evaporators are used for the concentration of liquid foods. The principal factors affecting the choice of an evaporation system are food quality, evaporation capacity, and energy/cost considerations. The food product quality depends primarily on the residence time–temperature combination in the evaporator. Evaporation capacity is related to the heat transfer rate, and energy utilization is improved by energy-saving evaporation systems (Hahn, 1986; Minton, 1986).

10.3.1 MATERIAL AND ENERGY BALANCES

The sizing of an evaporator is based on the estimation of the heat transfer surface area required for a given evaporation (and heat transfer) load. Material and energy balances, needed for sizing calculations, are estimated by the procedures discussed in Chapter 1 (Chen and Hernandez, 1987).

Figure 10.2 shows schematically the required quantities and process data for performing an elementary balance. The liquid to be concentrated is fed continuously to the top of a falling film evaporator of heating surface area A, which is heated with saturated steam, while the steam condensate is removed from the system. The vapor–liquid mixture from the bottom of the evaporator is separated in the separator (SP) into the vapors and the liquid product.

The following assumptions are normally made for preliminary process design: (a) The feed and the products enter and leave the evaporator as saturated liquids, i.e., at the boiling point for the given operating pressure. (b) The boiling point rise (BPR) can be neglected, which is a reasonable assumption for food materials, except for high sugar concentrations. e.g., above 60°Brix. (c) The heating steam and the steam condensates are saturated (at the condensation/boiling point). (d) Heat losses

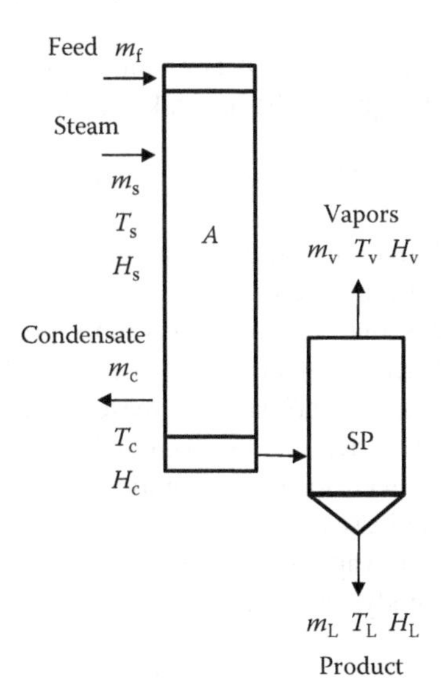

FIGURE 10.2 Material and energy balances in an evaporator unit: A, evaporator; SP, vapor–liquid separator.

to the environment are neglected. The heat losses are generally low, representing about 1%–3% of the total heat use, which could be reduced by insulation, but is not applied in practice.

$$\text{Overall balance,} \quad m_f = m_v + m_L \tag{10.4}$$

where m_f, m_v, and m_L (kg/s) are the flow of feed, vapors, and liquid product.

$$\text{Solids balance,} \quad m_f x_f = m_L x_L \tag{10.5}$$

where x_f and x_L are the mass fractions of solids in the feed and liquid product.

It is assumed that no solids are lost in the vapors (no entrainment).

$$\text{Steam balance,} \quad m_s = m_c \tag{10.6}$$

Equal flow rates of steam m_s and condensate m_s are assumed (kg/s).

The energy balances in evaporators are confined to enthalpy (heat) balances, since they represent the main energy changes in the system. Mechanical energy, used in pumping and vacuum equipment, is relatively small and can be neglected in preliminary design. For simplification, the feed is assumed to enter the evaporator at the boiling point, the steam condensate leaves the evaporator at the steam temperature, and that the vapors and liquid product leave at the boiling temperature. The BPE is neglected and the temperatures of water–vapor at a given temperature can be taken from the Steam Tables.

$$\text{Overall energy balance,} \quad m_f H_f + m_s \Delta H_s = m_v H_v + m_L H_L \tag{10.7}$$

In simplified evaporator calculations, it is assumed that the heat given up by the condensing steam q (W) is equal to the heat taken up by the vapors produced, i.e.,

$$Q = m_s \Delta H_s = m_v \Delta H_v \tag{10.8}$$

where ΔH_s and ΔH_v (kJ/kg) are the heat of evaporation of water at the steam T_s and boiling temperature ($T_b = T_f$), respectively.

Equation 10.8 indicates that the mass of vapors produced will be somewhat smaller than the mass of heating steam ($m_v < m_s$), since the heat of evaporation of water increases as the saturation pressure decreases ($\Delta H_v > \Delta H_s$ since $P_v < P_s$ and $T_v < T_s$, see Steam Tables). A detailed material and energy balance analysis would consider preheating of the feed to the evaporation temperature, subcooling of the steam condensate, and heat losses.

The basic calculation in evaporators is the estimation of the heat transfer area A (m^2) of the evaporation body, using the equation:

$$q = UA\Delta T \tag{10.9}$$

where $\Delta T = T_s - T_v$.

10.3.2 Short Residence–Time Evaporators

10.3.2.1 Film Evaporators

Evaporation of water from falling or rising films is used extensively in food evaporators, because of its advantages, simplicity of operation, and low equipment and operating cost. Figure 10.3 shows a diagram of the principles of operation for both falling and rising film evaporators (Moore and Hessler, 1965).

10.3.2.1.1 Falling Film Evaporators

In the falling film evaporators, the liquid film falls by gravity in the vertical evaporation surface (inside the tube or plate), while heat is transferred through the wall by condensing steam or water vapors. The mixture of liquid/vapors *L/V* leaves the bottom of the tube or plate and enters a vapor–liquid separator. The separated liquid concentrate is pumped out or to the next evaporation effect, while the water vapors are directed to a condenser or to heat the next effect (in multiple effect evaporator systems) (Moore and Pinkel, 1988).

In falling films, the minimum liquid flow rate per unit length of the feed surface, or "irrigation rate" Γ (kg/ms), is given by the empirical equation:

$$\Gamma_{min} = 0.008(\eta\sigma^3)^{1/5} \tag{10.10}$$

where

η(mPa s) is the viscosity
S (–) is the specific gravity related to water
σ (dyn/cm) is the surface tension of the liquid

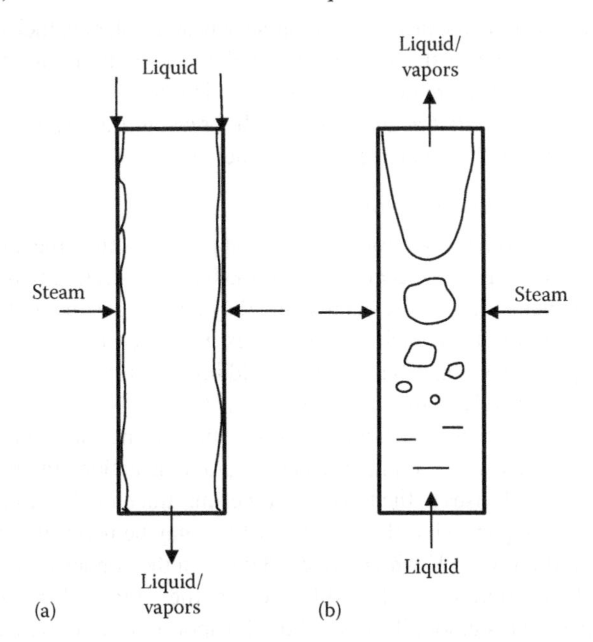

FIGURE 10.3 Principles of falling film (a) and rising film (b) evaporators.

Thus, the minimum flow rate of water 80°C to form a film ($\eta = 0.356$ mPa s, $S = 1$, $\sigma = 68$ dyn/cm) on a vertical surface will be, $\Gamma_{min} = 0.008 \times (0.356 \times 68^3)^{1/5} = 0.008 \times 10.23 = 0.08$ kg/ms. A food liquid film can be formed at a lower flow rate than pure water, because of lower surface tension, improving the heat transfer coefficient. Thus, for a liquid food with a surface tension of 34 dyn/cm, $\Gamma_{min} = 0.08 \times (34/68)^{3/5} = 0.052$ kg/ms.

The Reynolds number of the falling film is given by the simplified equation:

$$Re = \frac{4\Gamma}{\eta} \tag{10.11}$$

For water 80°C ($\eta = 0.356$ mPa s) at the minimum flow rate, the Reynolds number will be $Re = 4 \times 0.08 \times 10^3/0.356 = 900$, i.e., the flow is laminar.

Higher heat transfer coefficients h are obtained in the turbulent flow regime, i.e., at $Re > 2100$. The following empirical equation can be used to estimate the heat transfer coefficient of water films in the turbulent flow regime:

$$h = 9150\Gamma^{1/3} \tag{10.12}$$

The heat transfer coefficient h of falling liquid films for turbulent flow is also given by the general empirical equation,

$$h = 0.01(\varphi\ Re\ Pr)^{1/3} \tag{10.13}$$

where

$\varphi = (\lambda^3 \rho^2 g/\eta^2)$
$Re = 4\Gamma/\eta$
$Pr = C_p\, \eta/\lambda$

When SI units are used, the factor $\varphi^{1/3}$ has units of heat transfer coefficient [W/(m² K)].

The calculated heat transfer coefficient of the evaporation surface h_i from U depends strongly on the steam heat transfer coefficient h_s and the fouling resistance, which are difficult to predict accurately. For this reason, the experimental overall heat transfer coefficient U is more reliable in practice.

10.3.2.1.2 Rising Film Evaporators

The rising (climbing) film evaporators find fewer applications than the falling film systems, because of the longer residence-time and the higher operating temperatures and pressure drops, which require more energy, and they may be detrimental to the quality of heat-sensitive food liquids, such as fruit juices. However, the rising film systems do not require special feed distributors, they yield high heat transfer coefficients, and they do not foul as severely as the falling film units (Bourgois and LeMaguer, 1864).

In the rising film system, the liquid begins to boil in the tube, producing vapor bubbles of growing size as the liquid rises by natural convection, and finally forms a film on the walls, which rises to the top of the tube, entrained by the fast rising vapors. The *L/V* mixture is separated to the liquid *L*, which may be recirculated or removed as a product, and the vapors *V*, which are condensed in the condenser (Figure 10.3).

In a rising film evaporator, the liquid feed enters the bottom of the vertical tubes, and the water evaporates gradually, as the liquid–vapor mixture moves upwards. The evaporation surface should be covered completely with a rising liquid film, while the vapors flow upward as bubbles, plug, or stratified vapor–liquid mixture. At the top section of the tubes, the high-velocity vapors may entrain some liquid in the form of liquid droplets, reducing the product-side heat transfer coefficient.

The heat transfer coefficient in a rising film evaporator can be estimated by the following empirical equation:

$$Nu = 8.5\, Re^{0.2}\, Pr^{1/3}\, S^{2/3} \tag{10.14}$$

where S is the slip ratio, i.e., the ratio of vapor to liquid velocities in the evaporator tube. The numbers Nu, Re, and Pr are determined at a mean location, using mean velocities and property values.

Experimental values of overall heat transfer coefficients U in the evaporation of fruit juices in rising film evaporators may decrease from about 1.5 kW/(m^2 K) at the bottom (about 15°Brix) to nearly 1 kW/(m^2 K) at the top (about 60°Brix) of the evaporator tube. The liquid film velocity at the bottom and the top of the evaporator are respectively about 1.0 and 1.97 m/s. The respective vapor velocities are much higher, about 50 and 60 m/s.

As with falling films, the surface tension of the liquid is important in film formation. Food liquids, having surface tension lower than that of water, will cover the heating surface more effectively than pure water, resulting in higher heat transfer coefficients.

10.3.2.2 Long-Tube Vertical Evaporators

Most of the heat-sensitive food liquids, such as fruit juices and milk, are concentrated in long-tube vertical evaporators of 25–50 mm diameter and 4–10 m length. Falling film evaporators are more widely used than rising film units (Figure 10.3). Because of their length, many long-tube evaporators are often installed outside the plant building. Figure 10.4 shows schematic diagrams of the two types of long-tube vertical evaporators.

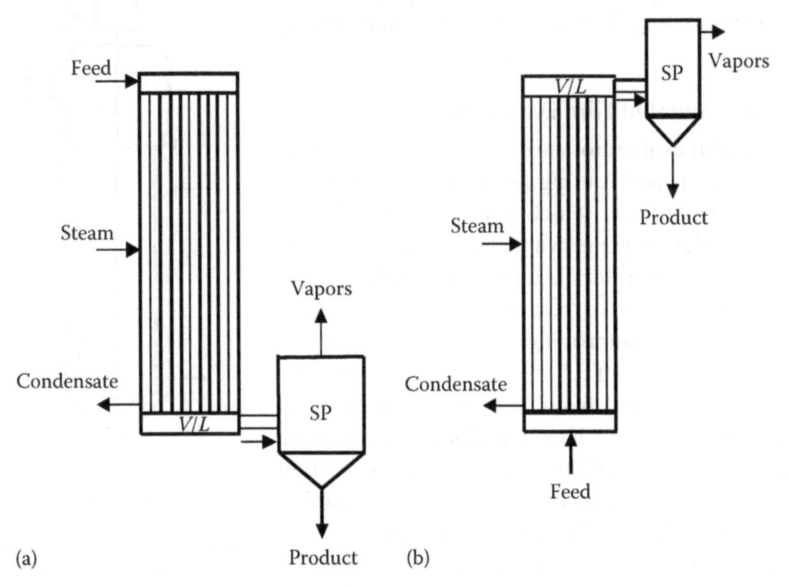

FIGURE 10.4 Diagrams of long-tube vertical evaporators: (a) falling film and (b) rising film. *V/L*, vapors/liquid mixture; SP, separator.

The falling film system is more popular, because of the high heat transfer coefficients, the low pressure drop, and the short residence-time (a few seconds). The preheated liquid feed must be distributed evenly at the top of the long-tubes and the vapor–liquid mixture exiting the bottom is separated in a centrifugal or baffled separator. The concentrated liquid is removed with an appropriate pump (positive displacement, if very viscous), and the vapors are condensed in a surface or mixing condenser, followed by a vacuum system (see Section 10.5).

Forced falling film evaporators are used for the concentration of fruit and vegetable pulps (e.g., tomato products). A pump is used to recirculate the partially concentrated product, improving heat transfer and reducing fouling of the evaporation surface.

The rising film evaporators do not need special feed distributors and they are less likely to foul, contrary to the falling film type. However, they operate at higher pressure drops, meaning that the liquid temperature at the bottom of the tube may be considerably higher than at the top. They also have longer residence-time. High vapor–liquid velocities (up to 100 m/s) develop at the exit of the tubes, due to the high evaporation rates in both types of evaporators.

A combination of rising and falling film concentrator (RFC) may have the advantages of both film evaporators.

10.3.2.3 Plate Evaporators

Plate evaporators, operating as falling film or combination RFC units, are similar in principle to the long tube systems. They have the advantage of shorter length and they are installed inside the plant building. They can also be disassembled and cleaned more easily than the tubular systems.

The plate evaporators are similar to the familiar plate heat exchangers (Chapter 8), with special designs for handling boiling viscous liquids and separating the water vapors produced during heating.

10.3.2.4 Agitated Film Evaporators

Agitated film evaporators are used in the processing of very viscous and fouling liquid foods, or suspensions of particulates, which cannot be handled in normal tubular or plate evaporators. The main feature of these evaporators is a rotor within the evaporator body (vertical or horizontal), which agitates the viscous fluid, improving heat transfer and preventing fouling (Figure 10.5). Various types of low wear rotors are used, fixed, scraping, or hinged. In the vertical units, the vapor–liquid separator and the motor can be placed either at the top or at the bottom of the unit. The agitated film units are often used in combination with tubular film evaporators, when a very high solids concentration is required. The evaporation surface of agitated film evaporators is limited, e.g., up to 10 m^2, due to mechanical limitation

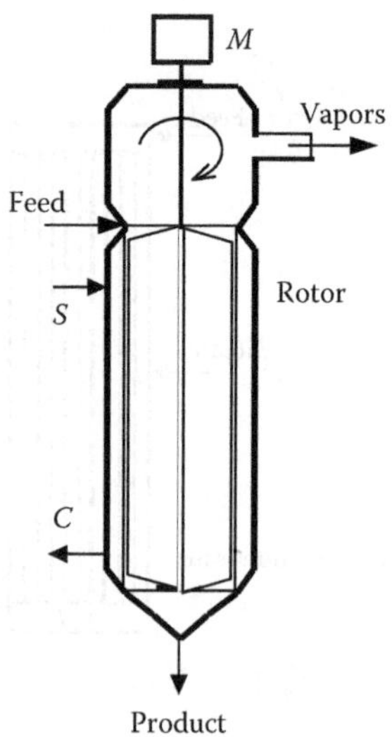

FIGURE 10.5 Diagram of agitated film evaporator with top vapor/liquid separator. *M*, motor; *S*, steam; *C*, condensate.

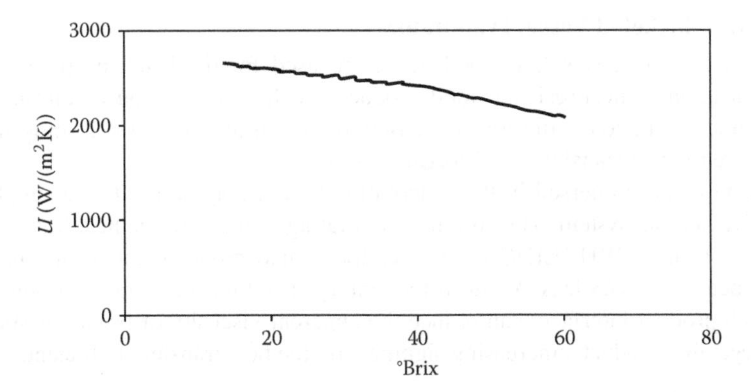

FIGURE 10.6 Overall heat transfer coefficients U in the evaporation of sucrose solutions at 100°C in an agitated film evaporator.

(a single tube with an agitator). By contrast, tubular (e.g., falling film) evaporators can be constructed with a large number of tubes, i.e., large evaporation surfaces.

The overall heat transfer coefficient U of an agitated film evaporator is generally high, depending on the rotational speed of the blades and the feed rate of the liquid product.

U values of 2.7–2.1 kW/(m^2 K) were obtained in the concentration of sucrose solutions at 100°C from 10°Brix to 60°Brix in an agitated film evaporator (Figure 10.6).

10.3.2.5 Centrifugal Film Evaporators

The heat transfer coefficients of liquid films can be increased in a centrifugal field, which increases the hydrodynamic and rheological processes of heat and mass transfer systems. Spinning core evaporators, with very short residence-time and high heat transfer coefficients, are suitable for concentrating very heat-sensitive food liquids.

The heat exchange surface consists or rotating concentric disks, similar to those used in centrifugal separators. The disks are double-walled and the heating medium (e.g., steam) flows inside the disks, while the product is distributed by nozzles on the lower external surface of the disks and climbs up during rotation. The product layer is less than 0.1 mm thick, the liquid holdup volume is less than 1.5 L, and the residence time in the evaporator is less than 1 s (Yanniotis and Kolokosta, 1996).

Heat transfer coefficients in the range of 2–10 kW/(m^2 K) were obtained in concentrating corn syrups (0°Brix–60°Brix), using a horizontal rotating disk at 200–1000 RPM. Industrial application of centrifugal film evaporators is limited, due to the high equipment and operating cost and the low evaporation capacity.

10.3.3 Long Residence–Time Evaporators

Heat-resistant foods, such as sugar solutions, syrups, tomato juice and tomato products, fruit jams and preserves, and salt solutions, can be evaporated in various types of evaporators, characterized by relatively long residence-times (several minutes or a few hours), high temperatures, and recirculation. High heat transfer coefficients are obtained by agitation and natural or forced recirculation of the fluid food. In addition, forced recirculation reduces fouling.

10.3.3.1 Jacketed Vessel Evaporators

Steam jacketed vessels (pans or kettles) are used for the batch evaporation and concentration of heat-resistant food products, such as tomato puree and ketchup. Mechanical agitation with scrapers is used to prevent fouling of the heating surface and increase heat transfer rate (Morgan, 1967).

Steam coils, immersed in the evaporating liquid, may be used as a simple and effective heating system. The coil may be rotating to increase the heat transfer rate. The coil heating WURLING evaporator for tomato products and fruit pulps was developed in the U.S.D.A. Western Laboratory. The high shear rates, developed at the coil–product interface, can reduce the apparent viscosity of pseudoplastic fruit and vegetable products, increasing significantly the heat transfer coefficient.

10.3.3.2 Short Tube Evaporators

The short tube evaporators (calandria) consist of a bundle or basket of short tubes, 2–3 m long, heated outside by steam, and immersed in the evaporating liquid. The liquid flows upward through the tubes by natural convection at velocities of about 1 m/s, and recirculates through the middle of the evaporator. Some boiling takes place within the tubes. Short tube evaporators are low-cost systems, effective for evaporation of low viscosity liquid foods, such as sugar solutions (sucrose processing plants).

10.3.3.3 Forced Circulation Evaporators

Figure 10.7 shows a schematic diagram of a forced circulation evaporator. The liquid is recirculated by a centrifugal pump through an external (or internal) heat exchanger at high velocity (3–5 m/s), where it is heated by condensing steam. Due to the high pressure drop, the liquid does not boil within the heat exchanger tubes, but it is flashed into the liquid–vapor separator, which is usually maintained in a

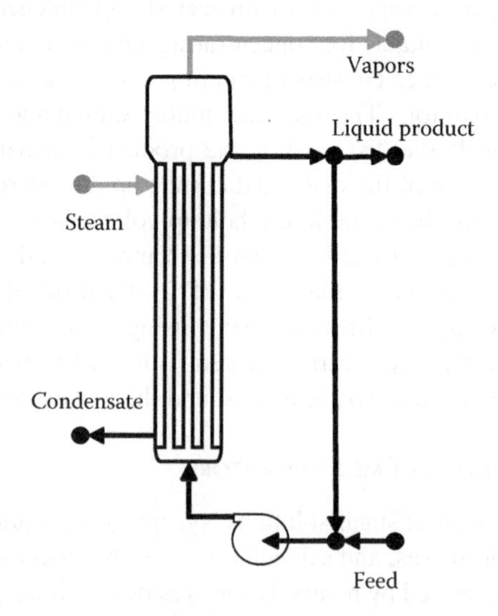

FIGURE 10.7 Diagram of a forced circulation evaporator. (Modified from Maroulis, Z.B. and Saravacos, G.D., *Food Process Design*, CRC Press, Boca Raton, FL, May 9, 2003, fig. 6.2.)

vacuum. High heat transfer coefficients are obtained, due to high liquid velocities, and fouling of the tubes is prevented. The residence time can be several minutes, and heat-resistant food liquids, such as sugar solutions are evaporated effectively.

10.4 ENERGY-SAVING EVAPORATION SYSTEMS

Evaporation and drying are the most energy-intensive unit operations of food processing. With increasing energy costs and concerns over the environmental impact of energy production, energy saving and utilization systems have been developed and applied in the industry. Energy utilization can be improved by low investment (fine tuning existing evaporators), moderate investment (modifying accessory equipment), or major investment (installing new energy-saving equipment).

The energy required for evaporation is used mainly to vaporize the water from the liquid food material. Theoretically, the evaporation of 1 kg of water requires slightly more than 1 kg of saturated steam, used as the heating medium, since the enthalpy (heat) of vaporization of water decreases as the pressure is increased. For example, evaporation of water at 100°C requires 2.26 MJ/kg and if saturated steam of 4 bar (absolute) pressure is used as the heating medium, the energy given up by its condensation will be 2.13 MJ/kg (Steam Tables). Thus, the steam economy (SE) in this operation will be SE = 2.13/2.26 = 0.94 kg water evaporated/kg steam (single effect operation). In this example, the liquid is assumed to enter and leave the evaporation unit thermally saturated (at the boiling point), and the steam condensate leaves the system at the saturation temperature (without subcooling) (Schwartzberg, 1977).

The steam economy of evaporators can be increased substantially, using various energy-saving systems, such as multiple effect and vapor recompression. Table 10.2 shows typical steam economies of industrial evaporators (ERDA, 1977).

The steam economy SE in a multiple effect evaporator system is approximately $SE = 0.85\,N$, where N is the number of effects.

10.4.1 MULTIPLE EFFECT EVAPORATORS

The multiple effect (ME) evaporation system is based on the repeated use of the water vapors from one evaporation unit (effect) to heat the next effect, which operates at a lower pressure. Thus, 1 kg of steam can evaporate more water, depending on the number of effects and the operating pressures.

TABLE 10.2
Steam Economies SE of Evaporator Systems

Evaporator System	SE (kg water/kg steam)
Single effect	0.90–0.98
Double effect	1.70–2
Triple effect	2.40–2.80
Six effect	4.6–4.9
Thermal recompressor	4–8
Mechanical vapor recompression	10–30

Thermodynamic considerations lead to the need for decreasing pressure (and temperature) from one effect to the next. For heat-sensitive liquid foods, the temperature in the first effect should not exceed 100°C, while the temperature in the last effect should not be lower than e.g., 40°C, in order to use cooling water at ambient temperature in the condenser of the last vapors. Assuming that the temperature difference ΔT in each effect is 10°C, the maximum number of effects in a food evaporation system should be about $N = 60/10 = 6$.

Figure 10.8 shows diagrammatically a three-effect rising film evaporator with forward feed operation (cocurrent flow of heating medium-steam/vapors and liquid). Forward feed evaporators are preferred because the feed (low concentration, low viscosity) is evaporated more efficiently at high temperatures, without serious fouling. In special cases, backward feed and parallel flow multiple-effect evaporators may be advantageous, compared to the forward feed system. In backward feed systems, the concentrated product is evaporated in the first effect at a higher temperature, obtaining a higher heat transfer coefficient. The dilute feed enters the last effect and it is evaporated at the lowest temperature of the system. Backward feed systems require pumps to transport the liquid product from the last to the first effect, against an increasing pressure.

The steam economy (SE) in a ME system is higher than (1) but less than the number of effects N,

$$SE = \sum \frac{m_{vi}}{m_s} \tag{10.15}$$

where

m_s is the steam consumption (kg/s)
m_{vi} is the evaporation rate (kg/s) of the (i) effect

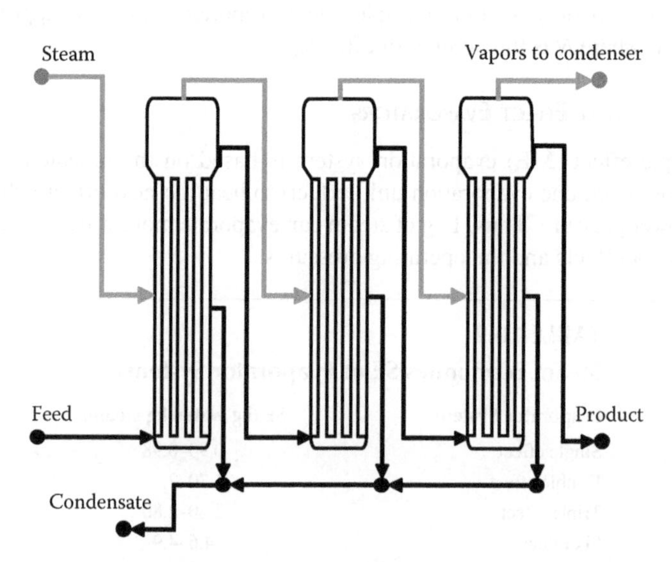

FIGURE 10.8 Schematic diagram of a three-effect, forward-feed rising film evaporator. (Modified from Maroulis, Z.B. and Saravacos, G.D., *Food Process Design*, CRC Press, Boca Raton, FL, May 9, 2003, fig. 6.4.)

The SE is estimated from material and energy balances around each effect and over the whole system. The boiling temperature (and pressure) in the last effect of a ME system is limited not only from consideration of the cooling water temperature in the condenser, but also from the high viscosity of the concentrated liquid food, which increases sharply as the temperature is lowered. High liquid viscosity means higher fouling and lower heat transfer coefficients, i.e., more expensive operation.

Simplified calculations of a three-effect evaporator are given in the Example 10.2.

The boiling point elevation (BPE) has a negative effect on the operation of a multiple effect evaporation system. In such a case, the vapors coming out of the vapor–liquid separator will be superheated by BPE degrees, but they will be condensed in the heater of the next effect at saturation temperature, losing the BPE superheat as available driving force ΔT. For most liquid foods, the BPE is usually small (about 1°C) and it can be neglected, except in very concentrated sugar solutions and juices (last effects).

In some food evaporation systems, more than one evaporator units are used in the last effect for more economical operation. The vapors coming from the previous effect are split into two or more parts, and they are used to heat two or more stages, operating at the same pressure of the last effect. Each stage is fed with concentrating liquid from the previous stage. A simplified two-effect, three-stage falling film evaporator, is shown in Figure 10.9. The feed enters the first effect (1), which is heated by steam with the condensate removed from the bottom. The vapors from the separator V_1 are split into two streams, V_{1a}, which heats the first stage of the second effect (2a), and V_{1b}, which heats the second stage (2b). The liquid product from the first effect L_1 is fed to the second effect (2a) and from there to (2b), from which it is removed as concentrated product. The vapors from the two stages of the second effect (V_{2a} and V_{2b}) are directed to a condenser.

In the evaporation of large quantities of aqueous nonfood solutions (e.g., water desalination) the number of effects in ME systems may be much higher than in food

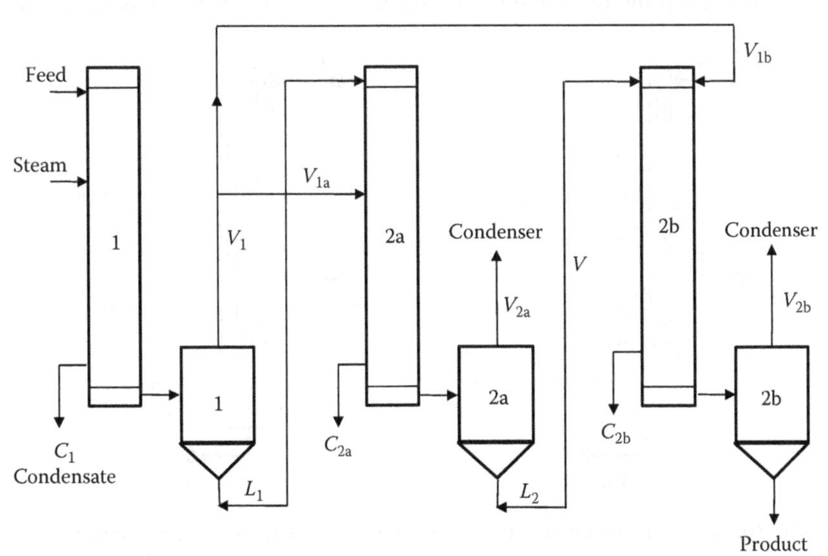

FIGURE 10.9 Diagram of a two-effect, three-stage falling film evaporator.

evaporators, because higher temperatures in the first effect can be used, and the temperature difference ΔT per effect can be smaller. Thus, ME systems with 8–12 effects may be used, achieving significant reduction of the cost of evaporation. In very large desalination plants (e.g., from sea water) the multiple stage flash (MSF) evaporation system is used with a large number of stages (25–50) and a small ΔT per effect.

10.4.2 Vapor Recompression Evaporators

Steam economies higher than those of multiple effect systems can be obtained by vapor recompression evaporators, in which the vapors from the evaporation unit are compressed and reused as a heating medium. Recompression is achieved by either thermal or mechanical compressors (Figure 10.10).

The thermocompressor system uses a steam ejector with high-pressure steam (7–10 bar) to increase the pressure and temperature of the water vapors and use the compressed mixture as the heating medium.

Material balances of the system indicate that part of the water vapors must be removed to the condenser for establishing an equilibrium balance in the system. The thermocompressor system is used when high pressure steam is available at a low cost. Steam economies SE of 4–8 can be achieved, higher than those of typical multiple-effects food evaporators.

Mechanical vapor recompression (MVR) evaporators are used more extensively than the thermocompressor system, because of their high steam economy (higher than 10) and the lower operating cost, especially when electrical power is available at low cost.

The vapors are compressed mechanically and they are used as the heating medium of the evaporator unit. A small amount of heating steam is added to the system to make up the condensate formed during compression of water vapors.

Centrifugal compressors are used to compress the water vapors by a ratio of 1.4–2.0, increasing the temperature difference ΔT by 5°C–20°C. More economical

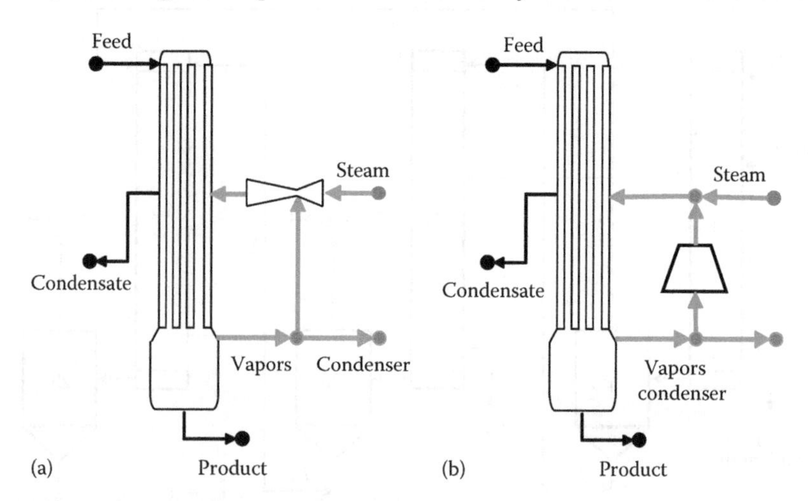

FIGURE 10.10 Vapor recompression evaporators. (a) Thermal; (b) mechanical. Falling film units, one-pass type. (Modified from Maroulis, Z.B. and Saravacos, G.D., *Food Process Design*, CRC Press, Boca Raton, FL, May 9, 2003, fig. 6.5.)

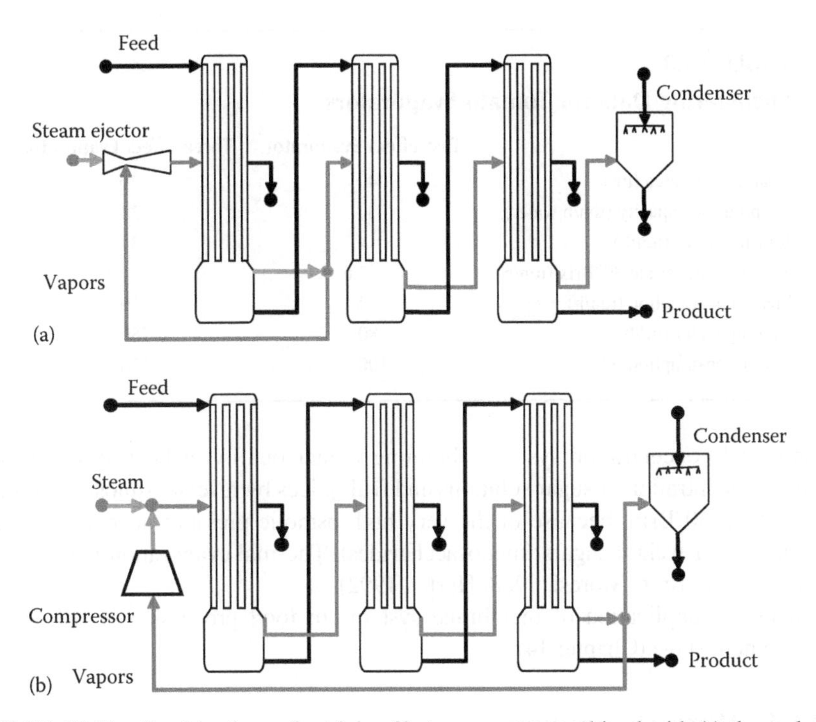

FIGURE 10.11 Combinations of a triple-effect evaporator combined with (a) thermal and (b) mechanical recompression. (Modified from Maroulis, Z.B. and Saravacos, G.D., *Food Process Design*, CRC Press, Boca Raton, FL, May 9, 2003, fig. 6.9.)

operation is obtained with turbo fans, which operate at a lower compression ratio, e.g., 1.2, which corresponds to a ΔT of about 5°C. Low ΔT values are applicable to falling film evaporators when there is no appreciable boiling point rise in the evaporator tube.

Combinations of multiple-effect and vapor recompression evaporators can be applied in food processing to reduce further the energy requirements. Figure 10.11 shows two evaporation systems: (a) A three-effect falling film evaporator, in which part of the vapors from the first effect is reused to heat the same effect, by thermal recompression through a steam ejector. The rest of vapors from the first effect are used to heat the second effect, and subsequently the third effect. The vapors from the third effect are condensed with cooling water. (b) A three-effect falling film evaporator, in which part of the vapors from the last (third) effect is reused to heat the first effect, by increasing the pressure (and the temperature) through a mechanical compressor. The rest vapors from the last effect are condensed with cooling water.

Typical engineering data on commercial evaporators of tomato products, given by equipment suppliers, are shown in Table 10.3.

10.4.3 Combined Reverse Osmosis/Evaporation

Removal of a large portion of water from dilute food liquids by membrane techniques (mostly reverse osmosis) followed by falling film evaporation would be an

TABLE 10.3
Engineering Data for Tomato Evaporators

	Two-Effect Evaporator	Three-Effect Evaporator
Heat transfer area (m^2)	40	120
Evaporation capacity (ton/h water)	11	33
Raw tomatoes (ton/h)	14	34
Tomato concentrate 30°Brix (ton/h)	2.6	7.4
Steam consumption (ton/h)	4	10
Cooling water (m^3/h)	80	250
Power consumption (kW)	100	120

economical concentration system, obtaining higher quality of heat-sensitive food fluids. Concentration of sugar solutions and fruit juices by reverse osmosis is limited to less than 50°Brix, because of the very high osmotic pressure developed by the low molecular weight sugars (monosaccharides). Thermal evaporation is needed to obtain higher °Brix (Moresi, 1988; Hartel, 1992).

Industrial application of membrane systems in food processing is still in the development stage (Chapter 14).

10.4.4 Water Desalination

Evaporation is the main desalination method for water desalination, followed by reverse osmosis (Chapter 14). Multiple effect evaporators, usually the falling film type, with a large number of effects (10–12) are used in medium-size applications to desalinate both brackish and sea water. Thinner tubes, made of heat-conductive metals, instead of the stainless steel tubes of the food evaporators, and higher operating temperatures can achieve very high overall heat transfer coefficients, e.g., 8 kW/(m^2 K). Vapor recompression evaporators with high heat economy are used in smaller installations. Scaling (fouling of the evaporation surface) is prevented by acid pretreatment.

For large capacities, the multiple-stage flash (MSF) evaporation system is used. It consists of a series of several heat exchangers (20–40), in which the feed water is preheated by the condensing vapors of flashing water. Small temperature differences are used between the two streams (e.g., 2°C–3°C), and capacities of about 20,000 m^3/day of desalted water are produced at the lowest desalination cost.

10.5 AUXILIARY EQUIPMENT

10.5.1 Vapor–Liquid Separators

The mixture of vapors–liquid, leaving the evaporator body, must be separated effectively into the concentrated liquid and the vapors, which are subsequently led to the condenser. The vapors may contain significant amounts of product in the form of droplets or foam, due to entrainment and foaming. Entrainment is caused by

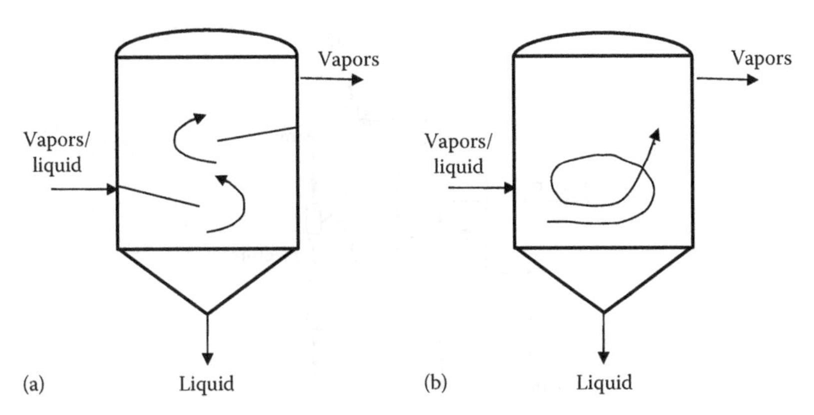

FIGURE 10.12 Vapor/liquid separators: (a) baffled and (b) centrifugal.

high-velocity vapors, produced by rapid evaporation. Foaming is caused by the presence of surface active gents in the liquid food. Special design of the vapor–liquid separators, or use of antifoam compounds can control foaming. Loss of product into the vapors and the condensate is undesirable for economic reasons and for environmental problems in the disposal of the condensate.

Two types of separators are used in industrial operations: (a) baffled and (b) centrifugal or cyclone separators (Figure 10.12) or cyclone separators and (b) baffle separators (Figure 10.12).

The velocity of the vapors entering the centrifugal separator should be lower than 100 m/s. The vapors enter tangentially the separator and they develop a swirling motion, which throws the liquid droplets to the walls of the separator. Sizing of the centrifugal separators is similar to the analysis of the mechanical cyclones, as discussed in Chapter 7. Larger separators may be needed in high-vacuum operation, such as in the last effect of a multiple effect evaporation system. The baffled separators are based on the change of the direction of vapor flow, due to mechanical obstacles (baffle plates). The liquid droplets are collected on the baffles, coalescing into a liquid, which flows by gravity downward.

10.5.2 Condensers

Vapors produced during evaporation are preferentially used as a heating medium in multiple effect evaporators or in vapor recompression systems. Some vapors may be used for preheating the feed to the evaporator. The remaining vapors are condensed in two types of condensers, i.e., surface and mixing condensers.

Condensers are part of vacuum maintaining systems, since they remove the water vapors, which otherwise would increase the evaporation pressure. Large diameter vapor pipes are required between the evaporator/separator and the condenser, for reducing the pressure drop, due to the high vapor velocity, in vacuum operation.

The surface condensers are actually shell and tube heat exchangers, cooled with cold water in the tubes, and the condensate is collected in the shell, preventing the pollution of the environment. The condensate may be used for the recovery of aroma components in a distillation column.

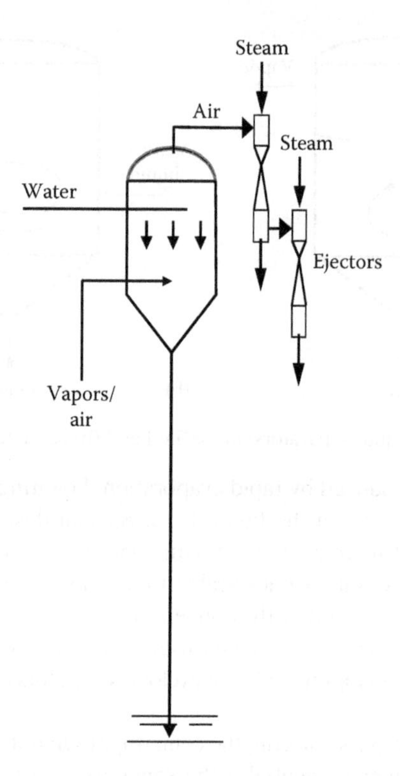

FIGURE 10.13 Barometric condenser and steam ejector vacuum system. Minimum height of the barometric column of water 10.33 m.

In the less expensive mixing condenser, the vapors are condensed by direct contact with cooling water, and the mixture is discharged to the environment. In vacuum operation, the condensate–water mixture is extracted from the system by either a centrifugal pump or a barometric condenser (Figure 10.13).

The barometric leg of the condenser should be at least 10.5 m, which is the liquid water column corresponding to atmospheric pressure (1.03 bar). The reduced pressure in the evaporator is achieved by condensation of vapors and the auxiliary action of the vacuum pump, which is necessary for removing any air leaks and noncondensable gases present in the fluid food.

10.5.3 Vacuum Systems

A vacuum maintaining system is necessary, in addition to the condenser, to remove any air leaks and noncondensable gases present in fluid food material. Two principal vacuum systems are used for food evaporators, i.e., the steam jet ejectors, and the liquid-ring vacuum pump, which are directly attached to the condenser.

The steam jet ejectors (Figure 10.13) remove the noncondensable gases by entrainment in high-pressure steam, flowing at high velocities in a specially designed ejector. Two or more jet ejectors (steam pressure about 7 bar) in series are used to produce high vacuum (down to 1 mbar or lower).

The liquid-ring pump is a centrifugal pump with a liquid ring, which seals the rotor chamber, and pumps out the air and condensable gases into the atmosphere (see Chapter 6). For high-vacuum operation, two-stage pump systems may be required. For normal evaporators, the liquid ring in the vacuum pump is water, but, for high-vacuum operation, oil and other low-vapor pressure liquids are used.

APPLICATION EXAMPLES

Example 10.1

Calculate the heat transfer coefficient in the evaporation surface h_i in a vertical falling film of water at 80°C at a flow rate of $\Gamma = 0.5$ kg/(m s). The overall heat transfer coefficient for evaporation of water in the same unit was determined experimentally as $U = 2$ kW/(m^2 K).

Viscosity of water at 80°C, $\eta = 0.356$ mPa s = 0.000356 Pa s, thermal conductivity $\lambda = 0.67$ W/(m K), density $\rho = 972$ kg/m^3 and $g = 9.81$ m/s^2.

Solution

Reynolds number, $Re = 4\Gamma/\eta = 4 \times 0.5/0.000356 = 5618$ and $Re^{1/3} = 17.4$.

The φ factor of Equation 10.13 becomes $\varphi = \dfrac{\left[(0.67)^3 \times (972)^2 \times 9.81\right]}{(0.000356)^2}$

$\varphi = 21.8 \times 10^{12}$ and $\varphi^{1/3} = 28{,}000$.

Prandtl number $Pr = C_p\,\eta/\lambda = (4100 \times 0.000356)/0.67 = 2.2$, and $Pr^{1/3} = 1.3$.

Therefore, $h_i = 0.01 \times 28{,}000 \times 17.4 \times 1.3 = 6.33$ kW/(m^2 K).

Equation 10.3 can be used to determine the heat transfer coefficient of the evaporation surface h_i from the experimental U value, assuming no fouling, $1/h_i = 1/U - x/\lambda - 1/h_s$. The thickness of the tube wall is $x = 3$ mm (2 in. tube, 10 gauge), and the thermal conductivity of the stainless steel $\lambda = 16$ W/(m K). The steam-side heat transfer coefficient is assumed to be $h_s = 10$ kW/(m^2 K). Therefore, $1/h_i = 1/2{,}000 - 0.003/15 - 1/10{,}000$, from which, $h_i = 4.8$ kW/(m^2 K).

The simplified equation for water films (Equation 10.13) yields the following heat transfer coefficient, for $\Gamma = 0.5$ kg/(m s), $h_i = 9150 \times (0.5)^{1/3} = 7.26$ kW/(m^2 K). It is seen that Equation 19.12 overestimates the heat transfer coefficient of boiling water. The general empirical Equation 12.12 yielded also higher heat transfer coefficients than the experimental values, presumably due to fouling of the evaporation surface. The experimental and calculated values of the heat transfer coefficients can be closer, if reliable fouling factors are considered for the evaporation system.

Example 10.2

Prepare a simplified design of a three-effect evaporator for the concentration of orange juice (OJ), using conventional calculations. A falling film LTV evaporator of 50 mm inner diameter and 10 m length will be used.

The evaporator will concentrate a feed $m_f = 5$ ton/h of orange juice 12°Brix to a concentrated orange juice (OJC) product of 65°Brix.

Solution

Assuming that the °Brix is equal to %TS (percent total solids), the OJC product will be equal to $m_p = m_3 = 5 \times (12/65) = 0.92$ ton/h of 65°Brix. Thus, the evaporation capacity of the evaporator system will be, $m_v = m_f - m_p = 5.00 - 0.92 = 4.08$ ton/h of water. The following assumptions are made to simplify the preliminary calculations: Feed forward triple-effect system. Feed enters the first effect and product leaves the last (third) effect at the corresponding boiling points, negligible boiling point elevation, and negligible heat losses. Saturated steam at 110°C (1.43 bar) is used to heat the first effect, and saturated condensates are removed from each effect. Pressure in the last effect 0.123 bar (boiling temperature 50°C). Thus, available overall temperature difference $\Delta T = 110 - 50 = 60$°C.

The whole evaporation system is assumed to be adiabatic, i.e., the heat transferred in each effect is identical:

$$Q_s = Q_1 = Q_2 = Q_3 \tag{10.16}$$

$$\text{or} \quad m_s \, \Delta H_s = m_{v1} \, \Delta H_1 = m_{v2} \, \Delta H_2 = m_{v3} \, \Delta H_3 \tag{10.17}$$

where

m_s, m_{v1}, m_{v2}, and m_{v3} (kg/s) are the flow rates of the steam, and the vapors from the first, second, and third effect, respectively

ΔH_s, ΔH_1, ΔH_2, and ΔH_s (MJ/kg) are the respective heats of evaporation of water

The heat transfer rate in each effect is given by the basic heat transfer Equation 10.2. Therefore, Equation 10.17 becomes

$$U_1 \, A_1 \, \Delta T_1 = U_2 A_2 \, \Delta T_2 = U_3 \, A_3 \, \Delta T_3 \tag{10.18}$$

where U_i, A_i, and ΔT_i represent respectively the overall heat transfer coefficient [W/(m^2 K)], heat transfer area (m^2), and temperature difference (K).

From the engineering and construction standpoint, the evaporator bodies of the three effects should be preferably identical, i.e., $A_1 = A_2 = A_3$, resulting in the equation

$$U_1 \, \Delta T_1 = U_2 \Delta T_2 = U_3 \Delta T_3 \tag{10.19}$$

The overall heat transfer coefficients in the three effects will decrease from the first to the third effect, due to the increase of concentration and viscosity of the food liquid. Therefore, according to Equation 10.19, the respective temperature differences will increase in the order $\Delta T_1 < \Delta T_2 < \Delta T_3$. Reliable data (experimental or industrial) of overall heat transfer coefficients U are needed for the design of evaporators. In this example of (OJ) evaporation, the U values of unfiltered (cloudy) apple juice (Figure 10.1) at corresponding °Brix can be used: $U_1 = 1.6$ kW/m^2 K, $U_2 = 1.4$ kW/m^2 K, and $U_3 = 0.7$ kW/m^2 K. An approximate (guessed) value for the intermediate heat transfer coefficient U_2 was assumed, since the concentration in the second effect is not known beforehand. Thus, Equation 10.19 becomes

$$1.6\Delta T_1 = 1.4\Delta T_2 = 0.7\Delta T_3 \tag{10.20}$$

The overall temperature difference ΔT is given by the equation:

$$\Delta T_1 + \Delta T_2 + \Delta T_3 = 60 \quad (10.21)$$

From the last two equations, it follows that, $\Delta T_1 = 13.6°C$, $\Delta T_2 = 15.4°C$, and $\Delta T_3 = 31°C$. Therefore, the boiling temperatures at the three effects will be $T_1 = 96.4°C$, $T_2 = 81°C$, and $T_3 = 50°C$.

It should be noted that the high temperature in the first effect (96.4°C) is sufficient to pasteurize the OJ and inactivate the pectic enzymes (stabilization of the juice).

From steam tables, the heats of vaporization of water at the three boiling temperatures will be $\Delta H_s = 2.23$ MJ/kg (110°C), $\Delta H_1 = 2.27$ MJ/kg (96.4°C), $\Delta H_2 = 2.31$ MJ/kg (81°C), and $\Delta H_3 = 2.38$ MJ/kg (50°C).

The flow rates of steam and water vapors in the three effects are calculated from the equations:

$$2.23m_s = 2.26m_{v1} = 2.31m_{v2} = 2.38m_{v3} \quad (10.22)$$

$$\text{and } m_{v1} + m_{v2} + m_{v3} = 4.08/3.6 = 1.13 \text{ kg/s} \quad (10.23)$$

from which, $m_s = 0.391$ kg/s, $m_{v1} = 0.385$ kg/s, $m_{v2} = 0.377$ kg/s, and $m_{v3} = 0.366$ kg/s, or $m_s = 1.41$ ton/h, $m_{v1} = 1.39$ ton/h, $m_{v2} = 1.36$ ton/h, and $m_{v3} = 1.32$ ton/h.

The steam economy of the three-effect evaporator will be

$$SE = \frac{(1.39 + 1.36 + 1.32)}{1.41} = \frac{10.99}{3.78} = 2.89$$

The heat transfer area of the first effect will be (Equation 10.2) $A = (0.391 \times 2230)/(1.6 \times 13.6) = 40\,m^2$, the same in the other two effects.

Assuming that tubes of 50 mm internal diameter and 10 m long are used, the required number of tubes per effect will be $N = 40/(3.14 \times 0.05 \times 10) = 25$. This is a medium size evaporator of water evaporation rate (capacity) 4 ton/h.

The concentration of the juice in the intermediate (second) effect of the evaporator x_1 is calculated by a material balance: $m_1 = m_f - m_{v1} = 5.00 - 1.39 = 3.61$ ton/h, and $x_1 = 12 \times (5/3.61) = 16.6$°Brix.

The vapor velocities at the exit of the falling film evaporator tubes are of importance to the design of the vapor–liquid separators. The vapor flow rates at the exit of the three effects will be, respectively, $m_{v1} = 0.385$ kg/s, $m_{v2} = 0.377$ kg/s, and $m_{v3} = 0.366$ kg/s. The vapor densities in the three effects, taken from the steam tables, will be: $\rho_{v1} = 0.54\,kg/m^3$, $\rho_{v2} = 0.243\,kg/m^3$, and $\rho_{v3} = 0.083\,kg/m^3$. The cross-sectional area of each tube will be $(3.14) \times (0.05)^2/4 = 0.002\,m^2$. Therefore, the exit vapor velocity in the three effects will be, respectively, $u_1 = 0.385/(25 \times 0.002 \times 0.54) = 14.25$ m/s, $u_2 = 0.377/(25 \times 0.002 \times 0.243) = 31$ m/s, and $u_3 = 0.366/(25 \times 0.002 \times 0.083) = 88$ m/s. The highest vapor velocity is found, as expected, in the last effect, which operates at the lowest pressure (highest vacuum) of the system. The vapor velocities are important in designing the vapor–liquid separators of the evaporator. It may be necessary to design a larger separator for the last effect in order to prevent liquid entrainment in the vapors.

Notes

1. The simplified solution of this example is based on assumed values of the overall heat transfer coefficients U in the three effects of the evaporator. The feed and product concentrations (°Brix) of orange juice are known, and so the corresponding U values can be taken from a reference material (in this case, unfiltered apple juice, Figure 10.1). However, the juice concentration in the intermediate (second) effect is not known, and a guess (approximation) is made, making the subsequent calculations a trial and error solution. The solution can be refined by repeating the calculations, using a new (improved) value of U, corresponding to the estimated concentration in the second effect. A spreadsheet solution in this example would simplify the calculations.
2. The third effect of the given evaporator operates at a large temperature difference, $\Delta T_3 = 31°C$ and a high juice concentration (65°Brix), conditions which favor fouling of the evaporator surface. The operation can be improved and be more economical by splitting the third effect into two stages, similar to the two-effect, three-stage system of Figure 10.9.
3. The calculated steam economy of the three-effect evaporator ($E = 2.89$) can be improved by adding one or two more effects to the system, or by using a mechanical vapor recompression system, compressing the vapors of the last effect to heat the first effect (Figure 10.11b). The latter alternative would be more economical, with the added advantage of eliminating the need for a large condenser for all the vapors of the last effect.

Example 10.3

Estimate the heat transfer area and the utilities requirements (steam, cooling water, and electric power) of a double-effect forced circulation evaporator to concentrate 10 ton/h tomato juice (TJ) of 6% TS to tomato juice concentrate (TJC) 32% TS. Saturated steam at 120°C is used to heat the first effect of the feedforward system, while the second effect operates at 60°C. Cooling water is available at 20°C for the condenser. Evaporator tubes of 20 mm inner diameter and 5 m length are used, and the heat transfer coefficients can be estimated from empirical equations, using literature data on physical and transport properties of TJ.

The feed enters the first effect at the boiling point, and the BPE and the heat losses from the evaporation system can be neglected.

Rheological data at mean evaporation temperatures: first effect, the partially concentrated (TJ) is considered a Newtonian fluid, $\eta = 2\,\text{mPa s}$; in the second effect, (TJC) 32% TS is considered a pseudoplatic fluid with rheological constants $K = 0.8\,\text{Pa s}^n$ and $n = 0.5$. Liquid densities $\rho_1 = 1050\,\text{kg/m}^3$, $\rho_2 = 1150\,\text{kg/m}^3$. Specific heats $C_{p1} = 3.9\,\text{kJ/(kg K)}$, $C_{p2} = 3.7\,\text{kJ/(kg K)}$. Thermal conductivities $\lambda_1 = 0.5\,\text{W/(m K)}$, $\lambda_2 = 0.5\,\text{W/(m K)}$ at mean evaporation temperatures.

Solution

The flow rate of the (TJ) into the evaporation system (feed) will be $m_f = 10$ ton/h or $m_f = 2.78\,\text{kg/s}$. The flow rate of the product (TJC) will be $m_p = m_2 = 10 \times (6/32)$ or $m_2 = 1.87$ ton/h $= 0.52\,\text{kg/s}$ of 32% TS. Thus, the evaporation capacity of the evaporator system will be, $m_v = 10.00 - 1.87 = 8.13$ ton/h or $m_v = 2.26\,\text{kg/s}$ of water.

For preliminary calculations, it is assumed that each effect evaporates the same amount of water, i.e., $m_{v1} = m_{v2} = 2.26/2 = 1.13$ kg/s. Thus, the flow rate of the partially concentrated product, leaving the first effect, will be $m_1 = 2.78 - 1.13 = 1.65$ kg/s, and its concentration will be $x_1 = 6 \times (2.78/1.65) = 10.1\%$ TS. The flow rate of (TJC) in the second effect will be 0.52 kg/s and its concentration $x_2 = 32\%$ TS. Approximate values of the heat transfer coefficients are needed in order for preliminary calculations, which can be refined later.

First effect: Assume a liquid velocity of $u_1 = 3$ m/s in the tubes. The Reynolds number will be $Re = (d\,u\,\rho)/\eta$, $Re_1 = (0.02 \times 3 \times 1050)/0.002 = 31{,}500$ (turbulent flow). Prandtl number, $Pr = (C_p\,\eta)/\lambda$, $Pr_1 = (3900 \times 0.002)/0.5 = 15.6$. Heat transfer coefficient, $(h\,d)/\lambda = 0.023\,Re^{0.8}\,Pr^{1/3}\,(\eta/\eta_w)^{0.14}$. To simplify the calculations, assume $(\eta/\eta_w)^{0.14} = 1$, $(0.02h_{i1})/0.5 = 0.023 \times (31500)^{0.8} \times (15.6)^{1/3}$, from which $h_{i1} = 5.70$ kW/(m^2 K).

Second effect: A higher velocity $u_2 = 6$ m/s is assumed, which is necessary to lower the apparent viscosity of the pseudoplastic fluid (TJC) and increase the heat transfer coefficient. The shear rate at the wall of the evaporator tube will be $\gamma = (8u)/d = (8 \times 6)/0.02 = 2400$, and the corresponding apparent viscosity of the (TJC), $\eta_\alpha = K\,\gamma^{n-1} = 0.8 \times (2400)^{-0.5} = 0.016$ Pa s. The Reynolds number becomes, $Re_2 = (0.02 \times 6 \times 1150)/0.016 = 8625$ (turbulent flow). The Prandtl number is $Pr = (3700 \times 0.016)/0.4$ or $Pr = 148$, and the heat transfer coefficient h_{i2}, $(0.02h_{i2})/0.4 = 0.023 \times (8625)^{0.8} \times (148)^{1/3}$, from which $h_{i2} = 4.27$ kW/(m^2 K).

The wall thickness of the 20 mm tubes is $x = 3$ mm and the thermal conductivity of the metal (stainless steel) $\lambda = 16$ W/(m K) = 0.016 kW/(m K).

It is assumed that the heat transfer coefficient of the steam side and the fouling factor of the evaporation surface are, respectively, $h_s = 10$ kW/(m^2 K) and $h_f = 6$ kW/(m^2 K).

For simplification, the overall heat transfer coefficient in tubes is assumed to be similar to that of plane surfaces. For detailed calculations, the equations for pipes/tubes given in Chapter 8 should be used. Thus, the overall heat transfer coefficients U in the two effects (1, 2) will be,

$$\frac{1}{U_1} = \frac{1}{h_{i1}} + \frac{1}{h_f} + \frac{x}{\lambda} + \frac{1}{h_s}$$

$$\frac{1}{U_1} = \frac{1}{5.7} + \frac{1}{6} + \frac{0.003}{0.016} + \frac{1}{10}$$

$$U_1 = 1.6\ \text{kW/(m}^2\,\text{K)}$$

$$\frac{1}{U_2} = \frac{1}{h_{i2}} + \frac{1}{h_f} + \frac{x}{\lambda} + \frac{1}{h_s}$$

$$\frac{1}{U_2} = \frac{1}{4.27} + \frac{1}{6} + \frac{0.003}{0.016} + \frac{1}{10}$$

$$U_2 = 1.45\ \text{kW/(m}^2\,\text{K)}$$

Note that the thermal resistance of the tube wall [0.187 (9 m^2 K)/kW] is significant (27%–30% of the total resistance of the system).

Normally, the heat transfer surfaces of the evaporator effects are assumed to be the same ($A_1 = A_2$), making the construction cost more economical.

The overall heat transfer rate in the two effects will be the same (adiabatic operation):

$$q = U_1 A_1 \Delta T_1 = U_2 A_2 \Delta T_2 \quad \text{or} \quad U_1 \Delta T_1 = U_2 \Delta T_2 \quad \text{and} \quad \Delta T_1 + \Delta T_2 = 120 - 60 = 60°\text{C}$$

From the equations $1.63\Delta T_1 = 1.45\Delta T_2$ and $\Delta T_1 + \Delta T_2 = 60$, it follows that $\Delta T_1 = 28.3°\text{C}$ and $\Delta T_2 = 31.7°\text{C}$.

The boiling temperatures of the juice in the two effects will be $T_1 = 91.7°\text{C}$ and $T_2 = 60°\text{C}$.

The corresponding heats of evaporation will be $\Delta H_1 = 2.28\,\text{MJ/kg}$ and $\Delta H_2 = 2.36\,\text{MJ/kg}$.

The evaporation rates in the two effects can be re-calculated, as follows.

$m_{v1} + m_{v2} = 2.26\,\text{kg/s}$, $2.28 m_{v1} = 2.36 m_{v2}$ and $m_{v1} = 1.15\,\text{kg/s}$, $m_{v2} = 1.11\,\text{kg/s}$, which are close to the assumed mean value of $m_{v1} = m_{v2} = 1.13\,\text{kg/s}$.

The common heat transfer area can be determined from the mean heat transfer rate in one effect, $q_1 = m_{v1}\,\Delta H_1 = U_1 A \Delta T_1$ or $1.15 \times 2.28 = (1.6 \times 28.3)$ A, from which $A = 58\,\text{m}^2$. The number of the 5 m evaporator tubes will be $N = 58/(3.14 \times 0.02 \times 5) = 185$.

Utilities

The amount of required heating steam ($T_s = 120°\text{C}$, $P_s = 1.98$ bar, $\Delta H_1 = 2.20\,\text{MJ/kg}$) will be $m_s = 1.15 \times (2.28/2.20) = 1.19\,\text{kg/s}$. The steam economy of the double-effect evaporation system will be SE = 2.26/1.19 = 1.9.

Cooling water is required for condensing the water vapors of the second effect. It is assumed that a water mixing condenser is used with a barometric leg to maintain the vacuum, and that the condensate–water mixture is removed at the saturation temperature of 60°C. The cooling water is available at 20°C and its specific heat is $C_p = 4.1\,\text{kJ/(kg K)}$. The amount of water m_w (kg/s) will be $(4.1 \times 40)m_w = m_{v2}\,\Delta H_2 = 1.11 \times 2360$ and $m_w = 16\,\text{kg/s} = 57.5$ ton/h.

Electrical power is required mainly for the mechanical pump used in the forced circulation evaporator. The pressure drop due to friction ($\Delta P_f/\rho$) in the two evaporator effects is the main energy requirement of the pump. It is calculated from the Fanning Equation 6.14 of Chapter 6 (Fluid Flow), $(\Delta P_f/\rho) = 4\,f\,(L/d)\,(u^2/2)$.

The friction factor f of the Fanning equation can be estimated from the Moody diagram of Figure 6.10 or from Equation 6.14 for turbulent flow of Chapter 6 (Fluid Flow), $f = 0.079\,Re^{-0.25}$. In the first effect, $Re_1 = 3150$ and $f = 0.079 \times (3150)^{-0.25}$ or $f_1 = 0.005$. In the second effect, $Re_2 = 3450$ and $f = 0.079 \times (3450)^{-0.25}$ or $f_2 = 0.008$.

First effect: $(\Delta P_f/\rho)_1 = 4 \times 0.005 \times (5/0.02) \times (3^2/2) = 22.5\,\text{J/kg}$. The flow rate (recirculation) of the product in the first effect will be $m_1 = [185 \times 3.14 \times (0.02)^2 \times 3 \times 1050]/4 = 183\,\text{kg/s}$. Pump power, $P_1 = 183 \times 22.5 = 4117\,\text{W} = 4.1\,\text{kW}$.

Second effect: $(\Delta P_f/\rho)_2 = 4 \times 0.008 \times (5/0.02) \times (6^2/2) = 144\,\text{J/kg}$. The flow rate (recirculation) of the product in the second effect will be $m_2 = [185 \times 3.14 \times (0.02)^2 \times 6 \times 1150]/4 = 400\,\text{kg/s}$. Pump power $= 400 \times 144 = 57.6\,\text{kW}$.

The pressure drop in the evaporator tubes due to friction will be: First effect, $(\Delta P_f/\rho)_1 = 22.5\,\text{J/kg}$ and $\Delta P_{f1} = 22.5 \times 1{,}050 = 23{,}625\,\text{J/m}^3 = 23{,}625\,\text{Nm/m}^3 = 23{,}625\,\text{Pa} = 0.24$ bar.

Second effect: $(\Delta P_f/\rho)_2 = 144$ J/kg and $\Delta P_{f2} = 144 \times 1150 = 165{,}600$ Pa = 1.65 bar.

Notice that a much higher pressure drop due to friction is developed in the second than in the first effect (1.65 bar versus 0.24 bar), due to the higher liquid velocity, which is needed to improve the heat transfer rate, and the higher apparent viscosity. This results in much higher power consumption (57.6 kW versus 4.1 kW) of the recirculation pump.

Example 10.4

Design a vapor recompression evaporator to concentrate 5 ton/h of nonfat (skimmed) milk from 12% to 30% TS. The evaporator will operate at a boiling temperature 80°C, heated with saturated steam and vapors at 100°C. The evaporator body consists of stainless tubes 25 mm inner diameter and 5 m long. To reduce fouling and obtain high heat transfer coefficients, a forced circulation evaporator system will be used and the velocity of the liquid will be 5 m/s. Milk is relatively stable at high temperatures and long residence-times. Figure 10.14 shows a diagram of the evaporator system. The viscosity of concentrated milk 30% at 80°C is $\eta = 10$ mPa s (approximately a Newtonian fluid), the specific heat $C_p = 3.6$ kJ/(kg K), and the thermal conductivity $\lambda = 0.4$ W/(m K). Density of the 30% TS milk, $\rho = 1100$ kg/m³.

Solution

The flow rate of the milk into the evaporation system (feed) will be $m_f = 5$ ton/h or $m_f = 1.38$ kg/s. The flow rate of the product (concentrated milk) will be $m_p = 5 \times (12/30) = 2$ ton/h = 0.56 kg/s. Thus, the evaporation capacity of the evaporator

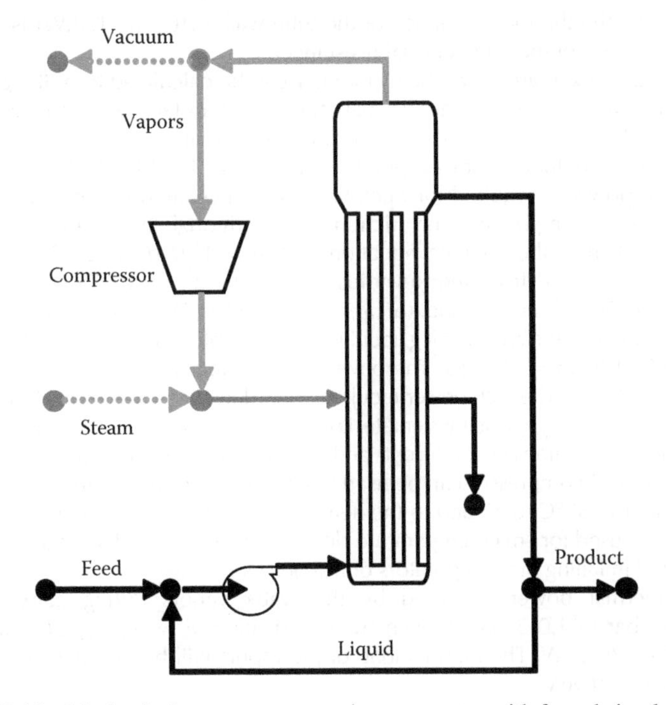

FIGURE 10.14 Mechanical vapor recompression evaporator with forced circulation.

system will be, $m_v = 5 - 2 = 3$ ton/h or $m_v = 0.82$ kg/s of water. The feed is assumed to enter the evaporator at the boiling point (80°C). The heating steam/vapors enter the evaporator and the condensate leaves the unit at the saturation temperature of 100°C.

It is assumed that all vapors produced in the evaporator are recompressed and are sufficient for maintaining the required evaporation rate. In practice, a small amount of external steam is needed to start the evaporation and make up any losses during the compression of the vapors. A vacuum system is required to maintain the low pressure constant in the evaporator.

An approximate value of the heat transfer coefficients is needed for preliminary calculations. A high velocity of $u = 5$ m/s is assumed in the evaporator tubes. The Reynolds number will be $Re = (d\,u\,\rho)/\eta$, $Re = (0.025 \times 5 \times 1100)/0.010 = 13{,}750$ (turbulent flow). Prandtl number $Pr = (C_p\,\eta)/\lambda$, $Pr = (3600 \times 0.010)/0.4 = 90$.

Heat transfer coefficient for turbulent flow, $(h\,d)/\lambda = 0.023 Re^{0.8}\,Pr^{1/3}\,(\eta/\eta_w)^{0.14}$. To simplify the calculations, assume $(\eta/\eta_w)^{0.14} = 1$. Then, $(0.025h_i)/0.4 = 0.023 \times (13{,}750)^{0.8} \times (90)^{1/3}$, from which $h_i = 3.36$ kW/(m² K).

The wall thickness of the 25 mm tubes is $x = 3$ mm and the thermal conductivity of the metal (stainless steel) $\lambda = 16$ W/(m K) = 0.016 kW/(m K).

It is assumed that the heat transfer coefficient of the steam side and the fouling factor of the evaporation surface are, respectively, $h_s = 10$ kW/(m² K) and $h_f = 6$ kW/(m² K).

For simplification, the overall heat transfer coefficient in tubes is assumed to be similar to that of plane surfaces. For detailed calculations, the equations for pipes/tubes given in Chapter 8 should be used. Thus, the overall heat transfer coefficient U in the evaporator will be $1/U = 1/h_i + 1/h_f + x/\lambda + 1/h_s$, $1/U = 1/3.36 + 1/6 + 0.003/0.016 + 1/10$, $U = 1.3$ kW/(m² K).

Note that the thermal resistance of the tube wall [0.167 (m² K)/kW] is significant (about 25% of the total thermal resistance).

The heat transfer area A of the evaporator can be calculated from the general equation, $q = UA\Delta T$, where the thermal load (heat transfer rate) is $q = m_v \Delta H_v = 0.82 \times 2300 = 1886$ kW and $A = 1886/(1.33 \times 20) = 71$ m².

The number of tubes in the evaporator will be $N = 71/(3.14 \times 0.025 \times 5) = 180$.

Electric power is required for operating the recirculation pump of the liquid and the vapor compressor. The friction energy through the evaporator tubes is calculated from the Fanning equation $(\Delta P_f/\rho) = 4f\,(L/d)\,(u^2/2)$. The friction factor f is calculated from the Reynolds number Re as $f = 0.079\,Re^{-0.25}$ or $f = 0.079 \times (13{,}750)^{-025} = 0.007$ and, $(\Delta P_f/\rho) = 4 \times 0.007 \times (5/0.025) \times (5^2/2) = 70$ J/kg.

The flow rate of the recirculating liquid in the evaporator will be, $m = 180 \times 3.14 \times (0.025)^2 \times 5 \times 1{,}100/4 = 486$ kg/s. The power of the pump will be $P = 70 \times 486 = 34{,}020$ W = 34 kW. The high power requirement is due to the high recirculation rate, needed to obtain a high heat transfer rate. Lower recirculation rates of the liquid would result in lower heat transfer coefficients and, therefore, in larger heat transfer area.

A centrifugal compressor can be used to compress the vapors from the evaporator separator (80°C, 0.47 bar) to the heating jacket (100°C, 1.03 bar). Fan compressors are used for small temperature differences, e.g., $\Delta T = 3$°C or 5°C, which are applied in falling film evaporators of fruit juices.

The thermal power absorbed by the water vapors during compression from 0.47 bar to 1.013 bar is given by the equation, $E_c = m_v\,C_{pv}\,\Delta T = 0.82 \times 1.88 \times 20 = 30.8$ kW. The thermal load of the vapors will be $Q = 0.82 \times 2300 = 1886$ kW = 1.89 MW.

Assuming that the overall (mechanical and electrical) efficiency of the compressor is 25% (Chapter 3, Themodynamics), the equivalent steam power supplied to the evaporator will be 30.8/0.25 = 123.2 kW. Steam economy, SE = 1886/123.2 = 15.3.

PROBLEMS

10.1 In the desalination of sea water to produce fresh water, multiple-effect evaporators are used. For economic operation, several effects (e.g., 10) are used, and thus the available temperature difference ΔT per effect is small, e.g., 5°C or less. *LTV* falling film systems are preferred.

In such installations, high overall heat transfer coefficients U are essential. Calculate the U value in a desalination falling film unit of 50 mm inner diameter tubes, assuming that the heat transfer coefficient in the evaporating side is 10 kW/(m^2 K) and in the heating side 12 kW/(m^2 K). Scaling (fouling) by salts can be prevented by acid treatment of the feed water. The tubes have a wall thickness of 3 mm and they are made of stainless steel [λ = 16 W/(m K)]. Calculate the percent thermal resistance of the tube wall.

It is suggested that the evaporator tubes be made of a copper alloy of thermal conductivity λ = 200 W/(m K) and the same dimensions as the SS tubes. Calculate the new U value and the % thermal resistance of the tube wall. Why are not such evaporator tubes used in food products?

10.2 Draw a flowsheet of three-effect feed forward rising film evaporator with two stages in the last effect, used to concentrate a fruit juice. What is the advantage of such a system over the normal three-effect evaporator?

10.3 Draw a flowsheet of a three-effect forced circulation evaporator for tomato juice, operated in the backward feed (opposite directions of flow of product and heating medium). What are the advantages and disadvantages of such system over the normal feed forward evaporator?

10.4 Prepare a preliminary design of a four-effect short tube evaporator for the concentration of 10 ton/h sugar (sucrose) solution from 12°Brix to 65°Brix (approximately 1°Brix = 1% TS). A forward feed evaporator system will be used, the first effect will be heated by saturated steam at absolute pressure 1.98 bar and the last effect will operate at an absolute pressure of 0.2 bar. Cooling water at 20°C is used to condense the water vapors of the last effect.

The evaporator bodies consist of short stainless steel tubes 30 mm inner diameter, 3 mm wall thickness, and 2 m long. The following empirical overall heat transfer coefficients can be used: U_1 = 1.7 kW/(m^2 K), U_2 = 1.3 kW/(m^2 K), U_3 = 1.0 kW/(m^2 K), and U_4 = 0.7 kW/(m^2 K).

Calculate: (1) The required heat transfer area and the number of tubes of the evaporator, assumed to be the same in all effects. (2) The steam economy of the evaporator. (3) The amount of cooling water, if the condensate is removed at the saturation temperature.

10.5 Repeat problem 4, assuming that the temperature difference ΔT in each effect is the same (meaning that the heat transfer areas are different).

LIST OF SYMBOLS

A	m^2	Heat transfer area
BPE	K	Boiling point elevation
C_p	kJ/(kg K)	Specific heat
d	m	Tube diameter
E_a	kJ/mol	Activation energy
FR	(m^2 K)/W	Fouling resistance
h	W/(mK)	Heat transfer coefficient
h	MJ	Enthalpy
K	Pa s^n	Rheological constant
L	m	Tube length
m	kg/s	Mass flow rate
n	—	Flow behavior index
Nu	—	Nusselt number ($h\ L/\lambda$)
Pr	—	Prandtl number ($C_p\ \eta/\lambda$)
q, Q	W	Heat transfer rate
R	J/(mol K)	Gas constant (8.31)
Re	—	Reynolds number ($d\ u\ \rho/\eta$)
SE	—	Steam economy
t	S	Time
T	°C, K	Temperature
TS	%	Total solids
u	m/s	Velocity
U	W/(m K)	Overall heat transfer coefficient
V	m^3	Volume
Γ	kg/(m s)	Film flow rate
η	Pa s	Viscosity
λ	W/(m K)	Thermal conductivity
ρ	kg/m^3	Density
σ	dyn/cm	Surface tension

REFERENCES

APV. 1987. *Evaporator Handbook.* Bulletin EHB-987, APV Crepaco, Rosemont, IL.

Bourgois, J. and LeMaguer, M. 1984. Modeling of heat transfer in a climbing film evaporator. III. Application to an industrial evaporator. *J. Food Eng*. 3: 39–50.

Chen, S.C. and Hernandez, E. 1997. Design and performance evaluation of evaporators. In: *Handbook of Food Engineering Practice*. K.J. Valentas, E. Rotstein, and R.P. Singh, eds. CRC Press, New York.

ERDA, 1977. *Upgrading Evaporators to Reduce Energy Consumption.* ERDA, U.S. Department of Commerce, Washington, DC.

Hahn, G. 1986. Evaporator design. In: D. MacCarthy, ed. *Concentration and Drying of Foods*. Elsevier Applied Science, London, U.K.

Hartel, R.W. 1992. Evaporation and freeze concentration. In: *Handbook of Food Engineering*. D.R. Heldman and D.B. Lund, eds. Marcel Dekker, New York.

Kessler, H.G. 1981. *Food Engineering and Dairy Technology*. Verlag A. Kessler. Freising, Germany.

Kessler, H.G. 1986. Energy aspects if food preconcentration. In: *Concentration and Drying of Foods*. D. MacCarthy, ed. Elsevier Applied Science, London, U.K.

Maroulis, Z.B. and Saravacos, G.D. 2003. *Food Process Design*. CRC Press, Boca Raton, FL.

Minton, P.E. 1986. *Handbook of Evaporation Technology*. Noyes Publ., Park Ridge, NJ.

Moore, J.G. and Hessler, W.E. 1963. Evaporation of heat sensitive materials. *Chem. Eng. Progr*. 59: 87–92.

Moore, J.G. and Pinkel, E.G. 1968. When to use single pass evaporators. *Chem. Eng. Progr*. 64: 29–44.

Morgan, A.I. 1967. Evaporation concepts and evaporation design. *Food Tech*. 21: 153.

Moresi, M. 1988. Apple juice concentration by reverse osmosis and falling film evaporation. In: *Preconcentration and Drying of Food Materials*. S. Bruin, ed. Elsevier, Amsterdam, the Netherlands, pp. 61–76.

Saravacos, G.D. 1970. Effect of temperature on the viscosity of fruit juices and purees. *J. Food Sci*. 25: 122–125.

Saravacos, G.D., 1974. Rheological aspects of fruit juice evaporation. In: *Advances in Preconcentration and Dehydration of Foods*. A. Spicer, ed. Applied Science Publ., London, U.K.

Saravacos, G.D. and Maroulis, Z.B. 2001. *Transport Properties of Foods*. Marcel Dekker, New York.

Saravacos, G.D. and Kostaropoulos, A.E. 2002. *Handbook of Food Processing Equipment*. Springer, New York.

Saravacos, G.D., Moyer, J.C., and Wooster, G.D. 1970. *Concentration of Liquid Foods in a Falling Film Evaporator*. New York State Agricultural Experiment Station, Cornell University, Bulletin No. 4, Geneva, NY.

Schwartzberg, H.G. 1977. Energy requirements in liquid food concentration. *Food Tech*. 31(3): 67–76.

Yanniotis, S. and Kolokotsa, D. 1996. Boiling on the surface of a rotating disc. *J. Food Eng*. 30: 313–325.

11 Drying Operations

11.1 INTRODUCTION

Food dehydration is a traditional method of food preservation, which is also used for the production of special foods and food ingredients, such as dry soups and food powders. In addition to the basic process engineering requirements, food dehydration must meet the strict standards for food quality, food hygiene, and food safety.

Removal of water from food materials is usually accomplished by thermal evaporation, which is an energy-intensive process, due to the high latent heat of vaporization of water. Part of the water in some "wet" food materials (e.g., solid food wastes) can be removed by inexpensive nonthermal processes, such as filtration and mechanical pressing or expression. Osmotic dehydration, requiring less energy than thermal drying, can be used for partial removal of water from foods. Novel membrane processes, such as ultrafiltration and reverse osmosis, requiring less energy can be used for preconcentration of liquid foods (Chen and Mujumdar, 2001).

Diverse drying processes and equipment are used in food processing, due to the difficulty of handling and processing solid materials, and the special requirements for the various food products. In addition, economics (investment and operation) is an important factor, especially for seasonal and low cost food products, such as fruits and vegetables. The equipment ranges from crude solar dryers to sophisticated spray dryers or freeze-dryers. In addition to process and equipment design, food dehydration involves processing technology for various food products. Modeling, simulation, and design are used in the engineering and economic analysis of food drying/dehydration processes.

The terms "drying" and "dehydration" are used interchangeably in process engineering, and in this book. However, in food science and technology, the term "drying" is traditionally used for thermal removal of water to about 15%–20% moisture (dry basis), which is approximately the equilibrium moisture content of dried agricultural products (e.g., fruits and grains) at ambient (atmospheric) air conditions. The term "dehydration" is traditionally used for drying foods down to about 2%–5%, e.g., dehydrated vegetables, milk, and coffee. The dehydrated foods usually require special packaging to protect them from picking up moisture during storage. The term "intermediate moisture foods" (IMF) is used for semimoist dried foods (fruits, meat, etc.) of 20%–30% moisture content (Barbosa and Vega-Mercado, 1996).

Recent advances in the application of transport phenomena, particle technology, and computer technology to food engineering have improved markedly the design and operation of food dryers (Saravacos and Kostaropoulos, 2002).

Most food products are dehydrated in convective dryers, in which air is used for heating the product and removing the evaporated water. In contact dryers, heat is transported to the product through the walls (e.g., shelves) of the equipment. In some dryers, heat may be transferred through radiation (infrared or microwave).

11.2 HEAT AND MASS TRANSFER

Most of the drying processes are carried out in an air stream (convective drying), and drying calculations are based on the properties of air–water vapor mixtures (psychrometrics) and the drying rate of the material. Specialized drying processes, such as contact drying and vacuum or freeze drying, require special treatment of heat and mass transfer, but the analysis of drying rate is the same.

11.2.1 HEAT AND MASS TRANSFER COEFFICIENTS

Heat and mass transport within the food materials (internal transport) controls the drying rate in most foods. The internal transport properties, i.e., mass diffusivity (moisture and solutes) and thermal conductivity/diffusivity affect strongly the drying rate.

Both mass and thermal transport properties are affected strongly by the physical structure (porosity) of the material, and to a lesser degree by the temperature and the moisture content (Chapter 5) (Saravacos and Maroulis, 2001).

Interphase (surface) heat and mass transfer is important in the early stages of drying, when the external drying conditions (air velocity, temperature, and humidity) have a decisive effect on the drying rate (Chapter 13).

The surface heat transfer coefficient h [W/(m^2 K)] in a drying operation is defined by the following equation:

$$q = h\Delta T \tag{11.1}$$

where

q (W/m^2) is the heat flux

ΔT (K) is the temperature difference between the heating medium and the heated surface of the material

The surface mass transfer coefficient h_M [kg/(m^2 s)] is defined by an analogous equation:

$$J = h_M \Delta Y \tag{11.2}$$

where

J [kg/(m^2 s)] is the mass transfer rate

ΔY (kg/kg db) is the difference of moisture content between the surface of the material and the bulk of the drying medium (air)

The mass transfer coefficient k_c (m/s), used in the literature, is related to the h_M [kg/(m^2 s)] by the following equation:

$$h_M = k_c \rho \tag{11.3}$$

where ρ(kg/m^3) is the density of the air.

For normal air-drying conditions (atmospheric pressure, temperature less than 100°C) the air density is approximately 1 kg/m^3 and the two coefficients become numerically equal (although with different dimensions).

$$h_M(\text{kg/m}^2\ \text{s}) = k_c(\text{m/s}) \tag{11.4}$$

The interphase heat and mass transfer coefficients are affected by the air velocity, the temperature, and the geometry of the transfer system. The transfer coefficients are correlated by empirical equations for various systems, using the dimensionless numbers Reynolds [$Re = (u\ \rho\ d)/\eta$], Prandtl [$Pr = (C_p\ \eta)/\lambda$], Nusselt [$Nu = (h\ d)/\lambda$], Schmidt [$Sc = \eta/(\rho\ D)$], and Sherwood [$Sh = (k_c\ d)/D$].

The following thermophysical and transport properties are used in these dimensionless numbers: density ρ(kg/m^3), specific heat C_p [J/(kg K)], velocity u (m/s), (equivalent) diameter d (m), thermal conductivity λ[W/(m K)], viscosity η(Pa s), mass diffusivity D (m^2/s), heat transfer coefficient h [W/(m^2 K)], and mass transfer coefficient k_c (m/s).

11.2.2 Heat and Mass Transfer Analogy

The Colburn analogy relates the heat and mass transfer coefficients, equating the dimensionless heat and mass transfer factors (j_H and j_M),

$$j_H = j_M \tag{11.5}$$

where

$j_H = St_H\ Pr^{2/3}$ and $j_M = St_M\ Sc^{2/3}$
$St_H = Nu/(Re\ Pr) = h/(u\ \rho\ C_p)$, Stanton number for heat transfer
$St_M = Sh/(Re\ Sc) = h_M/(u\ \rho) = k_c/u$, Stanton number for mass transfer

The Colburn heat and mass transfer analogy, applied to air/moisture systems at atmospheric pressure, yields

$$\frac{h}{h_M} = C_p \quad \text{or} \quad \frac{h}{k_c} = \rho C_p \tag{11.6}$$

For air at atmospheric pressure, approximately ρ = 1 kg/m^3 and C_p = 1000 J/(kg K). Therefore, the two coefficients are numerically identical, if expressed in appropriate units:

$$h[\text{W/(m}^2\ \text{K)}] = h_M[\text{g/(m}^2\ \text{s)}] = k_c(\text{mm/s}) \tag{11.7}$$

The Colburn analogies can be simplified by assuming that the heat and mass transfer factors (j_H and j_M) are functions of the Reynolds number only. Regression analysis of several literature data has yielded the following two empirical equations for convective (air) drying:

$$j_H = 1.04Re^{-0.45} \quad \text{and} \quad j_M = 23.5Re^{-0.88} \tag{11.8}$$

Figure 11.1 shows regression lines for the heat and mass transfer factors during the convective drying of corn and rice.

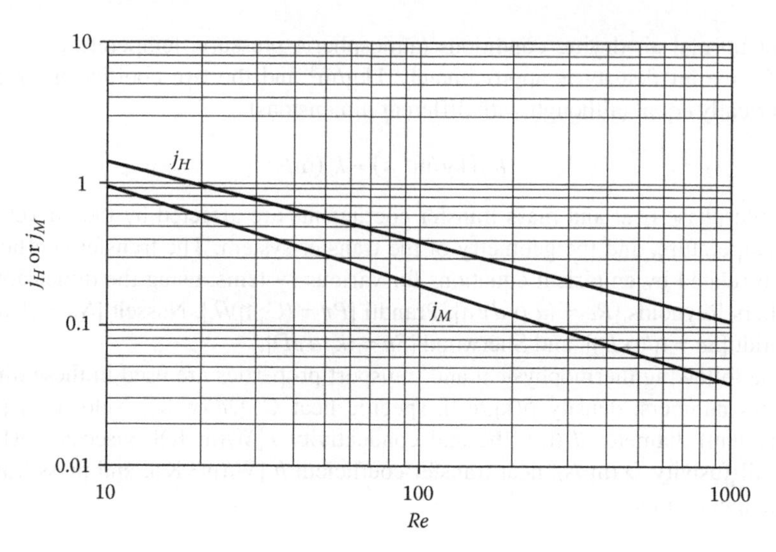

FIGURE 11.1 Regression lines of literature data on heat and mass transfer factors (j_H and j_M) for convective drying of corn and rice.

11.3 PSYCHROMETRICS

Psychrometrics is concerned with the determination of the thermodynamic properties of moist air, based on experimental data. The psychrometric model, consisting of a set of equations describing the thermodynamic relationships between these properties, is useful in process design (Maroulis and Saravacos, 2003).

11.3.1 WATER–AIR MIXTURES

Moist air is a dilute mixture of individual gases, dry air, and water vapor, each assumed to obey the perfect gas equation of state:

$$P_A V = \left(\frac{m_A}{M_A}\right) RT \tag{11.9}$$

$$P_V V = \left(\frac{m_V}{M_W}\right) RT \tag{11.10}$$

where

P_A and P_V (Pa) are the partial pressure of dry air and water vapor, respectively
V (m^3) is the total mixture volume
m_A and m_V (kg) are masses of air and water vapor
M_A and M_W (kg/mol) are the molecular weights of air and water
R [8.314 kJ/(kmol K)] is the universal gas constant
T (K) is the absolute temperature (note that 1 kJ = 1 kPa m^3)

The total pressure P of the mixture obeys the Dalton law:

$$P = P_A + P_V \tag{11.11}$$

The absolute humidity of the air, based on the dry air of the mixture (db = dry basis), Y_V (kg water/kg db), and the water activity a_w (–) are defined as follows:

$$Y_V = \frac{m_V}{m_A} \tag{11.12}$$

$$a_w = \frac{P_V}{P_S} \tag{11.13}$$

where P_S (bar) is the vapor pressure of water at saturation, which is calculated from the temperature T (°C) using the Antoine equation:

$$P_S = \exp\left[\frac{a_1 - a_2}{(a_3 + T)}\right] \tag{11.14}$$

where a_1, a_2, and a_3 are empirical adjustable constants,

$$a_1 = 1.19 \times 10,\ a_2 = 3.95 \times 10^3,\ \text{and}\ a_3 = 3.32 \times 10^2$$

The water activity in water vapor–air mixtures is identical to the fractional relative humidity RH or $a_w = (\%RH)/100$.

Equations 11.9 through 11.14 can be combined into the following psychrometric equation:

$$Y_V = \frac{m a_w P_S}{(P - a_w P_S)} \tag{11.15}$$

where m (–) is the water–air molecular weight ratio:

$$m = \frac{M_W}{M_A} \tag{11.16}$$

The humidity at saturation Y_S is obtained from Equation 11.15 when $a_w = 1$ and $Y_V = Y_S$:

$$Y_S = \frac{m P_S}{(P - P_S)} \tag{11.17}$$

The boiling temperature T_b (°C) is obtained from the Antoine equation 11.14, when $P_S = P$ (bar) and $T = T_b$.

$$T_b = \frac{-a_3 + a_2}{(a_1 - \ln P)} \tag{11.18}$$

The dew point temperature T_d (°C) is obtained by combining Equations 11.14 and 11.15 when $Y_S = Y$ and $T = T_d$:

$$T_d = \frac{a_2}{\left(a_1 - \ln\left[YP/(m+Y)\right]\right)} - a_3 \tag{11.19}$$

The dew point pressure P_d (bar) is obtained by combining Equations 11.14 and 11.15 when $Y_S = Y$ and $P = P_d$:

$$P_d = \exp\left[\frac{a_1 - a_2}{(a_3 + T)}\right] \tag{11.20}$$

The enthalpy of the moist air H (kJ/kg) is calculated from the equation:

$$H = C_{PA}T + Y_V(\Delta H_0 + C_{PV}T) + Y_L C_{PL}T \tag{11.21}$$

where

C_{PA}, C_{PL}, and C_{PV} [kJ/(kg K)] are the specific heats of air, liquid water, and water vapor, respectively

ΔH_0 (kJ/kg) is the latent heat of evaporation of water at 0°C

Y_V (kg/kg db) is the absolute humidity

Y_L (kg/kg db) is the liquid moisture (droplets) content of the air–water vapor mixture

Note that here the temperature T is in °C and not in K.

Generally, the total humidity Y refers to both water vapor Y_V and liquid water Y_L and

$$Y_V = \min(Y, Y_S) \tag{11.22}$$

$$Y_L = Y - Y_V \tag{11.23}$$

Equation 11.22 means that (at a given temperature) air cannot take up more than a certain amount of vapor. Liquid droplets then will precipitate due to oversaturation, and this is called the cloud or fog state.

From Equations 11.22 and 11.23 it follows that if the air–water vapor mixture is undersaturated ($Y < Y_S$) then $Y_V = Y$ and $Y_L = 0$, if saturated ($Y = Y_S$) then $Y_V = Y_S$ and $Y_L = 0$, if oversaturated ($Y > Y_S$) then $Y_V = Y_S$ and $Y_L = Y - Y_V$.

The adiabatic saturation temperature, which (for water–air mixtures) is equal to the wet-bulb temperature T_W, can be obtained from the following equation:

$$\frac{(Y_V - Y_W)}{(T - T_W)} = \frac{-C_p}{\Delta H_S} \tag{11.24}$$

where P_W and Y_W are the vapor pressure and absolute humidity at saturation at temperature T_W

$$P_W = \exp\left[\frac{a_1 - a_2}{(a_3 + T_W)}\right] \tag{11.25}$$

$$Y_W = \frac{mP_W}{(P - P_W)} \tag{11.26}$$

The specific heat of moist air C_p [kJ/(kg K)] and the latent heat of water vaporization ΔH_S (kJ/kg) in Equation 11.24 can be estimated from the following equations:

$$C_P = C_{PA} + Y_V C_{PV} \tag{11.27}$$

$$\Delta H_S = \Delta H_0 - (C_{PL} - C_{PV})T \tag{11.28}$$

Note that in psychrometric calculations of this chapter, T is in °C and P is in bar.

11.3.2 Psychrometric Charts

A psychrometric chart is a graphical representation of the thermodynamic properties of moist air. Its distinctive features are of practical value in solving engineering problems.

Psychrometric charts are plots of the model (equations) presented in this section. The numerical constants used in these equations are summarized in Table 11.1.

The basic psychrometric chart, used extensively in Process Engineering, is a plot of absolute humidity versus temperature at various water activities (Figure 11.2). Other charts include plots of humidity versus temperature at various pressures, humid enthalpy versus temperature at various absolute humidities, and humid enthalpy versus absolute humidity at various water activities (note that water activity = (%*RH*)/100) (Table 11.2).

The dew point temperature T_d is the temperature at which water vapor starts to condense out of the air, or it is the temperature at which air becomes completely saturated.

The wet bulb temperature is the temperature of a wet bulb thermometer that is in equilibrium with the air stream around the thermometer.

Adiabatic humidification of air is achieved in a closed system by conversion of the sensible heat of the air mixture into heat of evaporation of liquid water. The opposite process, adiabatic dehumidification, is the conversion of the heat of condensation of water into sensible heat of the air mixture. In both adiabatic humidification and dehumidification, no external heat transfer takes place.

TABLE 11.1
Psychrometric Data for Air–Water Vapor Mixture

R	8.314 kJ/kmol K	Ideal Gas Constant
m	0.622	Water–air molecular weight ratio
C_{PA}	1.00 kJ/kg K	Specific heat of air
C_{PV}	1.90 kJ/kg K	Specific heat of water vapor
C_{PW}	4.20 kJ/kg K	Specific heat of liquid water
ΔH_0	2.50 MJ/kg	Latent heat of water evaporation at 0°C
a_1	1.00×10^1	Antoine equation constants for water
a_2	3.99×10^3	
a_3	2.34×10^2	

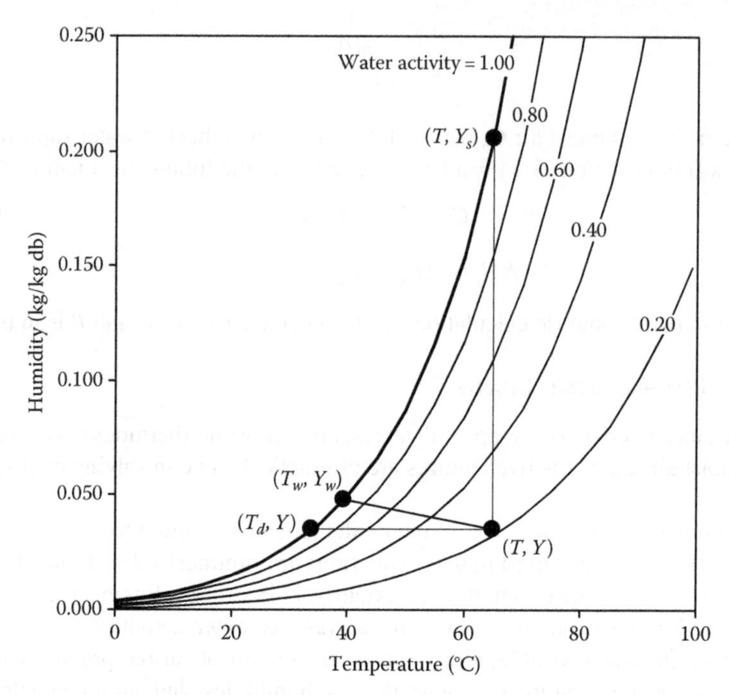

FIGURE 11.2 Psychrometric chart with characteristic temperatures and humidities. The corresponding numerical values are summarized in Table 11.2. (Modified from Maroulis, Z.B. and Saravacos, G.D., *Food Process Design*, CRC Press, Boca Raton, FL, May 9, 2003, fig. 7.18.)

TABLE 11.2
Numerical Values of Points on Psychrometric Chart of Figure 11.2

P	1 bar	Pressure
T	65°C	Temperature
Y	0.035 kg/kg db	Absolute humidity
a_w	0.214	(=21.4% *RH*)
T_d	34.0°C	Dew temperature
T_w	39.4°C	Wet bulb temperature
Y_s	0.206 kg/kg db	Saturation humidity at temperature T
Y_w	0.048 kg/kg db	Saturation humidity at temperature T_w

The specific volume of the air–water vapor mixtures V_a (m^3/kg) can be estimated from the psychrometric chart. It is about 0.85 m^3/kg db at ambient conditions and it increases at higher temperatures and higher humidities.

The psychrometric charts of the literature refer to the atmospheric pressure, i.e., $P = 1$ atm or $P = 1.013$ bar. For simplification, the psychrometric equations and diagrams presented in this book are based on a total pressure of 1 bar, i.e., 1.3% lower

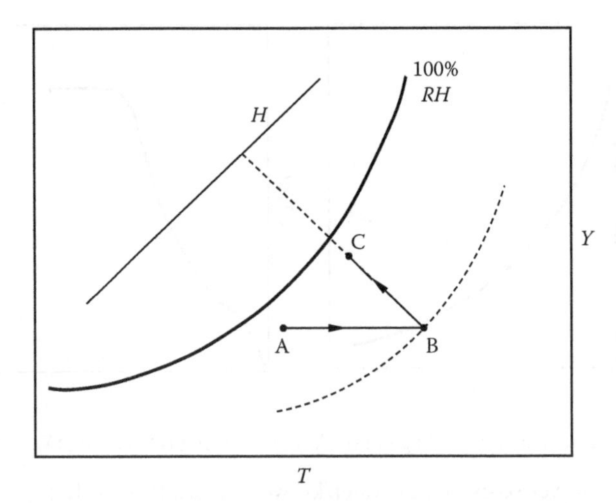

FIGURE 11.3 Air heating and adiabatic humidification on the psychrometric chart.

than 1 atm. For engineering design applications, the difference between the two calculations is considered negligible.

The basic psychrometric chart usually contains some additional data on the properties of moist air, such as specific volume (m^3/kg db) and enthalpy (kJ/kg db). Two useful psychrometric charts, one for lower temperatures (refrigeration and air conditioning) and another for temperatures up to 100°C (drying operations) are presented in the Appendix of this book. For temperatures higher than 100°C the analytical equations are recommended.

Figure 11.3 shows an example on the use of the psychrometric chart for an air heating and adiabatic cooling/humidification operation.

Moist air at point A is heated at constant absolute humidity Y until point B, which is at a lower water activity (relative humidity RH). The hot air is cooled and humidified adiabatically (constant enthalpy H) until point C is reached.

11.4 DRYING KINETICS

11.4.1 Drying Rates

The drying rates of food materials are usually determined experimentally, since it is very difficult to predict accurately the heat and mass transport rates theoretically. Drying tests are carried out on a layer of material, placed in an experimental dryer, which is operated under controlled conditions of temperature, air velocity, and humidity. The weight and the temperature of the sample are monitored as a function of time, obtaining the basic drying curve of moisture content dry basis X (kg/kg db) versus time t (s). The drying rate curve (dX/dt versus X) is obtained by differentiating graphically or analytically the original drying curve (Figure 11.4).

In drying and moisture sorption calculations, the moisture content of the material X is expressed on a dry basis (db) as kg water/kg db; and it is calculated from the

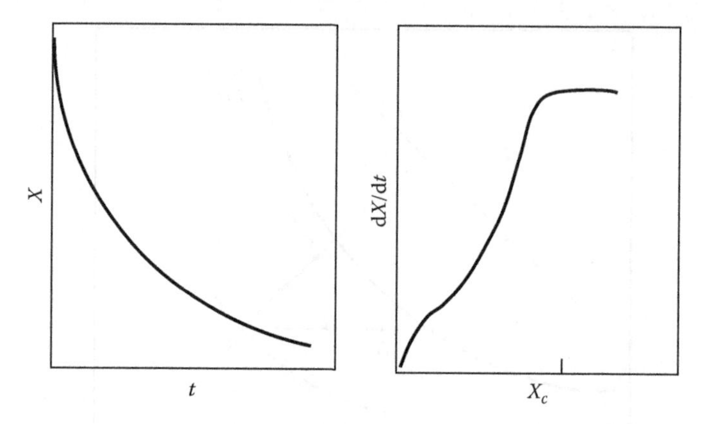

FIGURE 11.4 Experimental drying curve (X, t) and drying rate curve (dX/dt, X).

moisture content wet basis W (kg water/kg wet material) or % W (kg water/100 kg wet material) from the equation:

$$X = \frac{W}{(1 - W)} \quad \text{or} \quad X = \frac{(\%W)}{(100 - \%W)} \tag{11.29}$$

The drying rate curve may indicate a constant rate period, during which mass transfer from the surface of the material controls the drying process, depending on the initial moisture content and the external conditions (air velocity, temperature, and humidity). Short constant drying rate periods are observed in air-drying of food materials of high moisture content. However, most food materials do not show constant rate periods and they dry entirely in the falling rate period, during which mass transfer is controlled by the transport (mostly molecular diffusion) of water through the material to the surface of evaporation.

11.4.2 Drying Constant

The drying rate of a food material during the falling rate period can be expressed by the empirical equation of thin-layer drying:

$$\frac{dX}{dt} = -K(X - X_e) \tag{11.30}$$

where

X is the moisture content (dry basis) at time t
X_e is the equilibrium moisture content at infinite time
K (1/s) is the drying constant

Integration of Equation 11.30, yields the empirical drying equation, known also as the simplified Page equation:

$$\ln\left[\frac{(X - X_e)}{(X_o - X_e)}\right] = -Kt \tag{11.31}$$

The experimental drying data are usually plotted on semilogarithmic coordinates of $\log_{10}(R)$ versus t, where $R = (X - X_e)/(X_o - X_e)$, with a slope of $(-K/2.3)$. The equilibrium moisture content of the food material at the drying temperature (X_e) can be taken from the literature, determined experimentally, or estimated from empirical equations of the sorption isotherm of the material, such as the GAB equation (Chapter 3).

The empirical drying constant K depends on the material and the (dry bulb) temperature of the air. The drying constant of some food materials may change during the drying process, due to the significant changes in the physical structure of the material and, consequently, in the mass transport mechanism within the material. Water may be transported mainly by diffusion (liquid or vapor) or hydrodynamic/capillary flow.

When the slope of the drying ratio curve changes, two or more falling rate periods may be identified, obtaining two or more drying constants (K_1, K_2, ...). However, K may vary continuously with the moisture content X, and some empirical $K(X)$ relationship would be useful. As an alternative, the generalized Page equation may be used:

$$\ln\left[\frac{(X - X_e)}{(X_o - X_e)}\right] = -Kt^n \tag{11.32}$$

where n (–) is a characteristic constant.

The drying constant K increases exponentially with the temperature, and empirical models have been proposed for various materials.

11.4.3 Moisture Diffusivity

The water transport within most food materials can be expressed by the diffusion (Fick) equation, which is introduced in Chapter 2. The Fick equation, applied to the drying process, is based on the effective moisture diffusivity D, which is an overall transport property including molecular diffusion, and hydrodynamic or capillary flow of water within the material (Saravacos and Kostaropoulos, 2001).

The effective moisture diffusivity, or simply the moisture diffusivity, can be estimated from the slope ($-2.3K$) of the experimental drying ratio curve (Figure 11.5). For a slab (plate) drying from both flat surfaces, and exhibiting a constant slope of the drying ratio curve, the drying constant K is related to the diffusivity D (m^2/s) and the thickness L (m) of the material (slab or plate) by the diffusion-derived equation:

$$-K = \frac{\pi^2 D}{L^2} \tag{11.33}$$

Thus, if the effective moisture diffusivity D of the material is known at the given temperature and moisture content, the drying constant K of the material at the given thickness L of the slab can be estimated. For a spherical material, Equation 11.33 becomes

$$-K = \frac{\pi^2 D}{r^2} \tag{11.34}$$

where r (m) is the radius of the sphere.

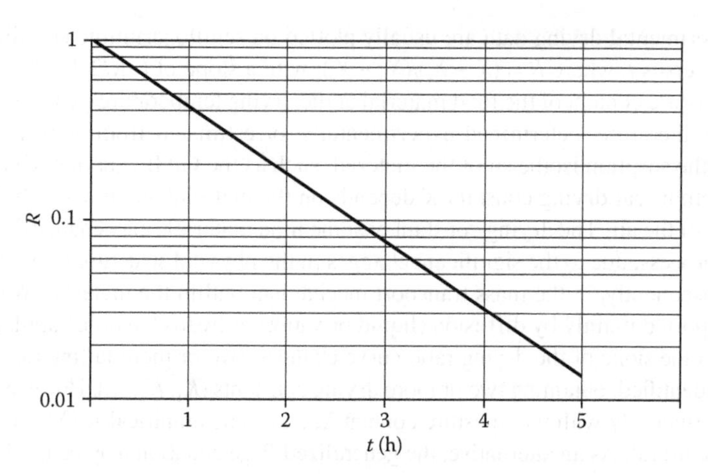

FIGURE 11.5 Drying ratio curve $\log_{10} R$ $[(X - X_e)/(X_o - X_e)]$ versus time (t); X_o, X_e, X, respective, moisture contents: initial, equilibrium, and after time (t).

Equations 11.33 and 11.34 are rough approximations, which are based on mass transport by diffusion, in which the rate is proportional to the square of thickness of the material. The diffusion mechanism may not be applicable to some food materials, due to their physical structure. For diffusion-controlled drying, the thickness (or diameter) of the material should be as low as possible, i.e., drying of thin layers or particles of small diameter is desirable. At the same time, higher effective diffusivities are required to achieve short drying times. Table 10.3 shows typical effective moisture diffusivities D and the energies of activation for diffusion E_D of some classes of food materials.

High energies of activation for diffusion E_D, meaning strong temperature effect, are characteristic of molecular diffusion in nonporous food materials, while low energies indicate vapor diffusion in porous materials. The energy of activation E_D (kJ/mol) is related to the temperature by the intergraded form of the Arrhenius equation:

$$\ln\left(\frac{D}{D_o}\right) = -\left(\frac{E_D}{R}\right)\left(\frac{1}{T} - \frac{1}{T_o}\right) \tag{11.35}$$

where

D, D_o (m²/s) are the moisture diffusivities at temperatures T, T_o (K), respectively
$R = 8.314$ kJ/(kmol K)

11.4.4 Empirical Model

The drying rate during the falling rate period is expressed by the kinetics model of Equation 11.30, which can be written as

$$\frac{dX}{dt} = -\left(\frac{1}{t_c}\right)\left(X - X_e\right) \tag{11.36}$$

where $t_c = 1/K$, the inverse of the drying constant with units (s) or (h).

For process design purposes, the drying rate constant t_c can be determined from experimental drying data obtained under conditions by varying the material size d (m), air velocity u (m/s), air temperature T (°C), and absolute humidity Y (kg/kg db):

$$t_c = c_0 d^{c_1} u^{c_2} T^{c_3} Y^{c_4} \tag{11.37}$$

where c_0, c_1, c_2, c_3, c_4 are adjustable empirical constants, depending on the material characteristics and drying equipment geometry.

The equilibrium material moisture content X_e depends on air temperature T and water activity a_w. Various empirical or semitheoretical models have been proposed, and the modified Oswin model is used here:

$$X_e = b_1 \exp\left[\frac{b_2}{(273+T)}\right]\left[\frac{a_w}{(1-a_w)}\right]^{b_3} \tag{11.38}$$

where b_1, b_2, and b_3 are adjustable empirical constants, depending on the material characteristics.

When drying conditions are kept constant, Equation 11.36 is integrated as follows:

$$t = -t_c \ln\left[\frac{(X - X_e)}{(X_o - X_e)}\right] \tag{11.39}$$

Equation 11.39 calculates the required drying time t to remove the moisture from the initial X_o to the final X material moisture content.

11.5 DRYING TECHNOLOGY

Food drying and food dehydration have developed mostly from practical and industrial experience, but during the last years food science and chemical (process) engineering are applied increasingly to analyze and improve this technology.

The major dehydrated food products are fruits and vegetables, dairy products (milk, whey), soluble coffee–tea, and soups. Fruit and vegetable dehydration has received special attention due to the diversity of the raw materials, the sensitivity of food products, the seasonal agricultural production, and the various drying processes and equipment used for these products (Baker, 1997).

11.5.1 PROCESS TECHNOLOGY

The main drying processes used for fruits and vegetables are sun drying, convective (air) drying, fluid bed drying, spray drying, and drum drying. Specialized drying processes include vacuum drying, freeze-drying, puff-drying, and foam-mat drying (Van Arsdel et al, 1973; MacCarthy, 1986; Greensmith, 1998).

Predrying treatments of fruits include slicing, treatment of fruits with special approved chemicals, such as sulfur dioxide and citric acid, and peeling and blanching of vegetables. Dipping of grapes in alkali solutions, containing ethyl oleate, increases substantially the drying rate. Sulfuring (gaseous SO_2 or sulfurous solutions) is used to preserve fruit color.

Dried and dehydrated foods are generally microbiologically stable, i.e., microbial growth is prevented by the low water activity ($a_w < 0.70$). Protective packaging and some approved additives may be required to preserve the quality of the product (color, flavor, structure).

IMF are dried to water activities of 0.90–0.70, corresponding to moisture contents 30%–20%, which prevent the growth of spoilage bacteria. The growth of yeasts and molds may be controlled by additives, such as sorbic acid. IMF foods are produced either by convective drying under mild conditions, or by osmotic dehydration.

Process and storage calculations require knowledge of the equilibrium moisture properties (moisture sorption isotherms) of the dried food materials. The quality of dehydrated foods, especially fruit products, is affected strongly by the retention of characteristic volatile aroma components during the drying operation. Some dehydration processes result in improved aroma retention, e.g., vacuum and freeze drying, spray drying, and osmotic dehydration.

Figure 11.6 shows a process block diagram for the air-dehydration of carrot slices, which is useful for material and energy balances, and for preliminary sizing of the process equipment.

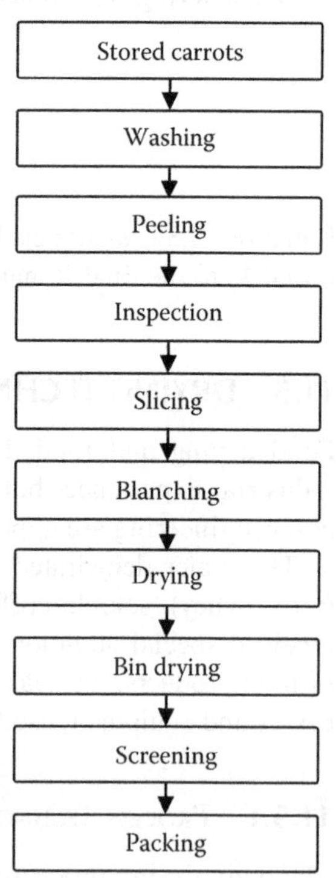

FIGURE 11.6 Block process diagram of carrot dehydration.

Osmotic dehydration is used to remove part of the free water in food materials by the osmotic action of sugar or salt solutions. It is based on the reduction of water activity of the food material, which prevents or delays the growth of spoilage microorganisms. It is operated at lower temperature and consumes less energy compared to thermal dehydration, which requires theoretically 2.3 MJ/kg and practically 3–6 MJ/kg water evaporated.

Osmotic dehydration is used to produce IMF or as a predrying step, removing about 50% of moisture. Mass transfer in osmotic dehydration involves the transport of water within the plant tissue, the transfer of water and natural solutes from the plant cells to the osmoactive solution, and the transfer of the osmoactive substance to the plant material. The transport mechanisms depend mainly on the physical structure of the food material. The diffusion (Fick) model is generally used for transport calculations, assuming that the driving force is a concentration gradient. The effective diffusivity D in osmotic treatment of apples is 10×10^{-10} m^2/s for water and 4×10^{-10} m^2/s for sucrose. Osmoactive substances are sugars (sucrose, corn syrups) and dextrins at about 65°Brix for fruits, and sodium chloride (about 15% TS) for vegetables and fish.

11.6 DRYER DESIGN

11.6.1 Design of Industrial Dryers

The design of industrial dryers is based largely on empirical knowledge, and several drying processes and dried products have been developed, using drying equipment specific for each class of products.

The thermophysical, transport, end equilibrium (isotherms) properties of the material are important in specifying the proper dryer and drying conditions. Drying rates (kinetics of moisture removal) are useful for preliminary estimation of the drying time of the specific material, which is taken as the mean residence time in the dryer.

The specifications of an industrial dryer should include the properties of the wet and dried material, the temperature sensitivity and water activity of the product, the (evaporative) capacity (kg water/h), the energy supply, and operating cost. Dryers are classified on the basis of type of operation (batch, continuous), type of feed (liquid, suspension, paste, granules, fibrous solids, porous solids), heating method (convection, contact, radiation, dielectric), and product sensitivity (vacuum, low temperature). The size of dryers can be small (up to 50 kg/h), medium (50–1000 kg/h), and large (above 1000 kg/h).

The capacity of the dryers can be expressed also as (kg) water evaporated per unit surface and unit time, which is very high in rotary dryers (about 50 kg/m^2 h) and low in tray dryers (about 1 kg/m^2 h). The cost of drying is an important factor in dryer design, especially for large volume products of relatively low value. Energy (fuel) is the major cost in drying operations, followed by capital and labor costs. The major energy use is for the evaporation of water (moisture), which varies considerably for the different dryers, from 3 MJ/kg water (spray dryers) to 6 MJ/kg (tray dryers). The energy efficiency of the dryers (ratio of the heat of evaporation to the heat input to the dryer) depends strongly on the type of dryer. It is higher in contact (40%–80%) than convective drying (20%–40%). Rotary dryers are more efficient than tray and spray dryers.

11.6.2 Selection of Industrial Dryers

Selection of the optimal dryer type and size should satisfy all process and product requirements at minimum cost. Recent progress in computer applications has resulted in a number of selection procedures, which facilitate the selection of the proper dryer for each particular application. Software programs and expert systems have been developed to select the proper dryer for a specific application.

In practice, directories of suppliers are used to select the dryer according to the field of application and the dryer characteristics, such as mode of operation (batch, continuous), convection heating (tray, belt, rotary dryers), contact heating, vacuum operation, and air-suspension dryers. Dryer selection depends also on the condition of the feed material, such as liquids, pastes, and size of solids: powders <0.5 mm, grains (0.5–5 mm), and pieces >5 mm (Kemp, 1999; VDMA, 1999).

Application of the diverse types of process dryers to food dehydration should take into consideration the unique requirements of processing of foods and biological products. Process engineering optimization of dryers based on equipment and operating cost alone is not sufficient. The strict product requirements for food dryers (organoleptic, nutritional, and functional) should be met at the lowest possible cost.

Preparation and pretreatment of raw food materials, especially fruits and vegetables, involving washing, peeling, slicing, blanching, and chemical treatment (e.g., sulfur dioxide, acids, salts, sugars) is an integral part of the dehydration flowsheet.

The various types of drying operations and equipment, used in commercial food processing, are shown in Table 11.4. Selected operating characteristics were taken from the literature.

11.6.3 Material and Energy Balances

Figure 11.7 shows diagrammatically a typical air dryer. It includes the material and energy flows, which are needed to formulate the overall material and heat balances.

The material enters in the dryer at a flow rate m_s (kg db/s), moisture content X_o (kg/kg db), and temperature T_o and exits at the same flow rate on dry basis m_s, moisture X (kg/kg db), and temperature T. Fresh air enters the dryer with a flow rate m_a (kg db/s), absolute moisture Y_o (kg/kg db), and temperature T_o and exits at the same flow rate m_a on dry basis, absolute moisture Y (kg/kg db), and temperature T. It is assumed that both streams, solids and air, leave the dryer in thermal equilibrium (at the same temperature), and that the thermal losses to the environment are negligible. The thermal energy input to the dryer is Q (kW).

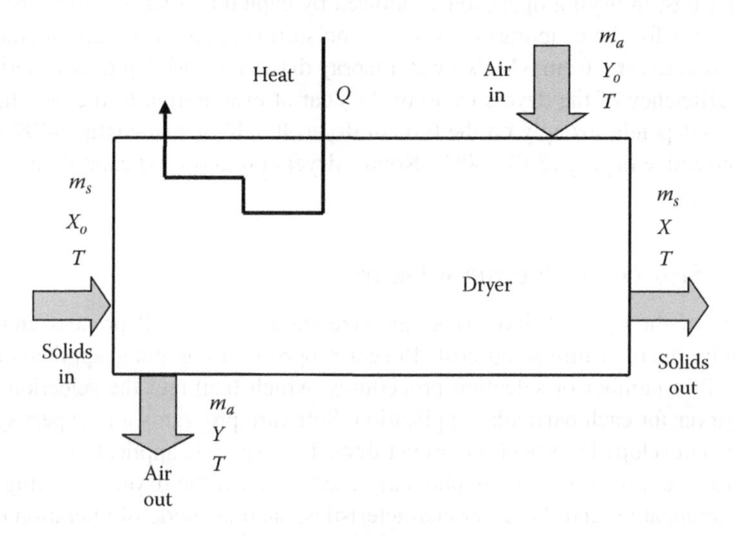

FIGURE 11.7 Diagram of a typical air (convective) dryer. (Modified from Maroulis, Z.B. and Saravacos, G.D., *Food Process Design*, CRC Press, Boca Raton, FL, May 9, 2003, fig. 7.22.)

The moisture balance in the dryer is expressed using the following equations, which indicate that the moisture removed from the solids is taken up by the air:

$$m_s(X_o - X) = m_a(Y - Y_o) \tag{11.40}$$

The total thermal load Q is obtained by an energy balance around the dryer

$$Q = H_a - H_{a0} + H_s - H_{s0} \tag{11.41}$$

where H_{a0}, H_a, H_{s0}, H_s, are the enthalpy of the air (a) and solids (s) streams at the dryer inlet and outlet, respectively:

$$H_a = C_{Pa}T + Y(\Delta H_o + C_{Pv}T) \tag{11.42}$$

$$H_{a0} = C_{Pa}T_o + Y_o(\Delta H_o + C_{PV}T_o) \tag{11.43}$$

$$H_s = (C_{Ps} + X\,C_{PL})T \tag{11.44}$$

$$H_{s0} = (C_{Ps} + X_oC_{PL})T_o \tag{11.45}$$

The energy balance of the dryer indicates that the thermal energy input to the dryer Q is used to heat the solids material and the fresh air, and to evaporate the material moisture.

The thermal efficiency of the dryer is defined as follows:

$$n = \frac{Q(\text{evaporation})}{Q} = \frac{\{m_s(X_o - X)[\Delta H_o - (C_{PL} - C_{Pv})T]\}}{Q} \tag{11.46}$$

The heat of evaporation of water at temperature T consists of the enthalpy of evaporation of water at 0°C (ΔH_o) plus the sensible heat of water vapors $C_{Pv}T$ minus the sensible heat of liquid water droplets $C_{PL}T$.

It should be noted that the temperature T is in °C and that the basis of enthalpy calculations H is 0°C. The heat of evaporation of water at 0°C is ΔH_o = 2.5 MJ/kg and the specific heat of liquid water and water vapors are, respectively, C_{PL} = 4.2 kJ/(kg K) and C_{Pv} = 1.9 kg/(kg K).

11.7 DRYING EQUIPMENT

11.7.1 Sun Drying

Large quantities of fruits, particularly grapes (raisins), apricots, figs, prunes, and dates are dried by direct exposure to sunlight in hot and dry climates. Coffee beans, cereal grains, and fish are also sun-dried prior to storage and preservation. Sun-dried fruits contain about 15%–20% moisture (wet basis), which is near the equilibrium moisture content at ambient air conditions, and they can be stored in bulk, without the danger of microbial spoilage.

Seedless (Sultana) grapes are usually pretreated by dipping in alkali solutions, containing vegetable oil or ethyl oleate, which increases the drying rate by increasing the moisture permeability of the grape skin. The grape bunches are spread in trays and dried by exposure to direct sunlight. The grapes may also be dried by hanging the bunches from a string, while they are covered by a transparent plastic cloth, which protects the product from adverse weather conditions. The sun drying time varies from 10 to 20 days, depending on the insolation (solar radiation). The ripe apricots are usually cut into halves before sun drying on trays, placed on the ground.

Dried fruits, especially figs and apricots, may require fumigation treatment with sulfur dioxide or other permitted insecticide during storage and also before packaging.

11.7.2 Solar Dryers

Solar drying is a form of convective drying, in which the air is heated by solar energy in a solar collector. Flat-plate collectors are used with either natural or forced circulation of the air. Figure 11.8 shows a simple solar dryer with a flat-plate solar collector connected to a batch tray dryer. The air movement is by natural convection, but addition of an electrical fan will increase considerably the collector efficiency and the drying rate of the product (Saravacos and Kostaropoulos, 2001).

Several types of solar collectors and drying systems have been proposed for drying various food and agricultural products, such as fruits, vegetables, and grains. The common flat-plate collector consists of a black plate, which absorbs the incident solar radiation, a transparent cover, and insulation material.

The incident solar energy (insolation) varies with the geographical location and the season of the year. A typical insolation for a hot climate would be 0.6 kW/m^2 with an average sunshine time of 7 h/day. This energy corresponds to about $0.6 \times 3600 =$ 2.16 MJ/h or 15 MJ/m^2 day. The evaporation of water at 40°C requires theoretically 2.4 MJ/kg and practically about 3 MJ/kg. Therefore, the mean evaporation rate of water will be about $2.16/3 = 0.72$ kg/m^2 h (intermittent operation 7 h/day).

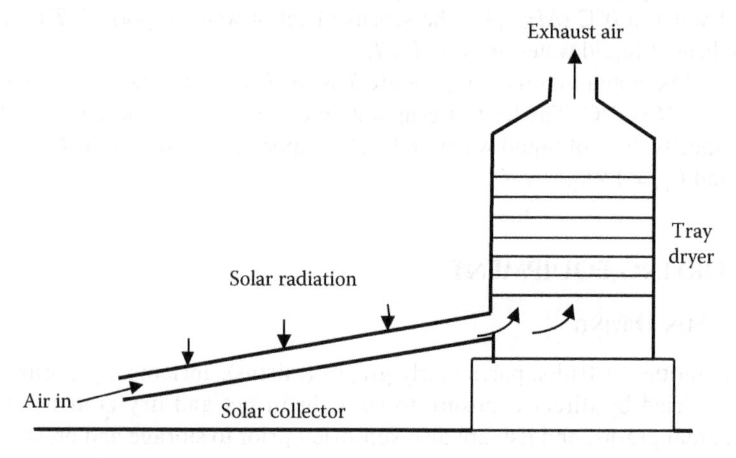

FIGURE 11.8 Simple solar dryer.

The relatively low intensity of incident solar radiation is a serious limitation for food drying applications, where large amounts of thermal energy are required for the evaporation of water.

Large surfaces of solar collectors are needed for drying significant amounts of food materials. For example, evaporation of 1000 kg/h of water (capacity of a typical mechanical convective dryer) would require about 1000/0.72 = 1400 m^2 of collector surface for a hot climate (intermittent operation 7 h/day). A larger surface would be required in a temperate zone. Solar drying is considered effective for relatively small drying operations for fruits, such as grapes and apricots, in high insolation regions.

Intermittent solar radiation (day–night) can be supplemented by the use of auxiliary energy, such as fuel or electricity. Thermal storage of solar energy can also be applied, using rock beds or water to absorb extra solar energy during the day, which can be used during the night or during cloudy weather.

Some other solar collectors, proposed for solar drying are (a) a low-cost tunnel collector 1 × 20 m connected to a tunnel dryer for drying a batch of 1000 kg of grapes; (b) a solar collector with V-grooves, attaining temperatures 50°C–70°C at 0.7 kW/m^2 insolation; and (c) an evacuated tubular solar collector (glass tubes 12.6 cm diameter and 2.13 m length), capable of heating the air to 90°C–110°C.

Solar collectors, integrated on the roof or the walls of a farm building can provide heated air for drying grain in a bin or silo.

11.7.3 Silo and Bin Dryers

Silo dryers are used for partial drying of large quantities of grains (wheat, corn, etc.) from moisture contents (wet basis) of about 25% (harvest) to 18% (storage). Hot air at 40°C–60°C is blown from the bottom of the fixed bed through the grains for several hours (Figure 11.9). Continuous tower dryers are more effective, using higher air temperatures (e.g., 80°C), while the grain slowly moves down.

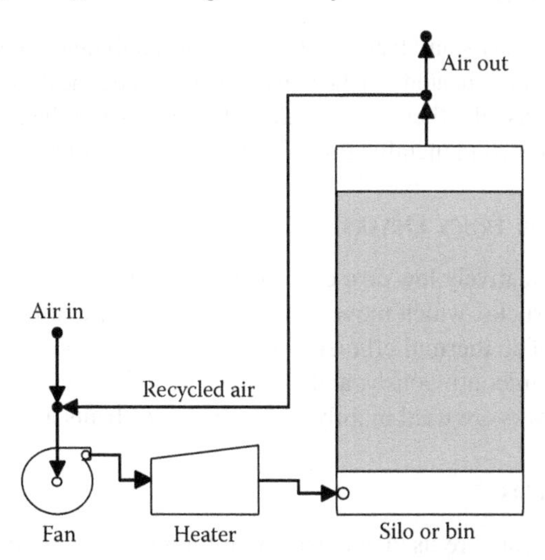

FIGURE 11.9 Batch silo or bin dryer. (Modified from Maroulis, Z.B. and Saravacos, G.D., *Food Process Design*, CRC Press, Boca Raton, FL, May 9, 2003, fig. 7.3.)

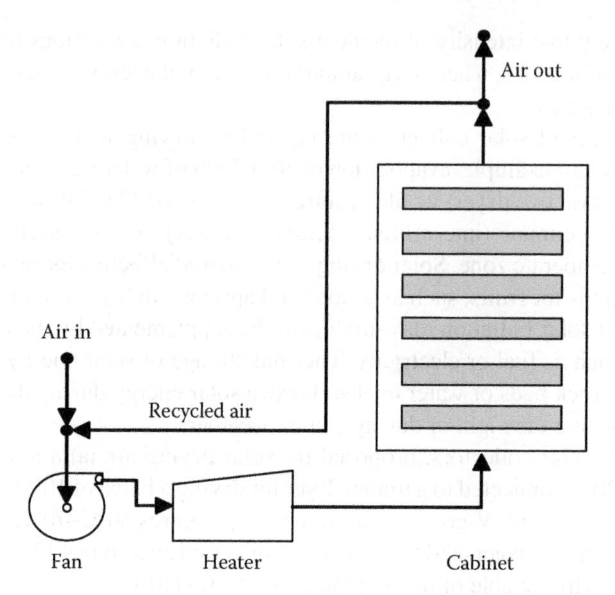

FIGURE 11.10 Diagram of a cabinet (tray) dryer. (Modified from Maroulis, Z.B. and Saravacos, G.D., *Food Process Design*, CRC Press, Boca Raton, FL, May 9, 2003, fig. 7.4.)

Bin dryers are similar but generally smaller than silo dryers and they are used as finish dryers of partially dried vegetables. Normal convective drying of vegetables reduces their moisture content to about 10%, and bin drying can bring it down to 2%–4%, which is necessary for preservation and storage. Bin dryers operate at relatively low temperatures with dry dehumidified air, blown upward.

11.7.4 Tray Dryers

Tray dryers are relatively small batch units for drying small quantities of food products (Figure 11.10). The air is heated in a heat exchanger outside the dryer, and it is usually recirculated to increase the thermal efficiency. The product in the form of pieces, particles, or pastes is placed in metallic trays, which are reused after the drying operation.

11.7.5 Tunnel or Truck Dryers

Tunnel dryers are relatively low cost constructions, with the product trays (pieces or pastes) loaded on trucks, which move slowly co-current or countercurrent to the hot air (Figure 11.11). The thermal efficiency of the dryer is improved by recirculation. The system runs semicontinuously, and the trays are loaded and unloaded manually. Tunnel or truck dryers are used mainly in the drying of fruits and vegetables.

11.7.6 Belt Dryers

Belt or conveyor dryers are used extensively in food processing for continuous drying of food pieces (Figure 11.12). The product, in the form of pieces, such as fruits and vegetables, is dried on a long perforated conveyor belt, which moves slowly

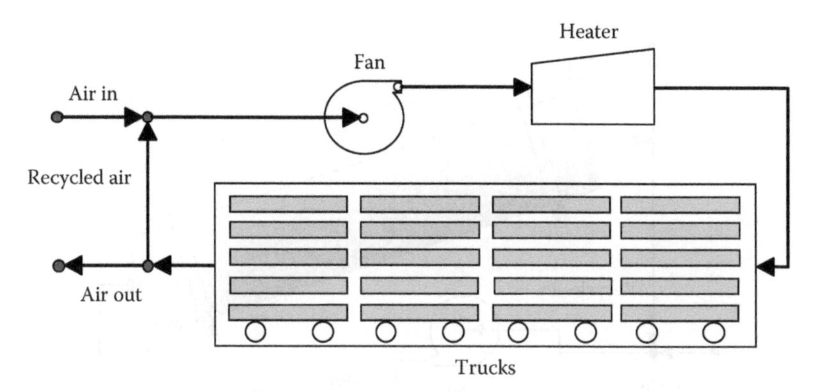

FIGURE 11.11 Diagram of a tunnel or truck dryer. (Modified from Maroulis, Z.B. and Saravacos, G.D., *Food Process Design*, CRC Press, Boca Raton, FL, May 9, 2003, fig. 7.5.)

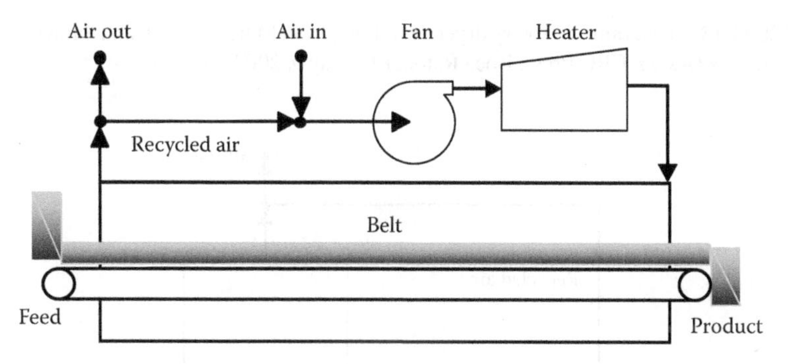

FIGURE 11.12 Diagram of a single-belt conveyor dryer. (Modified from Maroulis, Z.B. and Saravacos, G.D., *Food Process Design*, CRC Press, Boca Raton, FL, May 9, 2003, fig. 7.6.)

for the required drying time. The air is heated to the desired temperature in a heat exchanger or is mixed with the combustion gases of suitable fuels, and it is directed against the product in up- or down-flow. Long residence times are obtained using multibelt dryers (e.g., three belts), which run in opposite directions.

11.7.7 Rotary Dryers

The rotary dryers consist of an inclined long cylinder rotating slowly, while the material (grains, granules, powders) flows with the tumbling (cascading) action of the internal flights (Figure 11.13). The air is heated either in heat exchangers or by mixing with combustion gases of suitable fuel, e.g., natural gas. Rotary dryers are less expensive than belt dryers, but they cannot handle large food pieces, which may be damaged by mechanical abrasion during tumbling (Perry and Green, 1997).

11.7.8 Fluidized Bed Dryers

Fluidized bed dryers are used for fast drying of food pieces and particles that can be suspended in a stream of hot air (Figure 11.14). High drying rates are obtained due to high heat and mass transfer.

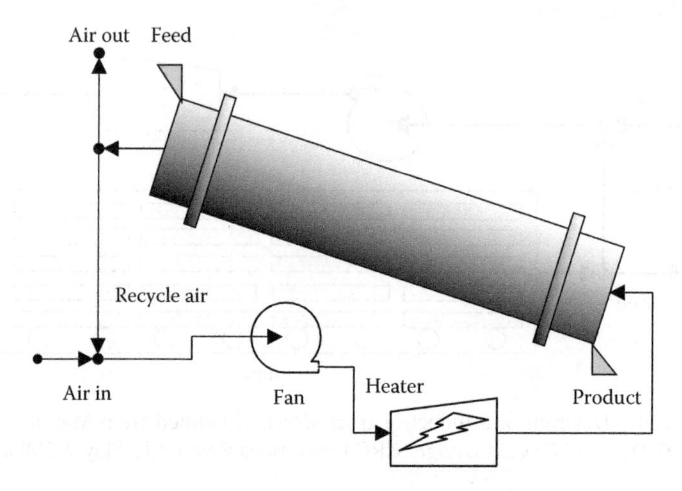

FIGURE 11.13 Diagram of a rotary dryer. (Modified from Maroulis, Z.B. and Saravacos, G.D., *Food Process Design*, CRC Press, Boca Raton, FL, May 9, 2003, fig. 7.7.)

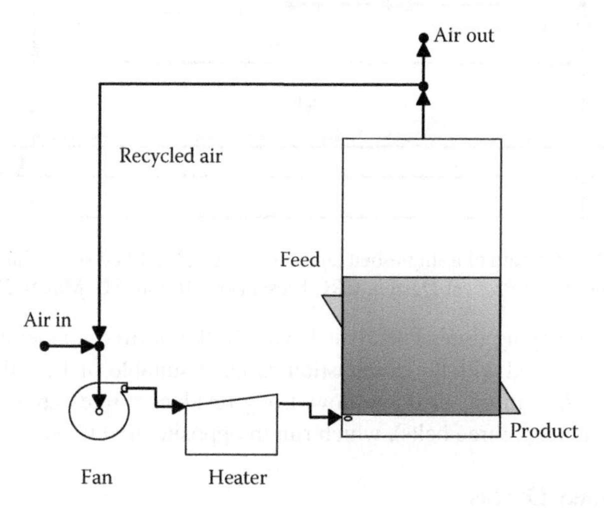

FIGURE 11.14 Diagram of a fluidized bed dryer. (Modified from Maroulis, Z.B. and Saravacos, G.D., *Food Process Design*, CRC Press, Boca Raton, FL, May 9, 2003, fig. 7.8.)

11.7.9 Spouted Bed Dryers

Spouted bed dryers are a special type of fluidized-bed equipment, in which the granular material is circulated vertically in a tall drying chamber. The heated gas enters as a jet at the center of the conical base of the vessel, carrying the granular material upward, which is dried partially and thrown to the annular space. The material in the bed moves slowly by gravity to the bottom, and the cycle is repeated continuously (Figure 11.15). Spouted bed dryers are suitable for granular materials larger than 5 mm, such as wheat grain.

11.7.10 Pneumatic or Flash Dryers

Pneumatic or flash dryers are used for fast and efficient drying of food particles that can be suspended and transported in the stream of heating air (Figure 11.16). The residence time in pneumatic dryers is much shorter than in fluidized-bed units.

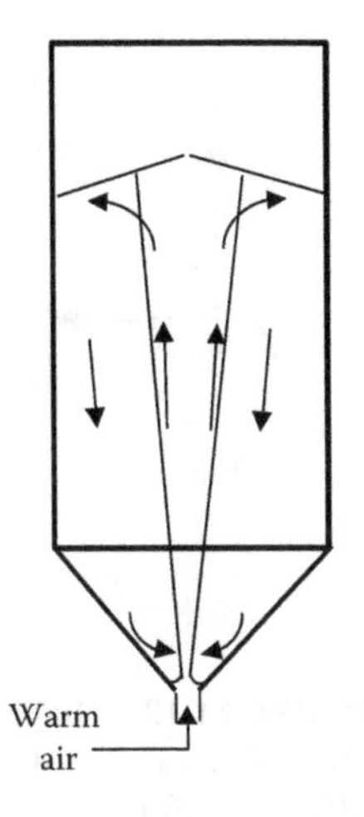

FIGURE 11.15 Diagram of a spouted bed dryer.

11.7.11 Spray Dryers

Spray dryers are used to dehydrate liquid foods or food suspensions into dry powders or agglomerates. The liquid feed is atomized in special valves (Chapter 7) and the droplets are dried by hot air as they fall in a large chamber (Figure 11.17). The flow of hot air can be cocurrent or countercurrent to the flow (fall) of the droplets and dried particles. The dryers are equipped with cyclone collectors and bag filters to collect the small particles from the exhaust air/gases, and prevent air pollution. Spray dryers are usually combined with agglomeration equipment, which produces food agglomerates of desirable quality (Masters, 1991).

The engineering of particles (liquid and solid) and agglomerates is discussed in the section of Mechanical Processing (Chapter 7). Pressure atomizers are preferred in spray drying because they produce droplets of approximately uniform size. The other two atomizers (centrifugal and pneumatic) produce a wider dispersion of droplet sizes, which dehydrate unevenly.

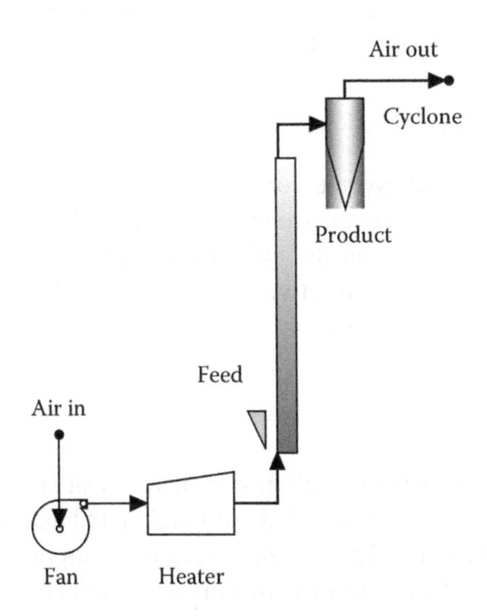

FIGURE 11.16 Diagram of a pneumatic dryer. (Modified from Maroulis, Z.B. and Saravacos, G.D., *Food Process Design*, CRC Press, Boca Raton, FL, May 9, 2003, fig. 7.9.)

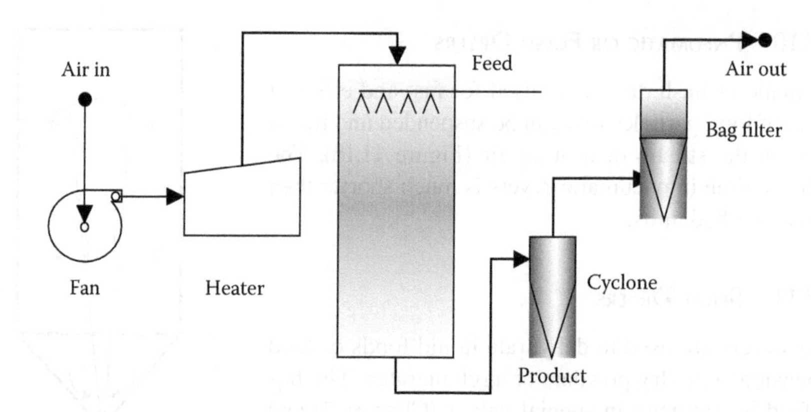

FIGURE 11.17 Diagram of a co-current flow spray dryer. (Modified from Maroulis, Z.B. and Saravacos, G.D., *Food Process Design*, CRC Press, Boca Raton, FL, May 9, 2003, fig. 7.10.)

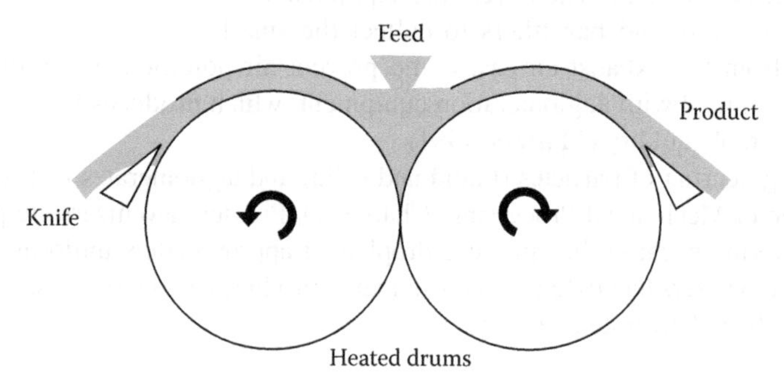

FIGURE 11.18 Diagram of a double drum dryer. (Modified from Maroulis, Z.B. and Saravacos, G.D., *Food Process Design*, CRC Press, Boca Raton, FL, May 9, 2003, fig. 7.11.)

11.7.12 Drum Dryers

Drum dryers are used to dehydrate concentrated liquid foods or suspensions and food pastes. They consist of one or two slowly rotating drums, heated internally by steam, with the product dried on the cylindrical surface (Figure 11.18). They are more efficient thermally than convective (air) dryers and they are operated either at atmospheric pressure or in vacuum.

11.7.13 Vacuum Dryers

Vacuum dryers are used for the dehydration of heat-sensitive food products, such as fruit juices. They operate at pressures of about 10 mbar and temperatures around 10°C (drying from the liquid state). They require vacuum pumping and low-temperature condensing equipment. Heat transfer is by contact to a heated shelf, infrared radiation, or microwaves. The product is dried either in trays or in a belt (Figure 11.19). Both batch and continuous operating systems are used.

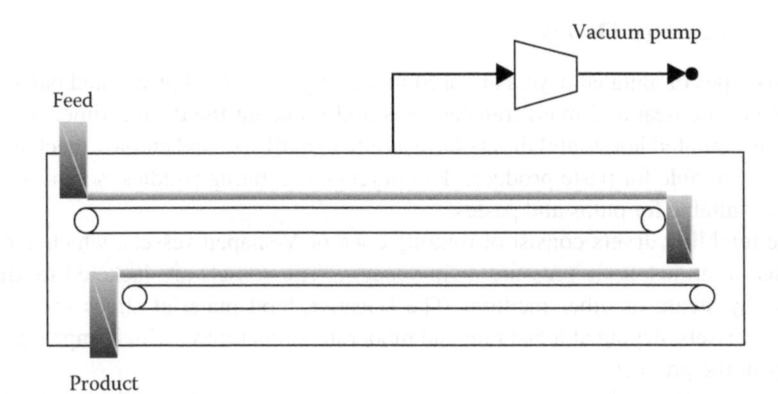

FIGURE 11.19 Diagram of a continuous belt-vacuum dryer. (Modified from Maroulis, Z.B. and Saravacos, G.D., *Food Process Design*, CRC Press, Boca Raton, FL, May 9, 2003, fig. 7.12.)

11.7.14 Freeze Dryers

Freeze dryers are the most expensive drying equipment, and they are justified economically only for drying certain expensive food products of unique quality, such as instant coffee. They are used mainly in the freeze drying of pharmaceutical products, which can afford the high cost (Liapis and Bruttini, 1995; Oetjen, 1999).

Freeze-dryers operate at pressures below 1 mbar and temperatures below −10°C (drying from the frozen state). The prefrozen product is placed in trays (Figure 11.20) and heated by contact, infrared radiation, or microwaves. Freeze-drying rate is limited by heat transfer to the drying surface. Batch freeze dryers are normally used, but there are some semicontinuous systems for large operations.

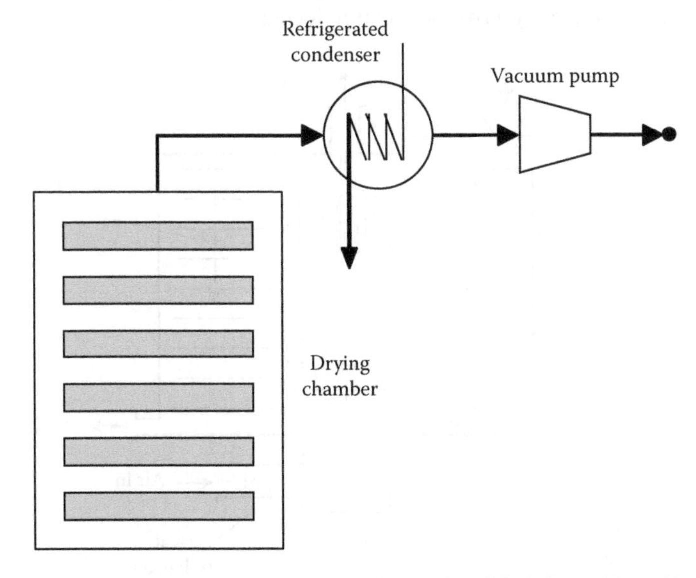

FIGURE 11.20 Diagram of a batch freeze-dryer. (Modified from Maroulis, Z.B. and Saravacos, G.D., *Food Process Design*, CRC Press, Boca Raton, FL, May 9, 2003, fig. 7.13.)

11.7.15 Agitated Dryers

Various types of agitated dryers are used for the drying of food pieces and particles, improving the heat and mass transfer rates and reducing the drying time. Among them, the agitated horizontal dryers have a rather small size and employ mechanical scrapers, suitable for paste products. Pan dryers use rotating paddles (scrapers) and they are suitable for pulps and pastes.

The tumbling dryers consist of rotating cone or V-shaped vessels, which can be operated at atmospheric pressure or in vacuum. The vessels are jacketed to allow heating by steam or other medium. The sensitive food material slides inside the rotating vessels, drying at a fast rate and moderate temperature, which improves the quality of the product.

The turbo dryer (Figure 11.21) is a special tray dryer with the particulate product flowing slowly down, following a helical path, while it is agitated by air blown countercurrently by two fans.

11.7.16 Microwave Dryers

Microwave (MW) and dielectric or radio frequency (RF) energy at 915 or 2450 MHz (Megacycles/s) are used to remove water from food materials at atmospheric or in vacuum. MW and RF energy heat the material internally, without the need of external convective or contact heat transfer. Water has a higher dielectric constant (about 8) than the other food components (about 2). Therefore, food materials of high moisture content absorb more MW or RF energy, facilitating the drying process. Free water can be removed more easily, because it absorbs more energy than adsorbed water (Chapter 8).

Internal absorption of the MW/RF energy by a wet material will increase its temperature and vapor pressure, creating a puffing effect on the product, and increasing the drying rate during convective or vacuum drying.

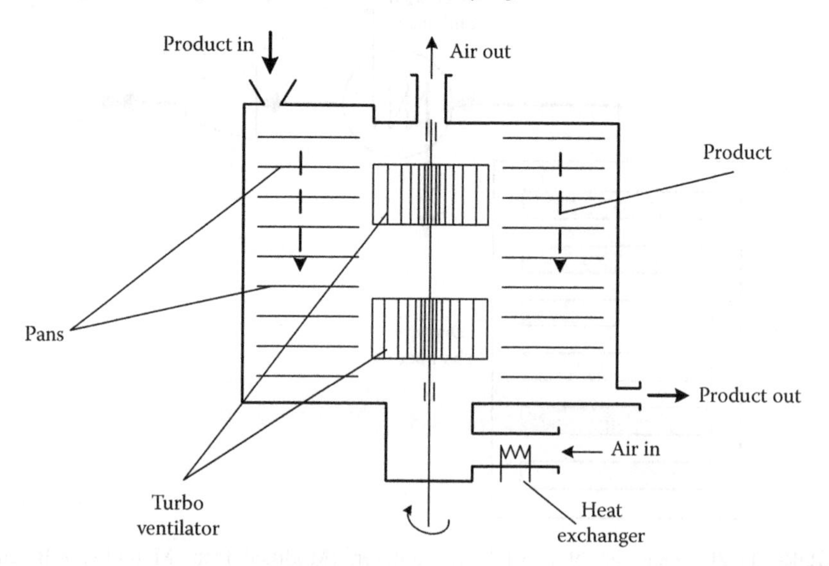

FIGURE 11.21 Diagram of a turbo dryer.

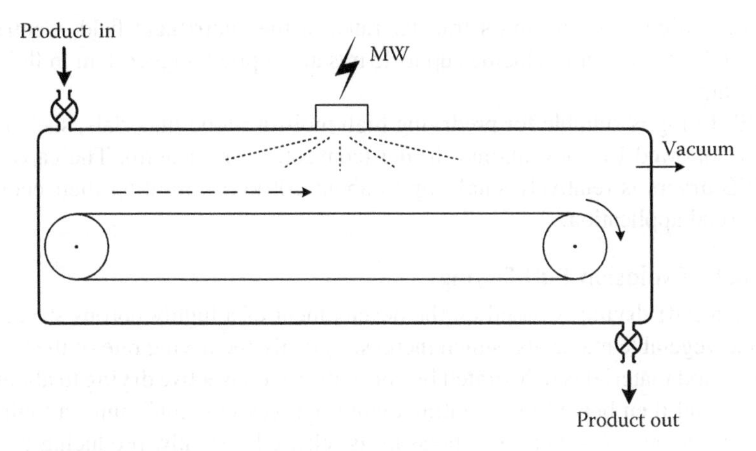

FIGURE 11.22 Diagram of a continuous MW-vacuum dryer.

MW energy improves the vacuum- and freeze-drying operations of food materials by better energy (heat) absorption within the product, or by the development of a porous structure (puffing) in the material, which increases substantially the effective moisture diffusivity.

RF drying is used in various post-baking systems, following the commercial fuel-heated oven, increasing the production rate of cookies, biscuits, etc. by 30%–50%. MW drying can be applied to pasta drying operations, reducing substantially the drying time of conventional hot-air drying, and improving the product quality.

Figure 11.22 shows a continuous MW-vacuum dryer, suitable for food products.

11.7.17 Impingement Dryers

Impingement jets of hot air (high speed) can be used in drying operations to increase the dying rate of food materials, due to the resulting higher heat and mass transfer coefficients. Nozzle design and geometry of the dryer-food material system are important factors in effective drying application. Specific energy consumption in impingement dryers is about 3.1 MJ/kg water evaporated.

11.7.18 Special Dryers

Several dryers have been tested and some of them have been applied in the food process industry. They are mostly batch food processing operations used to dry some sensitive food products, such as fruits and vegetables. Some of the special dryers are still in the development stage, and their commercial application will depend on the process economics and the acceptance of the new products by the consumers.

11.7.18.1 Centrifugal Dryers

The centrifugal fluid bed (CFB) dryers consist of a cylindrical vessel with perforated walls, which rotates horizontally at high velocity, and is heated by a cross flow air stream. Food material pieces move through the rotating cylinder and are dried fast, due to the high heat and mass transfer rates in the centrifugal field. Centrifugal forces of 3–15 *g*, and air velocities up to 15 m/s are applied, higher than in fluidized-bed drying.

CFB drying is suitable for predrying high moisture food materials, such as vegetables, followed by conventional drying (convective or vacuum). The capacity of the CFB dryers is relatively small (up to about 200 kg/h), limiting their economic commercial application.

11.7.18.2 Explosion-Puff Drying

Explosion-puff drying is based on the development of a highly porous structure in fruit and vegetable materials, which increases greatly the drying rate of the product. The wet food material is dehydrated by conventional convective drying to about 25% moisture and then heated in a rotating cylindrical vessel ("gun") until a high pressure is developed (2–4 bar). The pressure is released instantly, producing a puffed product, which is dried fast to the desired moisture content in a conventional dryer. The dehydrated porous product has improved rehydration properties, an important quality factor in many food products.

11.7.18.3 Foam-Mat Drying

Foam-mat drying is used in small scale for the drying of sensitive food products, such as concentrated fruit juices, fruit purees, and food slurries. The fluid food is foamed by incorporating a gas in a special mixer, using a foam stabilizer. The foamed material is applied as a thin film of 1.5 mm on a perforated tray or belt, and it is dried at moderate temperatures and air velocities. Very fast drying is achieved, e.g., 15 min at 70°C, and the product has a porous structure, which improves its rehydration properties. The operating cost of foam-mat drying is lower than vacuum drying, but, for commercial applications, large spray dryers are more cost effective.

11.7.18.4 Acoustic Dryers

Acoustic or sonic dryers can improve the drying rate of various food materials. Low frequency sound waves increase considerably the heat and mass transfer rates at the particle/air interface. The short drying times, achieved by sonic drying, improve the product quality, such as color, flavor, and retention of volatile aroma components. Food liquids of 5%–75% total solids have been dried to low moistures, e.g., citrus juices and tomato paste.

11.8 ENERGY EFFICIENCY IN DRYING

Energy represents the major part of the operating cost of industrial dryers (about 60%). Large amounts of energy are used for drying food and agricultural materials.

Energy is used mainly for the evaporation of free water, desorption of sorbed water, or sublimation of ice in freeze drying. Thermodynamically, evaporation of free water requires 2.26 MJ/kg at 100°C, 2.36 MJ/kg at 60°C, and 2.50 MJ/kg at 0°C. The heat of sublimation of ice at 0°C is 2.84 MJ/kg, and higher energies are required for desorption of water bound on food biopolymers. In addition, energy is required for sensible heating of the food material, the dryer, and the exhaust air, and for mechanical movement of the process air (operation of fans). Thus, the total energy consumption varies in the range of 3.2–4.5 MJ/kg water in continuous convective dryers. It is higher in batch dryers (4–6 MJ/kg water) and it may reach 10 MJ/kg water in vacuum and freeze dryers, where energy-consuming vacuum pumps and refrigerated condensers are needed (Strumillo and Kudra, 1987).

11.8.1 Heat Sources for Drying

The heat required for drying is provided mostly by direct heating with fuels (gas or oil) or saturated steam. In some small-size applications, heat is supplied in the form of hot water, infrared radiation, or microwave energy. Combustion of fuel (natural gas, liquefied petroleum gas [LPG], or fuel oil) is the most economical energy source for drying applications. Saturated steam and hot water are more expensive than direct fuel combustion because heat exchangers and condensers are needed, but they are preferred in some cases, when contamination of the food material with combustion gases may not be acceptable.

Natural gas, used for direct heating of the dryer, is suitable for agricultural and food drying. High temperatures are obtained, which increase substantially the drying rate and shorten the drying time. The heating value of natural gas is 37.2 MJ/m^3, propane 50.4 MJ/kg, LPG 46 MJ/m^3, and fuel oil 41.7 MJ/kg.

Direct heating by combustion of natural gas and LPG fuels produces significant amounts of water vapor in the flue gases. Stoichiometric calculations indicate that flue (combustion) gases would contain 19% by volume of water vapor, but in practice lower concentrations are obtained, due to the use of excess air. As a result, the capacity of the heating air/gases to remove water in the dryer is reduced, and a low moisture content may not be reached easily for some food materials, which may require finish-drying with low-humidity air, e.g., in bin dryers.

11.8.2 Heat Recovery

Recirculation of the exhaust air/combustion gases from the dryer can recover part of the heat rejected to the environment. Convective dryers use some form of recirculation, recovering only the sensible heat, since it is difficult to recover the latent heat of evaporation of water in the exhaust gases.

The theoretical thermal efficiency n of a convective dryer without recirculation is

$$n = \frac{(T_2 - T_3)}{(T_2 - T)} \quad (11.47)$$

where T, T_2, and T_3 are the ambient, inlet, and exit air temperatures, respectively.

The thermal efficiency of a dryer with recirculation is given by the following equation:

$$n = \frac{(T_2 - T_3)}{[(T_2 - T_3) + (1 - w)(T_3 - T)]} \quad (11.48)$$

where w (–) is the ratio of recirculated to total air flow.

It is evident that the thermal efficiency increases as the recirculation ratio *w* is increased. However, the amount of recirculation is limited by the increase in humidity of the air stream, which reduces the drying rate of the material. Part of the exhaust heat can be recovered by heating the inlet air to the dryer, using some type of heat exchanger, e.g., thermal wheel, pipe, or plate heat exchanger.

A number of efficient dryers have been developed, which can operate using less energy than conventional dryers. They operate mostly at high temperatures; however, they have not found wide applications yet in the food processing industry.

The superheated steam dryer is an energy-saving unit, which is used commercially to dry materials that can tolerate high temperatures, such as sugar beet pulp, and wastewater sludge. The superheated steam or "airless" dryer operates with superheated steam, which is heated by fuel gas and is recirculated in the dryer at about 100°C, removing the moisture from the wet product. The water vapors are condensed in a heat exchanger, which heats cold water to about 95°C. The dryer can operate at atmospheric pressure either batch or continuously. Stratification of the steam at the upper layer takes place, due to the difference in densities of air and steam, preventing their mixing in the dryer [(air) > ρ (steam)].

APPLICATION EXAMPLES

Example 11.1

Describe the physical processes and derive the equations of adiabatic humidification and wet bulb thermometry of air–water mixtures.

Solution

Figure 11.23 shows diagrams of the processes of adiabatic humidification and wet bulb thermometry.

The enthalpy H (kJ/kg db–dry basis) of air–water mixture is calculated from the following equation:

$$H = C_p(T - T_o) + Y\Delta H_s \tag{11.49}$$

where

T is the (dry bulb) temperature
T_o is a reference temperature, usually taken as $T_o = 0°C$
Y (kg/kg db) is the absolute humidity
ΔH_s is the heat of evaporation at the reference temperature (0°C), approximately $\Delta H_s = 2.5$ MJ/kg

The specific heat of the air–water vapor mixture C_p (kJ/kg K) is calculated from the specific heats of the components, using Equation 11.28. For engineering calculations C_p is estimated from the simplified equation:

$$C_p = 1 + 2Y \tag{11.50}$$

where the specific heat of dry air is taken as 1 kJ/(kg K) and of water vapor as 2 kJ/(kg K).

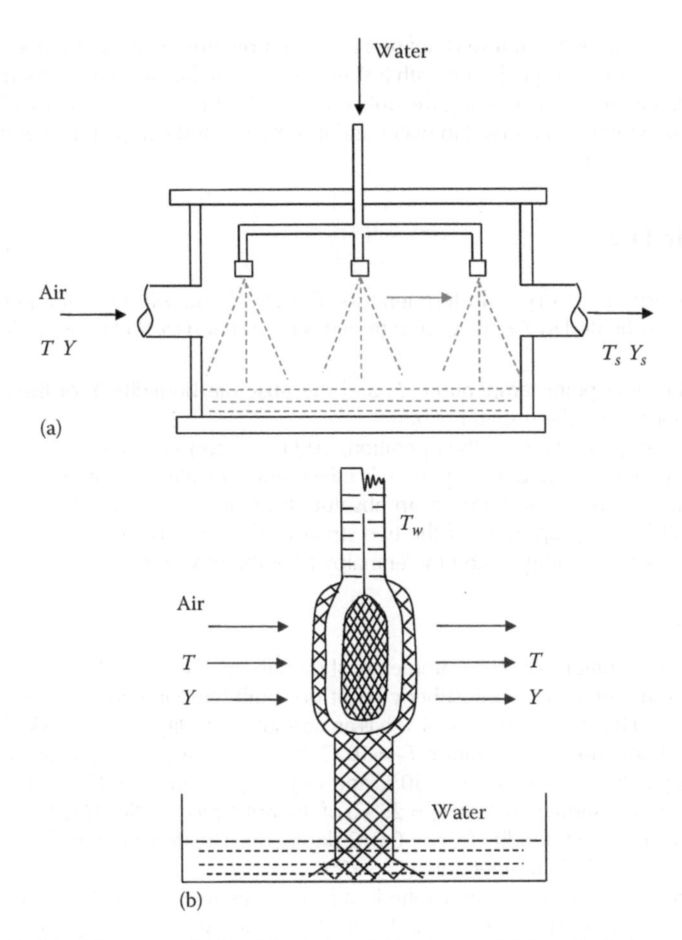

FIGURE 11.23 Diagrams of (a) adiabatic humidification and (b) wet bulb thermometer.

In the process of adiabatic humidification, the enthalpy of the air–water vapor mixture leaving the chamber must be equal to the heat of evaporation of water:

$$C_p(T - T_s) + Y\Delta H_s = Y_s\Delta H_s \tag{11.51}$$

or

$$\frac{(Y - Y_s)}{(T - T_s)} = \frac{-C_p}{\Delta H_s} \tag{11.52}$$

Equation 11.52 defines the saturation temperature T_s, and it is similar to Equation 11.24, which defines the wet bulb temperature T_w. Thus, for air–water vapor mixtures, the wet bulb temperature is equal to the saturation temperature ($T_w = T_s$). It should be noted that both Equations 11.24 and 11.52 represent straight lines on the psychrometric chart (Y, T) with a negative slope of ($-C_p/\Delta H_s$).

The wet bulb temperature is defined as the temperature of a wet thermometer, which is in thermal equilibrium with a stream of air flowing around the thermometer bulb (Figure 11.23). Usually, the bulb is covered with a special cloth (wick), the bottom of which is immersed in water and it is maintained completely wet during the measurement.

Example 11.2

Atmospheric air at dry bulb temperature T = 20°C and wet bulb temperature T_w = 14°C is heated to T = 50°C, and then it is humidified to 100% RH. Calculate

(a) The dew point temperature T_d and the absolute humidity Y of the atmospheric and the humid hot air.
(b) The required heat for the operation, in kJ/m^3 atmospheric air.
(c) The heated humid air of part (a) is mixed adiabatically with atmospheric air (same initial conditions) to an absolute humidity of 0.013 kg/kg db. What will be the proportion of the two streams, the temperatures (T, T_d, T_w), the absolute humidity Y, and the enthalpy H of the mixture?

Solution

(a) The operations of this example are shown in the Diagram 11.24 of the psychrometric chart. Atmospheric air at dry bulb temperature T_o = 20°C and wet bulb temperature T_w = 14°C has an absolute humidity Y_o = 0.0075 kg/kg db and dew point temperature T_d = 10°C. The air is heated at constant Y until (dry bulb) temperature T_1 = 50°C and wet temperature T_w = 25°C. Adiabatic humidification (constant T_w = 25°C) of the hot air to 100% RH will increase the absolute humidity to Y_1 = 0.0185 kg/kg db. The dew point will be equal to the wet bulb temperature (T_d = 25°C).
(b) The enthalpy of the atmospheric air increases from the initial conditions H_o = 39 kJ/kg db to H_1 = 66 kJ/kg db. Therefore, $\Delta H = H_1 - H_o = 66 - 39 =$ 27 kJ/kg db.

    The atmospheric air at the initial conditions has a specific volume of 0.84 m^3/kg db, and the heat required will be 27/0.84 = 32 kJ/m^3.
(c) The proportion of material between the two air streams can be calculated by material balance. If the mass fraction of the heated humid air is x, the atmospheric air mass fraction will be (1 – x). Moisture balance is expressed by the equation,

$$0.0185x + 0.0075(1 - x) = 0.0130$$

from which, x = 0.5, i.e., equal amounts of the two streams.

The enthalpy of the air mixture H is calculated as follows:

$$H = 66 \times 0.5 + 39 \times 0.5 = 52.5(1 - x) = 66 \times 0.5 + 50 \times 0.5 = 58.0 \text{ kJ/kg db}$$

On the psychrometric chart, the point (conditions D) of the mixture is found as the intersection of Y = 0.013 kg/kg db and H = 52.5 kJ/kg db. The same conditions (point D) of the mixture can be found as an intersection of the horizontal line Y = 0.013 and the line (AC) connecting the conditions (points) of the two components.

The air mixture (point D) has a (dry bulb) temperature $T = 22°C$, a wet bulb temperature $T_w = 20°C$, and a dew point $T_d = 18°C$.

Example 11.3

Estimate the time required to dry the following food materials from 80% to 10% wet basis. Drying will take place at a temperature of 60°C and the equilibrium moisture content is assumed to be zero ($X_e = 0$). Apply the simplified Page equation.

(a) Layer (slab) of porous starch material 1 cm thick; air-drying from both sides.
(b) Spherical particles 1 mm diameter of porous starch material; spray, flash, or fluidized-bed drying in a hot air stream.
(c) Spherical particles 1 mm diameter of nonporous starch/sugar material; spray, flash, or fluidized-bed drying in a hot air stream.

Solution

The moisture contents of the food material X (kg/kg db) before and after drying will be, $X_{o-} = 80/(100 - 80) = 4.0$ and $X = 10/(100 - 10) = 0.111$. The (effective) moisture diffusivities D of food materials given in Table 11.3 can be used in this example. The data refer to 30°C, and they must be converted to drying temperature (60°C), using the intergraded Arrhenius equation 11.36 and the respective activation energies.

For porous materials $D_o = 10 \times 10^{-10}$ m²/s, $E_D = 15$ MJ/kmol, $T_o = 303$ K, $T = 333$ K.

$$\ln\left(\frac{D}{D_o}\right) = -\left(\frac{E_D}{R}\right)\left(\frac{1}{T} - \frac{1}{T_o}\right) = -\left(\frac{15{,}000}{8.314}\right)\left(\frac{1}{333} - \frac{1}{303}\right) = 0.54$$

Therefore, $(D/D_o) = 1.7$ and $D = 17 \times 10^{-10}$ m²/s.

For nonporous materials $D_o = 1 \times 10^{-10}$ m²/s, $E_D = 40$ MJ/kmol, $T_o = 303$ K, $T = 333$ K.

$$\ln\left(\frac{D}{D_o}\right) = -\left(\frac{E_D}{R}\right)\left(\frac{1}{T} - \frac{1}{T_o}\right) = -\left(\frac{40{,}000}{8.314}\right)\left(\frac{1}{333} - \frac{1}{303}\right) = 1.4$$

Therefore, $(D/D_o) = 4$ and $D = 4 \times 10^{-10}$ m²/s.

TABLE 11.3
Typical Values of Effective Moisture Diffusivity D and Energy of Activation for Diffusion (E_D) of Food Materials (30°C)

Food Material	D, × 10^{-10} m²/s	E_D, kJ/mol
Highly porous	50	15
Porous	10	25
Nonporous starch/sugar	1	40
Nonporous protein/sugar	0.1	50

(a) According to Equation 11.33 the drying constant will be $-K = \pi^2 \times 17 \times 10^{-10}/0.01^2 = 9.86 \times 1.7 \times 10^{-5}$ 1/s, or $-K = 0.603$ 1/h. The material is dried from $X_o = 4$ to $X = 0.11$ kg/kg db moisture content. Assuming constant K value, the required drying time can be calculated from the Page equation 11.31, $t = [\ln(4/0.11)]/(0.603)]$ or $t = 5.97$ h, which is a reasonable drying time. It should be noted that the drying time could be reduced further by increasing the temperature or the bulk porosity, for example by puffing or extrusion.

(b) For the drying of spherical porous food particles of 1 mm diameter and $D = 17 \times 10^{-10}$ m²/s, the drying constant will be $-K = \pi^2 \times 17 \times 10^{-10}/(0.0005)^2 = 0.067$ 1/s. The drying time for $(X_o/X) = 4/0.11 = 36.4$ will be $t = [\ln(36.4)]/0.067 = 53.7$ s. This is a rather long residence time for a one-pass spray dryer. A recirculation spray dryer or a flash dryer may be used.

(c) For the drying of a nonporous starch/sugar spherical particle of 1 mm diameter, the diffusivity will be $D = 4 \times 10^{-10}$ m²/s, and the drying constant $-K = (9.86 \times 4 \times 10^{-10})/(0.0005)^2 = 0.015$ 1/s. The drying time will be $t = \ln(36.4)/0.015 = 240$ s = 4 min. This is a very long residence time for a spray or a flash dryer. A fluidized bed dryer is recommended.

Example 11.4

A convective air dryer is used to dry 1 ton/h of a food product from 50% to 10% moisture, wet basis. Atmospheric air at 25°C and 50% *RH* is heated to 95°C and passed adiabatically through the dryer, leaving at 70% *RH*. The air is heated by flue gases of LPG combustion.

Calculate

(a) The required amount of atmospheric air (m³/h)

(b) The required heat of drying in MJ/kg of removed moisture

(c) The required air and heat, if the dryer is operated with air recirculation; the air leaving the dryer at 70% *RH* is mixed with 50% atmospheric air (at 25°C and 50% *RH*) and is heated to 95°C, before reentering the dryer.

Solution

(a) The moisture content of the product before and after drying will be $X_1 = 50/50 = 1.00$ and $X_2 = 10/90 = 0.11$ kg water/kg solids db. The flow of dry food solids in the dryer will be $1000 \times 0.50 = 500$ kg/h, and the water removed during the drying operation will be $W = 500 \times (1.00 - 0.11)$ or $W = 445$ kg/h.

The dryer is assumed to operate by adiabatic humidification of the air (constant enthalpy).

The absolute humidities Y and enthalpies H of the air stream in the dryer, taken from the psychrometric chart, are as follows: atmospheric air T_o = 25°C, $Y_o = 0.01$ kg/kg db, $H_o = 51$ kJ/kg db; hot air entering the dryer T_1 = 95°C, $Y_1 = 0.01$ kg/kg db, $H_1 = 125$ kJ/kg db; air leaving the dryer T_2 = 40°C, $Y_2 = 0.034$ kg/kg db, $H_2 = 125$ kJ/kg db.

Moisture pick up in the dryer $\Delta Y = Y_2 - Y_1 = 0.034 - 0.01 = 0.024$ kg/kg db. Air required to remove 445 kg/h from the product in the dryer $445/0.024 = 18{,}541$ kg/h.

The specific volume of atmospheric air at 25°C and 50% RH is 0.86 m^3/kg. Therefore, the volume of atmospheric air required will be 18,541 × 0.86 = 15,945 m^3/h.

(b) The enthalpy increase of the air in the dryer will be ΔH = 125 − 51 = 74 kJ/kg db. The heat required for removal of water from the product in the dryer will be 18,541 × (74/1,000) = 1,372 MJ/h. Therefore, the heat required for drying will be 1,372/445 = 3.08 MJ/kg water.

It should be noted that the heat of evaporation of free water at the wet bulb temperature in the dryer (T_w = 35°C) is ΔH = 2.41 MJ/kg. Therefore, the thermal efficiency of the dryer would be (2.41/3.08) × 100 = 78%. The major portion of the energy loss is in the hot air leaving the dryer. Also, part of the removed water is moisture adsorbed on food components, which requires a higher energy than the heat of evaporation of free water.

(c) The dryer is assumed to operate adiabatically (constant wet bulb temperature T_w) with a mixture of 50% atmospheric air and 50% recycled air, heated to 95°C. Exit air at 70% RH.

The operation of the air dryer can be represented on the psychrometric chart, as shown in the diagram of Figure 11.24. The operation without recirculation,

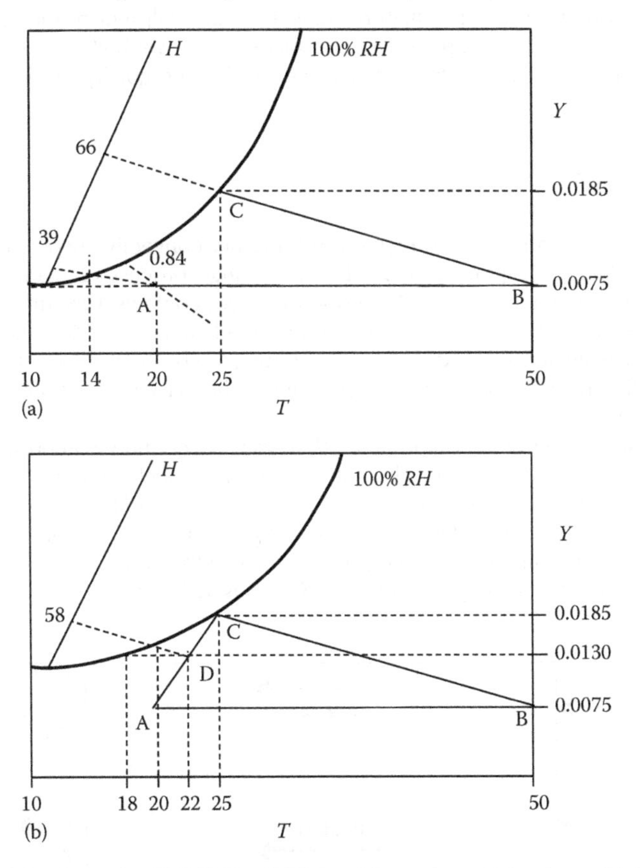

FIGURE 11.24 Operations of Example 2 on the psychrometric chart: (a) heating and humidification (ABC), (b) adiabatic mixing (ADC).

described in part (a) of this example, is represented by path (lines) ABC. The moisture pick up during this operation will be ΔY, the enthalpy increase will be ΔH, and the heat requirement will be $(\Delta H/\Delta Y)$. In this example, the simple (once through) operation requires $(\Delta H/\Delta Y) = 3.08$ MJ/kg water removed.

During recirculation, the absolute humidity of the air in the dryer increases gradually. If the air mixture entering the dryer is heated to the same (dry bulb) temperature (95°C) and if it leaves at the same humidity (70% *RH*), the moisture pick up ΔY and the required enthalpy ΔH will decrease gradually, as the recirculation continues. Due to the structure of the psychrometric chart, the ratio $(\Delta H/\Delta Y)$ tends to decrease, resulting in a reduction of heat requirement from 3.08 MJ/kg to about 2.80 MJ/kg, in this example. Thus, the thermal efficiency increases by air recirculation from $(2.41/3.08) \times 100 = 78\%$ to $(2.41/2.80) \times 100 = 86\%$.

The theoretical thermal efficiency n of an air convection dryer can also be estimated from the temperatures of the air at the entrance T_2 and the exit T_3 of the dryer, and the ambient temperature T, using Equation 11.47 or 11.48. In this example, for operation without air recirculation $T_2 = 95°C$, $T_3 = 40°C$, and $T = 25°C$. Equation 11.47 yields $n = (95 - 40)/(95 - 25) = 78\%$. For operation with 50% air recirculation ($w = 0.5$), $T_2 = 95°C$, $T_3 = 46°C$, and $T = 25°C$. Equation 11.48 yields $n = (95 - 46)/[(95 - 46) + 0.5 \times (46 - 25)] = 83\%$.

Operation of the dryer at high recirculation rates will reduce the drying rate and increase the drying time. It will also increase the wet bulb temperature and the residence time in the dryer, damaging possibly the quality of heat-sensitive products.

Example 11.5

Calculate the air and fuel requirements of a rotary convective dryer to dry 3 t/h of by-product peels/pulp, produced in an orange processing plant. The wet peels/pulp contain 15% TS and the dried product contains 10% moisture, wet basis. Because of the high evaporation duty, an efficient rotary dryer is selected. The dryer is heated by mixture of air and flue (combustion) gases of LPG fuel. A simplified diagram of the dryer, useful for preliminary calculations, is shown in Figure 11.25.

Atmospheric air at 20°C and 50% *RH* is mixed with flue (combustion) gases and enters the dryer at 100°C. The hot air flows in parallel and dries the material adiabatically (constant enthalpy), leaving the drier at 80% *RH*.

Calculate the required flow or air (m³/h) and LPG fuel (kg/h), (a) if the moisture in the flue (combustion) gases is neglected, and (b) if the moisture in the flue gases is considered. The heating value of the LPG fuel is 46 MJ/kg.

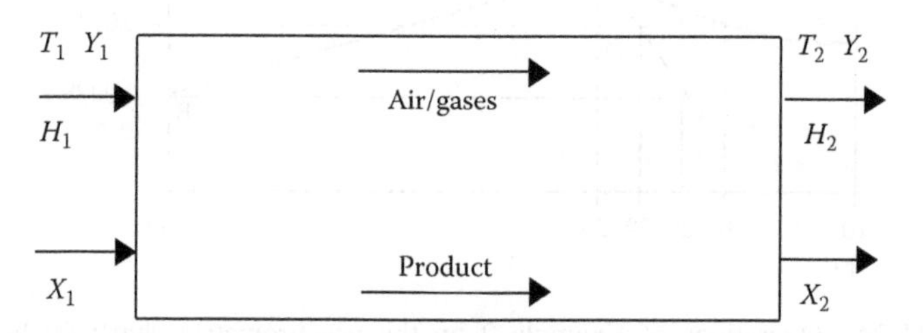

FIGURE 11.25 Diagram of the co-current rotary dryer of Example 11.5.

Solution

For approximate design estimation, a psychrometric chart is used to calculate the air and fuel requirements of the dryer.

(a) Assuming that atmospheric air at 20°C and 50% *RH* is used, its moisture content Y_o and enthalpy H_o, taken from the psychrometric chart (Appendix), will be, respectively, $Y_o = 0.0075$ kg/kg db and $H_o = 40$ kJ/kg db. The air is heated by mixing with the combustion gases to $T_1 = 100$°C and, neglecting the combustion water, the moisture content will remain constant at $Y_1 = Y_o = 0.0075$ kg/kg db, and the enthalpy will be $H_1 = 117$ kJ/kg db (Psychrometric chart). The dry air will be humidified adiabatically (constant H), as it passes through the dryer, leaving at 80% *RH* and air humidity $Y_2 = 0.035$ kg/kg db and the same enthalpy $H_1 = 117$ kJ/kg db.

The moisture pick up of the air in the dryer will be $\Delta Y = 0.0350 - 0.0075 = 0.0275$ kg/kg db, and the enthalpy difference will be $\Delta H = 117 - 40 = 77$ kJ/kg db.

The moisture content of the product before and after drying will be, respectively, $X_1 = 85/15 = 5.7$ and $X_2 = 10/90 = 0.11$ kg water/kg db. The dry solids flow rate in the dryer will be $3000 \times 0.15 = 450$ kg db/h. Therefore, the amount of water removed from the product in the dryer will be

$$W = 450 \times (5.7 - 0.11) = 2515 \text{kg/h}$$

The dry air needed to remove 2,515 kg/h water in the dryer will be 2,515/0.0275 = 91,454 kg db/h. Since the specific volume of the initial air (20°C and 50% *RH*) is 0.84 m^3/kg (psychrometric chart), the volumetric air flow in the dryer will be 91,454 × 0.84 = 76,821 m^3/h.

The total energy required for heating the air will be 91,454 × 77 = 7,042 MJ/h = 7,042/3,600 = 1.96 MW. If the heating value of the LPG is 46 MJ/kg, the amount of fuel gas needed will be 7,042/46 = 159 kg/h LPG.

(b) The presence of combustion water in the air/gas mixture may increase the requirements for process air and fuel. LPG consists of about 60% propane (44 kg/kmol) and 40% butane (58 kg/kmol).

The air requirement for the complete combustion of the LPG fuel is calculated as follows: Stoichiometric oxidation of 1 kmol of propane or butane to carbon dioxide and water requires 5 and 6.5 kmol of oxygen, respectively. In terms of mass, oxidation of 44 kg of propane requires 80 kg of oxygen, and 58 kg of butane requires 104 kg oxygen. Oxidation of 100 kg LPG mixture 60% propane and 40% butane will require 181 kg oxygen. Assuming that atmospheric air contains 20% of oxygen by mass, the amount of air required for oxidation of 100 kg of LPG will be 181 × 5 = 905 kg air or 9 kg air/kg LPG.

The amount of water vapor produced by the complete combustion (stoichiometric oxidation) of LPG is calculated as follows: 1 kmol (44 kg) of propane produces 4 kmol (72 kg) of water vapor. Also, 1 kmol (58 kg) of butane produces 5 kmol (90 kg) of water vapor. Thus, oxidation of 100 kg LPG mixture 60% propane and 40% butane will produce 132 kg water vapor or 1.32 kg water/kg LPG. The amount of combustion moisture introduced into the dryer will be 1.32 × 189 = 250 kg water/h, which is about 10% of the water evaporated from the product (2,515 kg water/h).

The combustion moisture will increase the moisture of the required air (estimated from the psychrometric chart) and may reduce its evaporative capacity ΔY. In order to maintain the desired evaporative capacity of the dryer, using the same (dry bulb) air temperature, the amount of air must be increased by about 10%. Thus, the amount of air will be 1.1 × 76,821 = 84,503 m^3/h. The required fuel consumption will be 1.1 × 159 = 175 kg LPG/h.

Example 11.6

Select a rotary convective dryer suitable for the drying operation of Example 11.5. Specify the dimensions of the dryer and estimate the power required.

Solution

The dryer should have an evaporative capacity of 2,515 kg water/h, requiring 80,000 m^3/h of atmospheric air, about 10% higher than calculated.

The size of the rotary dryer for this application can be selected from performance data of direct heated rotary dryers, found in the literature or provided by Equipment Suppliers.

The evaporative capacity of rotary dryers varies in the range of 30–100 kg water/m^2 h and the residence time is between 0.2 and 1 h (Table 11.4). Assuming an evaporation rate of 50 kg water/m^2 h, the required surface area of the dryer will be 2,515/50 = 50.3 m^2. A cylindrical dryer of internal diameter 1.80 m and length 10.0 m would have a wall surface area of A = 3.14 × 1.80 × 10.00 = 56.52 m^2, which is satisfactory for this application (about 10% oversize).

The cross-sectional area of this cylindrical dryer will be A = 3.14 × (1.80)^2/4 = 2.54 m^2 and the air velocity in the dryer u = 80,000/[(2.54) × (3,600)] = 8.7 m/s, which is a reasonable velocity for drying.

From the technical literature, the fan power of this dryer (to move the required air) would be about 60 kW and the motive power (to rotate the dryer) about 6 kW.

TABLE 11.4
Characteristics of Food Drying Operations and Equipment

Dryer Type	Product Form	Product Temperature, °C	Evaporative Capacity, kg/m^2 h	Residence Time
Sun drying	Pieces	Ambient	—	10–20 days
Bin or Silo	Pieces, grains	30–50	—	1–3 days
Tray	Pieces	40–60	0.2–2	3–10 h
Tunnel	Pieces	50–80	5–15	0.5–3 h
Conveyor belt	Pieces	50–80	5–15	0.5–3 h
Rotary	Grains, granules	60–100	30–100	0.2–1 h
Drum	Sheet	80–110	5–30	10–30 s
Fluid bed	Grains, granules	60–100	30–90	2–20 min
Pneumatic or flash	Grains, granules	60–120	10–100[a]	2–20 s
Spray	Powder	60–120	1–30[a]	10–60 s
Vacuum-freeze	Pieces	−10–20	1–7	5–24 h

[a] kg/m^3 h.

Notes

1. The psychrometric model equations, described in this chapter, can be used in calculations involving air–water vapor mixtures, instead of the psychrometric chart.
2. Complex drying problems, involving several calculations, can be solved using the computer Excel®/Speadsheet method described in Chapter 16.

PROBLEMS

11.1 The specific volume of air–water vapor mixtures V_a (m^3/kg) at atmospheric pressure can be calculated from the following equation:

$$V_a = 22.4\left(\frac{T}{273}\right)\left(\frac{1}{M_a}+\frac{Y}{M_w}\right) \tag{11.53}$$

where

T (K) is the (dry bulb) temperature
Y (kg/kg db) is the absolute humidity
M_a, M_w (kg/kmol) are the molecular weights of air and water, respectively

(a) Derive the above equation based on the perfect gas law.
(b) Calculate the specific volume of air at (30°C, 50% *RH*) and (70°C, 30% *RH*) and compare with the values obtained from a psychrometric chart.

11.2 Atmospheric air at 25°C and 50% *RH* is heated to 70°C and then it is humidified adiabatically until saturation. The saturated air is heated again to 70°C and humidified adiabatically to 80% *RH*. The humid air is heated again to 70°C.

(a) Show the process on the psychrometric chart.
(b) Estimate the moisture pick up by the air in kg/m^3 initial air.
(c) Estimate the enthalpy change of air in kJ/m^3 initial air.

11.3 The drying time constant t_c (h) of a food product can be estimated from the empirical model of Equation 11.37 $t_c = c_0\, d^{c1}\, u^{c2}\, T^{c3}\, Y^{c4}$ which is based on the material size d (m), air velocity u (m/s), air temperature T (°C), and absolute humidity Y (kg/kg db).

The five adjustable empirical constants were determined experimentally from drying tests of a food material as follows:

$$c_0 = 5000,\ c_1 = 1.40,\ c_2 = -1.65,\ c_3 = -0.25,\ \text{and}\ c_4 = 0.12$$

(a) Which are the two most important parameters of Equation 11.37?
(b) If the drying mechanism was entirely molecular diffusion, what would be the value of the empirical constant c_1?
(c) Estimate the drying time constant t_c and the drying constant K for the following parameters of Equation 11.37: $d = 0.015$ m, $u = 5$ m/s, $T = 60$°C, and $Y = 0.025$ kg/kg db.

11.4 Estimate the drying time t (h) for a food product from 90% to 10% moisture (wet basis) and equilibrium moisture content $X_e = 0.03$ kg/kg db. Under the drying conditions of this problem, the drying time constant t_c, estimated in problem (11.3) is applicable.

Use (a) the simple Page equation, and (b) the generalized Page equation with empirical constant $n = 0.65$.

11.5 A wet food product containing 90% moisture (wet basis) is dried in a stream of hot air of dry bulb temperature 60°C, wet bulb temperature 34°C, and air velocity 5 m/s. The product is placed in a tray of 1 m^2 flat surface and 5 cm depth, and it dried initially at a constant rate period until a moisture content of 50% (wet basis) is reached, followed by the falling rate period.

Calculate

(a) The heat and mass transfer coefficients during the constant rate period, assuming that Reynolds number is 20,000 and that the literature data on mass and heat factors are applicable.

The mean properties of air in the system are, density $\rho = 1$ kg/m^3, specific heat $C_p = 1$ kJ/kg K, and viscosity $\eta = 0.018$ mPa s.

(b) The drying rate during the constant rate period in kg/m^2 h.

(c) The duration of the constant rate period and the amount of water removed.

11.6 A fluidized bed dryer is used to dehydrate shelled peas from 75% to 6%, wet basis. The equilibrium moisture content is 3%, dry basis. The air temperature is 60°C and the air velocity is high enough to fluidize the pea bed and remove effectively the moisture from the dryer.

Calculate

(a) The drying time constant, assuming that the moisture transport in the peas is by molecular diffusion. Mean diameter of peas 8 mm, moisture diffusivity $D = 10 \times 10^{-10}$ m^2/s.

(b) The residence time in the fluidized bed dryer (drying time), using the simplified Page equation.

11.7 A solar dryer is designed for drying 1 t/day of apricots during the summer months, when solar radiation is high enough for this operation. The ripe fruit is cut into halves to increase the drying rate. It is normally dried from 80% to about 25% moisture, wet basis. The equilibrium moisture content of the product is 15%, wet basis.

The solar dryer would consist of a number of flat solar collectors, connected in parallel, through which atmospheric air, blown by a fan, is heated to a temperature of about 70°C and introduced to a batch tray dryer. The trays allow the flow of hot air though the product.

Calculate

(a) The minimum drying rate constant t_c and the drying constant K so that the desired drying of the product is achieved within 7 h of a day, during which the insolation (incident radiation of solar energy) is on the average 0.5 kW/m^2. The product is assumed to dry in the falling rate period and the simplified Page equation is applicable.

(b) The required surface area (m^2) of the solar collectors. The energy requirement in the air dryer is 3.5 MJ/kg water removed. The solar collector has a thermal efficiency of 75%.

11.8 There are considerable differences in energy efficiency of the various industrial dryers.

Explain why freeze-drying and vacuum drying require more than 5 kJ/kg water removed, contrasted to 3–4.5 kJ/kg water removed in convective air drying.

LIST OF SYMBOLS

a_w	—	Water activity
A	m^2	Surface area
C_p	kJ/(kg K)	Specific heat
d	m	Diameter
D	m^2/s	Mass diffusivity
E	kJ/mol	Energy of activation
g	m/s^2	Acceleration of gravity (9.81)
h	W/(m K)	Heat transfer coefficient
h_M	kg/(m^2 s)	Mass transfer coefficient
H	kJ/kg	Enthalpy
j_H	—	Heat transfer factor
j_M	—	Mass transfer factor
J	kg/(m^2 s)	Mass flux
k_c	m/s	Mass transfer coefficient
K	1/s	Drying constant
L	m	Length, thickness
m	kg/s	Mass flow rate
m	—	Ratio of mol. weights [M(air)/M(water)]
M	kg/kmol	Molecular weight
n	—	Efficiency
Nu	—	Nusselt number [$(hd)/\lambda$]
q, Q	W	Heat flow rate
P	Pa	Pressure
Pr	—	Prandtl number [$(C_p\eta)/\lambda$]
r	m	Radius
R	kJ/(kmol K)	Gas constant (8.314)
R	—	Moisture ratio, drying
RH	—	Relative humidity
Re	—	Reynolds number [$(d\ u\ \rho)/\eta$]
Sc	—	Schmidt number [$\eta/(\rho\ D)$]
Sh	—	Sherwood number [$(k_c\ d)/D$]
St	—	Stanton number [$Sh/(Re\ Sc)$]
t	s	Time

T	°C or K	Temperature
u	m/s	Velocity
U	W/(m^2 K)	Overall heat transfer coefficient
V	m^3	Volume
x	m	Wall thickness
X	kg/kg db	Moisture content, dry basis
W	—	Moisture content, wet basis (%)
ΔH	MJ/kg	Heat of evaporation
Δp	Pa	Pressure difference
ΔT	°C or K	Temperature difference
λ	W/(m K)	Thermal conductivity
ρ	kg/m^3	Density

REFERENCES

Barbosa-Canovas, G.V. and Vega-Mercado, H. 1996. *Dehydration of Foods*. Chapman & Hall, New York.

Baker, C.G.J. ed. 1997. *Industrial Drying of Foods*. Blackie Academic and Professional, London, U.K.

Chen, X.D. and Mujumbar, A.S. eds. 2008. *Drying Technologies in Food Processing*. Wiley-Blackwell, New York.

Greensmith, M. 1998. *Practical Dehydration*, 2nd edn. Woodhead Press, Cambridge, U.K.

Kemp, I. 1999. Progress in dryer selection techniques. *Drying Technol.* 17: 1667–1680.

Kiranoudis, C.T., Maroulis, Z.B., and Marinos-Kouris, D. 1994. Simulation of drying processes: An integrated computer-based approach. *Chem. Eng. Res. Des. Trans IChE.* 72(A3): 307–315.

Liapis, A.I. and Bruttini, R. 1995. Freeze drying. In: *Handbook of Industrial Drying*, A.S. Mujumdar, ed., 2nd edn., vol. 1. Marcel Dekker, New York.

MacCarthy, D., ed. 1986. *Concentration and Drying of Foods*. Elsevier Applied Science, New York.

Maroulis, Z.B. and Saravacos, G.D. 2003. *Food Process Design*. CRC Press, Boca Raton, FL.

Masters, K. 1991. *Spray Drying Handbook*, 5th edn. Longman, London, U.K.

Oetjen, G.W. 1999. *Freeze-Drying*. Wiley, New York.

Perry, R.H. and Green, D. 1997. *Chemical Engineers' Handbook*, 7th edn. McGraw-Hill, New York.

Saravacos, G.D. and Kostaropoulos, A.E. 2002. *Handbook of Food Processing Equipment*. Springer, New York.

Saravacos, G.D. and Maroulis, Z.B. 2001. *Transport Properties of Foods*. Marcel Dekker, New York.

Strumillo, C. and Kudra, T. 1987. *Drying: Principles, Applications and Design*. Gordon and Breach, New York.

Van Arsdel, W.B., Copley, M.J., and Morgan, A.I. eds. 1973. *Food Dehydration*, vols. 1–2. Avi Publ., Westport, CT.

VDMA. 1999. *Product Directory of Dryers and Drying Systems. Frankfurt/Main*, Verband Deutscher Maschinen und Apparatenbau (VDMA), Germany.

12 Refrigeration and Freezing Operations

12.1 INTRODUCTION

Refrigeration is used for the cooling or freezing of foods, mainly for preservation (extension of storage life) of fresh or processed foods by reducing the activity of microorganisms, enzymes, and chemical and biological reactions. Typical applications include the preservation of fresh products through precooling and the cold storage of fruits and vegetables, meat, and fish, and the freezing of meat, fish, and meals.

Cooling of liquid and solid foods is accomplished using various processes and equipment based on the principles and practices of heat transfer. Refrigerated foods are stored for several weeks at temperatures near 0°C in cold rooms or containers/trucks, sometimes in controlled atmosphere (Cleland, 1990).

Freezing of foods is based on the rapid crystallization of free water into ice, and storage of the frozen foods for several months at temperatures near −18°C. Frozen foods are usually packaged to prevent moisture loss and quality changes during storage (Heldman, 1992; Saravacos and Kostaropoulos, 2002).

The refrigeration and freezing equipment that is used for foods may be classified as (a) refrigeration producing equipment and (b) refrigeration using equipment.

In the mechanical compression system, low temperatures are created by the evaporation of refrigerants. A "pump" or compressor provides a pressure reduction over a refrigerant (which is a liquid evaporating at relatively low temperature), so that temperature reduction, through evaporation of the liquid, takes place.

12.2 COLD GENERATION

The basic methods for reducing temperature are (1) mechanical compression of refrigerants, (2) evaporation of cryogenic fluids, (3) melting of ice, and (4) vacuum evaporation. Mechanical compression is the most important method, followed by the evaporation of cryogenic fluids. Ice is used in fish preservation and in some smaller applications. Evaporation of free water under vacuum may be applied to cooling of some leafy vegetables (ASHRAE, 1993).

12.2.1 Mechanical Compression

The conventional mechanical refrigeration system (heat pump) consists basically of a compressor, a condenser, an expansion (throttling) valve, and an evaporator. The system is integrated with the necessary piping, control, and electrical components.

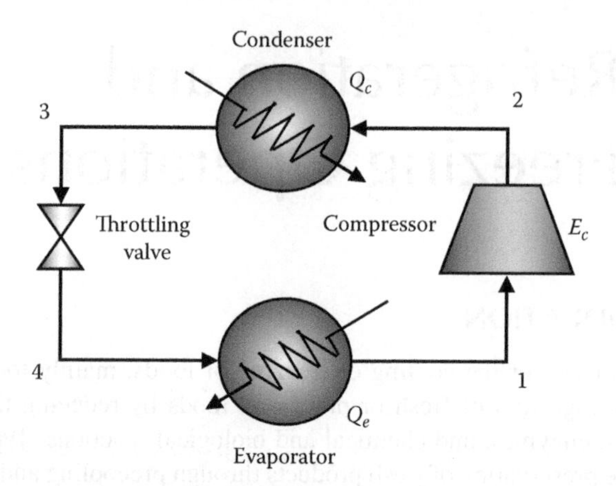

FIGURE 12.1 Mechanical compression refrigeration system. (Modified from Maroulis, Z.B. and Saravacos, G.D., *Food Process Design*, CRC Press, Boca Raton, FL, May 9, 2003, fig. 5.1.)

Refrigeration is produced by compression/expansion of a suitable refrigerant, such as ammonia or fluorinated hydrocarbons (Freons *R22* and *R134A*).

12.2.1.1 Refrigeration Cycles

The thermodynamic principles of compression i.e., the expansion processes of gases, are reviewed in Chapter 3. Figure 12.1 shows the diagram of a single-stage mechanical refrigeration system, while Figure 12.2 shows a thermodynamic diagram of pressure–enthalpy for a given refrigerant.

In compression refrigeration systems, special fluid refrigerants are used, which evaporate and condense within practical ranges of pressure and temperature.

The processes of evaporation (4–1) and condensation (2–3) are isothermic (constant temperatures T_e and T_c, respectively), while the expansion (throttling) process (3–4) is adiabatic or isenthalpic (constant enthalpy). The compression process (1–2) is considered isentropic (constant entropy S). The thermal (refrigeration) load Q_e (kW) is used to evaporate the liquid refrigerant F_r (kg/s) in the evaporator. The vapors are compressed to a higher pressure, absorbing mechanical power E_c (kW), and subsequently they are condensed in a water- or air-cooled condenser, releasing the thermal load of condensation, $Q_c = Q_e + E_c$.

The conventional (single stage) refrigeration cycle of Figure 12.2 is used for temperatures down to −20°C. Typical temperatures for single-stage refrigeration cycles are evaporator $T_e = -15$°C and condenser $T_c = 25$°C. In practical applications, the vapors leaving the evaporator are superheated by, e.g., 5°C before entering the compressor, and the liquid refrigerant is subcooled by, e.g., 5°C before expanding into the evaporator. The compressor operates more smoothly with superheated vapors than a mixture of saturated vapors and liquid droplets. Subcooling increases the refrigeration capacity.

The mechanical refrigeration cycle can be represented in a temperature (T)–entropy (S) diagram (Figure 12.3). In this diagram, the compression process (1–2) is isentropic (constant entropy S). The superheated vapors are cooled to the saturation

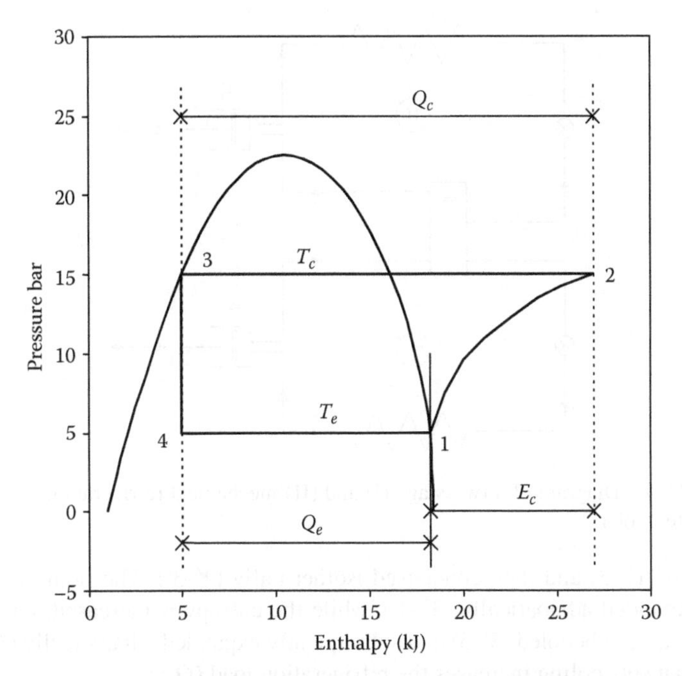

FIGURE 12.2 Pressure–enthalpy diagram of a single-stage refrigeration cycle. (Modified from Maroulis, Z.B. and Saravacos, G.D., *Food Process Design*, CRC Press, Boca Raton, FL, May 9, 2003, fig. 5.2.)

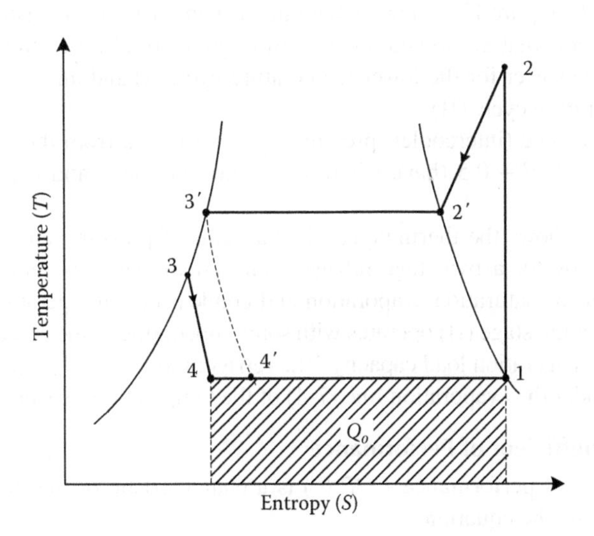

FIGURE 12.3 Temperature (T)–entropy (S) diagram of a single stage mechanical refrigeration system.

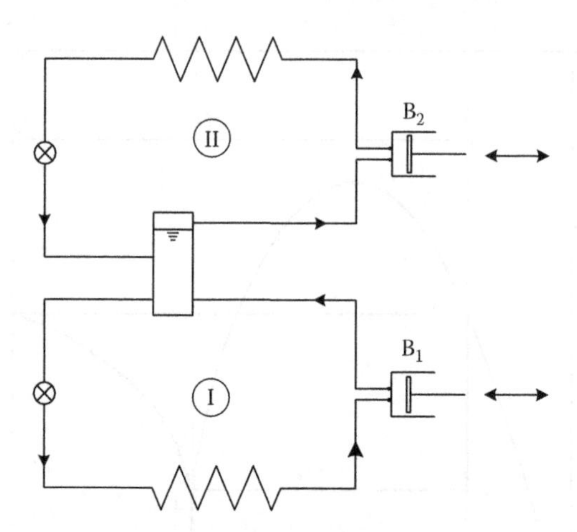

FIGURE 12.4 Diagram of a two-stage (I) and (II) mechanical refrigeration system joined with an intercooler.

temperature (2–3) and then condensed isothermally (2′–3′). The liquid condensate can be expanded adiabatically (3′–4′), while the entropy is increased. Or, the condensate can be subcooled (3′–3) and subsequently expanded adiabatically (3–4). It is evident that subcooling increases the refrigeration load (Q_o).

The refrigeration load of the process (Q_o) is the area under the evaporation line (4–1), calculated from the relation $Q_o = (\Delta H_e)(F_r)$, where ΔH_e (kJ/kg) is the enthalpy of evaporation, which is related to the entropy difference, ΔS (kJ/kg K) by the relationship $\Delta H_e = (\Delta S)\,T$, where T (K) is the absolute temperature. Entropy is discussed in Chapter 3.

A two-stage compression refrigeration cycle is used to obtain lower temperatures, e.g., $T_e = -40°C$. Figure 12.4 shows a diagram of such a system, consisting of two single-stage systems that are joined together through a closed-type intercooler, which serves as a condenser for the lower temperature cycle (I) and an evaporator for the higher temperature cycle (II).

The intermediate (intercooler) pressure P_m is estimated from the empirical relation, $P_m = (P_1\, P_2)^{1/2} + 0.5$ (bar), where P_1, P_2 are the lower and higher pressures, respectively.

Figure 12.5 shows the thermodynamic diagrams of pressure–enthalpy and temperature–entropy for a two-stage refrigeration system. The low temperature stage (I) operates at the saturation evaporation and condensation temperatures, while the higher temperature stage (II) operates with some subcooling of the condensate, which increases its refrigeration load capacity. The refrigeration load ($Q_o = Q_e$) is expressed by the area under the evaporation line (7–1) of the temperature–entropy diagram.

12.2.1.2 Coefficient of Performance

The coefficient of performance COP (–) is a characteristic of the thermodynamic cycle, defined by the equation

$$COP = \frac{Q_e}{E_c} \tag{12.1}$$

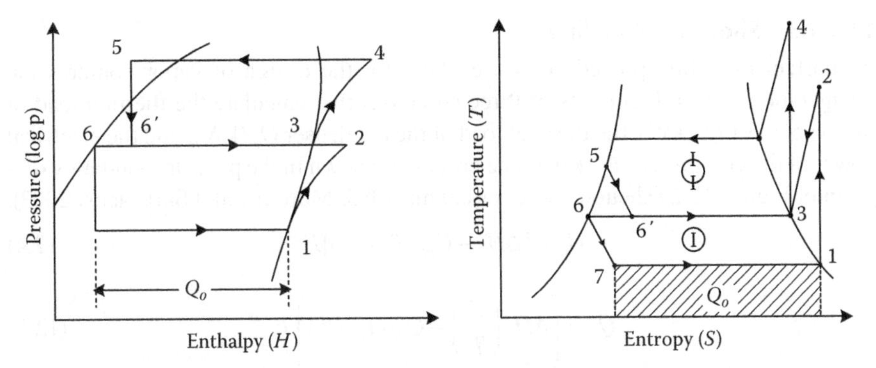

FIGURE 12.5 Enthalpy (*H*) and entropy (*S*) diagrams of a two-stage mechanical refrigeration system.

where

Q_e (kW) is the thermal load transported in the evaporator
E_c (kW) is the input mechanical power

In a thermodynamic cycle, operating between temperatures T_e and T_c (Figure 12.2), the maximum theoretical (Carnot) efficiency is given by the equation (Chapter 3),

$$COP = \frac{T_e}{(T_c - T_e)} \tag{12.2}$$

The *COP* in practice is smaller than the theoretical and is estimated from the equation,

$$COP = \frac{T_e}{(T_c - T_e)} - \frac{(C_{pL}T_e)}{\Delta H_r} \tag{12.3}$$

The first term of Equation 12.3 is the ideal Carnot coefficient of performance, which is independent of the refrigerant being used (Chapter 3). The second term contains the contribution of the refrigerant used.

It should be noted that *COP* is a dimensionless number with values higher than 1. A typical value *COP* = 4.5 means that for an input of mechanical power $E_c = 1\,\text{kW}$, the compressor transports a refrigeration load $Q_e = 4.5\,\text{kW}$ from the evaporator to the condenser, i.e., there is a net power gain of $4.5 - 1.0 = 3.5\,\text{kW}$.

The thermal loads to the evaporator (Q_e) and condenser (Q_c) represent the theoretical energy transferred, neglecting any losses to the environment or due to inefficient operation. The theoretical compression load (E_c) is lower than the mechanical power (*P*) by the efficiency (*n*), according to the relation, $P = E_c/n$. The mechanical efficiency is lower than 1.0, with a typical value $n = 0.80$.

12.2.1.3 Shortcut Modeling

A shortcut modeling procedure is used here for the design of vapor compression refrigeration cycles. It consists of three equations that calculate the thermal load of the evaporator Q_e (kW), the thermal load of the condenser Q_c (kW), and the electrical power of the compressor E_c (kW). The process is shown in the pressure–enthalpy diagram of Figure 12.2 (Shelton and Grossmann, 1985; Maroulis and Saravacos, 2003).

$$Q_e = [\Delta H_r - C_{pL}(T_c - T_e)]F_r \tag{12.4}$$

$$Q_c = \left[\Delta H_r\left(\frac{T_c}{T_e}\right) - C_{pL}(T_c - T)\right]F_r \tag{12.5}$$

$$E_c = \Delta H_r\left[\frac{(T_c - T_e)}{T_e}\right]F_r \tag{12.6}$$

where

T_e is the refrigerant evaporating temperature (K)
T_c is the refrigerant condensing temperature (K)
F_r is the refrigerant flow rate (kg/s)
E_c is the compressor power (kW)
ΔH_r is the latent heat of vaporization of refrigerant (kJ/kg)
C_{pL} is the specific heat of saturated liquid refrigerant [kJ/(kg K)]

In the English system of units, the refrigeration load is sometimes expressed in "tons of refrigeration," defined as the heat required to melt 1 short ton (2000 lb) of ice in 24 h. It can be shown that 1 ton (refrig) = 3.51 kW.

12.2.1.4 Refrigerants

The refrigerants should be nontoxic, nonflammable, noncorrosive, and chemically stable.

Table 12.1 shows the properties of three common refrigerants used in compression refrigeration, and two cryogenic fluids used in freezing operations of foods.

The hydrochlorofluorcarbons (*HFC*) *R22* (chlorodifluoromethane) and *R134A* (tetrafluorethane), called also Freons, are used in small and medium units, while *R717* (ammonia) is preferred in large installations, because of its high heat of evaporation. Refrigerant (Freon) *R12* (dichlorodifluoromethane), used extensively in the past, has recently been replaced by *HFC* (Freon) *R134A*.

Carbon dioxide and nitrogen are used as cryogens for fast freezing of food products.

Figure 12.6 shows vapor pressure–temperature diagrams of the three refrigerants (*R22*, *R124A*, and *R717*) and one cryogen (carbon dioxide), used mainly in food refrigeration and freezing operations. All refrigerants are gases at atmospheric pressure that can be liquefied at moderate pressures, and they can transfer heat from low to higher temperatures as the main components of the compression refrigeration systems (Table 12.2).

Thermodynamic diagrams of pressure–enthalpy of refrigerants, useful in refrigeration calculations, are published in the literature.

TABLE 12.1
Thermodynamic Properties of Some Refrigerants and Cryogens

Refrigerant	*BP* (°C)	ΔH_e (kJ/kg)	C_{pL} (kJ/(kg K))	T_c (°C)	P_c (Bar)
R22 (CHClF$_2$)	−41.0	234	1.30	96	50
R134A (CHFCF$_3$)	−26.5	215	1.50	101	40
R737 (NH$_3$)	−33.3	1350	4.35	133	113
Cryogen					
Carbon dioxide (CO_2)	−78.5	574	1.90	31	73
Nitrogen (N_2)	−197.0	200	2.00	−147	34

BP, boiling point (1 bar); ΔH_e, heat of evaporation (1 bar); C_{pL}, specific heat (liquid); T_c, P_c, critical point temperature and pressure.

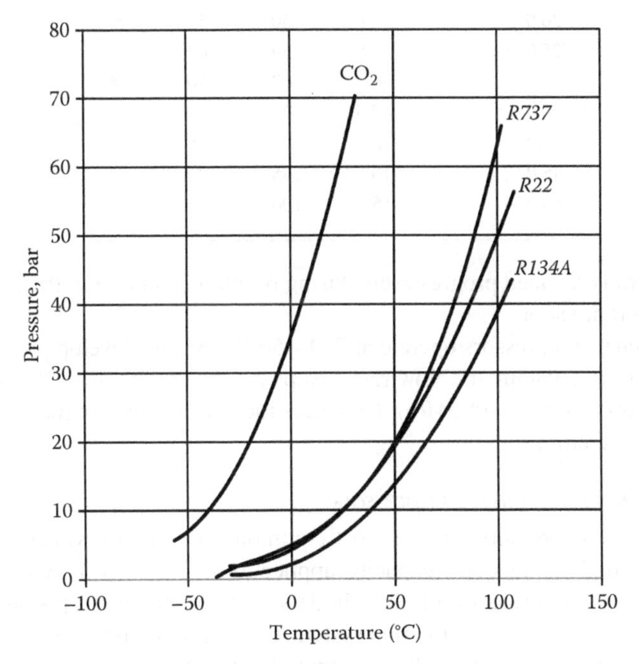

FIGURE 12.6 Vapor pressure of some refrigerants and cryogens.

12.2.1.5 Compressors

The compressors applied in refrigeration systems have similarities with the pumps and fans used in the transport of liquids and gases (Chapter 6). Five principal types of mechanical compressors are used in food refrigeration and freezing systems.

12.2.1.5.1 Rotary Compressors

Rotary compressors are used for very small to small capacities (refrigeration loads). They consist of a stator in which the rotor moves eccentrically, so that the space

TABLE 12.2
Vapor Pressures of Food Refrigerants

	Temperature (°C)			
Pressure (bar)	CO_2	*R717*	*R22*	*R134A*
0.4	—	−50	−58	−40
0.6	—	−40	−50	−35
1.0	—	−32	−40	−26
2.0	—	−18	−24	−10
4.0	—	−1	−6	9
6.0	−53	9	6	20
10.0	−46	25	24	31
12.0	−35	31	30	46
14.0	−31	36	37	52
16.0	−27	41	42	58
20.0	−19	50	51	63
25.0	−12	59	62	70
30.0	−6	67	70	76
35.0	0.5	73	78	85
40.0	6	75	85	100
45.0	10	85	90	—
50.0	15	100	94	—

between them is reduced progressively during rotation, resulting in the compression of the refrigerant vapor.

Small rotary compressors operate at 750–1500 RPM and develop pressures up to 5 bar. They have a volumetric flow rate (swept volume) of 30–40 m^3/h and a refrigeration load up to 30 kW (108 MJ/h). They have the disadvantage of the wear of stator wall during operation.

12.2.1.5.2 Reciprocating Compressors

Reciprocating compressors are the most common refrigeration systems. They are suitable for small to medium capacity applications (1–300 kW). A reciprocating compressor consists basically of (1) cylinders with reciprocating pistons, (2) inlet and outlet valves, and (3) a lubrication system. A compressor may have four or more cylinders, the larger number resulting in smoother pumping of the refrigerant.

The capacity of a compressor is defined by the volumetric displacement of the pistons V_p (m^3/h), which are given by the equation

$$V_p = 60AhNz \quad (12.7)$$

where

A (m^2) is the cross-sectional area of the cylinder
h (m) is the length of the stroke
N (RPM) is the speed of the crankshaft revolution
z (−) is the number of cylinders

For smooth operation, the piston velocity ($u = hN/30$) should not exceed 2.5 m/s and the revolution speed 1200–1750 RPM. The ratio of cylinder diameter (d) to the length of bore (h) should be (d/h) = 1.0–1.2. The volumetric piston displacement (V_p) in reciprocating compressors varies in the range of 100–250 m^3/h.

The volumetric efficiency of a compressor may be evaluated by the indicator diagram of pressure–volume, which may reveal the causes of efficiency reduction, such as leakage of valves, wet vapor, and dead space in the cylinder.

Lubrication of the moving parts of the compressor is important for efficient operation. In small compressors (<10 kW), oil splashing is used, while in larger machines, a gear oil pump may be installed.

12.2.1.5.3 Centrifugal Compressors

Centrifugal compressors are used for large refrigeration capacities (200–750 kW), such as the air conditioning of large installations. They are similar to the centrifugal pumps (Chapter 6) and they can handle large amounts of gas (>2000 m^3/h) at relatively low pressure differences (1.5–2 bar). High rotational speeds (>3000 RPM) are used to achieve large gas capacities.

The centrifugal compressor consists of a number of impeller wheels (usually two to four), which become smaller in the direction of gas flow, so that the pressure increases progressively from the suction to the discharge of the "pump."

12.2.1.5.4 Screw Compressors

Screw compressors are a recent development in refrigeration compressors, and they are used for medium to large refrigeration capacities (100–500 kW). They consist of two counter-rotating screws (rotors) in a steady stator. One screw is connected through a gearbox to the power source, driving the other during rotation.

Screw compressors operate at high rotational speeds (>3000 RPM). They have volumetric capacities 700–10,000 m^3/h and they can develop pressure differences of 7–20 bar. They do not have any valves and they can operate in a wide range of capacities. However, their fast operation causes wear to the screws and the bearings, which must be replaced.

12.2.1.5.5 Scroll Compressors

Scroll compressors, developed recently, are used for small to intermediate refrigeration and air conditioning capacities (1–100 kW). They consist of two interleaving scrolls (spirals), one of them rotating and the other stationary, which gradually compress the gas or vapors from the suction to the discharge. Normally, they do not use valves and are more efficient than other compressors of similar size.

12.2.1.6 Evaporators and Condensers

12.2.1.6.1 Evaporators

Evaporators are heat exchangers (Chapter 8), which are part of refrigeration cycles, and they are used in absorbing refrigeration load.

The refrigeration evaporators operate on two principles, the flooded and the dry expansion types, as shown in Figure 12.7. In the flooded type, the evaporator tubes are flooded with liquid refrigerant, while a float prevents the passing of liquid into

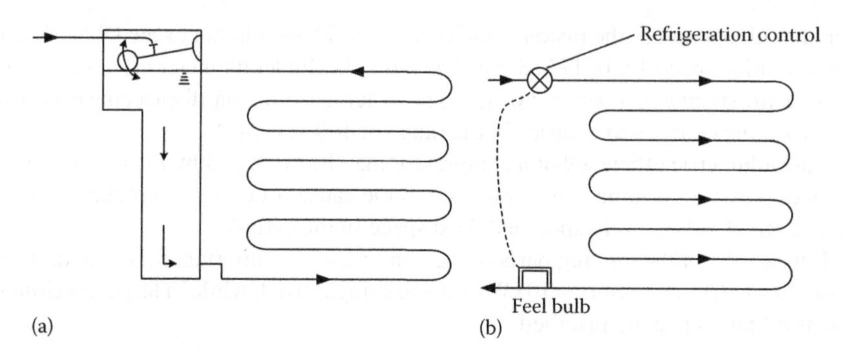

FIGURE 12.7 Types of refrigeration evaporators: (a) flooded and (b) dry expansion.

the compressor, allowing only the vapors to escape. In the dry expansion type, a valve control system allows the entrance of only the amount of liquid refrigerant, which can be evaporated completely in the evaporator.

Heat can be transferred from a product to the evaporator by direct contact, or by an intervening fluid (gas or liquid). The heat-exchanging surface can be a tube (coil), or a flat surface. Heat-exchanging tubes are either bare or finned. Fins, used outside or inside the tubes considerably increase the heat exchange surface area (Chapter 8). The heat transfer liquid medium to the evaporator can be cold water, brine, or a refrigerant liquid with a high boiling point (brine).

Outside-finned tubes are used when the inside heat transfer coefficient (h_i) is much smaller than the outside (h_o), as in using air in natural or forced convection on the outside and a liquid refrigerant in the inside. Tubes with inside fins may be used when $h_i < h_o$, and there is no need for fins when $h_i = h_o$. Blowers (fans) are needed to increase the air velocity and the heat transfer coefficients in evaporators with outside fins.

Examples of typical evaporators used in refrigeration applications:

1. *Double pipe evaporators*: They consist of two concentric tubes, with the product in the central tube in counterflow and the refrigerant in the annulus. The overall heat transfer coefficients of 300–900 W/(m^2 K) can be obtained. They are used for cooling liquid foods, such as fruit juices and wine.
2. *Shell and coil evaporators*: The liquid product flows through a coil, which is immersed in a closed shell containing the refrigerant. They are used in small-capacity applications for quick cooling.
3. *Shell and tube heat exchangers*: They are similar to the heating equipment discussed in Chapter 8. The product and the refrigerant can flow either in the tubes or in the shell. High overall heat transfer coefficients are obtained, e.g., 1 kW/(m^2 K). Large refrigeration loads can be obtained, up to 1 MW of water or brine cooling.
4. *Bath and tube evaporators*: The refrigerant, e.g., ammonia, flows in a bank of tubes immersed in a bath of the liquid to be cooled. The refrigerant tubes operate usually as a flooded evaporator, and the system is used for large amounts of liquids, such as water and brine. It is not suitable for direct cooling of food liquids because of poor sanitation.

5. *Baudelot evaporators*: They consist of several tubes placed in parallel one over the other, through which the refrigerant flows. The liquid product flows by gravity on the outside of the tubes from the top to the bottom forming a thin film, which improves heat transfer. Overall heat transfer coefficients up to 1 kW/(m^2 K) can be obtained.

12.2.1.6.2 Condensers

The condensers used in refrigeration systems are similar to the heat-exchange equipment described in Chapter 8. The condensation of refrigerant vapors requires the fast removal of large heat quantities, and high heat transfer coefficients and effective heat-exchange equipment are required.

1. *Tubular condensers*: They are similar to, e.g., shell and tube heat exchangers. The cooling water flows in the tubes, while the refrigerant condenses in the shell. The overall heat transfer coefficient is about 0.8 kW/(m^2 K). In order to reduce the water required, recycling of the used water is necessary through a cooling tower. In the cooling tower, water is sprayed from the top, while air is blown from the bottom. Fast cooling is achieved through the evaporation of water. The quantity of water sprayed is about 100 L/kWh heat removed.
2. *Evaporative condensers*: They consist of a bank of parallel tubes in which the refrigerant condenses. Cooling water is sprayed from the top of the tubes, while air is blown by fans from the bottom. The evaporative condensers require much less cooling water, e.g., 5 L/kWh, than the cooling towers.
3. *Falling film condensers*: They are similar to the falling film evaporators, discussed in Chapter 10. They consist of vertical tubes, 60 mm diameter and 3 m long, through which a film of water flows downward, while the refrigerant condenses on the outside. High overall heat transfer coefficients are obtained, e.g., 0.8–1.6 kW/(m^2 K).
4. *Air condensers*: The refrigerant vapors are condensed in a bank of tubes, which are cooled by air blown by fans through the bank. Units removing 150–350 kW refrigeration load require about 200–400 m^2 outside tube heat transfer area, with an overall heat transfer coefficient of 100 W/(m^2 K) and a mean temperature difference of 7°C.

12.2.1.7 Capacity Control

The capacity of refrigeration systems can be controlled by

1. The refrigerant flow is controlled in the evaporator by an expansion or throttling valve. The valve regulates the refrigerant flow so that it compensates the amount evaporated and maintains a constant pressure difference between the evaporator and the condenser (Figure 12.8). Automatic pressure valves can control the flow of liquid (a) or vapor (b) refrigerant. A float valve (c) can control the flow of the liquid refrigerant, as in flooded evaporators. Thermostatic valves and capillary tubes can control the flow of the refrigerant in small refrigeration units.

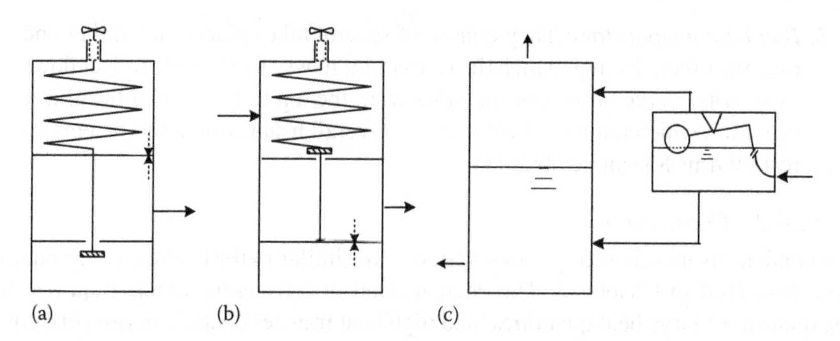

FIGURE 12.8 Pressure valves controlling refrigeration evaporators.

2. Control of compressor capacity: The compressor capacity can be controlled by (a) variation of the revolutions of the crank, (b) intervention in the opening and closing of the valves, (c) using a cylinder by-pass, and (d) altering the cylinder volume.
3. Control of cooling of the condenser: The capacity of the condenser can be controlled by changing the flow of cooling water, which controls the temperature and pressure of the condenser.

12.2.2 Cryogenic Liquids

Cryogenic liquids are used in fast cooling or freezing of foods at very low temperatures. They include liquid nitrogen and liquid/solid carbon dioxide, which cool or freeze the products by direct contact, either as vapors or liquids. The thermodynamic properties of cryogens and refrigerants are summarized in Table 12.1. The main application of cryogenic liquids is in food freezing, which is discussed in Section 12.4.

Fast cooling of fresh fruits and vegetables can be achieved using cryogenic nitrogen, which has the added advantage of creating an inert atmosphere for the stored cold product. It can be used in cooling and transporting food products by trucks, railcars, and ships.

The main advantages of cryogenic liquids are (1) high cooling or freezing speed (high capacity, better quality), (2) low product weight loss during cooling or freezing (0.1%–1.0%), (3) low initial capital (1/3 of that of mechanical systems), and (4) low maintenance cost (simple construction).

12.2.3 Ice Cooling

Ice is used commercially for cooling (a) fresh foods such as fish and vegetables, and (b) processed foods such as poultry and cut meat. The cooling effect of ice is produced by the melting of ice, which absorbs 334 kJ/kg ice. The cooling effect or refrigeration load of ice is expressed in kW or in tons of refrigeration (U.S.), which is equivalent to 3.51 kW.

Ice, for commercial cooling, is produced in ice manufacturing plants in the form of ice blocks, ice tubes, or ice flakes. Large amounts of ice are stored in fishing boats in some harbors before they sail into the ocean for the catching, cooling, and processing of fish.

12.2.4 Vacuum Cooling

Vacuum cooling is used for the removal of large quantities of heat by evaporation of water at low temperatures. It is applied for cooling foods with a large external surface containing free water, such as leafy vegetables. The food material is placed in a closed container, which can stand high vacuum, and steam ejectors are used to create the vacuum (Chapter 6). The temperature and pressure should not be lower than the triple point of water which is 0°C and 6.1 mbar. At lower pressures, created by mechanical pumps, ice is formed and the water is removed by sublimation.

The evaporation of water (*dm*/*dt*) (kg/s) at temperature *T* (K) can be calculated from the equation

$$\left(\frac{dm}{dt}\right) = \left[\frac{mC_p}{\Delta H}\right]\left(\frac{dT}{dt}\right) \tag{12.8}$$

where

m (kg) is the mass of free water
$C_p = 4.18\,\text{kJ/(kg K)}$ is the specific heat of water
ΔH (kJ/kg) is the heat of evaporation of liquid water at temperature T

The heat of evaporation of water for cooling near 0°C is assumed constant, $\Delta H = 2.5\,\text{MJ/kg}$.

The evaporation rate can also be calculated from the perfect gas equation

$$\left(\frac{dm}{dt}\right) = \left[\frac{273}{(T+273)}\right]\left(\frac{P}{1}\right)\left(\frac{18}{22.4}\right)Q \tag{12.9}$$

where

T (°C) is the temperature
P (bar) is the pressure of the gas phase
Q (m³/s) is the capacity of the vacuum system, which is assumed to be constant

The pressure in Equation 12.9 can be expressed by the Antoine equation $P = \exp[A - B/(C+T)]$, where A, B, C are the constants for water in appropriate units, yielding

$$\left(\frac{dm}{dt}\right) = 0.80\left[\frac{273}{(T+273)}\right]\left[\exp\left[\frac{A-B}{C+T}\right]\right]Q \tag{12.10}$$

Combining Equations 12.8 and 12.10 yields

$$\left(\frac{dt}{dT}\right) = \frac{(mC_p)}{\{0.80Q\Delta H\,[273/(273+T)]\,\exp[A - B/(C+T)]\}} \tag{12.11}$$

Integration of the complex differential Equation 12.11 results in the time (t) required to cool a given mass of water (m) from temperature T_1 to T_2, using a vacuum system of capacity Q.

12.3 COOLING OF FOODS

12.3.1 Cooling Load

Cooling or chilling of foods at temperatures near 0°C is used to extend the shelf life of fresh or processed foods. The products are stored usually in insulated refrigerated storage rooms, trucks, or ships.

The cooling or refrigeration load q (kW), or the heat removed from the product upon cooling is calculated from the equation

$$q = mC_p\Delta T \tag{12.12}$$

where

m (kg/s) is the mass of the product per unit time (product feed)
C_p (kJ/kg K) is the specific heat
ΔT (K) is the temperature difference

Note that ΔT (K) = ΔT (°C).

The specific heat of a composite (heterogeneous) product can be estimated from the empirical "mixing rule" equation

$$C_p = x_w C_{pw} + x_c C_{pc} + x_f C_{pf} + x_s C_{ps} \tag{12.13}$$

where

x_w, x_c, x_f, x_s are the mass fractions
C_{pw}, C_{pc}, C_{pf}, C_{ps} are the specific heats of the water, carbohydrate, fat, and salt components of the food product

The enthalpy change of fruits and vegetables ΔH (kJ/kg) can be estimated from the empirical equation,

$$\Delta H = (1 - x)\Delta H_L + xC_{ps}\Delta T \tag{12.14}$$

where

ΔH_L is the enthalpy change of the juice
x, C_{ps} are the mass fraction and specific heat of the dry matter (nonjuice dry solids), respectively

Thermophysical and transport properties of foods, useful in refrigeration and freezing, are reviewed in Chapter 5. Typical values of specific heats of foods used in refrigeration and freezing calculations are the following:

- High-moisture foods above freezing ($T > 0$°C): 3.5–3.9 kJ/kg K
- High-moisture foods below freezing ($T < 0$°C): 1.8–1.9 kJ/kg K
- Dehydrated foods: 1.3–2.1 kJ/kg K
- Fat ($T > 40$°C): 1.7–2.2 kJ/kg K
- Fat ($T < 40$°C): 1.5 kJ/kg K

The enthalpy of some important classes of food, such as fruits and vegetables, meats, and fats, is given in the Riedel diagrams (literature) as a function of moisture content and temperature. In the same diagrams, the percentage of unfrozen water can be estimated.

12.3.2 Cooling Equipment

Conventional heat transfer equipment is applied for the cooling of liquid foods. Special equipment is used for the cooling of various solid food products.

12.3.2.1 Liquid Foods

The cooling of liquid foods is normally carried out in conventional heat exchangers, similar to those described in Chapter 8. Typical cooling units are

- The plate heat exchanger
- The shell and tube heat exchanger
- The scraped surface heat exchanger
- The shell and coil
- The jacketed vessels

12.3.2.2 Solid Foods

12.3.2.2.1 Hydrocooling

Hydrocooling is effected by contacting solid foods with cold water, resulting in a fast cooling of the product. The cooling process can be improved significantly by increasing the relative motion of solids/water, which results in high heat transfer coefficients at the water/solid interface. Hydrocooling is applied mainly to fresh fruits and vegetables, where cooling may be combined with washing of the product, prior to processing. The cooling water is recirculated through heat exchangers, and it may be chlorinated to prevent microbial growth. Hydrocooling equipment includes immersion and spraying, rotating (tumbling) drums, and tunnel systems.

12.3.2.2.2 Air Cooling

Air cooling equipment for solid foods is similar to air (convective) dryers (Chapter 11). Low-temperature tunnels are similar to drying tunnels, with the food product moving in trucks, in racks, or on conveyors in counterflow with cold air blown from fans at air velocities near 1 m/s. The fans and cooling coils are placed on the upper part or on the sides of the tunnel. Higher air velocities reduce the cooling time but they may increase moisture losses from the product by evaporation. Evaporation losses may be reduced by using higher relative humidity in the tunnel.

12.3.2.2.3 Vacuum Cooling

Leafy vegetables and liquid foods can be cooled rapidly by applying vacuum near to 6 mbar (near the freezing point of the product). Evaporation of free water results in a very fast cooling of the product (see Section 12.2.4, Vacuum Cooling).

12.4 FOOD FREEZING

Freezing of (free) water takes place at a constant temperature (0°C) until all liquid water is crystallized into ice, and the temperature of the ice decreases, as the temperature is reduced. Due to the presence of water-soluble components, such as sugars and organic acids or salts, freezing of foods is a gradual process and it is not completed until the temperature drops considerably below 0°C. To assure good freezing, fruits and vegetables are frozen to −1°C (0°F), and ice cream and fish at −25°C (IIR, 1972; Singh and Sakar, 2006).

12.4.1 FREEZING TIME

The freezing of food products involves the crystallization of free water into ice and the lowering of temperature to about −18°C so that the product can be preserved for several months without any appreciable microbial, enzymatic, or quality deterioration (Singh and Heldman, 2005).

The speed of freezing u (m/s), defined as the speed of movement of the "frozen front" between the unfrozen and frozen product, should be high enough, e.g., $u > 5$ cm/h, for economic operation and preservation of food quality.

The freezing point of food materials depends on the composition (moisture content), e.g., meat −1.0°C, fish −2.0°C, peaches −1.4°C, and cherries −4.5°C. The fraction of frozen water in frozen foods depends on the temperature and the moisture content, and the type of product, varying from 50% (bread) to 90% (fruit juice) at −18°C.

The freezing time t_f (s) is useful in the design calculations of freezing processes and freezing equipment. It is estimated by the conventional Plank equation, which is derived from basic heat transfer and freezing analysis (Cleland and Valentas, 1997),

$$t_f = \left[\frac{(\rho \Delta H_L)}{(T_f - T_a)}\right]\left[\left(\frac{2Pr}{h}\right) + \left(\frac{4Rr^2}{\lambda}\right)\right] \tag{12.15}$$

where

ρ is the density of the frozen product (kg/m³)
ΔH_L is the latent heat of crystallization of ice (334 kJ/kg)
T_f is the freezing point (°C)
T_a is the temperature of the freezing medium (°C)
r is the half-thickness (radius) of the product (m)
h is the convective heat transfer coefficient [W/(m² K)]
λ is the thermal conductivity of the frozen product [W/(m K)]

P, R = shape factors:

sphere, $P = 1/6$, $R = 1/24$
infinite plate, $P = 1/16$, $R = 1/24$
infinite cylinder, $P = 1/16$, $R = 1/24$

The following Pham equation, which is a modified form of the Plank equation, is more accurate and it can be used for various shapes of solid foods

$$t_f = \frac{1}{E}\left[\left(\frac{\Delta H_1}{\Delta T_1}\right) + \left(\frac{\Delta H_2}{\Delta T_2}\right)\right]\left[\left(\frac{r}{h}\right) + \left(\frac{r^2}{2\lambda}\right)\right] \quad (12.16)$$

where

E is the shape factor related to heat transfer:

sphere: $E = 3$,

infinite cylinder: $E = 2$

infinite plate (slab): $E = 1$

$\Delta T_1 = 0.5\ (T_i + T_{fm}) - T_a$

$\Delta T_2 = T_{fm} - T_a$

T_i is the initial product temperature (°C)

T_{fin} is the initial freezing temperature (°C)

T_{fm} is the mean initial freezing temperature (°C)

T_a is the temperature of the freezing medium (°C)

$\Delta H_1 = \rho C_{pu}\ (T_i - T_{fin})$, volumetric enthalpy for the prefreezing period (kJ/m³)

$\Delta H_2 = \rho \Delta H_L + \rho C_{pf}\ (T_{fm} - T_{fin})$, volumetric enthalpy of phase change (kJ/m³)

ΔH_L is the latent heat of freezing of the product (kJ/kg)

C_{pu} is the specific heat of the unfrozen product [kJ/(kg K)]

C_{pf} is the specific heat of the frozen product [kJ/(kg K)]

$T_{fm} = 1.8 + 0.263\, T_{fin} + 0.105\, T_a$

The shape factor E (dimensionless) is estimated from the following empirical equation:

$$E = 1 + \left[\frac{(1 + (2/B_i))}{\left(\beta_1^2 + (2\beta_1)/B_i\right)} + \frac{(1 + (2/B_i))}{\left(\beta_2^2 + ((2\beta_2)/B_i)\right)}\right] \quad (12.17)$$

where

$\beta_1 = A/(\pi\ r^2)$

$\beta_2 = 3V/(4\ \pi\beta_1 r^3)$

$B_i = (\mathrm{h\ r})/\lambda$, Biot number

r is the half-thickness (radius) (m)

V is the volume (m³)

A is the smallest cross-sectional area of the product (m²)

h is the convective heat transfer coefficient, [W/(m² K)]

λ is the thermal conductivity of the frozen product, [W/(m K)]

For irregular shapes, the calculation of the E value is difficult. In this case, experimentally determined shape E factors may be found in the literature, e.g., fish $E = 1.8$, beef leg $E = 1.3$, and lamb leg $E = 2.0$.

The Pham equation is more accurate than the original Plank equation because it takes into consideration the prefreezing and postfreezing periods.

12.4.2 Heat Load

The total heat load q (kW) to be removed in a freezing process from a product of mass (m) per unit time (kg/s) includes lowering the product temperature from the ambient (initial) T_i to the mean freezing temperature T_o (°C), removing the latent heat of crystallization of ice ΔH_L (334 kJ/kg), and lowering the product temperature to the temperature of the frozen product T_f (°C), according to the equation,

$$q = m[C_{pu}(T_i - T_o) + \Delta H_L + C_{pf}(T_o - T_f)] \tag{12.18}$$

where C_{pu} and C_{pf} [kJ/(kg K)] are the specific heats of the unfrozen and frozen product, respectively.

12.4.3 Freezing Equipment

The freezing equipment is selected according to the type of product (liquid, solid), the freezing medium (air, cold surface, liquid), and the type of processing (batch, continuous).

12.4.3.1 Air Freezing Equipment

Air is used in freezing food in tunnels, conveyor belts, and fluidized bed equipment. In all cases, air is blown counter-currently to the product flow and, depending on the freezing method, it is blown horizontally or vertically to the product. In tunnel freezing, the horizontal blow method prevails. In fluidized beds, air is blown upward vertically, and in belt freezing, both blowing methods are used. Since the specific heat of air is low, large air quantities are required for freezing.

12.4.3.1.1 Tunnel Freezers

Tunnel freezing is used in a wide range of products, such as meat cuts or minced products, whole poultry, and half-beef carcasses. It consists of an insulated room with one door, or with two doors for a continuous operation. Air temperatures −30°C to −40°C and air velocities 3–6 m/s are used. Three loading systems of a tunnel are used: (a) the push-through system, in which for each new trolley coming into the tunnel, a trolley with frozen product gets out; (b) the rack system, which is applied to the freezing of carcasses; and (c) the chain drive system, in which the trolleys are pulled by a chain, in and out of the tunnel. The freezing time depends on the size and thermal conductivity of the product, lasting usually 1.5–6 h. The fans are installed usually above the trolleys, and the heat exchangers are on both sides. The capacity of a tunnel is in the range of 1.5–4.0 ton/h.

12.4.3.1.2 Belt Freezers

This equipment consists of belts moving through a cold air steam (Figures 12.9 through 12.12). The belts are either straight or curved, made of steel or plastic material,

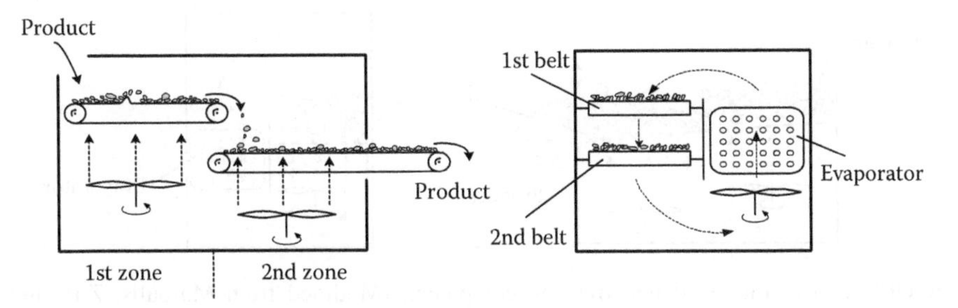

FIGURE 12.9 Straight-belt freezing equipment. (Modified from Maroulis, Z.B. and Saravacos, G.D., *Food Process Design*, CRC Press, Boca Raton, FL, May 9, 2003, fig. 5.5.)

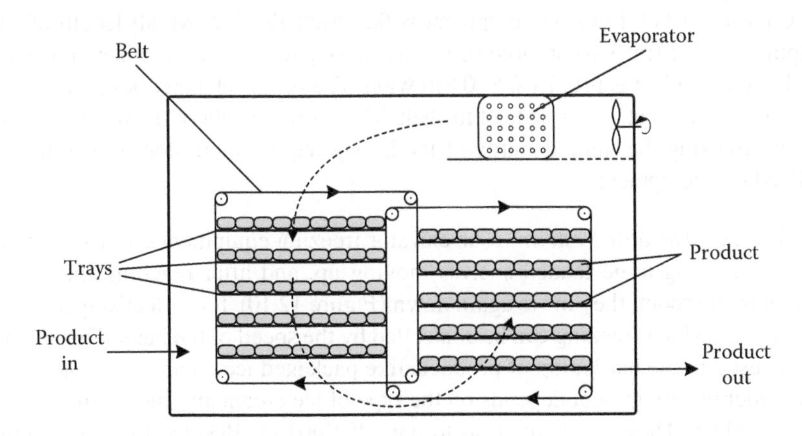

FIGURE 12.10 Elevator belt freezing equipment.

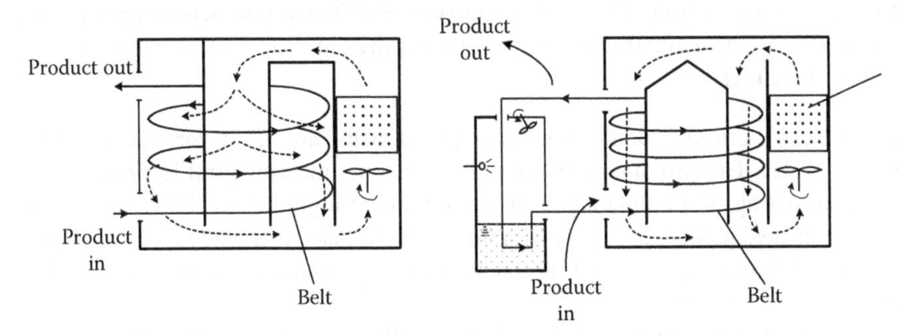

FIGURE 12.11 Spiral belt freezing equipment. (Modified from Maroulis, Z.B. and Saravacos, G.D., *Food Process Design*, CRC Press, Boca Raton, FL, May 9, 2003, fig. 5.6.)

allowing air to pass through, and they are suited for freezing sensitive and relatively large or heavy pieces of food, such as apple slices, cauliflower, and strawberries.

12.4.3.1.2.1 Straight Belts In some cases, the straight belts are separated into zones (Figure 12.9). In the first zone, the air recirculates fast, causing a surface freezing of the product (crust freezing). The freezing of the product is completed in the second zone. Buckles of the belt cause turning over of the product, contributing to more

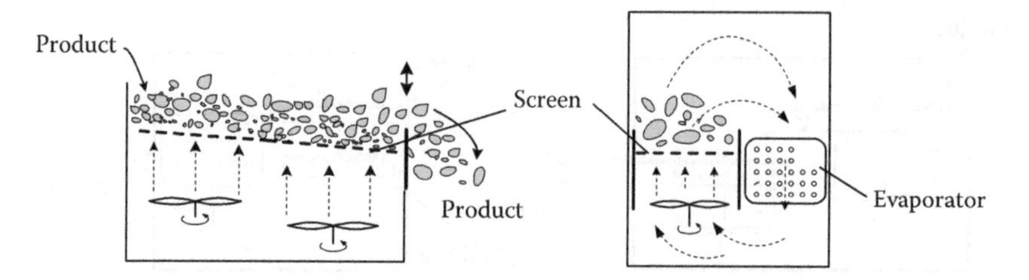

FIGURE 12.12 Fluidized bed freezing equipment. (Modified from Maroulis, Z.B. and Saravacos, G.D., *Food Process Design*, CRC Press, Boca Raton, FL, May 9, 2003, fig. 5.7.)

even freezing. Strawberries can be frozen in about 12 min, fish fillets in about 20 min. The capacity of belt freezing equipment is 0.2–6 tons/h. The overall length of such equipment is 5–13 m. Two or more belts may move parallel to each other, at different speeds. A single belt is usually 0.5–0.8 m wide. The heat exchanger lies in a separate part of the equipment, on the side of the belt. The overall height of the freezer is about 5 m. For freezing the same quantity of food, they require more floor space than the fluidized bed equipment.

12.4.3.1.2.2 Elevator System The elevator freezing equipment consists of parallel belts carrying large loaded shelves moving up, and after reaching the highest position in the room, they move again down (Figure 12.10). It is effectively an elevator system in which freezing can be controlled by the speed of the belts. This method is often used in the hardening of products like packaged ice cream. The capacity of such hardening equipment depends on the type of ice cream and the desired texture of the product. They can harden up to e.g., 20,000 L/h. Besides freezing control, through the speed of the belt it is also possible to load and empty the shelves at different positions, enabling the parallel freezing of different products or packages of different size. The method saves room but it requires more energy in comparison to flat belt structures.

12.4.3.1.2.3 Curved Belts Curved (spiral) belts are used for saving space (Figure 12.11). The two main types are the spiral and the semispiral freezing equipment, consisting of a combination of curved and straight belts. The spiral type is quite often used in the freezing of hamburgers, fish sticks, and ready meals. It is also used in the hardening of frozen products. The combined type is normally used in hardening frozen products.

The length of the curved belts can be up to 300 m, and the width is about 4–7 m. Air is blown horizontally or vertically through the product, which moves around a cylindrical core. The air is cooled in finned heat exchangers, placed in a separate room. Large equipment may freeze continuously more than 5 tons/h. The main advantages are the continuous, gentle product transport, and the possibility of parallel freezing of products differing in size or in packaging.

12.4.3.1.3 Fluidized Beds

The fluidized bed method (Figure 12.12) is an individual quick freezing (IQF) method, used in freezing small, whole, or cut pieces or food (diameter up to about

3 cm and length up to about 12 cm), such as peas, french fries, sliced or cut carrots, beans, mushrooms, etc. The food pieces are frozen individually as they are suspended in the air that freezes them quickly. The equipment consists of an inclined screen, fans (usually radial) blowing air upward through the perforated bottom, and heat exchangers cooling the air to about −40°C. The product is frozen quickly, because of high heat transfer coefficients. Examples of freezing time are: peas, 3–5 min, and french fries and strawberries, 9–13 min. The product layer over the screen depends on the product, e.g., 3–25 cm (usually about 12 cm). In proper design, the weight loss of the product is less than 1.5%–2%. The capacity of fluidized bed freezing equipment varies between about 0.5 and 10 tons/h.

12.4.3.2 Cold Surface Freezers

Figure 12.13 shows a cold surface freezer, which consists of several double-walled plates, in which the refrigerant circulates. Food is placed between the plates, which press the food lightly by means of a hydraulic system (0.06–0.1 bar) to reduce air pockets between the cooling surface and for packaging and increasing heat conduction. The plates can be horizontal or vertical. Vertical plates are used in freezing fish in ships, because they require less free headroom. The number of parallel plates can be 5–20, their spacing is up to 7 cm, and their surface is 1.5–2.0 m^2. Plate equipment is used in the freezing of whole fish, fish fillets, pieces of meat (e.g., chops) products packed in rectangular packages, and liquid slurries. The last product is frozen in plastic bags, hanging between vertical plates. The capacity of plate equipment is 6–13 tons/24 h. The refrigeration capacity of large units is about 75 kW. The freezing time of a 5 cm fish is about 1.15 hT. The main advantage of plate freezers is their high specific capacity, which is about four times higher than that of freezing tunnels.

12.4.3.3 Heat Exchanger Freezers

Heat exchangers, such as the scraped surface and the agitated vessel, used in cooling viscous liquid foods can also be used in freezing processes.

12.4.3.4 Frozen Pellets

Food liquids are used in the production of frozen pellets of foods, such as dairy products, liquid egg, fruit pulps, sauces, and vegetable purees. The liquid foods are frozen between two parallel moving metallic belts. The corrugated lower belt gives

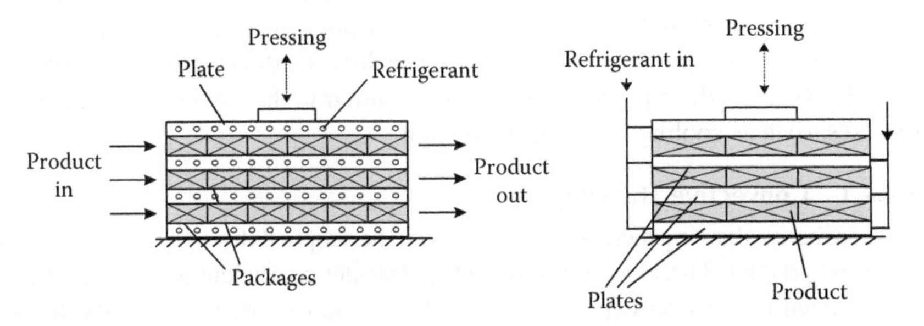

FIGURE 12.13 Diagram of a cold surface freezer. (Modified from Maroulis, Z.B. and Saravacos, G.D., *Food Process Design*, CRC Press, Boca Raton, FL, May 9, 2003, fig. 5.8.)

the shape of the pellets, while the upper belt is flat. A liquid refrigerant, such as a propylene glycol/water mixture, is sprayed on the external sides of both belts.

12.4.3.5 Cryogenic Liquids

Cryogenic liquids, such as liquid nitrogen and liquid carbon dioxide, can be used in freezing processes. They cause rapid freezing of the product due to high heat transfer coefficients.

In cryogenic nitrogen freezing, straight belts or spiral belts running around one or two cylindrical cores are normally used. For small quantities, a batch process chamber can be used. The straight belt, up to 12 m long and 1.5–2.5 m wide, has three zones for precooling, freezing, and equilibration. In the first zone, the product is precooled down to about −70°C through gas of just-evaporated liquid N_2. In the second zone, liquid nitrogen (−196°C) evaporates, while it is sprayed directly onto the food. Due to the high heat transfer coefficient of liquid N_2 during its evaporation, up to 2.3 kW/(m^2 K), the product is frozen very fast reaching a surface temperature of about −190°C. In the third zone, the temperature of the product is equilibrated by air blown on its surface. The capacity of this equipment can be up to about 1.5 ton/h.

Immersion in liquid nitrogen freezes the food products within 15–100 s, depending on their size. However, since not all foods resist the freezing shock caused by the sudden immersion in such a low temperature, the method is restricted to products such as berries, shrimp, and diced meat and fruits. In direct contact, liquid N_2 may remove 330 kJ/kg from a product being frozen. The liquid N_2 consumption is about 1–1.5 kg/kg product. In some cases, to reduce N_2 consumption, the products processed by cryogenic nitrogen are frozen only on their surface. The final freezing is carried out in other equipment (e.g., belt freezers) or in cold stores at −20°C.

Carbon dioxide (CO_2) has a triple point at −56.6°C and 5.28 bar. The operation of the freezing equipment is based on cryogenic CO_2 (dry ice). When liquid CO_2 is released in to the atmosphere, half of it becomes dry ice and the other half vapor at the equilibrium temperature of −78.5°C. Spraying of the product with CO_2 dry ice is done at the entrance in the freezing cabinet in order to have time for sublimation. The overall cost of CO_2 freezing is comparable to that of N_2 freezing.

12.4.4 Thawing of Frozen Foods

The commercial thawing of frozen foods is important in connection to catering. Thawing can be accomplished by convective, vacuum, or contact heating, and by electrical methods. The convective methods include methods using air, water, or steam. In all cases, the equipment used is very similar to that of other heat exchange processes, such as cooling, heating, and drying.

12.4.4.1 Convective Thawing

In all convective thawing methods, the heat has to be supplied through the external surface of the product. Therefore, the surface heat transfer coefficient is very important.

The speed of thawing depends also on the thermal conductivity of the thawed product, since heat has to pass through the melted zone of the product to the thermal center of the frozen food. Air blast thawing is slower than water or steam thawing, because of the lower heat transfer coefficients. However, this method is used often, because it requires less capital and is applicable to all products. The thawing tunnel is very similar to those used in freezing, cooling, or drying.

Water thawing equipment is similar to the hydro-cooling of fresh foods.

Steam can be also used in the thawing of frozen foods. However, although this method is very fast, it has the disadvantage of steam condensation to the thawed product.

12.4.4.2 Vacuum Thawing

Vacuum thawing with water vapor has the advantage of a rapid rate, which is caused by high vapor diffusivity and high vapor pressure potential. Vacuum thawing is a batch process and the thawing equipment has a relatively small capacity (less than 2 tons). However, since the thawing rate is increased, a greater number of operating cycles is possible per day. The thawing equipment consists basically of a vacuum vessel loaded with trays of layers of frozen product, which are stacked on trolleys.

12.4.4.3 Contact Thawing

Double walled surfaces are used for transferring heat to the product that has to be thawed. This method is applied when the frozen food is available in small pieces. Semicylindrical jacketed vessels, equipped with a screw propeller in the center for agitating the frozen food, can thaw about 2.5 tons/h.

12.4.4.4 Electrical Thawing

Two methods of electrical thawing are used, dielectric and microwave (MW) thawing. The heat generated depends on the electrical frequency and the electrical properties of the food, such as the dielectric constant (ε', efficiency to accumulate electrical energy) and the loss factor (ε'', electrical energy that can be transformed to heat (Chapter 8).

In dielectric heating, frequencies about 10 MHz are used, while in MW heating, the standard frequencies are 915 MHz and 2.45 GHz. Table 12.3 shows typical dielectric properties of foods and packaging materials.

In the dielectric heating process, the frozen product is placed between plates or electrodes, which are connected to a source supplying alternating high-frequency

TABLE 12.3
Dielectric Properties of Food and Packaging Materials (2.45 GHz, 25°C)

Material	Dielectric Constant (ε')	Loss Factor (ε'')	Penetration (m)
Potato	64	14	5×10^{-3}
Beef	50	15	9×10^{-3}
Water	77	11	16×10^{-3}
Ice (−3°C)	3	0.03	10
Polyethylene	2	0.001	26
Glass	6	0.005	10

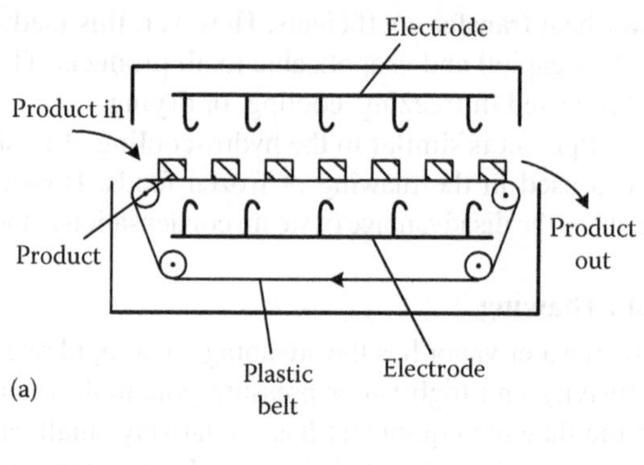

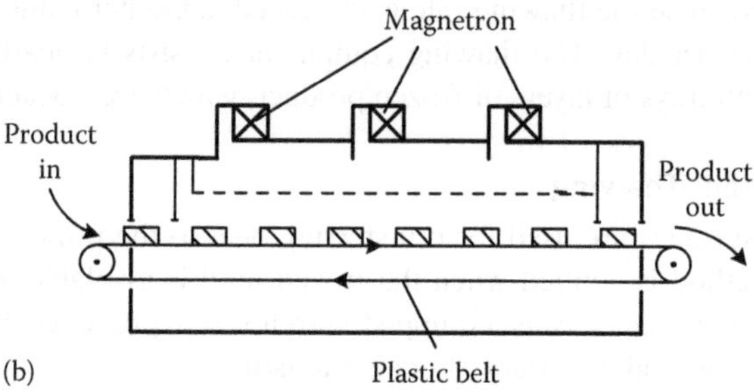

FIGURE 12.14 Electrical thawing equipment: (a) dielectric and (b) microwave (MW).

voltage. As electricity flows through the product, alternating from one plate to the other, heat is generated, thawing the food. Thawing of large blocks of frozen food (e.g., fish) with voids is facilitated in trays filled up with water.

Dielectric equipment consists of a rubber or plastic conveyor belt, which carries the food between electrodes or plates connected to electrical current (Figure 12.14). The electric power of each unit may be about 15–30 kW with a thawing capacity of about 1 ton/h. The thawing time depends on the size and the product; usually it is less than 1 h.

The industrial MW equipment consists of a belt transporting the food to a chamber in which magnetrons supply the electromagnetic energy. Metallic trays or packaging materials are not used in MW heating. For an even distribution of the electromagnetic energy, fans are used in the thawing chamber (Figure 12.14).

The main advantage of both electrical methods is the fast heating rate, which reduces weight loss during thawing, compared to conventional heating. Due to shorter time, there is less danger of contamination during thawing.

12.5 COLD STORAGE OF FOODS

12.5.1 Cold Storage Rooms

Refrigerated foods are stored for a short or long time in special insulated storage rooms, which are kept at a constant low temperature, using mechanical refrigeration. The refrigerated storage conditions of temperature and relative humidity for various foods are described in the literature (IIR, 1967; Rao, 1992).

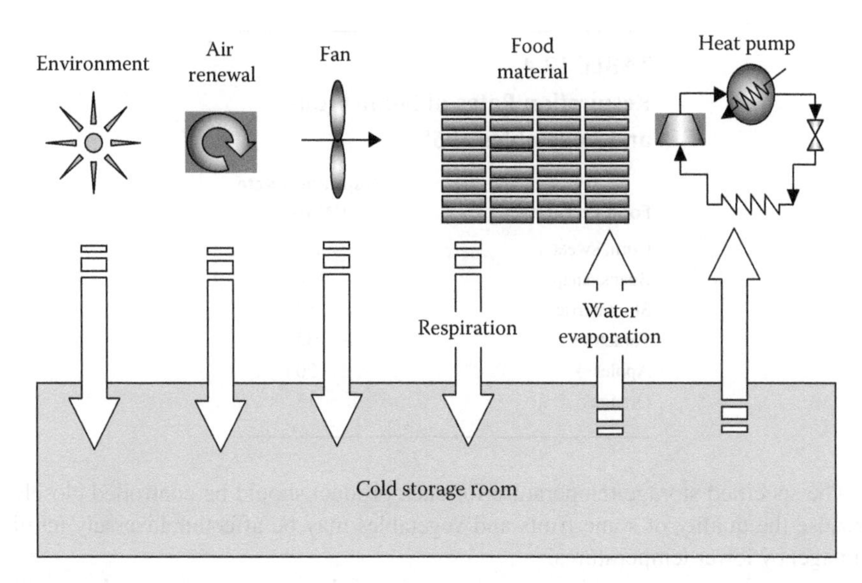

FIGURE 12.15 Diagram of energy flow in a cold storage room. (Modified from Maroulis, Z.B. and Saravacos, G.D., *Food Process Design*, CRC Press, Boca Raton, FL, May 9, 2003, fig. 5.14.)

The refrigeration load of cold storage rooms involves calculations for the removal of the following heat loads (Figure 12.15):

- Sensible heat of the fresh product
- Heat of respiration of the fresh product (W/ton)
- Heat produced by the equipment (fans, blowers, etc.)
- Heat due to leakage (thermal losses)
- Heat produced by personnel, light, etc.

In several applications, a secondary refrigeration system (brine), e.g., ethylene glycol solution, is used to transfer the heat load from the room/product to the primary (compression) refrigeration system, which is usually installed in a separate room.

High humidity may be required during cold storage of several foods. Humidification of the air can be achieved by various humidifiers, which can disperse water in the form of vapor or small droplets (nebulizers).

12.5.2 Cold Storage Conditions

The storage life of fresh fruits and vegetables is extended at low temperatures in the range of 0°C–15°C, depending on the variety of the product. High relative humidities in the range of 80%–99% are required in order to prevent the loss of product weight (moisture) during storage. The storage room should be insulated, and refrigeration losses should be minimized (Figure 12.15).

TABLE 12.4
Respiration Rates of Some Fruits and Vegetables at 5°C

Food Product	Respiration Rate (W/ton)
Corn, sweet	230
Beans, snap	100
Strawberries	50
Tomatoes	35
Apples	20
Oranges	15

The specified storage temperature for each product should be controlled closely because the quality of some fruits and vegetables may be affected adversely (chill damage) by lower temperatures.

A significant amount of heat may be produced during storage by respiration of some fruits and vegetables, which if not removed properly, can damage the quality of the product. The respiration rate is expressed in (W/ton) and it increases almost linearly with the storage temperature (Table 12.4). It varies over a wide range, e.g., 120 W/ton for sweet corn and 15 W/ton for oranges. Different food products stored together in the same room should be compatible, in terms of aroma, flavors, and ethylene evolved.

Fresh food products of animal origin (meat), fish, and prepared dishes should be stored at temperatures just above freezing (near 0°C) and high % RH to prevent moisture loss. Such foods may have a short storage life of 3–7 days (Table 12.5).

TABLE 12.5
Storage Conditions of Some Fruits and Vegetables

Food Material	Temperature (°C)	Humidity (% RH)	Storage Life
Apples	2	90–95	1–6 months
Apricots	0	90	2 weeks
Bananas	15	90	2–4 weeks
Beans, green	5	90	3 weeks
Cabbage	0	95–99	2–3 months
Carrots	0	95–99	2 months
Grapes	0	90	2–4 weeks
Lettuce	1	95–99	2 weeks
Onion	0	80	2–3 months
Peaches	0	95–98	2–4 weeks
Pears	0	90–95	2–5 months
Oranges	2	85–90	3–4 months
Potatoes	4	90–95	5–8 months
Tomatoes	10	90	1–2 weeks

12.5.3 CA Storage

Controlled atmosphere (CA) or Modified Atmosphere (MA) storage rooms, containing less oxygen and more carbon dioxide than normal, can improve the storage life of fresh foods. CA storage rooms are airtight structures, maintaining a low temperature, e.g., 3°C, a high humidity, e.g., 90% RH, and a modified composition of oxygen and carbon dioxide. Under these conditions, fresh fruits and vegetables, and some packaged meals can be stored for an extended time without significant spoilage or quality deterioration.

Oxygen in atmospheric air (21%) is reduced to about 5% by catalytic combustion of propane or methane. The oxygen content in the CA room may be reduced to about 3% by equilibration with stored food products.

Carbon dioxide is maintained constant at about 3% by chemical or physical scrubbing of the room's air. Chemical scrubbing is affected by absorption in alkali solutions, while physical scrubbing is based on adsorption by activated carbon or molecular sieves. The physical adsorbents may reduce the excessive ethylene content of the air in the storage of some fruits.

APPLICATION EXAMPLES

Example 12.1

A single stage refrigeration system using refrigerant R134A is operated between evaporator and condenser temperatures $T_e = -20°C$ and $T_c = 30°C$, respectively. The refrigeration load to the evaporator is $Q_e = 20\,kW$. The saturated refrigerant vapors are compressed following a constant entropy (S) line of the pressure–enthalpy diagram. The condensed liquid is expanded adiabatically to the evaporator without subcooling.

(a) Show the refrigeration operation on a pressure–enthalpy diagram of R134A.
(b) Calculate the required flow rate of the refrigerant.
(c) Calculate the coefficient of performance (COP) and the required power of the compressor.

Solution

(a) Figure 12.16 shows a semilogarithmic plot of pressure–enthalpy for refrigerant *R134A*. The thermal load Q_e evaporates the refrigerant at a constant temperature ($T_e = -20°C$) following line (4–1) with the heat of evaporation $\Delta H_r(4-1) = 140\,kJ/kg$. The saturated vapors are compressed following the constant entropy (S) line (1–2). The superheated vapors are cooled following line (2–2′) and condensed at a constant temperature ($T_c = 30°C$) following line (2′–3). The saturated liquid is expanded directly into the evaporator (no subcooling) following the constant enthalpy (adiabatic) line (3–4).

(b) The flow rate (F_r) of the refrigerant will be $F_r = 20/\Delta H_r(4-1) = 20/140 = 0.143\,kg/s$ or $F_r = 515\,kg/h$. Note that a relatively large refrigerant (*R134A*) flow rate is required because of its low heat of evaporation.

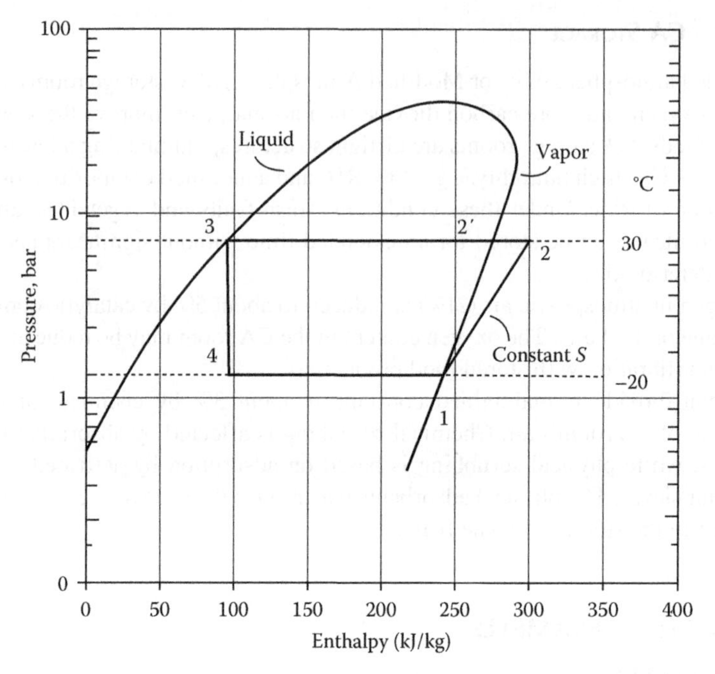

FIGURE 12.16 Diagram of refrigeration cycle of *R134A* for Example 12.1.

(c) The coefficient of performance of the refrigeration system is calculated from the equation,

$$COP = T_e/(T_c - T_e) - (C_{pr}T_e)/\Delta H_r = 253/50 - (1.5 \times 253)/215 = 5.10 - 1.76 = 3.34.$$

The maximum (theoretical) *COP* in this example is $COP = T_e/(T_c - T_e) = 5.10$.

The compressor power will be $E_c = Q_e/COP = 20/3.34 = 6.0\,\text{kW}$. The thermal load to the condenser will be $Q_e = 20.0 + 6.0 = 26\,\text{kW}$. Note that the compressor increases the thermal load to the condenser by 30%.

Example 12.2

A two-stage (I, II) refrigeration system using refrigerant *R717* (ammonia) is used to transfer a thermal load of 20 kW from an evaporator temperature of −40°C to a condenser operated at 30°C. The intermediate temperature of the evaporator/condenser (intercooler) is −10°C. The saturated refrigerant vapors from evaporators (I) and (II) are compressed following constant entropy (*S*) lines of the pressure–enthalpy diagram. The condensed liquid is expanded adiabatically to the evaporator without subcooling.

(a) Show the refrigeration operation on a pressure–enthalpy diagram of *R717*.
(b) Calculate the required flow rate of the refrigerant, the coefficient of performance (*COP*), and the required power of the compressor for both refrigeration cycles.

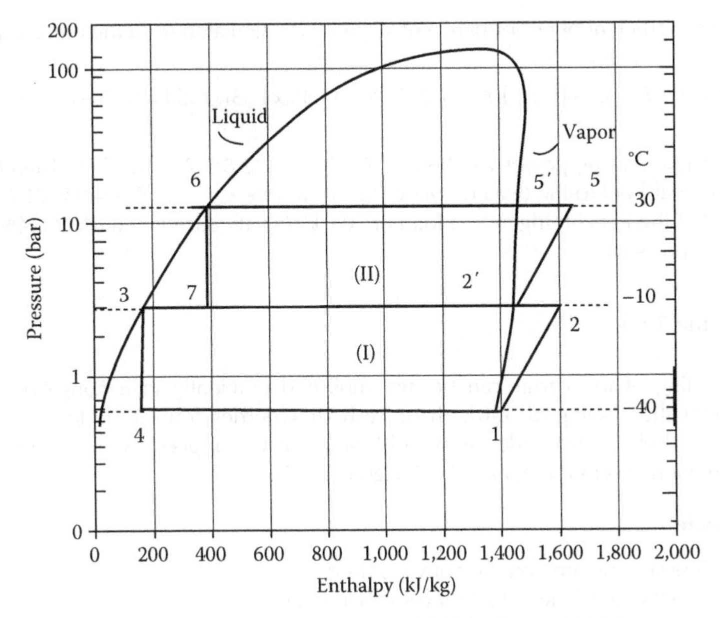

FIGURE 12.17 Diagram of refrigeration cycle of *R717* for Example 12.2.

Solution

(a) Figure 12.17 shows a semilogarithmic plot of pressure–enthalpy for refrigerant *R717*. The thermal load Q_e evaporates the refrigerant of stage (I) at a constant temperature ($T_e = -40°C$) following line (4–1) with the heat of evaporation $\Delta H_r(4–1) = 1200\,kJ/kg$. The saturated vapors are compressed following the constant entropy (S) line (1–2). The superheated vapors are cooled following line (2–2′) and condensed at a constant intermediate temperature ($T_i = -10°C$) following line (2′–3). The saturated liquid is expanded directly into the evaporator (no subcooling) following the constant enthalpy (adiabatic) line (3–4).

The saturated vapors of stage (II) are compressed following the constant entropy line (2′–5), the superheated vapors are cooled following line (5–5′), condensed at constant temperature 30°C following line (5′–6), and expanded adiabatically, following line (6–7).

(b) The flow rate (F_r) of the refrigerant *R717* in stage (I) will be $F_r = 20/\Delta H_r(4–1) = 20/160 = 0.0167\,kg/s$ or $F_r = 60\,kg/h$. Note that the flow rate of refrigerant *R717* (ammonia) is relatively low because of its high heat of evaporation. The coefficient of performance of refrigeration stage (I) is calculated from the equation

$$COP = T_e/(T_c - T_e) - \left(C_{pr}T_e\right)/\Delta H_r = 233/30 - (4.35 \times 233)/1350 = 7.77 - 0.75 = 7.02$$

The compressor (I) power will be $E_c = C_e/COP = 20/7.02 = 2.8\,kW$. Therefore, the condenser (I) thermal load will be $Q_c = 20.0 + 2.8 = 22.8\,kW$. The refrigerant flow rate (F_r) in the evaporator of stage (II) will be $F_r = 22.8/\Delta H_r(2'–7) = 22.8/1050 = 0.022$ or $F_r = 78\,kg/h$. The evaporator load is denoted as C_e or Q_e.

The coefficient of performance of stage (II) is calculated from the equation

$$COP = T_e/(T_c - T_e) - (C_{pr}T_e)/\Delta H_r = 263/40 - (4.35 \times 263)/1350 = 6.58 - 0.85 = 5.73$$

The compressor (II) power will be $E_c = C_e/COP = 22.8/5.73 = 4.0\,\text{kW}$. Therefore, the thermal load to the (water) condenser (II) will be $Q_c = 22.8 + 4.0 = 26.8\,\text{kW}$. Note that the initial refrigeration load of 20.0 kW is increased by 6.8 kW (34%) by the compressors.

Example 12.3

Air cooling of fresh fruit can be accomplished efficiently in a conveyor belt. Consider the cooling of 500 kg/h of fresh strawberries from 32°C to 2°C on a conveyor belt 1.00 m wide, using cold air at −8°C, supplied by a compression refrigeration unit of refrigerant *R717* (Figure 12.18).

Calculate

(a) The cooling time on the conveyor belt
(b) The length of the belt, if its velocity is 0.5 m/s
(c) The thermal load to the evaporator (Q_e) of the *R717* refrigeration system
(d) The required power of the compressor, if the coefficient of performance is $COP = 3.5$
(e) The required flow rate of cooling water in the condenser, if its temperature at the entrance and exit is 15°C and 25°C, respectively

Technical Data

Density of fruit, $\rho = 1000\,\text{kg/m}^3$
Size of fruit (spherical diameter), $d = 3\,\text{cm}$
Specific heat of fruit, $C_p = 3.80\,\text{kJ/(kg K)}$
Specific heat of water, $C_p = 4.20\,\text{kJ/(kg K)}$
Heat transfer coefficient, condenser, $U = 0.75\,\text{kW/(m}^2\,\text{K)}$
Evaporator temperature, $T_e = -18°\text{C}$
Condenser temperature, $T_c = 35°\text{C}$
Belt velocity, $u = 0.5\,\text{m/min}$

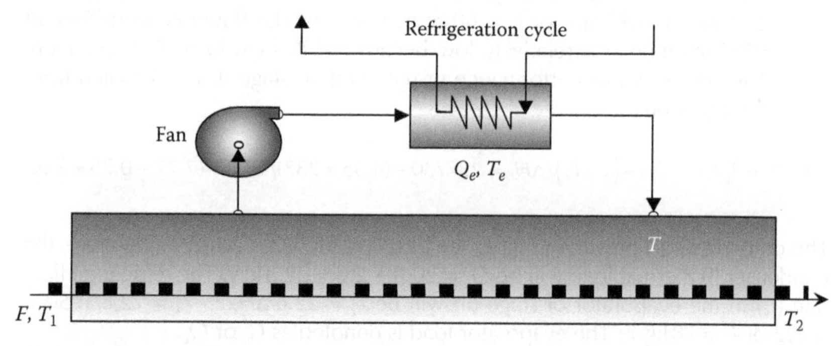

FIGURE 12.18 Fruit cooling on a conveyor belt.

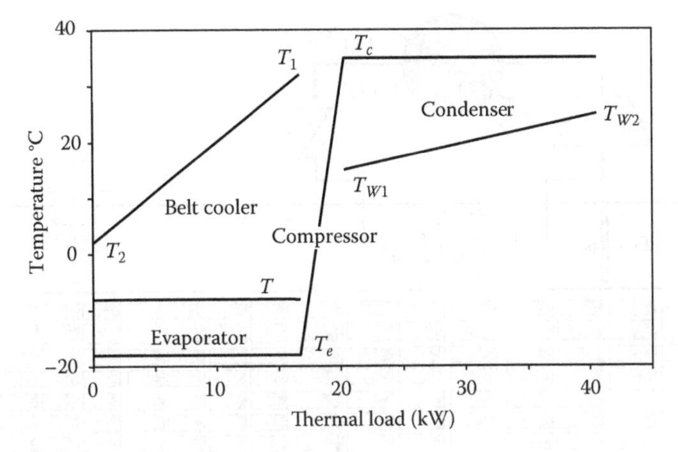

FIGURE 12.19 Temperature–thermal load diagram of Example 12.3.

Void fraction of fruit on belt, $\varepsilon = 0.50$
Load depth of fruit on belt, $z = 10\,\text{cm}$
Thermal diffusivity of fruit, $\alpha = 1.18 \times 10^{-7}\ \text{m}^2/\text{s}$

Solution

(a) The cooling time t (s) for a single fruit of diameter d (m) is given by the unsteady-state heat conduction equation, $t = t_c \ln[(T_1 - T)/(T_2 - T)]$, where $t_c = d^2/(\pi^2\alpha)$ is the cooling time constant (s), and α (m²/s) is the thermal diffusivity (Chapter 2). In this example $\alpha = 1.18 \times 10^{-7}\ \text{m}^2/\text{s}$. $T = -8°\text{C}$ is the temperature of the cooling air, and $T_1 = 32°\text{C}$, $T_2 = 2°\text{C}$ are the initial and final temperatures of the fruit, respectively.

Thus, $t_c = (0.03)^2/[(3.14)^2 \times 1.18 \times 10^{-7}] = 770\,\text{s} = 0.215\ \text{h}$, and

$$t = 0.215 \ln[(32+8)/(2+8)] = 0.215 \times 1.386 = 0.298\ \text{h or}\ t = 18\ \text{min}.$$

(b) The length of the conveyor belt will be $L = u\,t = 0.5 \times 18 = 9\,\text{m}$.

(c) The thermal load of the evaporator (Q_e) is equal to the heat removed from the fruit on the cooling belt, $Q_e = m\,C_p\,(T_1 - T_2) = (500/3600) \times 3.80 \times 30 = 15.8\,\text{kW}$.

(d) The power of the compressor will be, $E_c = 15.8/3.5 = 4.5\,\text{kW}$.

(e) The thermal load of the condenser will be, $Q_c = Q_e + E_c = 15.8 + 4.5 = 20.3\,\text{kW}$. The required flow rate of water F_w (kg/s) is calculated from the equation

$$Q_c = F_w C_p (T_1 - T_2) \quad \text{or} \quad F_w = Q_c/[C_p(T_1 - T_2)] = 20.3/(4.2 \times 10) = 0.483\ \text{kg/s} = 1740\ \text{kg/h}.$$

Figure 12.19 shows a diagram of the thermal loads for Example 12.3.

Example 12.4

Fresh fruits and vegetables are preserved in cold storage rooms for several months. Figure 12.20 shows a diagram of the main components of a typical cold storage room.

The food material of mass M (tons) is kept in an insulated room for the appropriate time t (h/y). The cold room is rectangular with dimensions D (m) × L (m),

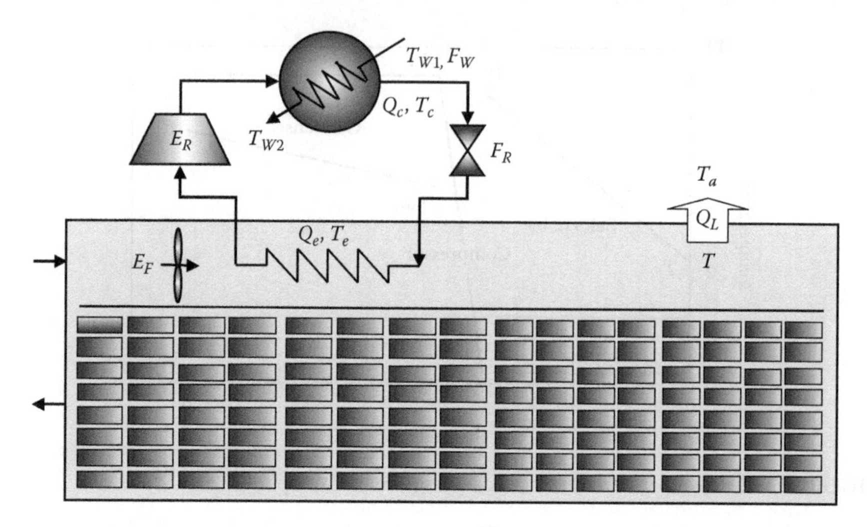

FIGURE 12.20 Diagram of a cold storage room. (Modified from Maroulis, Z.B. and Saravacos, G.D., *Food Process Design*, CRC Press, Boca Raton, FL, May 9, 2003, fig. 5.13.)

height Z (m), and volume V (m^3). The stored material is packed in boxes so that a void (empty) fraction ε (–) is created, which permits circulation of the cold air.

The cold storage room is kept at temperature T (°C) by circulation of cold air with superficial velocity u (m/s) via a fan of power E_f (kW). The circulated air is cooled in the evaporator of a heat pump (compression refrigeration system). The refrigerant is evaporated at temperature T_e (°C) in the evaporator, removing a thermal load Q_e (kW). It is compressed by a compressor of power E_c (kW), then it is liquefied at temperature T_c (°C) in the condenser of thermal load Q_c (kW) via cooling water, and finally it is expanded through a valve to the refrigerant evaporator.

Thermal energy Q_L (kW) is transferred into the room through the insulated wall from the environment, which is at a higher temperature T_a (°C) than that of the cold room T (°C). Thermal energy Q_A (kW) is also introduced through the air renewal. The fan also adds a thermal load Q_F (kW). The food material generates thermal energy Q_R (kW) due to respiration and loses thermal energy Q_W (kW) due to water, which is evaporated during the storage.

Application to Apple Storage

Process Data

$M = 360$ t Mass of apples
$\varepsilon = 0.85$ Packing void (empty) fraction of room
$T_a = 35$°C Ambient air temperature
$T = 2$°C Cold room temperature
$T_{W1} = 15$°C Cooling water inlet temperature
$T_{W2} = 25$°C Cooling water outlet temperature
$T_c = 35$°C Refrigerant condensing temperature
$T_e = -8$°C Refrigerant evaporating temperature
$t_{ar} = 6$ h Required time for air removal
$L = 38$ m Cold room length

$D = 15$ m Cold room width
$Z = 5$ m Cold room height
$u = 1$ m/s Superficial air velocity

REFRIGERANT *R717* (AMMONIA)

Property Data

$\rho_s = 850$ kg/m^3 Product density
$\rho_a = 1$ kg/m^3 Air density
$C_{pw} = 4.20$ kJ/(kg K) Water specific heat
$C_{pa} = 1.00$ kJ/(kg K) Air specific heat
$C_{pr} = 4.35$ kJ/(kg K) Refrigerant specific heat (liquid *R717*)
$\Delta H_w = 2500$ kJ/kg Water heat of evaporation (0°C)
$\Delta H_r = 1350$ kJ/kg Refrigerant *R717* heat of evaporation

Heat Transfer Coefficients

$U_L = 0.15$ W/(m^2 K) Overall heat loss coefficient through the wall
$U_E = 0.25$ kW/(m^2 K) Heat transfer coefficient at the refrigerant evaporator
$U_C = 0.75$ kW/(m^2 K) Heat transfer coefficient at the refrigerant condenser

Storage Data

$w = 0.50$ kg/(t day) Water loss rate
$r = 20$ W/t Heat of respiration
$t_{ar} = 6$ h Air renewal time

CALCULATIONS

Apples in Storage Room

The volume of stored apples is $V_a = 360/0.85 = 423$ m^3. The given dimensions of the cold stage room are assumed to measure the internal length ($L = 38$ m), width ($D = 15$ m), and height ($Z = 5$ m). Therefore, the volume of the room will be $V_a = 38 \times 15 \times 5 = 2850$ m^3. The empty volume of the room is $2850 - 423 = 2427$ m^3. Therefore, the void (empty fraction) of the room will be $\varepsilon = 2427/2850 = 0.85$, which agrees with the given data (ε).

It should be noted that the void fraction ($\varepsilon = 0.85$) refers to the volume of the room that is not occupied by apples, including space between the apples, containers, intermediate spaces, and corridors. The void fraction of bulk stored apples is much smaller, which can be estimated from the bulk density (ρb) and the material density (ρ) of the apples, according to the equation, $\varepsilon = 1 - (\rho b/\rho)$. For the normal values of $\rho = 850$ kg/m3 and $\rho b = 600$ kg/m3 (literature), it follows that $\varepsilon = 1 - (600/850) = 0.30$.

The external surface area of the room, which is subject to heat transfer (A), is calculated from the external dimensions of the room, assuming that the thickness of the walls and the roof is equal to 25 cm:

$$A = (38.50) \times (15.50) + 2 \times (38.50) \times (5.25) + 2 \times (15.50) \times (5.25) = 1163.75 \text{ m}^2$$

The heat transfer between the room floor and the ground is neglected.

Heat losses (heat input) in the room:

Heat transfer from the environment into the room

$$Q_L = AU\Delta T = (1163.75) \times (0.15) \times (35 - 2) = 5.76 \text{ kW}$$

Air renewal, $Q_A = \rho_a C_{pa} V (T_a - T)/t_{ar} = 1 \times 1 \times 2850 \times 33/(6 \times 3600) = 4.3\,\text{kW}$

Heat generated by the fan; assume the approximate value is, $Q_F = 5\,\text{kW}$. A detailed calculation of Q_F requires data on superficial air velocity in the room and friction factor (f).

Heat of respiration $Q_R = Mr = 360 \times 20 = 7.2\,\text{kW}$

Total heat input $Q_{IN} = Q_L + Q_A + Q_F + Q_R = 5.76 + 4.30 + 5.00 + 7.20 = 22.26\,\text{kW}$.

Note that the heat of respiration (7.20 kW) is the highest energy input in the system.

Cold generation (heat output) of the room:

Cold generation by water evaporation; $Q_W = wM\Delta H_w = 0.50 \times 360 \times 2500/(24 \times 3600)$ and $Q_W = 5.2\,\text{kW}$. Required cold generation by the heat pump (compression refrigeration system) for energy balance, $Q_{OUT} = Q_{IN} - Q_W = 22.26 - 5.20 = 17.1\,\text{kW}$.

The heat output is absorbed by the refrigeration evaporator, $Q_{OUT} = Q_e = 17.1\,\text{kW}$.

The surface area of the evaporator A_e (m^2) is calculated from the equation

$$Q_e = A_e U_e (T - T_e) \quad \text{or} \quad A_e = (17.1)/(0.25 \times 10) = 6.84 \text{ m}^2$$

The refrigerant flow rate F_R (kg/s) in the evaporator is calculated from the enthalpy balance equation, $Q_e = [\Delta H_r - C_{pr}(T_c - T_e)]F_R$ or $F_R = (17.1)/(1350 - 4.35 \times 43)$, $F_R = 0.015\,\text{kg/s}$ or $F_R = 53\,\text{kg/h}$.

The coefficient of performance (COP) of the refrigeration system in this example is given by Equation 12.3

$$COP = T_e/(T_c - T_e) - (C_{pr} T_e)/\Delta H_r = 265/43 - (4.35 \times 265)/1350$$
$$= 6.20 - 0.85 = 5.35.$$

The maximum (theoretical) COP in this example is, $COP = T_e/(T_c - T_e) = 6.2$.
The compressor power will be $E_c = C_e/COP = 17.1/5.35 = 3.2\,\text{kW}$.
The thermal load of the condenser will be $Q_c = Q_e + E_c = 17.1 + 3.2 = 20.3\,\text{kW}$.
The flow rate of cooling water, F_W (kg/s), in the condenser will be calculated from the equation, $Q_e = F_W C_{pw} (T_{w2} - T_{w2})$ and $F_W = (20.3)/(4.20 \times (25 - 15) = 0.48\,\text{kg/s} = 1.74\,\text{m}^3\text{/h}$.
Surface area of condenser $A_c = Q_c/(U_c\, \Delta T_{LM})$, where $\Delta T_{LM} = (\Delta T_1 - \Delta T_2)/\ln(\Delta T_1/\Delta T_2)$, $\Delta T_1 = (35 - 15) = 20$ and $\Delta T_2 = (35 - 25) = 10$, $\Delta T_{LM} = 10/\ln(2) = 10/0.69 = 14.5°\text{C}$.
$A_c = (20.3)/(0.75 \times 14.5) = 1.86\,\text{m}^2$.
The results of this example are shown in Figure 12.21.

Example 12.5

Blanched green peas are to be frozen in a fluidized bed system. Calculate the size of the system and the required utilities for freezing 0.5 tons/h of the product (refrigerant *R717*).

Figure 12.22 shows a diagram of the proposed fluidized bed freezer for the green peas.

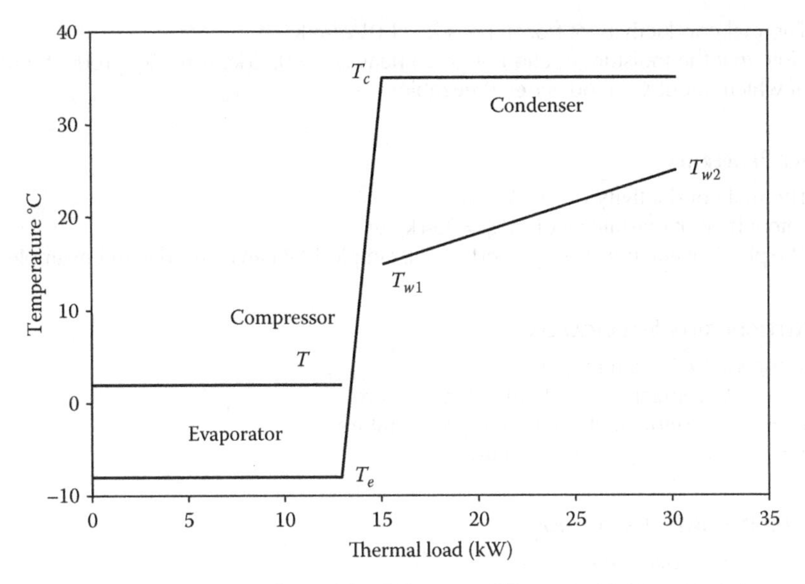

FIGURE 12.21 Temperature–thermal load diagram of Example 12.4.

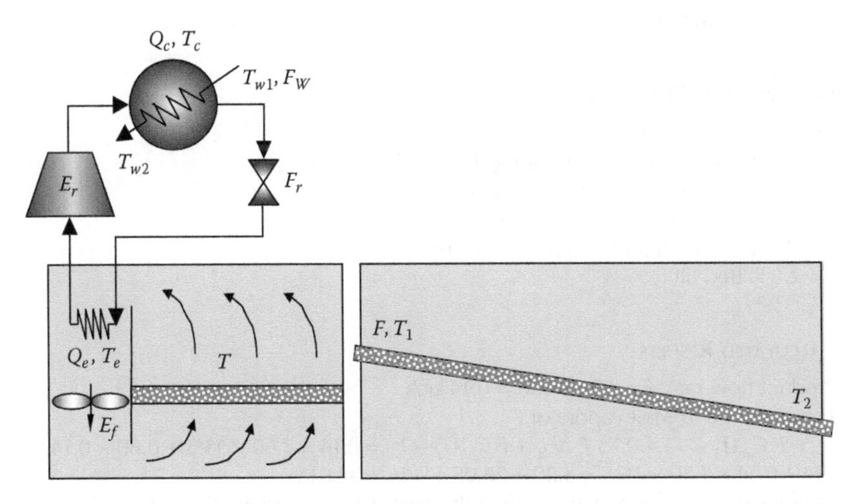

FIGURE 12.22 Diagram of the fluidized bed freezer of Example 12.5. (Modified from Maroulis, Z.B. and Saravacos, G.D., *Food Process Design*, CRC Press, Boca Raton, FL, May 9, 2003, fig. 5.20.)

Physical Properties of Green Peas

Freezable (free) water fraction $x = 0.60$
Density $\rho_s = 1000\,\text{kg/m}^3$
Specific heat of the unfrozen product $C_{pu} = 3.60\,\text{kJ/(kg K)}$
Specific heat of the frozen product $C_{pf} = 1.80\,\text{kJ/(kg K)}$
Initial freezing temperature $T_f = -1.5°\text{C}$

Thermal conductivity of frozen peas $\lambda = 1.0\,W/(m\,K)$
Note that the moisture fraction of green peas is $x = 0.75$ kg water/kg product, out of which about $x = 0.60$ is free (freezable).

Ice Properties

Thermal conductivity $\lambda = 1.5\,W/(m\,K)$
Latent heat of crystallization $\Delta H_L = 334\,kJ/kg$
The physical and transport properties of Example 12.4 can be used in this example.

Refrigeration Specifications

Refrigerant *R717* (ammonia)
$T_c = 35°C$ Refrigerant condensing temperature
$T_e = -35°C$ Refrigerant evaporating temperature
$T = -25°C$ Cooling air temperature

Fluidized Bed Specifications

$A_b = 1.35\,m^2$ Bed area
$D = 1$ m Bed width
$L = 1.35$ m Bed length
$Z = 0.10$ m Loading depth
$d = 2r = 6$ mm Particle diameter

Fluidized Bed Operation

$F = 0.5$ ton/h Material flow rate
$T_1 = 10°C$ Initial temperature of food
$T_2 = -18°C$ Final temperature of food
$T = -25°C$ Cooling air temperature
$h = 20\,W/(m^2\,K)$ Heat transfer coefficient
$s = 2.8\%$ Bed tilt

Calculated Results

Product flow rate, $F = 500/3600 = 0.14$ kg/s.
Thermal load to the evaporator,
$Q_e = F\,C_{pu}(T_1 - T_f) + 0.75\,F\,\Delta H_L + F\,C_{pf}\,(T_f - T_2) = 0.14 \times 3.60 \times 11.5 + 0.60 \times 0.14 \times 334 + 0.14 \times 1.80 \times 16.5 = 5.80 + 28.05 + 4.16 = 38.0\,kW$
Coefficient of performance (*COP*) of the compression refrigeration system
$COP = T_e/(T_c - T_e) - (C_{ps}\,T_e)/\Delta H_r = 238/70 - (4.35 \times 238)/1350 = 3.40 - 0.77 = 2.63$.
Note that the *COP* of the low temperature heat pump of this example is much lower than the *COP* of the refrigeration heat pump, operated at higher temperatures (Example 12.4).
Compressor energy $E_c = Q_e/COP = 38.0/2.63 = 14.45\,kW$.
Heat load to the condenser $Q_c = Q_e + E_c = 38.0 + 14.45 = 52.45\,kW$.
The surface area of the refrigeration evaporator A_e is calculated from the equation,
$Q_e = A_e U_e \Delta T$ and $A_e = 52.45/(0.25 \times 10) = 21.0\,m^2$.

The refrigerant flow rate F_R (kg/s) in the evaporator is calculated from the enthalpy balance equation, $Q_e = [\Delta H_r - C_{pr}(T_c - T_e)]F_R$ or $F_R = (52.45)/[1350 - 4.35 \times 70]$, $F_R = 0.05$ kg/s or $F_R = 180$ kg/h.

The flow rate of cooling water F_W (kg/s) in the condenser will be calculated from the equation, $Q_c = F_W C_{pw}(T_{w2} - T_{w2})$ and $F_W = 52.45/[4.20 \times (25 - 15)] = 1.25$ kg/s $= 4.5\,m^3/h$.

Surface area of condenser $A_c = Q_c/(U_c\,\Delta T_{LM})$, where $\Delta T_{LM} = (\Delta T_1 - \Delta T_2)/\ln(\Delta T_1/\Delta T_2)$, $\Delta T_1 = (35 - 15) = 20$ and $\Delta T_2 = (35 - 25) = 10$, $\Delta T_{LM} = 10/\ln(2) = 10/0.69 = 14.5$.

$A_c = (52.45)/(0.75 \times 14.5) = 4.82\,m^2$.

The mean freezing time of spherical green peas, assuming a mean diameter of 6 mm is calculated from the Plank Equation 12.15

$$t_f = [(\rho\,\Delta H_L)/(T_f - T_a)]\;[(2P_r/h) + ((4Rr^2/\lambda)] \qquad (12.15)$$

$t_f = [(1000 \times 334000)/23.5] \times [(2 \times 1 \times 0.003/6 \times 20) + (4 \times 1 \times 0.000009/24 \times 1]$ and $t_f = 731\,s = 0.2\,h$.

The freezing time ($t_f = 0.2$ h) is equal to the residence time in the fluidized bed of the freezer. The product holdup of the bed will be equal to $F\,t_f = 500 \times 0.2 = 100$ kg peas.

If the void fraction of the bed is assumed to be $\varepsilon = 0.5$, the volumetric flow rate of peas will be $V = 500/(1000 \times 0.5) = 1.0\,m^3/h$. The product hold up will have a volume of 100/0.5 L or $0.2\,m^3$. The loading depth is assumed to be $Z = 0.1$ m, and the bed surface will be $A = 0.2/0.1 = 2\,m^2$. If the width of bed is assumed to be $D = 1.0$ m, the length (flow path) L will be $L = 2.0$ m. The cross section of the pea flow will be $A = (1.0\,m) \times (0.1\,m) = 0.10\,m^2$, and the flow velocity of the peas in the bed will be $u = V/A = 1.0/0.10 = 10.0$ m/h.

The fan power (E_f) can be calculated from relevant data on the air flow rate and the pressure drop through the fluidized bed. An approximate value of $E_f = 10$ kW is suggested for this example. Thus, the total power required will be $E = E_c + E_f = 14.45 + 10.0 = 24.45$ kW. Figure 12.23 shows a diagram of the thermal loads of Example 12.5.

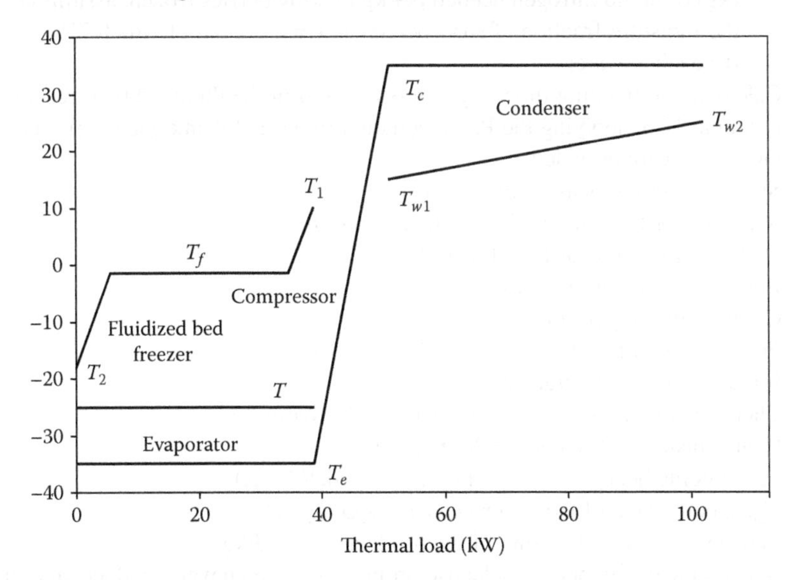

FIGURE 12.23 Temperature–thermal load diagram for Example 12.5.

PROBLEMS

12.1 Show that the unit I ton of refrigeration used in the United States is equal to 3.51 kW. Note that in this case 1 (short) ton is equal to 2000 lb.

12.2 Show that, for a given set of evaporator/condenser temperatures in a heat pump, the coefficient of performance (*COP*) decreases slightly, when the heat of evaporation of the refrigerant decreases, or when its specific heat increases.

12.3 A cold storage room requires a total refrigeration load of 30 kW, which is provided by a refrigeration cycle using refrigerant *R717* (ammonia) operating between pressures 3 bar (evaporator) and 15 bar (condenser).

Calculate the *COP* and the condenser load of the system using thermodynamic data for the refrigerant from the literature.

12.4 Calculate the freezing time of a pizza 3 cm thick being frozen in a cold air stream at −20°C, and using the Plank equation. The frozen pizza has a freezing point of −1.0°C, density $\rho = 1000\,kg/m^3$, and thermal conductivity $\lambda = 1$ W/(m K). The heat transfer coefficient in the cold air is $h = 20\,W/(m^2\,K)$. The flat surface of the pizza is much higher than the thickness (an infinite plate).

12.5 Calculate

(a) the freezing time of strawberries, 2.0 cm spherical diameter, in a fluidized bed of cold air at −20°C. The freezing point of strawberries is −2.0°C, the density $\rho = 1100\,kg/m^3$, and the thermal conductivity of the frozen product $\lambda = 1.2$ W/(m K). The heat transfer coefficient is $h = 25\,W/(m^2\,K)$. Use the simplified Plank equation.

(b) The freezing time of the above strawberries in liquid nitrogen at −150°C with a heat transfer coefficient of $h = 1.0\,kW/(m^2\,K)$. Estimate the amount (kg) of liquid nitrogen needed per kg of strawberries frozen, assuming that the moisture fraction of strawberries is $x = 0.80$, out of which 75% is free (freezable) water.

12.6 Calculate the freezing time of green peas, used in the fluidized bed freezing of Example 12.5, applying the Pham equation (modified Plank equation). The following data are provided:

Size of spherical peas, radius $r = 4$ mm
Shape factor for spheres $E = 3$ (Phan equation)
Density of green peas $\rho = 1000\,kg/m^3$
Initial freezing temperature $T_f = -1.5$°C
Cooling air temperature $T_a = -25$°C
Final temperature of frozen peas $T_{final} = -18$°C
Initial product temperature 10°C
Thermal conductivity of frozen peas $\lambda = 1.0$ W/(m K)
Heat transfer coefficient $h = 20\,W/(m^2\,K)$
C_{pu}, specific heat of unfrozen product = 3.6 kJ/(kg K)
C_{pf}, specific heat of frozen product = 1.8 kJ/(kg K)
ΔH_L, latent heat of freezing of the product = 250 kJ/kg

12.7 Discuss the difference in dielectric properties between water and ice, and their effect on the electrical thawing of frozen foods.

12.8 Calculate the shape factor and the freezing time of a circular pizza 30 cm diameter and 3 cm thick, using the Pham equation (modified Plank equation), applying the spreadsheet analytical method.

Data

Product half-thickness $r = 1.5\,\text{cm} = 0.015\,\text{m}$

Surface area $A = 3.14 \times (15)^2 = 706.5\,\text{cm}^2 = 0.0706\,\text{m}^2$

Volume $V = 3.14 \times (15)^2 \times (0.03) = 0.0021\,\text{m}^3$

Density of product $\rho = 1000\,\text{kg/m}^3$

Heat transfer coefficient $h = 50\,\text{W/(m}^2\,\text{K)}$

Thermal conductivity if frozen product $\lambda = 1.0\,\text{W/(m K)}$

T_i, initial product temperature = 10°C

T_{fin}, initial freezing temperature = −1.5°C

T_a, temperature of freezing medium = −25°C

C_{pu}, specific heat of unfrozen product = 3.60 kJ/(kg K)

C_{pf}, specific heat of frozen product = 1.80 kJ/(kg K)

ΔH_L, latent heat of freezing of the product = 250 kJ/kg

T_{fm}, mean initial freezing temperature = $1.8 + 0.263\,T_{fin} + 0.105\,T_a$, °C

$\Delta T_1 = 0.5\,(T_i + T_{fm}) - T_a$, °C

$\Delta T_2 = T_{fm} - T_a$, °C

ΔH_1, volumetric enthalpy for the precooling period = $\rho C_{pu}\,(T_i - T_{fin})$, kJ/m³

ΔH_2, volumetric enthalpy of phase change = $\rho \Delta H_L + \rho C_{pf}\,(T_{fm} - T_{fin})$, kJ/m³

LIST OF SYMBOLS

A	m²	Surface area
Bi	Biot number	[(h r)/λ]
BP	°C	Boiling point
CA	—	Controlled atmosphere
COP	—	Coefficient of performance
C_p	kJ/(kg K)	Specific heat
d, D	m	Diameter
E	kJ	Enthalpy
F	kg/s	Flow rate
G	m/s²	Acceleration of gravity (9.81)
G	kg/(m² s)	Mass flux (u ρ)
h, H	m	Height
H	W/(m² K)	Heat transfer coefficient
j_H	—	Heat transfer factor [St Pr$^{2/3}$]
HFC	—	Hydrochlorofluorocarbons
L	m	Length
M	kg	Mass
M	g/s	Mass flow rate (G A)
N	—	Efficiency
N	1/min	Speed of revolution (RPM)
Nu	—	Nusselt number [(h d)/λ]

q, *Q*	kJ	Heat quantity
P	Pa, bar	Pressure
R	m	Radius
Re	—	Reynolds number [(d u ρ)/η]
RH	%	Relative humidity
R22	—	Chlorodifluoromethane
R124A	—	Tetrafluorethane
R717	—	Ammonia
S	kJ/K	Entropy
T	s	Time
T	°C or K	Temperature
U	m/s	Velocity
U	$W/(m^2 K)$	Overall heat transfer coefficient
V	m^3	Volume
X	m	Thickness
X	—	Mass fraction
ΔH	kJ/kg	Heat of phase change (evaporation, melting)
Δp	Pa	Pressure difference
ΔT	°C or K	Temperature difference
η	Pa s	Viscosity
λ	W/(m K)	Thermal conductivity
ρ	kg/m^3	Density

REFERENCES

ASHRAE. 1993. *ASHRAE Handbook-Fundamentals-Refrigeration*. American Society of Heating, Refrigeration and Air-Conditioning Engineers, Atlanta, GA.

Cleland, A.C. 1990. *Food Refrigeration Processes: Analysis, Design and Simulation*. Elsevier, London, U.K.

Cleland, A.C. and Valentas, K.J. 1997. Prediction of freezing time and design of food freeze. In: *Handbook of Food Engineering Practice*. K.J. Valentas, E. Rotstein, and R.P. Singh, eds. CRC Press, New York.

Heldman, D. 1992. Food freezing. In: *Handbook of Food Engineering*. D.R. Heldman and D.B. Lund, eds. Marcel Dekker, New York.

IIR. 1967. *Recommended Conditions for Cold Storage of Perishable Produce*. International Institute of Refrigeration, Paris, France.

IIR. 1972. *Recommendations for Processing and Handling of Frozen Foods*. International Institute of Refrigeration, Paris.

Maroulis, Z.B. and Saravacos, G.D. 2003. *Food Process Design*. Marcel Dekker, New York.

Rahman, S. 2009. *Food Properties Handbook*, 2nd edn. CRC Press, New York.

Rao, M.A. 1992. Transport and storage of food products. In: *Handbook of Food Engineering*. D.R. Heldman and D.B. Lund, eds. Marcel Dekker, New York.

Saravacos, G.D. and Kostaropoulos, A.E. 2002. *Handbook of Food Processing Equipment*. Kluwer Academic/Plenum Publ., New York.

Shelton, M.R. and Grossmann, I.E. 1985. A shortcut procedure for refrigeration systems. *Comput. Chem. Eng*. 9(6): 615–619.

Singh, R.P. and Sarkar, A. 2005. Thermal properties of frozen foods. In: *Engineering Properties of Foods*, 3rd edn. M.A. Rao, S.S.H. Rizvi, and A.K. Datta, eds. CRC Press, New York.

13 Mass Transfer Operations

13.1 INTRODUCTION

Mass transfer operations are used in food processing for physical separation of components from liquids or solids, for recovering valuable products, or for removing undesirable food or nonfood components. The controlling transport mechanism is mass transfer at the molecular level (Chapter 2) (Treybal, 1980; Cussler, 1997).

Table 13.1 lists some mass transfer operations used in food processing in which the controlling mechanism is mass transfer within the liquid or solid phase and at the phase boundaries. Distillation and extraction find more industrial applications than the other operations (King, 1982; Perry and Green, 1867).

Mass transfer operations are based on two fundamental physical processes, i.e., phase equilibria and mass transfer. Both processes are controlled by molecular forces and they are evaluated and predicted by molecular dynamics or empirical correlations (Prausnitz et al, 1994; Geankoplis, 1993).

Phase equilibria indicate the final concentrations of the components of two phases after long enough time to reach thermodynamic equilibrium in which the activity (or partial vapor pressure) of each component is equal in both phases.

Phase equilibria are calculated by equating the partial pressure of the various components in the two phases. The partial pressure of a component in ideal liquid solutions is proportional to its concentration and vapor pressure (Raoult's law). In nonideal solutions, the partial pressure is also proportional to its activity coefficient, which is usually higher than 1. Aqueous solutions of volatile components have very high activity coefficients, making easier the removal of food volatiles during distillation (Saravacos, 2005; Saravacos and Maroulis, 2001).

Mass transfer rates in separation processes are important for the quick attainment of thermodynamic equilibrium. Mass transfer in gases and liquids is fast, due to the molecular motion and, in liquids, it can be enhanced by mechanical mixing or high flow velocity (turbulence) (Schweitzer, 1988).

Mass transfer in solid foods is slow, and it is controlled by molecular diffusion or other transport mechanisms (Chapter 2). The equipment used in mass transfer operations has been developed mostly in the chemical and petrochemical industries. Mass transfer operations involving gases and liquids, are important in operations such as distillation and absorption. Solid/fluid operations, such as leaching, adsorption, and ion exchange have been developed from empirical and industrial experience, and they are specific for a given separation system and product (Gekas, 1992).

Equipment used in food mass transfer operations is similar to chemical engineering equipment, with special requirements for hygienic (sanitary) design and corrosion due to the presence of free water in food products.

TABLE 13.1
Mass Transfer Operations (Separations) in Food Processing

Operation	Basis of Operation	Applications
Distillation	Volatility	Recovery of volatiles
Extraction/leaching	Solubility	Oil from seeds, sugar from beets
Absorption	Solubility	Absorption of O_2, CO_2
Adsorption/ion exchange	Sorption capacity	Removal of components
Crystallization	Solubility	Salt, sugar

The design of distillation and extraction is described in more detail in this chapter, since they constitute the most important mass transfer operations used in food processing.

13.2 DISTILLATION

The design and operation of distillation equipment requires vapor/liquid equilibrium data for the estimation of the theoretical separation stages (van Winkle, 1967).

13.2.1 VAPOR/LIQUID EQUILIBRIA

In ideal solutions, the partial pressure p_i of a component i in the liquid phase is given by Raoult's law:

$$p_i = x_i p_i^o \tag{13.1}$$

where

x_i is the mole fraction (molar concentration) of i
p_i^o is the vapor pressure of pure i at the given temperature

The p_i^o is taken from Tables or is calculated from the Antoine equation as a function of the temperature.

The mole fraction x_i of a component i of mass fraction w_i (kg/kg) in a liquid mixture is calculated from the following equation:

$$x_i = \frac{(w_i/M_i)}{\sum (w_i/M_i)} \tag{13.2}$$

where M_i (kg/kmol) is the molecular weight of i. Note that $\sum x_i = 1$ and $\sum w_i = 1$. The same equation is used to calculate the mole fraction of the vapor phase y_i.

13.2.1.1 Activity Coefficients and Relative Volatility

Aqueous solutions of food volatile components are highly nonideal, and the partial pressure p_i is given by the following equation (Reid et al, 1967):

$$p_i = \gamma_i \, x_i \, p_i^o \tag{13.3}$$

where ($\gamma_i > 1$) is the activity coefficient of i in the mixture.

The vapor phase in food systems can be considered as ideal, i.e., the Dalton law is applicable:

$$p_i = y_i P \tag{13.4}$$

where

y_i is the mole fraction (molar concentration) in the vapor phase
P is the total pressure

In food systems, the vapor phase is considered as ideal, since most food processing operations are carried out at atmospheric pressure or in vacuum. Nonideal gas phases characterize special high-pressure operations (e.g., $P > 10$ bar), such as supercritical fluid extraction.

At equilibrium, the partial pressure of a component i is the same in both phases, i.e.,

$$y_i P = \gamma_i x_i p_i^o \tag{13.5}$$

$$y_i = \left(\frac{\gamma_i p_i^o}{P} \right) x_i = K_i x_i \tag{13.6}$$

where ($K_i = y_i/x_i$) is the partition coefficient of component i between the two phases, which is directly proportional to the activity coefficient γ_i. The activity coefficients are preferred in most distillation applications, since they can be calculated and correlated, using computer techniques.

The relative volatility of component i to component j is defined by the following equation:

$$\alpha_{ij} = \left(\frac{K_i}{K_j} \right) = \left(\frac{(y_i/x_i)}{(y_j/x_j)} \right) = \left(\frac{y_i}{y_j} \right) \left(\frac{x_j}{x_i} \right) \tag{13.7}$$

For ideal mixtures

$$\alpha_{ij} = \frac{p_i^o}{p_j^o} \tag{13.8}$$

For nonideal mixtures at relatively low pressures (ideal vapor phase)

$$\alpha_{ij} = \left(\frac{\gamma_i}{\gamma_i}\right)\left(\frac{p_i^o}{p_j^o}\right) = \left(\frac{K_i}{K_j}\right) \tag{13.9}$$

The vapor/liquid equilibrium data at a constant pressure are usually plotted in y–x diagrams, according to the equilibrium equation:

$$y_i = \frac{(\alpha_{ij}\, x_i)}{[1+(\alpha_{ij}-1)\, x_i]} \tag{13.10}$$

13.2.1.2 Nonideal Mixtures and Azeotropes

The activity coefficients of the various components of a nonideal mixture are determined experimentally or correlated in semiempirical or empirical relations. The activity coefficient γ_i is determined by measuring the concentrations of the vapor and liquid phases y_i, x_i and by applying Equation 13.3. The activity coefficients are strong functions of concentration of the liquid mixtures, reaching unity (1) at very low and very high concentrations ($x_i = 0$ and $x_i = 1$).

Empirical correlations are used to correlate the activity coefficients, such as the van Laar (two-parameter) and the Wilson (n-parameter) equations. The Wilson equation and its modifications are suited for computer calculations of multicomponent mixtures. The generalized correlations for multicomponent equilibrium Universal Quasi-Chemical (UNIQUAC) and Universal Function Activity Contribution (UNIFAC) are based on structural and thermodynamic parameters of the components of the liquid mixture.

Phase equilibrium data on multicomponent mixtures of chemicals are available in Data Banks, such as the DECHEMA and the AIChE collections. Equilibrium data are often available in the form of the partition coefficient ($K_i = y_i/x_i$) in various Tables and diagrams of the literature.

A useful application of partition coefficients K_i is the estimation of the bubble and dew points for mixtures at a constant pressure. The bubble point of a liquid mixture is the temperature at which the liquid begins to boil, and the following equation applies:

$$\sum K_i x_i = 1 \tag{13.11}$$

The dew point of a vapor mixture, at a constant pressure, is the temperature at which the mixture begins to condense, and the following equation applies:

$$\sum \left(\frac{y_i}{K_i}\right) = 1 \tag{13.12}$$

The highly nonideal mixture ethanol/water, forms an azeotropes, i.e., the mole fraction of a component is identical in both vapor and liquid phases, and the equilibrium

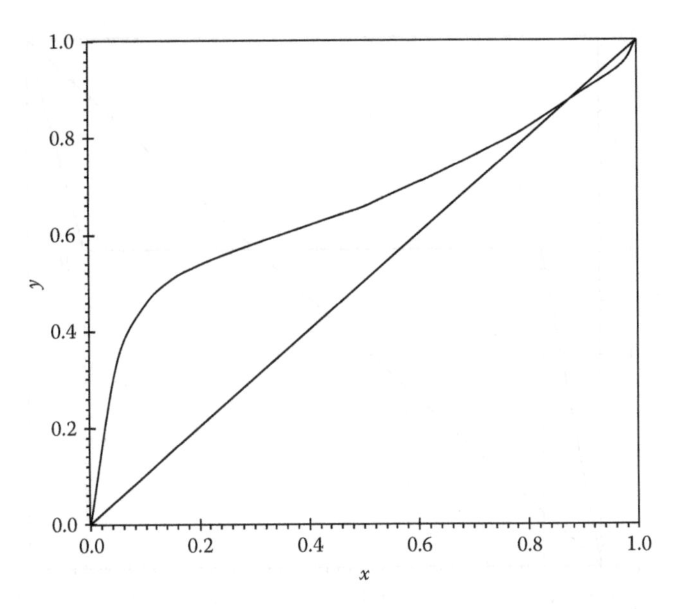

FIGURE 13.1 Vapor/liquid equilibrium (y, x) of ethanol/water at atmospheric pressure.

curve of Equation 13.9 crosses the diagonal ($y = x$). The ethanol/water azeotrope at atmospheric pressure is 95% ethanol by volume, with a boiling point of 78.1°C. This corresponds to a mole fraction of $y_i = x_i = 0.894$ (Figure 13.1).

Partially soluble components, such as organic aroma compounds in water, form azeotropes, with the equilibrium line becoming horizontal and crossing the diagonal. At low concentrations, the partially soluble components are very volatile in water.

In a partially soluble system, two layers are formed, a lower water layer W, saturated with the organic component A, and an upper layer of the organic component A, saturated with water W. At equilibrium, the partial pressure of each component is the same in the double layer liquid and in the vapor space above.

The relative volatility of a partially soluble component A in the water layer can be calculated, assuming an ideal vapor phase. The following approximate equation is derived for the relative volatility α_{AW} of component A in water W:

$$\alpha_{AW} = \left(\frac{x_A^A}{x_A^W}\right)\left(\frac{p_A^o}{p_W^o}\right) \tag{13.13}$$

where, x_A^A, x_A^W are the mole fractions of A in phases A and W, respectively, and p_A^o, p_W^o are the vapor pressures of (pure) components A and W at the equilibrium temperature.

Equation 13.13 predicts that the relative volatility of a slightly soluble component in water $\left(x_A^W \ll x_A^A\right)$ can be high, even for high-boiling components, i.e., compounds of vapor pressure lower than that of water ($p_A^o < p_W^o$).

Figure 13.2 shows the vapor/liquid equilibrium diagram of a partially miscible aroma component A in water W. Line OA represents the solubility of A in W, while

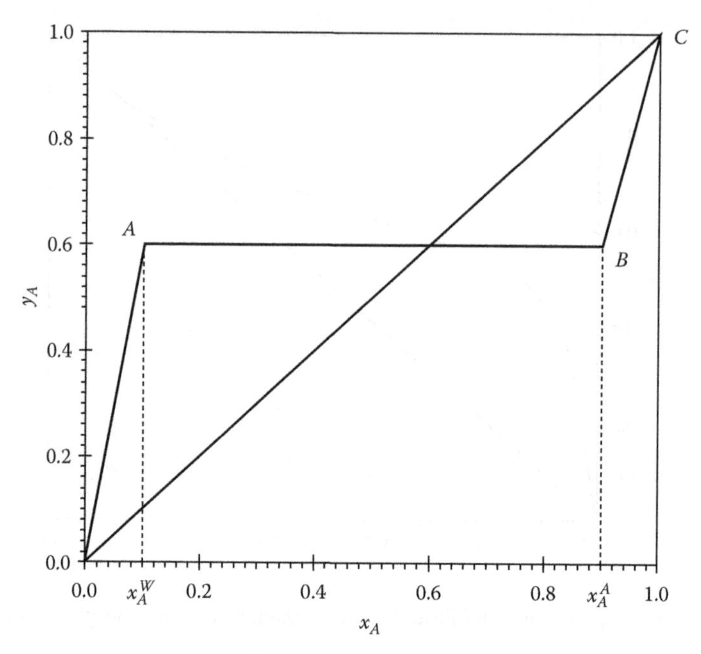

FIGURE 13.2 Equilibrium diagram of partially soluble components.

line BC shows the solubility of W in A, according to the Henry law $y_A = H_A x_A$ and $y_W = H_W x_W$, respectively. The Henry law is discussed in Section 13.4.

13.2.1.3 Volatile Food Aromas

The volatile components (aromas) of foods are organic compounds, usually partially soluble in water, which form highly nonideal aqueous solutions. The high activity coefficients of these compounds at very low (ppm) concentrations (infinite dilution activity coefficients) vary widely, depending on their molecular structure and the interactions with the water and the other food components (notably sugars).

The activity coefficients γ_i in aqueous solutions of some alcohols and esters, related to food aroma, range from about 4 (ethanol) to above 1000 (ethyl butyrate). They increase considerably, when the sugar concentration is increased, due to a "salting out" effect (Saravacos and Moyer, 1968).

In engineering applications, the relative volatilities α of aroma components A to water W are more useful than the activity coefficients. Some useful values of the relative volatility α_{AW}, determined from experimental activity coefficients γ_i data, are shown in Table 13.2.

Partially water-soluble compounds, such as ethyl acetate and ethyl butyrate, are much more volatile than water soluble components, such as ethanol. Methyl anthranilate, an aroma component of Concord grapes, partially soluble in water, has a relative volatility in water of about 4.0, although it has a very low vapor pressure (boiling point 266.5°C at atmospheric pressure).

In the distillation of wine and ethanol-fermentation products, the volatility of aroma components decreases, in general, as the ethanol concentration is increased.

TABLE 13.2
Relative Volatilities (α_{AW}) of Volatile Aroma Compounds in Water

	Relative Volatility (α_{AW})		
Aqueous Solution	**Water**	**15% Sucrose**	**60% Sucrose**
Ethanol	8.3	9.0	14.5
n-Propanol	9.5	10.0	18.5
n-Butanol	14.0	15.0	43.0
n-Amyl alcohol	23.0	25.0	105.0
Hexanol	31.0	34.0	195.0
Ethyl acetate	205	265	985
Ethyl butyrate	643	855	6500

The changes in volatility are important in the distillation of ethanol-fermentation liquors for the production of brandy and other alcoholic drinks. Higher alcohols in the fermented liquors are undesirable, because of their flavor, and must be separated from the distilled product in the distillation column.

The partition coefficient or the volatility of aroma compounds is high at low ethanol concentrations, i.e., when water is the main component of the mixture. These compounds are more soluble in ethanol, which reduces their activity coefficients and relative volatilities. At higher ethanol concentrations, e.g., above 50% mole fraction, the partition coefficient K drops below 1 and the compound becomes less volatile than water in the liquid mixture.

13.2.2 Simple Distillations

13.2.2.1 Differential Distillation

Differential distillation is a simple batch distillation in which the vapor and liquid composition change continuously. It is similar to the traditional laboratory distillation, consisting of a boiler (still), a condenser, and heating/cooling units (Figure 13.3).

The analysis of differential distillation is based on the assumption of instantaneous vapor/liquid equilibrium between the two phases, although the compositions of liquid and vapor change throughout the process.

At any time, the still contains a component i in the liquid residue L (kmol) of composition x_i (kmol/kmol), which is in equilibrium with the vapor phase of composition y_i (kmol/kmol), according to the material balance equation:

$$y_i \mathrm{d}L = \mathrm{d}(Lx_i) \quad (13.14)$$

Integrating Equation 13.14 between residue quantities L_1 and L_2 yields the Rayleigh equation:

$$\ln\left(\frac{L_1}{L_2}\right) = \int_{x_i^1}^{x_i^2} \left[\frac{dx_i}{(y_i - x_i)}\right] \quad (13.15)$$

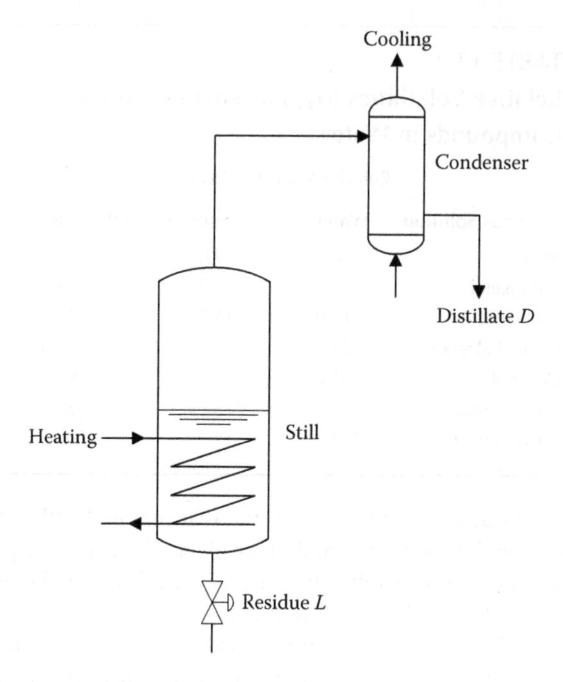

FIGURE 13.3 Diagram of differential distillation.

If the relative volatility α in a binary mixture is constant, the y_i, x_i equilibrium is expressed by Equation 13.10 and the Rayleigh equation becomes

$$\ln\left(\frac{L_1}{L_2}\right) = \left[\frac{1}{(\alpha-1)}\right]\left\{\ln\left(\frac{x_1}{x_2}\right) + \alpha\ln\left[\frac{(1-x_1)}{(1-x_2)}\right]\right\} \tag{13.16}$$

The Rayleigh equation for a binary mixture (A, B) of constant relative volatility α is also written as

$$\ln\left(\frac{A}{A_0}\right) = \alpha\ln\left(\frac{B}{B_0}\right) \tag{13.17}$$

where

A_0, B_0 are the initial quantities (kmol) of the two components

A, B are the quantities of the two components at the end of the distillation

13.2.2.2 Steam Distillation

The presence of water vapor and other inert components changes the thermodynamic equilibrium in a mixture, facilitating the separation of the desired components. Steam distillation is used to separate small quantities of high-boiling compounds.

Water is assumed to be partially soluble with organic components, forming two phases with the other compounds. Organic compounds of high boiling can be removed and recovered at relatively low temperatures. Application of steam distillation includes removal of off-flavors from vegetable oils and essential oils from fruit materials.

The amount of steam W (kmol) required to distill A (kmol) of an organic compound is given by the following equation:

$$W = A\left[\frac{p_w}{(Ep_A)}\right] = A\left[\frac{(P - p_w)}{(Ep_A)}\right] \tag{13.18}$$

where

p_A (bar) is the vapor pressure of component A
p_w (bar) is the vapor pressure of water
P (bar) is the total pressure
E (0.60–0.95) is the efficiency of distillation

According to Equation 13.18, large quantities of steam W are required to distill a given quantity of component A. The steam requirement is reduced by operating at higher temperatures or in vacuum.

13.2.3 Fractional Distillation

13.2.3.1 Distillation Columns

Fractional distillation is used widely in the separation of volatile components of a mixture. Batch or preferably continuous distillation columns are used, consisting of stripping and fractionation (enrichment) sections, and heating (reboiler) and condensing auxiliaries (Figure 13.4). The column may have a larger diameter in the upper (rectifying) section, because of the higher vapor flow rates. Uniform diameter columns are used more often (AIChE, 2000; Maroulis and Saravacos, 2003).

The distillation column of Figure 13.4 separates a binary mixture F (kmol), volatile concentration x_F, into a distillate (top product) D (kmol), concentration x_D, and a residue (bottom product) B (kmol), concentration x_B.

The feed F is preheated with steam S, enthalpy Q_F, and enters near the middle of the column. It splits into liquid, flowing downward, and vapor, flowing upward.

In the stripping section the liquid L' (kmol), consisting of the liquid feed and the liquid from the rectifying section L, flows downward, countercurrently to the vapors V' (kg) rising from the reboiler. At the bottom of the column, the liquid L' is split into the residue (bottom) product B (kmol), concentration x_B, and the vapors V' (kmol), produced in the reboiler, which is heated by steam S, enthalpy Q_F.

In the rectifying section, the vapors V (kmol), consisting of the vapors from the stripping section V' and any vapors from the feed, flow upward, countercurrently to the reflux L.

The vapors V from the top of the column are condensed with cooling water W and split into the reflux L and the distillate product D, concentration x_D.

In some applications, single stripping columns are used, with the liquid entering the top of the column and the exhausted "heavy" liquid removed from the bottom.

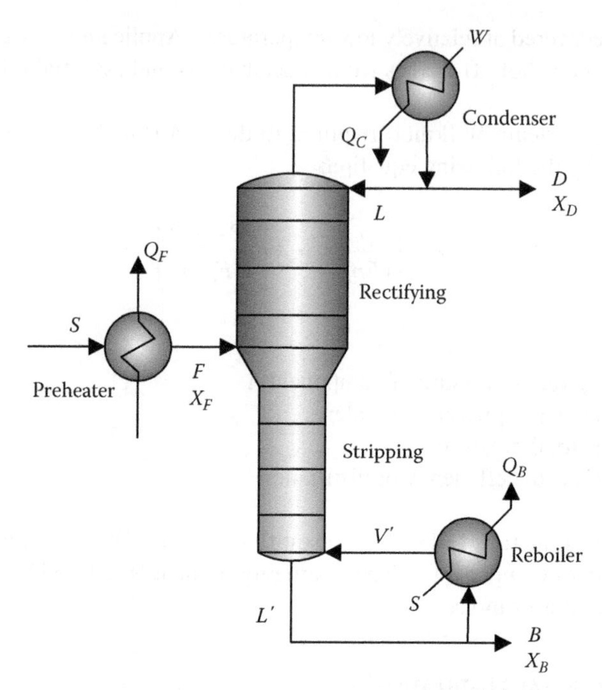

FIGURE 13.4 Diagram of a fractional distillation column. (Modified from Maroulis, Z.B. and Saravacos, G.D., *Food Process Design*, CRC Press, Boca Raton, FL, May 9, 2003, fig. 9.1.)

Instead of vapors produced in the reboiler, "live" steam may be used for stripping the volatile components, which are removed from the top of the column, without further fractionation.

Both stripping and rectifying sections contain plates or trays, each one of which acts as a vapor/liquid equilibrium stage. Since thermodynamic equilibrium is not possible to be reached in a tray, liquid/vapor contactor, and the number of actual trays, for a given separation, is always larger than the number of equilibrium (theoretical) stages.

Some distillation columns, usually of low capacity, operate in continuous vapor/liquid contact, without discrete separation stages (trays). These units are known as packed columns or towers, and they are discussed in Section 13.4.2.3.

13.2.3.2 Equilibrium Stages

The number of theoretical stages in a distillation column is estimated either graphically or by analytical methods. In both methods, vapor/liquid equilibria of the system are required. Developed for the distillation of binary mixtures, they have been modified and extended to multicomponent systems.

Most of the industrial distillation columns are operated as continuous units, although there are some batch columns, used in small-scale operations. The food processing industry uses distillation columns of medium to small size, some of them batch-operated.

In most distillation columns, a total condenser is used, i.e., all the vapors coming out of the first (top) tray are condensed, and the liquid condensate is split into two streams, the distillate product and the reflux, which is returned to the column. In some columns, a partial condenser may be used, with the vapors coming out of the condenser as a product in equilibrium with the liquid condensate, which is returned to the column as the reflux. In this case, the condenser acts as an additional separation stage of 100% efficiency.

13.2.3.2.1 Graphical Methods

The theoretical (equilibrium) stages of distillation are usually determined graphically by the McCabe–Thiele diagram, which is also used for the calculation of separation stages of absorption and extraction operations.

Figure 13.5 shows a McCabe–Thiele diagram for the separation of an ethanol/water mixture, based on vapor/liquid equilibrium data at the given pressure (atmospheric). The basic assumption of this method is the constant molar flow (kmol/h) of liquid L and vapors V in the two sections of the column.

Material balances on the fractionating and stripping sections of the column result in the following equations, respectively:

$$y_n = \left(\frac{L}{V}\right) x_{n-1} + \left(\frac{D}{V}\right) x_D \tag{13.19}$$

$$y_m = \left(\frac{L'}{V'}\right) x_{m-1} - \left(\frac{B}{V'}\right) x_B \tag{13.20}$$

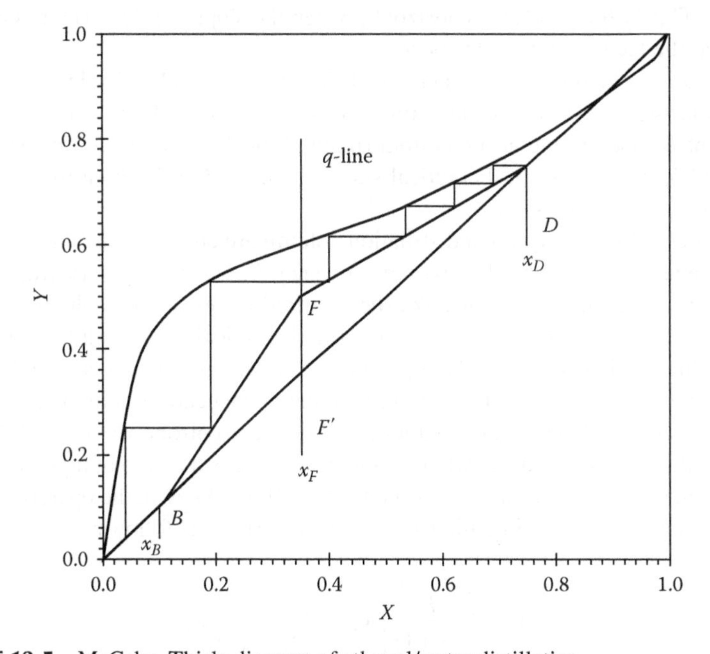

FIGURE 13.5 McCabe–Thiele diagram of ethanol/water distillation.

where

L, V, and L'/V' are the liquid/vapor flow rates of the fractionating and stripping sections

D, B are the flow rates of the distillate and the residue (bottoms), respectively (Figure 13.4)

The flow rates are expressed in molar units, i.e., kmol/h.

For constant molar flows, the material balance equations are represented by the straight lines DF and BF on the McCabe–Thiele diagram. Thus, line DF of the fractionating section is plotted from point D (x_D) on the diagonal with slope L'/V', and line BF is plotted from point B (x_B) on the diagonal with slope L/V. The lines intersect at point F (x_F), corresponding to the composition of the feed.

In the operating lines, the composition of the vapors y_i leaving a stage i is related by material balances to the composition of the liquid x_{i-1} coming from the previous (above) stage i–1, and (in the fractionating section) the compositions of the distillate x_D or (in the stripping section) the residue x_B. At the same time, the composition of vapors y_i and liquid x_i leaving the same stage (i) are in thermodynamic equilibrium.

The two operating lines intersect at point F, which corresponds to the feed concentration x_F. The line $F'F$ is known as the "q-line," and it represents the thermal condition of the feed. It is plotted from point F' with a slope of $q/(q-1)$. The values of q are the following: $q = 1$ (saturated liquid, i.e., liquid at its boiling point), $q = 0$ (saturated vapors), $0 < q < 1$ (vapor/liquid mixture), $q < 0$ (superheated vapors), and $q > 1$ (subcooled liquid).

In the example of Figure 13.4, the feed enters the column as a saturated liquid, i.e., $q = 1$, and the "q-line" has a slope $q/(q-1) = \infty$, i.e., it is a vertical line plotted from point F'. The "q-line" becomes horizontal when the slope is $q/(q-1) = 0$, i.e., when $q = 0$ and the feed is saturated vapors.

The number of theoretical stages N is determined graphically by constructing rectangular steps between the operating and the equilibrium lines, starting from the top (point D) and ending at the bottom (point B) of the column. In the example of Figure 13.5, the number of theoretical stages is nearly $N = 5$ (about three fractionation and two stripping stages).

The actual plates (trays) of a distillation column are equal to the number of theoretical stages N divided by the column efficiency E, which may vary from 50% to 75%. Thus, for an efficiency of 65%, the required plates for the column of Figure 13.5 will be equal to 5/(0.65) = 7.7. While the theoretical stages can be a fractional number, the number of actual plates should be rounded off, in this case eight plates.

As shown in Figure 13.5, the maximum ethanol concentration in distillation columns of aqueous ethanol/water solutions will be the azeotrope $y = x = 0.894$.

The reflux ratio R in a distillation column is defined as the ratio of liquids $R = L/D$, which returns to the column from the condenser. If R is known, the operating line of the fractionating section (Equation 13.19) can be written as follows:

$$y_n = \left[\frac{R}{(R+1)}\right] x_{n-1} + \left[\frac{1}{(R+1)}\right] x_D \tag{13.21}$$

When the two operating lines of a distillation column intersect F on the equilibrium line y_n, x_n, the number of theoretical stages becomes infinite, and the mixture cannot be separated by the particular column. This limiting operating condition is known as the minimum reflux ratio R_{min}. The opposite condition is the total reflux ($R = \infty$), in which the slope of the operating line becomes equal to (1), coinciding with the diagonal line. In the total reflux operation, the number of stages becomes minimum (N_{min}). For economic reasons, the distillation columns are operated between the two extremes of total reflux and minimum reflux, usually at $R = (1.1–1.5)\ R_{min}$.

The McCabe–Thiele diagram is applicable when the operating lines are straight lines (equal molar flows from stage to stage). However, in many nonideal mixtures, including the system ethanol/water, the molar flows are variable, due to significant differences in the enthalpy of the mixtures. For such systems, the Ponchon-Savarit graphical method or analytical stage to stage methods can be used. The Ponchon-Savarit diagram is an enthalpy-concentration diagram for both the liquid and the vapor phases. The number of theoretical stages is estimated by joining the equilibrium enthalpies of the two phases. The Ponchon-Savarit method is applied to the Solvent Extraction operations in this chapter (Example 13.3), where the enthalpy is replaced by the solvent.

13.2.3.2.2 Analytical Methods

The theoretical stages in distillation and other separation processes can be estimated by analytical methods. They are useful when a large number of stages is involved, such as when the operating lines are very close to the equilibrium lines, and in calculations of multicomponent systems, where the graphical methods are difficult to apply. The analytical calculations can be included in computer calculation packages.

The most common analytical method is the Fenske–Underwood–Gilliland calculation method, which involves the determination of the minimum number of theoretical stages N_{min} at total reflux, the minimum reflux ratio R_{min}, with infinite number of theoretical stages and the number of stages or theoretical plates N for the given finite reflux ratio R. This method is used in binary and multicomponent mixtures for the separation of two "key" components, i.e., the light and heavy "keys."

The minimum number of theoretical stages N_{min} is estimated by the Fenske equation for a binary mixture of relative volatility α into a distillate (top) product x_D and a residue (bottom) product x_B at total reflux $R = \infty$:

$$N_{min} = \log \frac{[x_D(1 - x_B)/x_B(1 - x_D)]}{\log(\alpha)} \tag{13.22}$$

The Underwood equations estimate the minimum reflux ratio R_{min}, for the same separation:

$$\frac{(ax_F)}{(\alpha - \theta)} + \frac{(1 - x_F)}{(1 - \theta)} = 1 - q \tag{13.23}$$

$$R_{min} + 1 = \frac{(ax_D)}{(\alpha - \theta)} + \frac{(1 - x_D)}{(1 - \theta)} \tag{13.24}$$

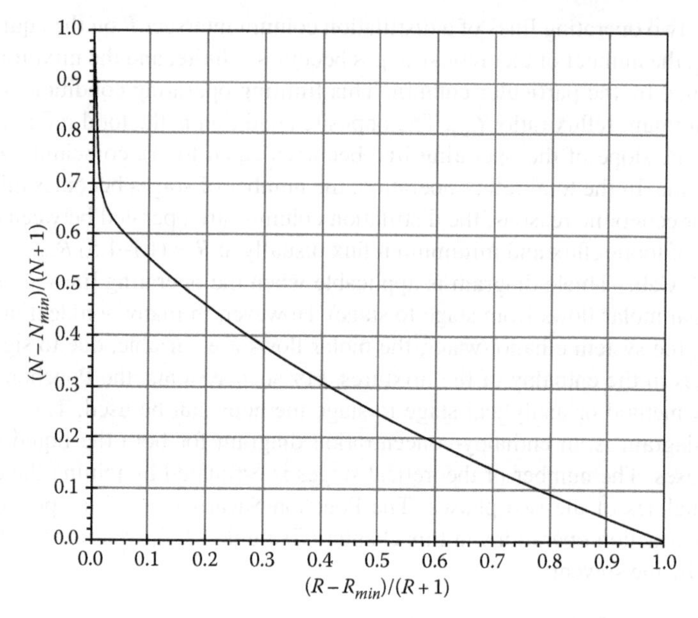

FIGURE 13.6 Gilliland diagram for the number of theoretical stages N.

where
- x_F is the mole fraction of the feed
- θ is an empirical parameter ($1 < \theta < \alpha$) connecting Equations 13.23 and 13.24

The Underwood equations can be extended to n components (multicomponent systems).

The limiting operating conditions N_{min}, R_{min} are used in the empirical Gilliland diagram $\{(N - N_{min})/(N + 1)$ versus $(R - R_{min})/(R + 1)\}$ to estimate the number of stages N at a given reflux ratio R. The Gilliland diagram (Figure 13.6) is given by the empirical equation:

$$\frac{N - N_{min}}{N + 1} = 1 - \exp\left\{\left[\frac{1 + 54.1\,\Psi}{1 + 117.2\,\Psi}\right]\left[\frac{\Psi - 1}{\Psi^{0.5}}\right]\right\} \tag{13.25}$$

where $\Psi = (R - R_{min})/(R + 1)$.

Economic analysis shows that the optimum reflux ratio is in the range of (1.1–1.5) R_{min}, and the corresponding optimum number of stages is in the range of (1.5–2)N_{min}.

13.2.3.3 Column Efficiency

The actual number of trays N_T in a distillation column is estimated from the number of equilibrium (theoretical) stages N, according to the following equation:

$$N_T = \frac{N}{E_o} \tag{13.26}$$

where E_o is the overall column efficiency, which for distillation columns varies from 0.50 to 0.75.

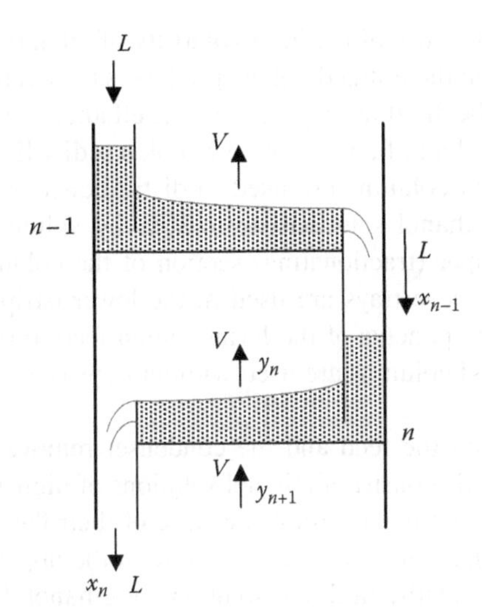

FIGURE 13.7 Sieve tray hydraulics in a distillation column.

The efficiency of distillation columns is determined by detailed mass transfer and fluid flow analysis of the vapor/liquid mixture in a tray. Empirical correlations and diagrams are used for approximate estimations. The efficiency is inversely proportional to the product of liquid viscosity and relative volatility ($\eta\ \alpha$) of the key component. The viscosity is known to have a negative effect on mixing, while high relative volatilities increase the escaping tendency of the volatile components, reducing the residence time and the efficiency of the tray.

The distillation trays or plates must provide adequate residence time and mixing for the liquid and the vapors to approach thermodynamic equilibrium, and subsequently separate the two phases for the next vapor/liquid stage contact. Sieve (perforated) trays are commonly used, while bubble-cup, valve trays, and other specialized designs are applied to some distillations. Sieve trays are preferred in the distillation of fermented food liquids, particularly in the stripping section of the column, because they can handle the food suspensions in a better manner, without the food being fouled and plugged.

Figure 13.7 shows the diagram of a distillation sieve tray. The liquid L flowing down through the down-comers from tray to tray spreads on each (sieve) tray and it comes in contact with the vapors, rising from the bottom tray. The liquid is held on the sieve by the up-flowing vapors through the tray perforations. Valve trays can hold the liquid without flowing through (dripping), even without upward vapor flow.

13.2.4 Food Distillations

13.2.4.1 Ethanol Distillation

Simple (differential) distillation of ethanol from aqueous is applied to produce various food drinks, such as brandy. In simple distillation, the vapors are assumed to be instantaneously in equilibrium with the residue liquid. Thus, the separation is considered to be affected by one theoretical stage, limiting the separation of the volatile components. Because of the high volatility of ethanol in dilute water solutions, its concentration in the initial distillate product is relatively high (30%–40% by volume). However, as the distillation proceeds, the ethanol concentration decreases. In order to obtain a product of about 50% ethanol, a redistillation is necessary.

Fractional distillation columns are used to distill aqueous fermentation liquors (including wine) into ethanol solutions up to 95% by volume (azeotrope) and an aqueous waste. The upper (fractionating) section of the column contains valve or sieve trays, while only sieve trays are used in the lower (stripping) section, which can handle the solid components of the fermentation feed to the column. In small-size applications, packed columns are used without trays (Packed Towers section of this chapter).

A sidestream between the feed and the condenser removes the fusel oils from the column. The fusel oils consist of ethanol solutions of higher alcohols (e.g., butanol, amyl alcohol), which are undesirable because of their flavor and toxicity in the ethanol product. The fusel oils tend to concentrate in the upper part of the column, because they are more volatile in dilute solutions of ethanol, but they become less volatile at higher ethanol concentrations.

The stripping section of the column is usually heated by direct (live) steam at the bottom of the column, eliminating the need of a reboiler (Figure 13.8). The live steam column is more economical than the conventional reboiler column.

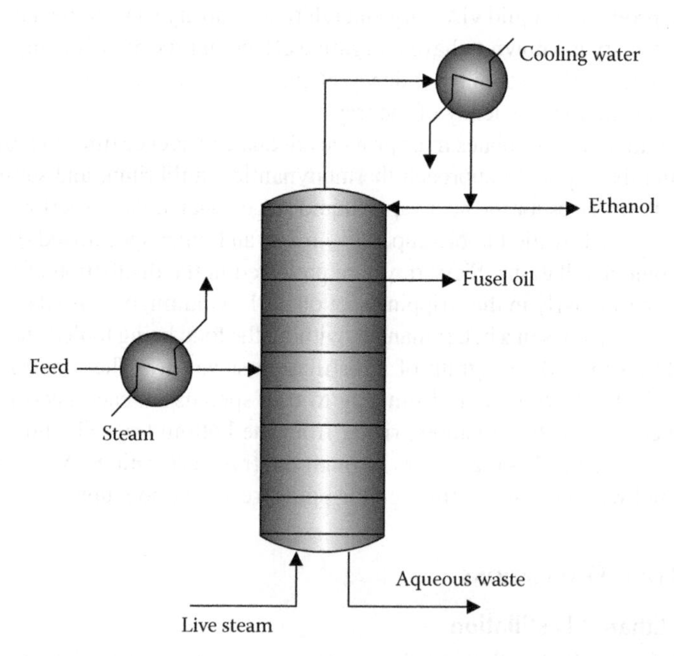

FIGURE 13.8 Distillation of ethanol from aqueous solution, live steam heating. (Modified from Maroulis, Z.B. and Saravacos, G.D., *Food Process Design*, Marcel Decker, New York, May 9, 2003, fig 9.10. CRC Press.)

13.2.4.2 Aroma Recovery

Recovery of aroma or volatile components by distillation is important in the processing of fruit juices and other fluid foods, and it is applied in connection with evaporation and dehydration operations.

The fruit aromas consist of volatile organic compounds, such as esters, alcohols, and oxygenated compounds, which are lost or changed during evaporation, drying, and other physical and thermal processes. Most of these compounds are partially soluble in water and, therefore, they have high relative volatilities. They are present at very low concentrations (less than 0.1%), and the presence of soluble sugars increases their volatility in the aqueous food product.

Aroma recovery can be accomplished by stripping/distillation operations, supercritical fluid extraction, pervaporation (membrane separation), solvent extraction, and solid/fluid adsorption. The design of aroma recovery systems is based on vapor/liquid equilibria of aroma components/water and on distillation theory and practice.

A conventional essence (aroma) recovery system, used in food processing, is shown in Figure 13.9. The fruit juice is stripped of the volatile aroma in a falling film

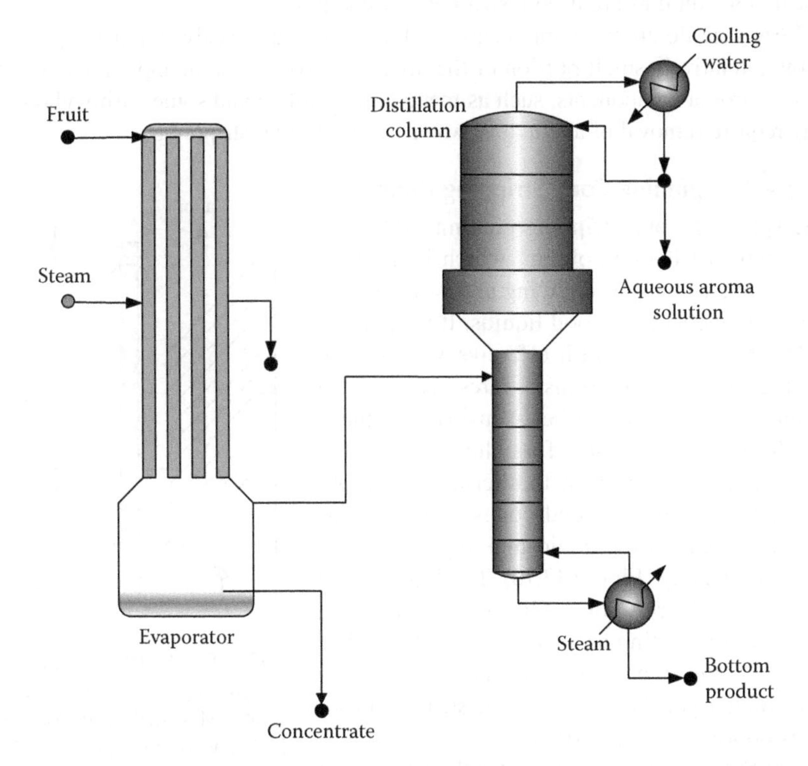

FIGURE 13.9 Simplified diagram of an essence recovery system from fruit juice. (Modified from Maroulis, Z.B. and Saravacos, G.D., *Food Process Design*, CRC Press, Boca Raton, FL, May 9, 2003, fig. 9.11.)

evaporator, heated with saturated steam. The vapors from the vapor/liquid separator are introduced into the distillation column, while the stripped juice is concentrated further by evaporation. The distillation column separates the mixture into a distillate (aqueous aroma solution) and a bottom product. The concentration of the volatile components in the aqueous aroma solution may be about 100–200 times higher than in the juice.

The separation of complex aroma mixtures is based on the separation of a typical "key" component from water, which is difficult to remove from the aqueous solution, i.e., it has a relative volatility lower than the other aroma components. Thus, by recovering this "key" component, most of the aroma components will be recovered in the distillate. As an example, the characteristic flavor component methyl anthranilate of Concord grapes with a relative volatility $\alpha = 3.4$ is used as a "key" aroma component in designing an aroma recovery system for grape juice. The design of such a column would be similar to the design of ethanol distillation, described in detail in this chapter. The column efficiency E in water-rich distillations is relatively low (50%–60%), due to the difficulty of mixing water/vapors effectively. Typical columns of aroma recovery contain about 15 trays of 1 m diameter and 7.5 m high (distance between trays about 0.5 m). In small- to medium-size operations, packed distillation columns are used instead of tray columns.

Very volatile aroma components, such as esters, are easily stripped by evaporating a relatively small portion of the juice, e.g., 10%–25% in apple juice. Higher boiling aroma components, such as methyl anthranilate and some carboxylic acids may require removal of about 40% water for substantial stripping.

13.2.4.3 Spinning Cone Stripping Column

The spinning cone stripping column (SCSC) is a special distillation column, which is used for recovering aromas or removing undesirable volatile components from food liquids. It is operated at low temperatures with effective vapor/liquid mixing and high mass transfer rates, without heat damaging the quality of heat-sensitive products. The SCSC unit consists of a column, about 1 m diameter and 3 m high, with alternating stationary and rotating truncated cones, which act as contacting stages for liquid and vapors flowing countercurrently (Figure 13.10). The liquid is fed at the top of the column and live steam is injected at the bottom, acting as a stripping medium. The volatile components are condensed in refrigerated condensers at the top, and the stripped product is obtained at the bottom.

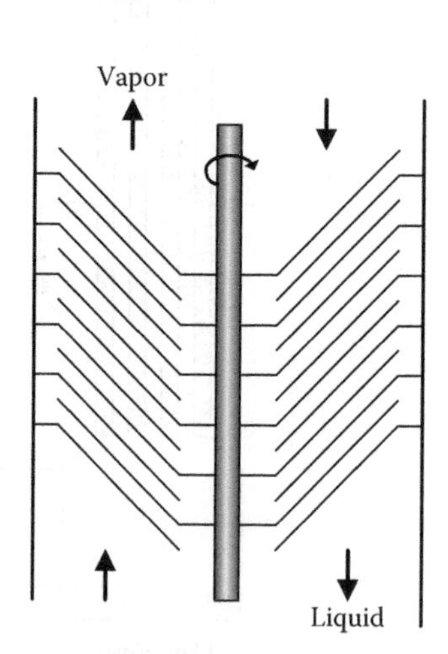

FIGURE 13.10 Diagram of a spinning cone stripping column. (Modified from Maroulis, Z.B. and Saravacos, G.D., *Food Process Design*, CRC Press, Boca Raton, FL, May 9, 2003, fig. 9.12.)

The SCSC unit can be used for the removal of undesirable volatiles from fruit juices, e.g., sulfur dioxide from grape juice, and it is suited for the preparation of alcohol-free (nonalcoholic) wines, which contain less than 0.5% ethanol. The SCSP dealcoholization process first involves the recovery of the volatile aromas (esters, etc.) in a low-temperature column ($T < 30°C$), and second, the stripping of ethanol at a higher temperature (about 35°C), resulting in a stripped wine. The volatile aromas, recovered in the first stripping, are returned to the stripped wine, obtaining a full-flavored nonalcoholic product.

13.2.5 Molecular Distillation

Molecular distillation is applied to the separation of high-boiling and heat sensitive food components, such as flavors, vitamins, and monoglycerides. The mixture is first separated from the volatile components by normal vacuum evaporation or distillation.

Molecular distillation is an expensive process, requiring very high vacuum (about 0.01 mbar), which is achieved with special vacuum pumps. For small, laboratory-size applications, an oil rotary (ballast) pump, followed by an oil diffusion pump may be used. For pilot plant and industrial applications, a rough vacuum pump, e.g., steam jets or liquid ring pump is used to reach a vacuum (absolute pressure) of about 10 mbar. The high vacuum required is obtained by using a rotary dry (oil) pump, followed, if necessary, by an oil diffusion pump.

The molecular still consists basically of a film evaporator, which is surrounded by a condenser. The high-boiling compounds are evaporated and condensed very fast on the condenser surface, which is located very close to the evaporator (short path distillation). The short distance between evaporator–condenser is smaller than the mean free path of the molecules at the prevailing pressure and temperature of the still, so that the distilling components quickly reach the condenser.

13.3 SOLVENT EXTRACTION

Extraction operations include solvent extraction of liquids or solids and leaching of solids. The analysis and design of extraction is similar to distillation operations. Empirical correlations are used for liquid/liquid equilibria, and experimental equilibrium (solubility) data are essential for liquid/solid systems. The equilibrium stages are estimated by approximate methods, and the extraction/leaching equipment is more specialized than the generalized distillation systems (Walas, 1988).

13.3.1 Liquid/Liquid and Liquid/Solid Equilibria

Liquid/liquid extraction is based on the thermodynamic equilibrium between two partially soluble phases, the extract and the raffinate, between which the solute is distributed. The equilibrium is expressed by the empirical equation:

$$Y = KX \tag{13.27}$$

The concentrations of the solute in the extract Y and raffinate X phases are usually expressed as mass fractions of the mixture (kg/kg), not mole fractions (used in

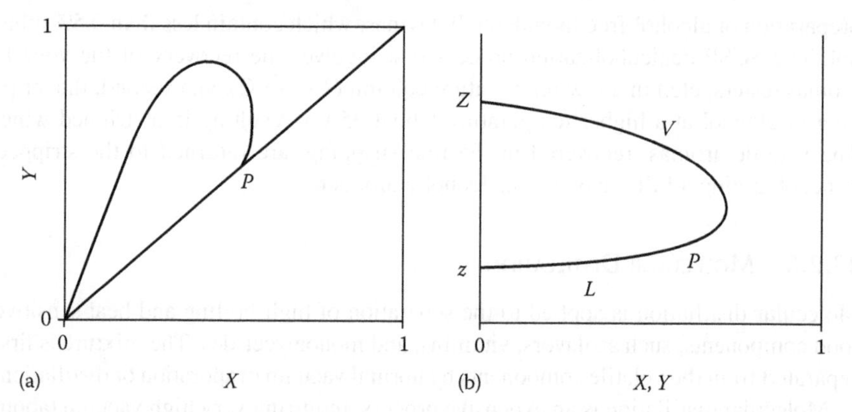

FIGURE 13.11 McCabe–Thiele (a) and Ponchon-Savarit (b) diagrams for liquid/liquid extraction for a partially soluble system. *P*, plait point.

distillation). The partition or distribution coefficient *K* of the solute in the extract and the raffinate is analogous to the volatility of vapor/liquid equilibria, and it is related to the activity coefficients γ of the solute in the raffinate γ_r and the extract γ_e:

$$K = \frac{\gamma_r}{\gamma_e} \tag{13.28}$$

where $K > 1$, or $\gamma_r > \gamma_e$.

The activity coefficients of various liquid systems are correlated in empirical relations, such as the Wilson, the UNIQUAC, and the UNIFAC equations. Liquid/liquid mixtures, used in solvent extraction, are nonideal systems with high *K* values, facilitating the separation in a relatively small number of theoretical equilibrium stages.

The liquid/liquid equilibrium data are presented usually in the modified McCabe–Thiele, and the Ponchon-Savarit diagrams (Figure 13.11). The orthogonal (right angle) triangular diagrams are sometimes also used.

The modified McCabe–Thiele diagram for liquid/liquid extraction uses as coordinates the mass fractions of the solute *B* in the extract *Y* and the raffinate *X*, defined by the following equations:

$$Y = \frac{B}{A+B} \text{ extract phase}$$
$$X = \frac{B}{A+B} \text{ raffinate phase} \tag{13.29}$$

where

A (kg) is the amount of the residue
B (kg) the amount of solute in the mixture

The two phases (extract and raffinate) have a common equilibrium point, called the "plait" point.

The Ponchon-Savarit diagram considers the amount of solvent S (kg) in the system, in addition to components A and B. The role of the solvent in extraction is similar to the enthalpy (heat content) in distillation operations. The ordinates of the Ponchon-Savarit diagram for liquid/liquid equilibria are

$$Z = \frac{S}{A+B} \text{ for the extract}$$
$$z = \frac{S}{A+B} \text{ for the raffinate} \tag{13.30}$$

Equilibrium "*tie*" lines connect the solvent V and the raffinate L phases.

Liquid/solid equilibria are determined experimentally, assuming that the solute B is completely soluble in the solvent S, and that it is not sorbed on the inert solid. Thus, equilibrium is expressed by the simple equation:

$$Y_e = X \tag{13.31}$$

where

Y_e is the mass fraction of the solute B in the extract
X is the mass fraction of the solute in the liquid remaining in the pores of the solids, after reaching equilibrium

The liquid/solid equilibrium data can be plotted on modified McCabe–Thiele and Ponchon-Savarit diagrams, in a similar manner with the liquid/liquid equilibria. In the modified McCabe–Thiele, the coordinates are the mass fractions of the solute B in the extract Y and the residue X, respectively, as defined by Equations 13.29.

The ordinates of the modified Ponchon-Savarit diagram for liquid/solid equilibria are the concentrations of the inert solids A in the solid (underflow) and in the extract (overflow), defined by the equations:

$$z = \frac{A}{B+S} \text{ underflow}, \quad Z = \frac{A}{B+S} \text{ overflow} \tag{13.32}$$

where A, B, S are the amounts (kg) of inert solids, solute, and solvent, respectively.

For the normal extraction case of an inert solid that is completely insoluble in the solvent, the overflow line coincides with the X-axis, i.e., $Z = 0$. The equilibrium "tie" lines are vertical lines, which facilitate the graphical construction of the equilibrium stages between the equilibrium and the operating lines (Figure 13.12).

13.3.2 Equilibrium Stages

Similar to distillation, more than one equilibrium contact stages are needed in an extraction separation process. While in liquid/liquid extraction the separation stages can be arranged in a continuous column, special contact equipment is used in liquid/solid extraction and leaching, such as batteries of static beds, operated semicontinuously.

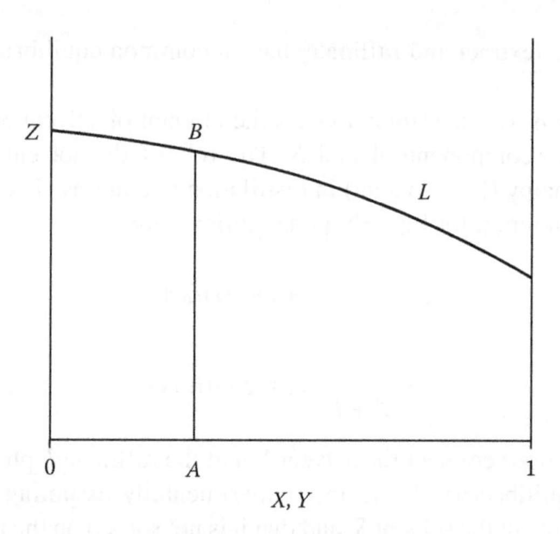

FIGURE 13.12 Modified Ponchon-Savarit diagram for liquid/solid extraction. Insoluble inert solid. Vertical line *AB*, tie line.

The number of theoretical stages in liquid/liquid (*L/L*) extraction is determined using the modified McCabe–Thiele (dilute solutions) or Ponchon-Savarit diagrams (concentrated solutions).

In the McCabe–Thiele diagram, the equilibrium line is plotted, using literature or experimental data, and the operating line is constructed from material balances of the given system, in analogy to distillation. In extraction, the solvent V and the raffinate L flows are taken as mass flow rates (kg/h) and the concentrations Y, X as mass fractions. In dilute (low-concentration) solutions, straight equilibrium, and operating lines facilitate the graphical construction. The equilibrium line ($Y = KX$) is plotted from the origin of the coordinates with a slope of K. The operating line representing the overall material balance in a countercurrent column becomes a straight line with a slope of L/V:

$$Y_1 - Y_{n+1} = \left(\frac{L}{V}\right)(X_0 - X_n) \tag{13.33}$$

where

X_0, X_n are the solute concentrations in the raffinate

Y_1, Y_{n+1} are the solute concentrations in the extract at the entrance (stage 1) and exit (stage n) of the column, respectively

In multistage separations of dilute solutions, the number of equilibrium stages N is more easily determined analytically by the Kremser equation:

$$N = \frac{\log[R(1-E)+E]}{\log(E)} \tag{13.34}$$

where

$E = (KV)/L$ is the extraction factor

$R = (X_n - Y_{n+1}/K)/(X_0 - Y_1/K)$ is the extraction ratio

The liquid/liquid extractors have lower efficiencies (20%–30%) than the distillation columns, because of difficulties in mixing the liquid phases, i.e., inefficient mass transfer. The efficiency can be improved by thorough mixing, using agitation or vibration.

In liquid/solid extraction (leaching), a series (battery) of vessels can be considered as a continuous countercurrent equilibrium operation, similar to a countercurrent liquid/liquid column. The extract V flows countercurrently to the liquid (usually water) residue L, while the inert solids remain fixed in the beds. The solvent gradually removes the solute from the liquid residue, reaching equilibrium at each stage (sufficient residence time, mechanical mixing).

The number of equilibrium stages in liquid/solid extraction can be determined graphically, using the modified McCabe–Thiele diagram (dilute solutions) or the Ponchon-Savarit diagram (concentrated solutions). Graphical construction of the stages is facilitated by the assumption that the equilibrium line coincides with the diagonal line ($Y = X$).

For low concentrations of the solute in the inert solid and in the extract, a straight operating line is obtained, assuming constant flows of liquid underflow X and extract V. The number of equilibrium stages N can be calculated from the Kremser equation (13.34). The efficiency of the solid/liquid extractors is high, about 90%–95%, due to the thorough mixing and long residence time in each stage.

13.3.3 Mass Transfer Considerations

Solvent extraction is generally controlled by mass transfer of the solutes from the material to the transfer interface and then to the solvent. Mass transfer within liquid and solid materials is assumed to take place by molecular diffusion, and the transport rate is expressed by the effective diffusivity D, which is an overall transport coefficient, based on the diffusion (Fick) equation.

The diffusivity in liquids is related to the molecular (particle) size of the solute and the viscosity of the mixture, or the molecular solute/solvent interactions. Typical values of D of solutes in dilute water solutions (25°C) are, sodium chloride, 12×10^{-10} m²/s; sucrose, 5×10^{-10} m²/s; and lactalbumin, 0.7×10^{-10} m²/s.

Diffusivity data in solids are more variable, depending strongly on the microscopic and macroscopic structure (homogeneous, porous, fibrous, etc.) of the material. Prediction of D in solids is difficult, and experimental data are required for each material at given conditions. Typical data on D of importance to extraction are oil (soybean flakes)/hexane, 1×10^{-10} m²/s, and coffee solubles (coffee beans)/water, 1×10^{-10} m²/s. The D of solutes increases when the porosity of the solid is increased. Large molecules, such as lipids and proteins have smaller diffusivities, compared, e.g., to sugars.

The diffusion rate of the solutes can be increased by various mechanical and hydrothermal pretreatments of the solid foods, e.g., slicing, flaking, and steam injection. Slicing of beets and fruits and flaking reduce the thickness (diffusion path) of the material, without serious damage of the cells, which would release cell components into the solution and make it difficult for the solid/liquid separation. Heating by steam or water modifies the cellular structure, increasing the diffusivity in oil extraction.

Recovery of oil from oilseeds containing above 22% oil is accomplished by mechanical expression, followed by leaching with an organic solvent, such as hexane, chlorinated hydrocarbons, and alcohols. Mechanical pressing can remove up to 90% of the oil in the seeds. Direct solvent extraction is applied to oilseeds containing less than 22% oil, such as soybeans (Aguilera, 1999).

Sugar beets are cut into long, thin slices and heated to 50°C–60°C to denature the proteins and increase the diffusivity of sucrose, without leaching the nonsugar components.

Water extraction of soluble solids from roasted coffee beans is facilitated by the high porosity of the beans, developed during thermal treatment (roasting). In the processing of decaffeinated soluble coffee, the diffusion of caffeine in water controls the process.

13.3.4 Solvent Extraction Equipment

Multistage countercurrent extraction in liquid/liquid (*L/L*) systems is performed in extraction columns, which are similar to the stripping section of a standard distillation column. The columns are analyzed either as tray (plate) columns or packed towers.

Industrial liquid/liquid solvent extraction columns include the sieve tray, the rotating disc (RTC), the Oldshue (paddle agitation), the Scheidel (packing and agitation), the pulsed columns, the Graesser rotary raining cup extractor, and the centrifugal extractor.

Extraction (leaching) of solutes from solids is usually carried out in static (fixed) beds of particulate material, prepared to the appropriate size and physicochemical condition to facilitate mass transport of the solute. The bed of material is enclosed in vertical cylindrical vessels, which can stand high pressure operation of volatile solvents. The solvent is fed at the top, while the extract can be recirculated through the solids, before it is removed from the system.

In the leaching of sugar beets with water, 10–14 extraction cells are used, each with a volume of 4–12 m^3. Leaching time is 60–100 min and 110 kg of sugar solution 12–16°Brix is produced per 100 kg beets. In the processing of instant coffee, five to eight extraction columns are used with cycle time 1/2–1 h. Hot water at 175°C is used in ratios 7/2 to 5/1 to solids, and the extracts contain 25%–30% soluble solids.

Figure 13.13 shows a semicontinuous battery of extraction columns (cells) for soluble coffee. Six-extraction columns (percolators) are filled with ground coffee of appropriate size to facilitate diffusion. The columns are operated under pressure (about 20 bar), and they are equipped with special valves for filling and removing solid particles. Hot water at about 175°C is fed to column (1), which contains the most-spent coffee grounds, and the lean extract is reheated and fed to column (2), and so on. Finally, rich extract is removed from the last column (6), which contains fresh grounds. The final extract, containing about 25%–30% soluble solids, is concentrated in an evaporator and spray-dried into granular soluble coffee. When exhausted, column (1) is emptied and refilled with fresh grounds, while the hot water is fed to column (2), and the extract from column (6) is fed to column (1). The cycle is repeated by emptying and refilling column (2), and so on.

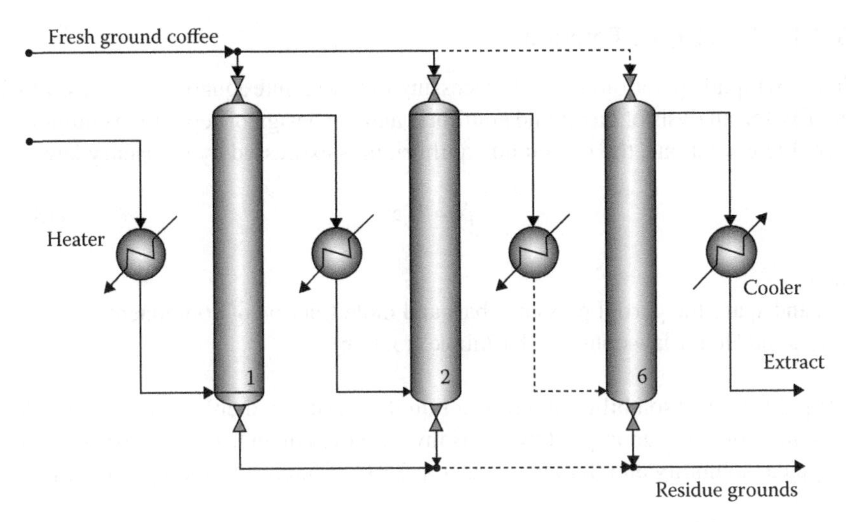

FIGURE 13.13 Semicontinuous extraction columns for soluble coffee. (Modified from Maroulis, Z.B. and Saravacos, G.D., *Food Process Design*, CRC Press, Boca Raton, FL, May 9, 2003, fig. 9.13.)

Special liquid extraction equipment for solids, includes the following:

- The Bollmann bucket elevator with perforated buckets carrying each about 40 kg of particulate material in a solvent bath, contained in a closed vessel.
- The Hildebrand extractor, in which the solids are transported with screw conveyors through three sections in countercurrent flow with the solvent/extract.
- The Bonotto multitray tower extractor in which the trays rotate while the solid particles are scraped and discharged from tray to tray.
- The Rotocel extractor, which consists of about 18 wedge-shaped cells in a rotating shell enclosed in a stationary tank. Fresh solvent is charged to the last cell and the drained solutions are pumped countercurrently to each cell in series. The Rotocel is used in the leaching of sugar beets and oilseeds.
- The Screw extractor is used for leaching sugar beets, where the beet pieces move countercurrently at a slope against a hot water solution.

13.4 GAS/LIQUID ABSORPTION

Gas absorption and stripping are physical separation processes, in which a gas component of a mixture is absorbed or stripped in a liquid, using single- or multistage operations, similar to distillation and liquid/liquid extraction. Absorption and stripping of oxygen and carbon dioxide in liquid foods are the most important systems in food processing (aerobic fermentations, deaeration of liquid foods, carbonation of beverages). Absorption of sulfur dioxide in some fluid foods (fruit juices, wine) is also of interest.

13.4.1 Gas/Liquid Equilibria

Most gas/liquid operations in food processing involve dilute aqueous solutions, which simplify the analysis of gas/liquid equilibria and the design of separation equipment.

In dilute solutions, the gas/liquid equilibrium is expressed by the Henry law:

$$p_i = H_i x_i \tag{13.35}$$

where

p_i and x_i are the partial pressure (bar) and mole fraction of component i
H_i is the Henry law constant, bar/(mole fraction)

Equilibrium or solubility of gases in liquids is expressed either as the Henry law constant H_i or, more appropriately, by its inverse ($1/H_i$) in units of (mole fraction)/bar.

Typical solubility data for gases/water at 25°C of interest to food processing are

Oxygen, $1/H_i = 2.3 \times 10^{-5}$ mole fraction/bar $= 2.3 \times 32 \times 10^{-5}/18 = 4 \times 10^{-5}$ g/(g water bar) or 40 ppm/bar in oxygen atmosphere 1 bar. The solubility of atmospheric oxygen (20% oxygen) at atmospheric pressure ($P \approx 1$ bar) in water is about $40/5 = 8$ ppm.
Carbon dioxide, $1/H_i = 8 \times 10^{-4}$ (mole fraction)/bar $= 8 \times 44 \times 10^{-4}/18 = 1960$ ppm/bar, or 0.2% by weight in carbon dioxide atmosphere 1 bar.
Nitrogen, $1/H_i = 1.2 \times 10^{-5}$ mole fraction/bar $= 1.2 \times 28 \times 10^{-5}/18 = 1.8 \times 10^{-5}$ g/g water bar or 18 ppm/bar in nitrogen atmosphere 1 bar. The solubility of atmospheric nitrogen (80% nitrogen) at atmospheric pressure in water becomes $18 \times (4/5) = 14.4$ ppm.

In food processing (low pressure), the gas phase is considered as ideal, and the Dalton law predicts the partial pressure p_i of component i as $p_i = y_i P$, where y_i is the mole fraction of component i in the gas phase, and P is the total pressure. Thus, the equilibrium relationship of Equation 13.35 becomes

$$y_i = m_i x_i \tag{13.36}$$

where $m_i = H_i/P$.

The gas/liquid equilibrium constant (m_i) is equivalent to the distribution coefficient K_i of the vapor/liquid and liquid/liquid phase equilibria.

13.4.2 Absorption Operations

13.4.2.1 Agitated Vessels

Absorption and stripping of gases in solutions are carried out in agitated vessels and in columns/towers. In the agitated vessels, equilibrium (one stage) is approached by vigorous mixing. Agitated vessels are used for the transfer of oxygen in aerobic fermentations, and their construction and power requirements are discussed in Chapter 7 (Mixing). Multistage countercurrent absorption is carried out either in tray columns or in packed towers.

13.4.2.2 Tray Columns

Operation of absorption tray columns is similar to the operation of distillation and liquid/liquid tray columns. For dilute solutions, common in food applications, the gas (G) and liquid (L) flow rates in countercurrent columns are considered constant, and overall material balance in the column simplifies to a straight operating line:

$$y_1 - y_2 = \left(\frac{L}{G}\right)(x_1 - x_2) \tag{13.37}$$

where

y_1, y_2 are the mole fractions of the component in the gas at the inlet (bottom) and outlet (top) of the column, respectively

x_1, x_2 are the corresponding mole fractions in the liquid at the bottom and the top of the column

The number of equilibrium stages N for absorption can be determined graphically, using the McCabe–Thiele diagram, in which both equilibrium and operation are represented by straight lines with slopes m and L/G, respectively. The number of stages can also be calculated analytically using the Kremser equation, similar to extraction:

$$N = \log\frac{[R_a(1-1/A)+1/A]}{\log(A)} \tag{13.38}$$

where

$R_a = (y_1 - mx_2)/(y_2 - mx_2)$, absorption ratio

$1/A = (m\ G/L)$, absorption factor

If pure liquid is used at the top of the column ($x_2 = 0$), then absorption ratio $R_a = (y_1/y_2)$.

An equation similar to 13.38 can be used for the calculation of N of a stripping column, substituting R_a with $R_s = (x_2 - y_1/m)/(x_1 - y_1/m)$, stripping ratio, and $1/A$ with $A = L/(m\ G)$, stripping factor.

The efficiency of the tray columns, used for absorption and stripping operations, is generally lower than the efficiency of the distillation and extraction columns (about 10%–15%), due to the poor mixing of the gas with the liquid and inefficient mass transfer.

13.4.2.3 Packed Towers

Absorption of gases in liquids is often carried out in packed towers, which use various packing materials to affect gas/liquid contact and mass transfer, instead of the tray columns. Packed towers are continuous separation systems, the separating capacity of which is measured by the number of theoretical transfer units NTU instead of the number of theoretical stages N of the multistage systems.

In dilute solutions, common in food systems, the height of absorption or stripping towers z is equal to the product of the number of theoretical transfer units NTU times the height of transfer unit HTU:

$$z = (NTU)(HTU) \tag{13.39}$$

In absorption towers, the number of transfer units *NTU* is estimated from the integral

$$NTU = \int_{y_2}^{y_1} \left[\frac{dy}{(y - y_e)}\right] \quad (13.40)$$

where

y_1, y_2 are the mole fractions of the gas component at the inlet (bottom) and outlet (top) of the tower, respectively

y_e is the mole fraction of the gas in equilibrium with the liquid phase ($y_e = mx$)

In absorption towers, the height of a transfer unit *HTU m* is estimated from the following equation:

$$HTU = \frac{G}{(K_g \alpha P)} \quad (13.41)$$

where

G is the gas flow rate [kmol/(m^2 s)]

K_g is the overall mass transfer coefficient [kmol/(m^2 s Pa)]

α is the specific surface of the packing (m^2/m^3)

P is the total pressure (Pa)

For stripping towers, the following analogous relationships are used:

$$NTU = \int \left[\frac{dx}{(x_e - x)}\right] \quad \text{and} \quad HTU = \frac{L}{(K_l \alpha \rho_M)} \quad (13.42)$$

The integral $\int[dx/(x_e - x)]$ is estimated from x_2 to x_1, where x_2, x_1 are the liquid mole fractions of the component *i* at the inlet (top) and outlet (bottom) of the stripping tower, respectively, and x_e is the mole fraction in the liquid, which is in equilibrium with the gas phase entering the tower ($x_e = y/m$). For clean stripping gas $y_2 = 0$ and $x_e = 0$ at the bottom of the tower. L is the liquid flow rate [kmol/(m^2 s)], K_l is the liquid overall mass transfer coefficient (m/s), α is the specific surface of the packing (m^2/m^3), and ρ_M is the molar liquid density (kmol/m^3).

The overall mass transfer coefficients K_g and K_l, based on the gas and liquid phase, respectively, are characteristic parameters of the mass transfer system, related to the geometry and flow conditions of the equipment, and to the physical and transport properties of the materials involved.

In dilute solutions, the number of theoretical transfer units in a countercurrent absorption tower can be estimated by the Colburn equation, which is analogous to the Kremser equation for trays:

$$NTU = \frac{\ln[R_a(1 - A) + A]}{[1 - (1/A)]} \quad (13.43)$$

where the absorption ratio R_a and the absorption factor A are defined as in the Kremser equation for absorption columns (13.38). The same equation (13.43) can be used for estimating the *NTU* of packed stripping towers, substituting R_a with the stripping ratio R_s and A with the stripping factor $1/A$, as defined in the Kremser equation (13.38).

The height of a packed tower Z can also be determined from the following equation:

$$Z = (N)(HETP) \tag{13.44}$$

where

N is the number of equilibrium (theoretical) stages of a multistage countercurrent column of equivalent separating capacity with the absorption tower

HETP (m) is the height of an equivalent theoretical plate

The number of plates N is more easily determined with methods developed in distillation, and it can replace *NTU*, provided that the data on *HETP* are available. In some systems, *NTU* and *HTU* may coincide with N and *HETP*, respectively. Experimental and operating data on *HTU* and *HETP* of various packing materials have been correlated empirically with the flow conditions of absorption and stripping equipment.

The packing materials include porcelain, metal and plastic materials in the form of rings, cylinders, and saddles. The effectiveness of the packing materials depends on their specific surface α (m^2/m^3) and the packed porosity ε, expressed by the packing factor $F = \alpha/\varepsilon^3$ in units (m^2/m^3). The *HETP* of packed materials ranges from 0.3 to 0.5 m.

13.4.2.4 Gas Scrubbers

Gas scrubbers are applied to remove small particles and undesirable gases from industrial exhaust gases, for the purpose of reducing environmental pollution. They are usually installed after the mechanical cyclone collectors, which are efficient for removing particles larger than 1–5 μm (Section 7.7).

Cyclone scrubbers are mechanical cyclones in which an absorbing liquid (water or aqueous solutions) is sprayed from several nozzles in a central manifold. The gas comes into intimate contact with the liquid and it leaves the cyclone near equilibrium, equivalent to one-stage operation.

The ejector-venturi scrubbers use high-velocity jets of water, which create a suction and absorb the gas, in a parallel-flow operation. Large quantities of water are required, which should be disposed without creating pollution problems.

Scrubbers are necessary to reduce air pollution in some food processing plants. Certain food processing operations produce undesirable gases and volatiles, which cause air pollution (mainly offensive odors) in the area surrounding the food plant. Food processes involved in air pollution include air drying, solvent extraction, refining of edible oils, coffee processing, fermentation, and baking. Air scrubbers, usually installed after dust collection equipment (cyclones or air filters), can remove most of the offensive gases from the exhaust streams.

13.5 ADSORPTION AND ION EXCHANGE

Adsorption and ion exchange operations are used to adsorb solute components from liquids or gases with the purpose of clarifying the fluid of unwanted materials or recovering valuable components. The separation is based on physical adsorption on solid adsorbents or ion exchange resins, which are regenerated for repeated use.

Commercial adsorbents include activated carbon, silica gel, activated alumina, and molecular sieves, while ion exchangers include cation and anion exchange resins. Batch adsorption in fixed beds is used in most applications. Desorption of the adsorbed components is accomplished by washing the fixed bed or by increasing the temperature. Regeneration of the ion exchange beds is achieved by washing with salt or alkali solutions.

13.5.1 Adsorption Equilibria and Mass Transfer

The adsorption capacity of a solid adsorbent is determined by the amount of the solute component as a function of its partial pressure or concentration in the fluid phase. Empirical equations are used to express this relationship, such as the Freundlich equation, which for a gas/solid system becomes

$$w_i = Kp_i^n \tag{13.45}$$

where

w_i is amount of adsorbed component (kg/kg adsorbent)
p_i (Pa) is the partial pressure of the solute component in the gas phase
K, n are characteristic constants of the system. For systems favoring adsorption, $n < 1$

For liquid/solid adsorption the partial pressure is replaced by the concentration C (kg/m^3) of the solute component in the liquid phase.

More complex sorption relationships, such as the Langmuir, the *B.E.T.*, and the *G.A.B.* equations, have been used to express the fluid/solid equilibria in food systems.

Fluid/solid equilibrium data are usually presented as component mass fraction w_i versus relative pressure p_i/p_o, where p_o is the vapor pressure of component i at the given temperature of the system. These plots are known as the sorption isotherms of the solute component/solid adsorption system, and they are used extensively in food science and engineering (Chapter 3).

The volumetric mass transfer rate from a gas phase to a solid adsorbent J [kg/(m^3 s)] is given by the following equation:

$$J = k_G \alpha (p_i - p_{ie}) \tag{13.46}$$

where

p_i (Pa) is the partial pressure of component i
p_{ie} (Pa) is its equilibrium concentration
k_G [kg/(m^2 s Pa)] is the mass transfer coefficient of the gas phase
α (m^2/m^3) is the interfacial area

The volumetric mass transfer from a liquid phase becomes

$$J = k_L\alpha(C_i - C_{ie}) \tag{13.47}$$

where
- C_i, C_{ie} (kg/m^3) are the concentrations of component i in the liquid and at equilibrium, respectively
- k_L (m/s) is mass transfer coefficient in the liquid phase
- α (m^2/m^3) is the interfacial area

The mass transfer coefficients can be estimated from empirical correlations of the mass transfer factor j_M:

$$j_M = aRe^m \tag{13.48}$$

The mass transfer factor is defined by the following equations:

$$j_M = \frac{(k_G P)}{G} \quad \text{or} \quad j_M = \frac{(k_L \rho)}{L} \tag{13.49}$$

where
- G, L [kg/(m^2 s)] are the gas or liquid flow rates, respectively
- P (Pa) is the total pressure
- ρ (kg/m^3) is the liquid density

The Reynolds number Re is defined as $Re = (dG)/\eta$ or $Re = (dL)/\eta$, where d(m) is the particle diameter and η (Pa s) is the fluid viscosity.

Physical properties of the adsorbents, which affect mass transfer and adsorption capacity are the particle diameter d, bulk density ρ_b, bulk porosity ε, and specific surface area α. Typical values of these properties for commercial adsorbents are

$$d = 1-4 \text{ mm}, \quad \rho_b = 500-800 \text{ kg/m}^3, \quad \varepsilon = 0.25-0.35, \quad \alpha = 1000-3000 \text{ m}^2/\text{m}^3.$$

Regression analysis of published data on mass transfer coefficients in food systems has yielded the following average values of the constants of Equation 13.48, $a = 1.11$ and $m = -0.54$. Thus,

$$j_M = 1.11Re^{-0.54} \tag{13.50}$$

13.5.2 Adsorption Equipment

Fixed vertical beds of porous and granular adsorbents are used for the adsorption of components from fluids. The fluid flows from the top down through the bed, while regeneration is carried out by upward flow of the regenerant solution.

Figure 13.14 shows a diagram of an adsorption bed with typical breakthrough curve.

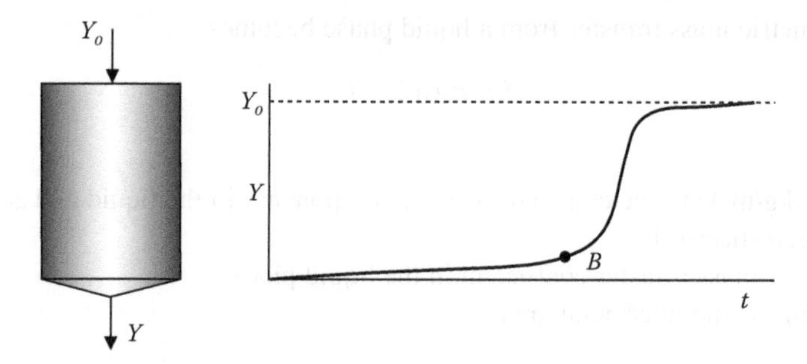

FIGURE 13.14 Breakthrough curve in an adsorption column.

A fluid of initial solute concentration Y_o is fed to the top of the bed, and the concentration of the effluent Y is recorded. Initially, all the adsorbent in the bed is active, and the effluent concentration is nearly zero. After some time of operation, the concentration Y starts to rise sharply at the "break" point B. After all the adsorbent is saturated with the solute, the effluent concentration reaches asymptotically the initial solute concentration Y_o.

The operation of a fixed adsorption bed is similar to the operation of packed absorption towers. In differential form, the change in concentration of the component in the gas phase Y over the height (z), according to Equations 13.39 through 13.41, will be

$$\frac{\partial Y}{\partial z} = \left(\frac{1}{HTU}\right)(Y - Y_e) \tag{13.51}$$

where

Y, Y_e are the concentrations of the solute in the bed at bed depth z and at equilibrium, respectively

HTU is the height of one transfer unit, defined as, gas phase, $HTU = G/(k_G\alpha)$; liquid phase $HTU = L/(k_L\rho)$

The height of the bed is estimated by integrating Equation 13.51.

13.5.3 Ion Exchange Equipment

Ion exchange separations are based on the exchange of cations and anions from a solution or food liquid with the ions of an ion exchange resin.

Commercial ion exchange columns have diameters up to 4 m and bed heights of 1–3 m. Sufficient free space above the bed is required for bed expansion during operation and regeneration, which can exceed 50%. The particle size of the ion exchange resins ranges from 0.3 to 0.8 mm. The resin beds are supported by a distributor at the bottom.

The ion exchange rate is affected by the mass transfer resistances of the resin. At low solute concentrations, which is the case of most food applications, diffusion in the film of liquid/resin interphase controls the transfer rate.

In the case of water softening, the capacity of ion exchange resin is about 2 meq/g of resin, which corresponds to the removal of 0.2 kg of calcium carbonate/kg of resin (note that molecular weight of $CaCO_3$ = 100). The bulk density of the ion exchange resins varies from 600 to 900 kg/m^3. Liquid flow rates in the ion exchange beds range from 14 to 18 m^3/(m^2 h).

Regeneration of the cation exchange resins is usually accomplished by upward flow of sodium chloride solutions, while alkali solutions are used to regenerate the anion exchange resins. Mixed beds, consisting of two layers of cation and anion exchange resins of differing bulk densities, are regenerated by salt and alkali solutions, introduced from the top and the bottom, while the waste solutions are removed from the middle of the bed.

The operating cycle of an ion exchange system includes the following: (1) passing of the process stream through the bed for the proper time, (2) rinsing the bed with water and recovering any remaining valuable solution, (3) backwashing of the bed with water to remove accumulated materials and reclassify the particle size distribution, (4) regeneration of the bed for the proper time, and (5) rinsing of the bed with water to remove any remaining regenerant.

13.5.4 Food Applications of Adsorption

13.5.4.1 Water Treatment

Adsorption in activated carbon beds is used to remove odors, chlorine, and other undesirable compounds from drinking and process water. The spent adsorbent is regenerated by controlled burning in special furnaces.

Ion exchange is used for softening and demineralization of drinking, process, and steam boiler water. Water softening and removal of carbonates is accomplished using two cation exchange beds, one weak acid and a second strong acid resin. Complete demineralization of water with simultaneous removal of silicates can be accomplished with four columns, in the following order: strong acid, weak alkali, strong alkali, and mixed bed. Regeneration is accomplished with acid (hydrochloric acid) and alkali (sodium hydroxide).

Special ion exchangers are used to remove some specific mineral ions from drinking and process water, which may be toxic or radioactive, such as nitrates, lead, barium, strontium, and cesium.

Figure 13.15 shows a system of cation/anion exchange columns for water softening.

13.5.4.2 Recovery of Valuable Components

Recovery of proteins from food and biotechnological solutions can be accomplished with special ion exchange resins. Depending on the pH of the liquid, the proteins behave either as cation or anion molecular structures, and thus anion or cation exchange resins can be used for their recovery. The adsorbed proteins are recovered from the column either by altering the pH or increasing the ionic strength.

Industrial enzymes, such as amylase, can be recovered from fermentation liquids or from food materials with special ion exchange resins (combinations of strong anion and strong cation resins).

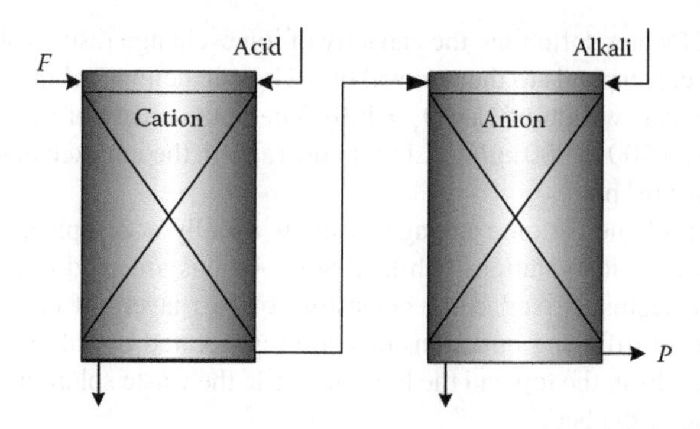

FIGURE 13.15 Water softening by a cation/anion exchange column system. *F*, feed water; *P*, product water.

13.5.4.3 Removal of Undesirable Components

Adsorption and ion exchange operations are used to remove undesirable components from food liquids, as a pretreatment step during processing or as a final step to improve the quality of the food product. Typical examples are decolorization of sugar solutions and liquid foods, decaffeination of soluble coffee, demineralization of dairy products and fruit juices, and debittering of citrus juices.

Demineralization of food liquids, such as cheese whey, with ion exchange resins can remove undesirable ions, such as Na, K, Mg, Cl, phosphate, citrate, and lactate. A system of strong cation, followed by a weak anion exchanger may be used, followed by regeneration with strong acid and alkali.

Demineralization of cane, beet, and hydrolyzed sugar solutions with ion exchange resins is used to clarify the sugar solutions before further processing by evaporation, followed by crystallization.

Bitter components of citrus (grapefruit and navel orange) juices, such as limonin and naringin, should be removed from the processed product. The juices are first clarified by centrifugation or membrane ultrafiltration (UF), and then debittered in beds of special ion exchange resins, made of divylbenzene polymer. The bed removes practically all bitter components and it is regenerated with a weak alkali solution. The debittered clear juice is combined with the separated fruit pulp to make the regular cloudy citrus juice.

Ion exchange resins and adsorption materials, used in food processing, must be nontoxic and approved by national food health authorities.

13.6 CRYSTALLIZATION FROM SOLUTION

Crystallization from solution is a process operation used to separate and recover various solutes from solutions by evaporation of the solvent. Most crystallizations in food processing take place from aqueous solutions, e.g., sugar and salt. Crystallization of ice in freeze-concentration of fruit juices is treated in Chapter 14. As in the other mass transfer operations, crystallization from solution is based on phase equilibria (solubility) and mass transfer rate (Mullin, 1993; Hartel, 2002).

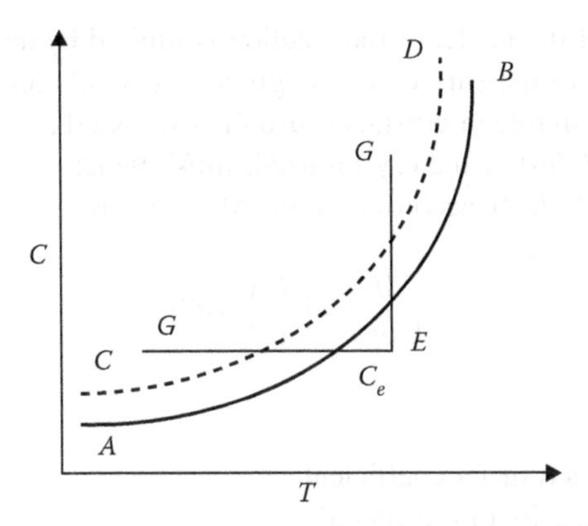

FIGURE 13.16 Solubility (C)–temperature (T) diagram. (AB), saturation; (CD), supersaturation.

13.6.1 Solubility in Crystallization

The solubility C of a solute in water is expressed in crystallization calculations as a mass ratio, i.e., kg solute/kg of water, and it increases significantly with the temperature (Figure 13.16).

Below the saturation line AB the solution is undersaturated. By cooling a solution from point E at constant concentration C (no evaporation), the solution can cross the supersaturation line CD and reach a supersaturation state G, where fast crystallization may start.

A similar condition may be created by concentrating the solution at constant temperature T, e.g., by evaporation, when a supersaturation point G may be reached, starting fast crystallization. The area between saturation and supersaturation is the unstable state.

The supersaturation of a solute is expressed by the difference $\Delta C = C - C_e$, where C_e is the equilibrium concentration (on the saturation line).

13.6.2 Nucleation and Mass Transfer

Crystallization from solution starts with nucleation, followed by crystal growth. Nucleation (formation of several microscopic nuclei) is either heterogeneous (foreign particles) or homogeneous (particles of the same material). The rate of homogeneous nucleation N [nuclei/(m^3 s)] is given by the following empirical equation, based on reaction-rate kinetics:

$$N = k(\Delta C)^i \tag{13.52}$$

The nucleation constants k, i depend on the geometry of the system and the agitation rate. The nucleation can be enhanced by adding small crystals (about 1 μm size) to the saturated solution, as in the seeding of sugar crystallizers.

Crystal growth of the nuclei of the solution is limited by mass transfer from the bulk of the solution to the surface of the growing crystals (molecular or turbulent diffusion). The growth rate of crystals is usually expressed as the rate of increase of a linear dimension L (m) of the crystal dL/dt (m/s), which is related linearly to the supersaturation ΔC (kg/m^3), according to the McCabe law:

$$\left(\frac{dL}{dt}\right) = \left(\frac{K}{\rho}\right)(\Delta C) \tag{13.53}$$

where

K (m/s) is the mass transfer coefficient
ρ (kg/m^3) is the density of the crystals

The crystal growth rate is a function of the mass transfer coefficient K, which increases at high agitation rates and lower viscosity of the solution. The growth of crystals is affected by the presence of foreign substances in the solution, resulting in the production of crystals of special sizes and shapes, e.g., in the crystallization of sodium chloride.

The crystallization kinetics (nucleation and crystal growth) is investigated in the laboratory and the pilot plant, using the mixed suspension mixed product removal (MSMPR) continuous crystallizer. In analyzing the operation of the (MSMPR) crystallizer, in addition to the usual material balance and mass transfer rates, the population balances of the system are considered.

13.6.3 Commercial Crystallizers

Commercial crystallizers are classified according to the method by which the supersaturation is achieved, such as cooling or evaporation. The crystallizing suspension is called "magma," while the saturated solution, remaining after removing the crystals, is known as "mother liquor."

Cooling crystallizers consist of a cooling/separation system, resembling the forced circulation evaporator (Chapter 10), with the heater replaced by a cooling (shell and tube) heat exchanger. In the draft tube baffled (DTB) crystallizers, the recirculation is carried out in a draft tube, which is installed in the crystallizer. Crystallization by cooling of viscous solutions is accomplished in scraped surface heat exchangers (Chapter 8), e.g., in margarine and ice cream production.

Evaporative crystallizers are similar to the forced circulation evaporators (Chapter 10) with an additional crystallization vessel below the vapor/liquid separator for the growth of the crystals (Oslo crystallizer). Simultaneous evaporation and cooling crystallizers operate without external heat exchangers, with the cooling effect provided by vacuum evaporative cooling of the saturated solution.

Production of large crystals can be obtained by recirculation of the magma within the crystallizer, and removal of the small crystals by dissolving in an outside vessel. Large crystals are produced in a crystallization column at the bottom of the crystallizer (elutriation leg).

13.7 FOOD PACKAGING

Mass transfer is important in the transmission of water vapor and other vapors and gases through packaging materials and in the transport of moisture and other food components within the package. Transmission properties (mainly permeability) characterize the polymeric materials (plastic films and paper), whereas mass transfer within the package relates to all packages, including metal containers.

Because permeation of gases depends on both solution and diffusion, the state of the polymer is very important. At temperatures above the glass transition temperature T_g (rubbery polymers), gases quickly reach equilibrium with the polymer surface, and diffusion through the polymer controls the transport process. At temperatures below T_g (glassy state), true equilibrium is not reached easily.

The permeability (PM) of packaging films to gases and moisture is measured using the permeation Equation 13.54. The driving force for gases is the partial pressure difference Δp of the penetrant, while for water vapor, it is the humidity difference ΔY. Measurements of the permeability reflect both the solubility S and the diffusivity D of the penetrant into the film, according to the relationship $PM = SD$.

$$J = \frac{[(PM)\Delta p]}{\Delta z} = \frac{[(DS)\Delta p]}{\Delta z} \tag{13.54}$$

PM, permeability, has units kg/(m s Pa) or g/(m s Pa)
J [kg/(m^{2}s)] is the mass flux through the film
Δp (Pa) is the pressure difference
z (m) is the film thickness
D (m^2/s) is the molecular diffusivity
$S = C/p =$ kg/(m^3 Pa) is the solubility of the gas in the film

Standard methods are used to measure the water vapor transmission (WVT) and the gas transmission (GT). Various units of permeability are used traditionally in Packaging, related to the established methods of measurement (Chapter 5).

Table 13.3 shows typical permeabilities of packaging and food films, useful in the design and applications of packaging operations. The elements of Food Package Engineering are introduced in Chapter 15.

TABLE 13.3
Permeabilities *PM* of Water Vapor in Packaging Films

Film or Coating	$PM \times 10^{10}$, g/(m s Pa)
HDPE	0.002
LDPE	0.014
PVC	0.041
Cellophane	3.70
Protein films	0.1–10
Polysaccharide films	0.1–1.0
Lipid films	0.003–0.1
Chocolate film	0.11

Differential permeability of packaging materials is important in maintaining a controlled atmosphere within the package during storage. Thus, polymer films may permit the partial removal of carbon dioxide while maintaining the water vapor in the package during the storage of packaged fresh fruits and vegetables.

Transport processes of food components within the package may affect the quality of the stored food product. Transfer of moisture from high- to low-moisture regions may cause agglomeration (caking) of hygroscopic food powders. Transfer of oxygen may cause oxidation reactions and loss of nutrients and sensory quality.

APPLICATION EXAMPLES

Example 13.1

A distillation column is used to separate the ethanol from a fermentation solution, containing 4% of ethanol, into distillate and bottom products containing 80% and 0.1% (molar basis) ethanol, respectively. The feed to the column will be 8 ton/h. The mean relative volatility of ethanol/water in the column may be taken as $\alpha = 3.5$.

Estimate the required number of equilibrium (theoretical) stages, the number of sieve plates (trays), and the dimensions of the column. Column efficiency is 60%. Use the simple analytical method and make the needed simplifying assumptions.

The column is assumed to operate at atmospheric pressure with steam heating the reboiler and water cooling the total condenser.

The feed is assumed to enter the column as saturated liquid ($q = 1$). The molar flow rates of liquid and vapors in the stripping and enriching sections of the column are assumed constant, although ethanol/water is a nonideal mixture.

Note that the maximum concentration of the distillate at atmospheric pressure is the azeotrope of 0.896 mole ethanol.

Solution

The minimum number of theoretical stages N_{min} at total reflux will be, for $x_D = 0.80$, $x_B = 0.001$ (Equation 13.22)

$$N_{min} = \log\left\{\frac{[(0.80) \times (0.999)]/[(0.001) \times (0.20)]}{\log(3.5)}\right\} = 6.7$$

The minimum reflux ratio R_{min} for infinite number of stages is estimated from the Underwood Equations 13.23 and 13.24, for $x_F = 0.05$, $q = 1$ and $\alpha = 3.5$

$$\frac{(3.5 \times 0.05)}{(3.5 - \theta)} + \frac{(1 - 0.05)}{(1 - \theta)} = 0$$

and

$$R_{min} + 1 = \frac{(3.5 \times 0.80)}{(3.5 - \theta)} + \frac{(1 - 0.8)}{(1 - \theta)}$$

from which $\theta = 3.11$ and $R_{min} = 6.1$.

Assume that the reflux ratio is $R = 1.2\ R_{min} = 1.2 \times 6.1 = 7.32$. Then $(R - R_{min})/(R + 1) = (8.4 - 7.0)/9.4 = 0.135$. From the Gilliland diagram or from the Gilliland correlation (Equation 13.25), we obtain $(N - N_{min})/(N + 1) = 0.45$. For $N_{min} = 6.1$, $N = 11.9$.

Assuming a column efficiency of 60%, the number of trays of the column will be $N_T = 11.9/0.60 = 18.8$ or $N_T = 20$.

The column diameter and height are estimated from material balances in the column.

The feed to the column of 8 ton/h must be converted to molar flow. Material balance of the feed gives, $F = 46E + 18W = 8000\ kg/h$, where E and W are the kmols of ethanol and water with molecular weight 46 and 18, respectively. The feed contains 5% molar ethanol and 95% molar water, and $W = (95/5)E = 19E$. Therefore, $46E + 18 \times 19E = 8000$, and $E = 20.6$ kmol/h and $W = 391.7$ kmol/h, or $F = 20.6 + 391.7 = 412.3$ kmol/h.

Material balances (overall and ethanol) over the whole column:

$F = D + B$, or $412.3 = D + B$ and $(0.05) \times 412.3 = (0.80)D + (0.001)B$, from which $D = 25.3$ kmol/h and $B = 387.0$ kmol/h.

The column diameter is calculated on the basis of vapor flow in the upper (enriching) section, which is usually higher than in the lower (stripping) section. For the reflux ratio $R = (L/D) = 7.32$, the liquid flow in the enriching section will be $L = 7.32 \times 25.3 = 185.2$ kmol/h. The vapor flow in this section will be $V = L + D = 185.2 + 25.3 = 210.5$ kmol/h. Therefore, the liquid/vapor ratio in this section (slope of the upper operating line) will be $L/V = R/(R + 1) = 6.3/7.3 = 0.88$.

Assume that the mean molar concentration of ethanol in the enriching section is 50% with a corresponding mean molecular weight of $0.5 \times 46 + 0.5 \times 18 = 32$. Therefore, the mass flow of the vapors will be $V = 210.5 \times 32 = 6{,}736$ kg/h.

The vapor density in the enriching section at a mean temperature of 90°C is taken approximately as $\rho = 0.4\ kg/m^3$. Therefore, the volumetric flow rate of vapors will be $V = 6{,}736/0.4 = 16{,}840\ m^3/h$ or $V = 14{,}887/3{,}600 = 4.68\ m^3/s$.

The vapor velocity in distillation columns is selected on the basis of the flooding (maximum) velocity, which for this system can be taken as 1.5 m/s. The vapor velocity is taken as 80% of the flooding velocity, or $0.8 \times 1.5 = 1.2$ m/s. Therefore, the column cross sectional area (for vapor flow) will be $A = 4.68/1.2 = 3.9\ m^2$, and the column diameter will be $d = (4A/\pi)^{1/2} = (4 \times 3.9/3.14)^{1/2} = 2.22$ m.

Sieve (perforated) trays are selected, since they are inexpensive and they can be cleaned more easily than other complex tray arrangements. The distance between the trays is taken empirically for this system as equal to 50 cm. Therefore, the column height will be equal to $20 \times 0.5 = 10.0$ m.

Notes

1. As discussed earlier in this chapter, the mixture ethanol/water is highly nonideal and the mean volatility used in this Example is a rough approximation. The graphical method of McCabe-Smith takes into consideration the variation of relative volatility, but it assumes constant molar flow, which may be not true in this case. The Ponchon-Savarit (PS) graphical method is more accurate, because it is not based on these restrictions. The PS is applied in Example 13.3 (Extraction).
2. The design of the reboiler and the condenser of the column is similar to conventional procedures, applied to shell and tube heat exchangers (Chapter 8).

Example 13.2

Apply the analytical and graphical methods in determining the height of a countercurrent liquid/liquid extraction column to remove 85% of the caffeine from an aqueous solution of 4% caffeine, using an immiscible organic solvent. What will be the approximate diameter of an extraction column for treating 6 ton/h aqueous solution?

The equilibrium constant (partition coefficient) of the solvent/solution is m = 2.5, and the liquid/solvent mass flow ratio is constant, $L/V = 1.5$.

Graphical Method

For dilute systems, straight equilibrium and operating lines can be assumed,

$y_e = m\,x$ or $y_e = 3x$ (equilibrium) and $y = 1.5x$ (operating). The feed composition is $x_2 = 0.04$ and the product $x_1 = 0.006$ (mass fractions).

Figure 13.17 shows a diagram of the graphical method McCabe–Thiele, applied to the extraction operation of this example.

The number of equilibrium stages is estimated by a step-by-step graphical construction between the operating and the equilibrium lines. The construction starts at the intersection of the feed $x_2 = 0.04$ with the operating line and ends when concentration $x_1 = 0.006$ is reached. The number of equilibrium stages is about $N = 2.5$.

It should be noted that the graphical estimation of the equilibrium stages, in general, becomes difficult to apply at very low concentrations of the solvent, when the operating and equilibrium lines approach together near the zero point. The analytical method is more accurate in such applications.

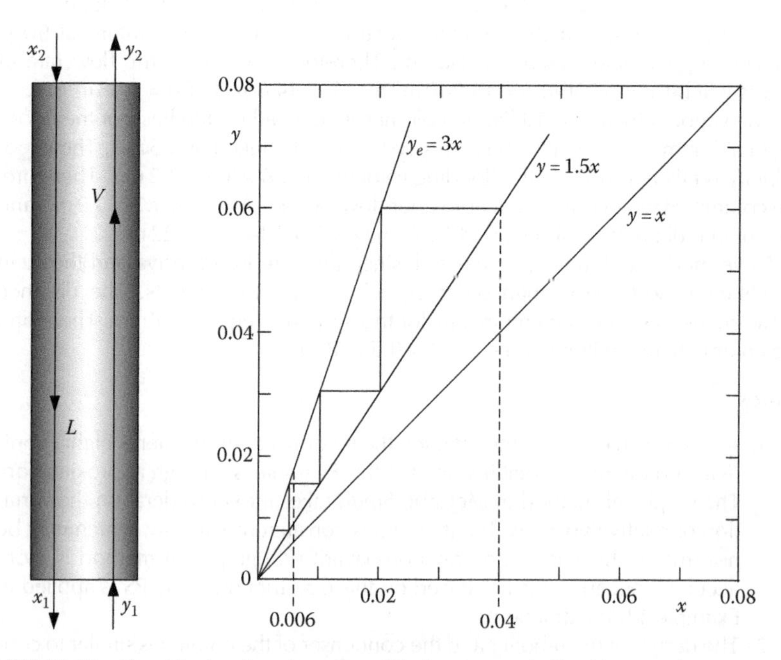

FIGURE 13.17 Graphical estimation of equilibrium stages in liquid extraction.

Analytical Method

The Kremser equation (13.38), modified for stripping, is applied to the extraction process in dilute solutions for calculating the number of equilibrium stages (N):

$N = \log[R_s(1 - A) + A]/\log(1/A)$, where $R_s = (x_2 - y_1/m)/(x_1 - y_1/m)$ is the stripping ratio and $1/A = (m\ V)/L$ is the stripping factor, x_2, x_1 and y_2, x_1 are the mass fractions of the solute in the solution and the solvent, respectively, at the top (2) and bottom (1) of the column, and m is the equilibrium constant. If the solvent enters the bottom of the column free of solute, then $y_1 = 0$ and the stripping ratio becomes $R_s = (x_2/x_1) = 0.04/0.006 = 6.7$.

The stripping factor in this column will be $1/A = (m\ V)/L = 3/1.5 = 2$, and $A = 0.5$.

$$N = \log\{6.7(1 - 0.5) + 0.5\}/\log(2) = \log(3.85)/\log(2) = 1.95, \text{ approximately } 2.$$

The number of equilibrium stages, calculated by the Kremser equation (13.2), is somewhat less than the stages found by the graphical method (2.5). In practice an integer number of plates (trays) is used, after considering the efficiency of the extraction column.

Column Dimensions

The efficiency of liquid/liquid extraction columns depends on the mass transfer between the immiscible phases (solvent/residue), and it can be increased by agitation, such as the rotating (RDC) column. Assuming an efficiency of 30%, the number of plates will be $2.5/0.3 = 8$. If the distance between the plates is taken as 30 cm, the height of the column will be $8 \times 0.3 = 2.4$ m.

The diameter of the column will depend on the mass flow rate (kg/h) of the solution L through the trays. Sufficient residence time should be allowed on the trays for good contact mixing, and mass transfer between the two liquid phases. Assuming a mean velocity of the solution in the column $u = 10$ cm/min or 6 m/h and a density $\rho = 1000$ kg/m, the cross sectional area A can be calculated from the mass balance, $Au\rho = 5000$ kg/h, from which, $A = (5000)/(1000 \times 6) = 0.83\ m^2$. The column diameter d will be, $d = [(4 \times 0.83)/3.14]^{1/2} = 1.03$ m or $d = 1$ m.

Example 13.3

Extraction of vegetable oil from soybean flakes can be achieved in a series of extraction cells, in countercurrent flow with an organic solvent. The soy flakes contain 20% oil, 10% moisture, and 70% solids. It is desired to extract 90% of the oil from the flakes.

Calculate the required number or theoretical (equilibrium) stages, if 2 ton of solvent are used for 1 ton of flakes.

The following data and simplifying assumptions are considered:

The oil is completely soluble in the solvent, while the soy solids and the contained moisture are completely insoluble. Hexane is a typical solvent for soy oil.

Experimental liquid/solid equilibria data of liquid/solid extraction are obtained by mixing and equilibrating the solvent with the original solids at various proportions. The component to be extracted is determined in the liquid above the solids, overflow Y [kg (component)/kg (solution)], and in the liquid trapped in the pores of the solids, underflow X [kg (component)/(kg solution)].

Equilibrium data of this Example are given in Table 13.4. Here, the value $X = 0$ corresponds to the initial solids [underflow, entering the first stage (1) of the system ($z = 80/20 = 4.0$)], while $X = 1$ corresponds to the underflow leaving the last stage N ($z = 2.0$). The overflow line coincides with the horizontal abscissa ($Z = 0$).

TABLE 13.4
Liquid/Solid Equilibrium Data for Soy Flakes/Oil/Solvent

X = kg (Oil)/kg (Solution)	kg (Solution)/kg (Inert Solids)	z = kg(Inert Solids)/kg (Solution)
0	0.25	4.00
0.25	0.27	3.70
0.50	0.30	3.30
0.75	0.48	2.80
1.00	0.50	2.00

Figure 13.18 shows the underflow equilibrium data as z [kg (inert solids)/kg (solution)] versus X [kg (oil)/kg (solution)]. Note that kg (solution) = kg (oil) + kg (solvent).

The liquid/solid extraction system of this example operates at high concentrations of oil and solvent, and it cannot be considered as a dilute mixture, where simplified methods of analysis can be used, such as the analytical method. The Ponchon-Savarit and the triangular diagrams are more suitable in such cases. The Ponchon-Savarit method is preferably applied in this Example.

The Ponchon-Savarit diagram for extraction is constructed with coordinates X [kg (oil)/kg (solution)] and z [kg (inert solids)/kg (solution)].

In this example, 1 kg of soy flakes contain 0.2 kg oil, 0.1 kg moisture, and 0.7 kg of solids. For simplification, the sum of soy solids and the adsorbed moisture (0.8 kg) is considered as "inert solids" (IS), which are insoluble and they do not interfere with the extraction process. The residue (extracted flakes) will contain $0.1 \times 0.2 = 0.02$ kg oil.

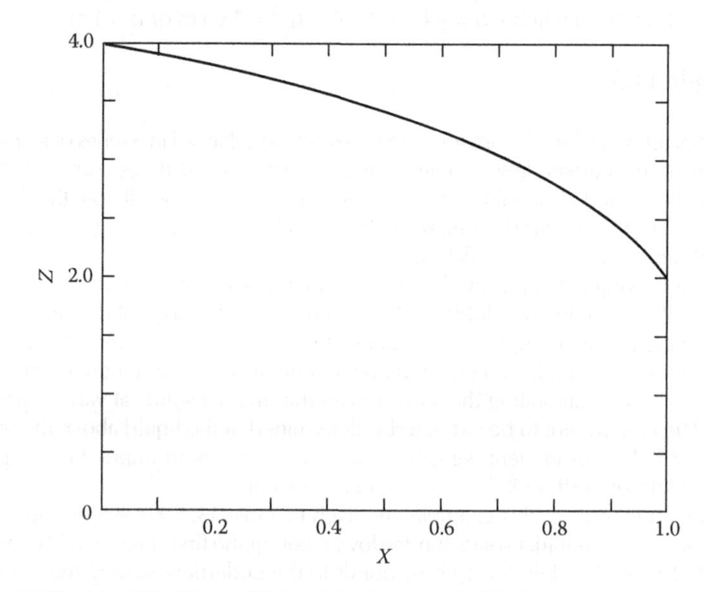

FIGURE 13.18 Liquid/solid equilibrium of oil/solution/solids in soy flakes.

Figure 13.19 shows the application of the Ponchon-Savarit diagram for the estimation of the equilibrium stages required for the extraction of oil from soy flakes.

A series of N stages are operated in counterflow of the flakes and the solvent solution. The fresh flakes enter as underflow the first stage (1), where they are mixed with solution overflow from stage (2). After mixing and approaching to equilibrium, the solution (oil/solvent) leaves stage (1) as overflow product. The counter flows of solution and product continue until stage N is reached, which is fed with pure solvent, while the oil-extracted flakes plus the adhering solution leave the system.

The overflow oil concentration increases from $Y_{N+1} = 0$ at the entrance of the N stage to the maximum value of oil concentration Y_1 in the overflow of stage (1). At the same time, the inert solids concentration in the underflow

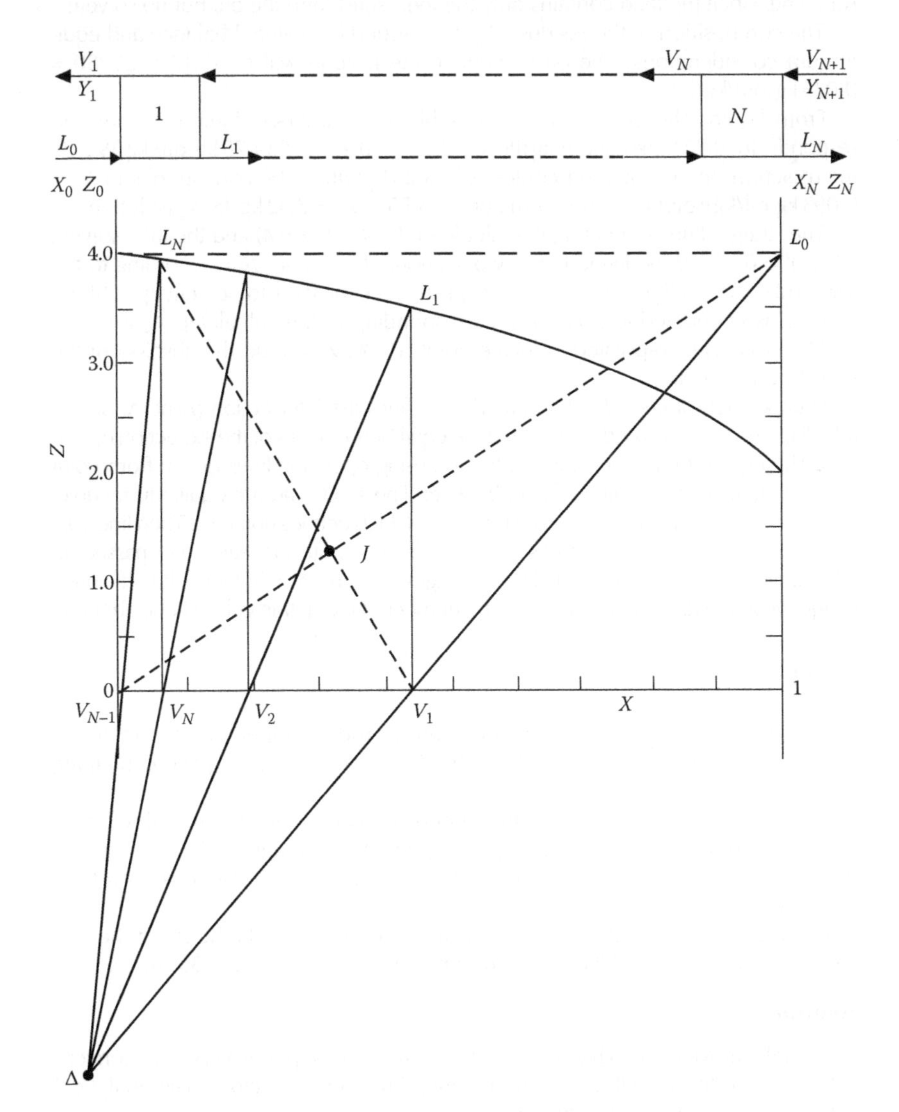

FIGURE 13.19 Ponchon-Savarit diagram of liquid/solid extraction of oil from soy flakes.

decreases from the maximum value of z_0 at the entrance of stage (1) to the minimum z_N at the exit of stage N.

In the Ponchon-Savarit (PS) diagram of Figure 13.19 the underflow L curve (z, X) represents the liquid/solid equilibrium data of Figure 13.18, while the overflow V line (Z, Y) coincides with the abscissa ($0XY$), since the inert solids are insoluble in the solvent ($Z = 0$).

The underflow feed L_0 (z_0, X_0) enters stage (1), where $z_0 = 4.0$ kg (IS)/kg (solution) or 4.0 kg (IS)/kg (oil), where (*IS*) stands for inert solids. The flakes enter the extraction system solvent-free ($X_0 = 1$), and they leave the last stage N as L_N (z_N, X_N). The overflow enters stage N as V_{N+1} (oil-free, $Y = 0$) and leaves stage (1) as V_1 ($Y = Y_1$). Point L_0 ($X_0 = 1$ and $z_0 = 4.0$) of the underflow on the PS diagram lies above the equilibrium line, since the feed contains only the inert solids and the oil, but no solvent.

The composition of the residue (X_N, z_N) is found by material balance and equilibrium considerations. The oil content of the residue will be $(0.10 \times 20)/80 = 0.025$ kg (oil)/kg (IS).

From liquid/solid equilibrium data (Table 13.4), 0.25 kg (oil)/kg (solution) corresponds to 0.27 kg (solution)/kg (IS) or $0.25 \times 0.27 = 0.0675$ kg (oil)/kg (IS). By interpolation, the residual oil content of 0.025 kg (oil)/kg (IS) corresponds to $X_N = 0.093$ kg oil/kg (solution), and, from curve 13.E3.1, $z_N = 3.92$ kg (IS)/kg (solution).

The straight line connecting the inlet feed (L_0, $X = 1$, $z = 4$) and the inlet solvent (V_{N+1}, $Y = 0$, $Z = 0$) is divided into two sections (JV_{N+1}) and (JL_0), according to the lever rule $(JV_{N+1})/(JL_0) = L_0/V_{N+1} = 1/2$ (given ratio of feed to solvent quantities). Point (J) is connected to point (L_N), corresponding to the calculated $X_N = 0.093$, and line (JL_N) is extrapolated to locate point (V_1, Y_1, $Z = 0$) on the abscissa of the equilibrium diagram.

Extrapolation of lines (L_0V_1) and (L_NV_{N+1}) defines the intersection point (Δ) on the (PS) diagram, which is used to construct the equilibrium stages of the extraction operation. The equilibrium stages are constructed using operating lines drawn from point (Δ), and equilibrium (tie) lines drawn between line $z = 0$ and the equilibrium curve. Thus, a vertical tie line is drawn from point (V_1), which defines point (L_1). Then line (ΔL_1) is drawn, defining point (V_2), and so on, until final point (L_N) is reached or passed. In this example, the number of equilibrium stages is nearly $N = 3$. Since the efficiency of liquid/solids extractors is very high, the number of extraction cells will be about 3.

Example 13.4

In wine processing, grape juice is preserved at room temperature by treatment with SO_2. It is desired to strip 95% of the SO_2 from a sulfured juice containing 1000 ppm SO_2.

The stripping of the juice will be carried out with countercurrent flow of air in a tray column, operated at atmospheric pressure, at a ratio of $(L/G) = 4.0$. The solubility of SO_2 in the grape juice is given by the Henry constant $(1/H) = 0.02$ mole fraction/bar.

Calculate the required number of equilibrium stages and the height of the column, if its efficiency is 20% and the distance between the trays is 35 cm.

Solution

The graphical McCabe–Thiele method is difficult to apply at very low concentrations, since the equilibrium and operating lines are very close. The analytical method is more suitable for this case.

The Kremser equation (13.38) for stripping is written as

$N = \log[R_s(1 - A) + A]/\log(1/A)$, based on the stripping ratio (R_s) and the stripping factor ($1/A$), defined as, $R_s = (x_2 - y_1/m)/(x_1 - y_1/m)$, and $(1/A) = (mG/L)$.

Here, (x_2, x_1) are the mole fractions of SO_2 in the liquid at the top and bottom of the column, respectively, and (y_1) is the mole fraction of SO_2 in the stripping air at the bottom of the column. The equilibrium constant is defined as $m = H/P$, where H (bar/mole fraction) is the Henry constant and P (bar) is the total pressure. The Henry constant is defined by the Henry law, $p_i = H_i x_i$, where p_i (bar) is the partial pressure and x_i is the mole fraction of component (i). The ratio of quantities of liquid (juice) to gas (air) is given as $(L/G) = 4.0$.

In this example, the air enters the bottom of the column clean ($y_1 = 0$) and the stripping ratio becomes $R_s = (x_2/x_1) = 95/5 = 19$. The Henry constant is $H = (1/0.020) = 50$ bar, the equilibrium constant $m = 50$, and the stripping factor is $(1/A) = (mG)/L = 50/4 = 12.5$.

Therefore, $N = \log[19 \times (1 - 0.08) + 0.08]/\log(12.5) = \log(17.56)/\log(12.5) = 1.13$.

The number of trays (20% efficiency) will be $N_T = 1.13/0.20 = 5.6$ or 6 trays, and the height of the column will be $h = 6 \times 0.35 = 2.10$ m.

Example 13.5

Vacuum or inert gas deaeration of food liquids is used to remove dissolved oxygen, which may damage the quality of the product. Oxygen absorption by water is indicative of oxygen absorption in aqueous food products.

Determine the equilibrium dissolved oxygen in water (ppm) at 20°C in the following conditions: (a) In contact with 100% Oxygen; (b) In contact with atmospheric air (20% O_2); (c) In a vacuum of 1 mbar; and (d) In an atmosphere of 100% N_2.

The solubility of oxygen in water at 20°C is given as $(1/H) = 2.3 \times 10^{-5}$ mole fraction/bar.

Solution

The solubility of oxygen in water follows the Henry law $p = Hx$, where p (bar) is the partial pressure, H (bar/mole fraction) is the Henry constant, and x is the mole fraction. The solubility is given as $(1/H) = 2.3 \times 32 \times 10^{-5}/18 = 4 \times 10^{-5}$ g/(g water) (bar). It is evident that the solubility of oxygen depends on the pressure above the water solution.

(a) The solubility of oxygen in water at atmospheric pressure (1.013 bar) and 100% O_2 is $40 \times 1.013 = 40.5$ ppm.
(b) The solubility of atmospheric oxygen (20% oxygen) at atmospheric pressure ($P = 1.013$ bar) is $40.5 \times 0.2 = 8.1$ ppm.
(c) The solubility of oxygen in absolute pressure 1 mbar will be $8.1/1013 = 8$ ppb, where 1 ppb = 0.001 ppm. It is assumed that the vacuum atmosphere contains 20% oxygen.
(d) According to the Henry law, the solubility of oxygen in a nitrogen atmosphere (0% oxygen) will be zero ($x = 0$). Thus, an inert atmosphere (100% nitrogen), is an effective method of reducing the dissolved oxygen from aqueous solutions. It should be noted, however, that dissolved oxygen in food liquids may not reach equilibrium with the surrounding atmosphere, due to physicochemical interactions with food components.

PROBLEMS

13.1 A high-boiling organic component *A* is partially soluble in water, forming two phases. The water phase contains 5% of component *A* and 95% of water *W* in molar basis.

Estimate the relative volatility α_{AW} at 30°C if the vapor pressures at this temperature are 42 mbar for *W*, and 5 mbar for *A*. Water is considered completely insoluble in *A*.

13.2 A large volume of food processing wastewater is stored outdoors before it is disposed to the environment. The wastewater contains 0.1% by weight of a volatile compound, which has a relative volatility $\alpha = 10$. Estimate the percentage (%) of water evaporation required in order to remove 95% of the volatile compound into the surrounding atmosphere.

13.3 Brandy liquor, containing 50% ethanol by volume, is produced by simple (differential) distillation of wine. The wine contains 12% ethanol by volume and two successive simple distillations are required. In the first distillation, 95% of the ethanol is removed, while the second distillation is stopped when the distillate reaches 50% (vol.) ethanol.

Calculate the (kmol) of brandy produced per 100 kmol of wine.

Density of wine 970 kg/m^3, 50% ethanol 900 kg/m^3. Molecular weight ethanol 46, water 18. Wine is assumed to consist of only ethanol and water.

13.4 Steam distillation is used to deodorize an edible oil which contains an undesirable compound *A* of vapor pressure 300 mbar at 100°C. Estimate the amount of steam required *W* (kmol) to remove 1 kmol of *A* at pressure 1 bar, if the efficiency of the system is 85%.

13.5 In normal distillation, a total condenser is used, which condenses all the vapors leaving the top plate of the column, and the condensate is split into reflux and product. In some distillations a partial condenser is used, with the liquid condensate returning to the column as the reflux and the vapors recovered as the product.

Show in a diagram the top of a column with a partial condenser. Why such a system is more effective in distillation than the normal total condenser?

13.6 The Ponchon-Savarit *PS* graphical method is more suitable for the determination of equilibrium stages in extraction and in distillation of concentrated solutions than the McCabe–Thiele and analytical methods (Example 13.3). In the *PS* diagram the "lever-arm rule" is often applied to facilitate calculations. The lever-arm rule is expressed by the relationship $L/V = VJ/JL$ in a straight line *LJV* connecting *L* and *V* and intermediate point *J*.

Prove the lever rule on a *PS* diagram, using material and energy balances.

In extraction, *L* (*z*, *X*) is the underflow and *V* (*z*, *Y*) is the overflow, where z = kg (solids)/kg (solution), and *X*, *Y* = kg (component)/kg (solution) in the underflow and overflow, respectively (normally $z = 0$ in the overflow).

In distillation, *L* (*H*, *x*) is the liquid and *V* (*H*, *y*) is the vapor, where *H*(kJ/kmol) is the enthalpy and (*x*, *y*) are the mole fractions of the liquid and vapor, respectively.

13.7 The combustion gases of a steam boiler contain 5% CO_2 which can be removed by washing with water. An absorption column with sieve trays, operated at atmospheric pressure and 30°C, may be used for this purpose. Determine the number of equilibrium stages for the removal of 95% of CO_2, using countercurrent flow of water to gases $L/G = 2.0$. Solubility of CO_2 in water, $(1/H) = 8 \times 10^{-4}$ mole fraction/bar. Henry equation, $p = Hx$, where p (bar) is the partial pressure of the gas, and x (–) is the mole fraction in the liquid phase.

13.8 Industrial water contains 1000 ppm of ammonia, which should be removed down to 20 ppm before it is reused. Ammonia can be stripped from water in a tray column, operated at atmospheric pressure and 30°C with countercurrent flow of atmospheric air. The column will be operated at a liquid/gas ratio of $L/G = 2.0$. The solubility of ammonia in water at 30°C is given by the inverse of Henry constant $(1/H) = 0.3$ (mole fraction)/bar.

Calculate the required number of equilibrium stages and the height of the column, if its efficiency is 25% and the distance between the trays is 35 cm.

LIST OF SYMBOLS

A	m^2	Surface area
α_w	—	Water activity
B	kmol	Bottom product
C	kg/m^3, $kmol/m^3$	Concentration
C_p	kJ/(kg K)	Specific heat
d	m	Diameter
D	m^2	Diffusivity
D	kmol	Distillate
E	kJ/kmol	Energy of activation
E	—	Efficiency
F	kmol	Feed mixture
G	kg/h	Gas flow rate
H	$W/(m^2\ K)$	Heat transfer coefficient
H	Pa/(mole fraction)	Henry's constant
H	$(m^3\ Pa)/kg$	Henry's constant
HETP	m	Height of equivalent theoretical plate
HTU	m	Height of a transfer unit
J	$kg/m^2\ s$, W/m^2	Mass or heat flux
j_M	—	Mass transfer factor
k	1/s	Reaction constant
K	—	Partition coefficient, Equilibrium constant
K_G	$kmol/(m^2\ s\ Pa)$	Overall mass transfer coefficient, gas phase
K_L	m/s	Overall mass transfer coefficient, liquid phase
L	kmol	Residue
L	kmol	Liquid flow
L	kg	Liquid underflow
m	—	Equilibrium constant

m	kg	Mass
M	kg/kmol	Molecular weight
N	—	Number of theoretical stages
NTU	—	Number of transfer units
p, P	Pa	Pressure
(PM)	kg/(m s Pa)	Permeability
R	m	Radius
Re	—	Reynolds number ($du\rho/\eta$)
q	—	Thermal state of distillation feed
Q	kJ	Enthalpy
R	kJ/(kmol K)	Gas constant (8.31)
R	—	Reflux ratio
S	kg/(m^3 Pa)	Solubility (C/p)
S	kg	Steam
t	s	Time
T	K	Temperature
u	m/s	Velocity
V	kmol	Vapor flow
V	kg	Solvent overflow
x	—	Mole fraction, liquid phase
X	kg/kg	Composition, liquid phase
y	—	Mole fraction, vapor phase
Y	kg/kg	Composition, vapor phase
z	m	Distance
z, Z	—	Solvent ratio (extraction)
α	—	Relative volatility
γ	—	Activity coefficient
ε	—	Bulk porosity (void fraction)
η	Pa s, kg/(m s)	Viscosity
λ	W/(m K)	Thermal conductivity
ρ	kg/m^3	Density

REFERENCES

AIChE. 2000. *Distillation in Practice* CD-ROM. American Institute of Chemical Engineers, New York.

Aquilera, J.M. and Stanley, D.W. 1999. *Microstructural Principles of Food Processing and Engineering*, 2nd edn. Aspen Publ., Gaithersburg, MD.

Cussler, E. 1997. *Diffusion Mass Transfer in Fluid Systems*, 2nd edn. Cambridge University Press, Cambridge, U.K.

Geamkoplis, C.J. 1993. *Transport Processes and Unit Operations*, 3rd edn. Prentice Hall, New York.

Gekas, V. 1992. *Transport Phenomena of Foods and Biological Materials*. CRC Press, New York.

Hartel, R.W. 2002. *Crystallization in Foods*. Kluwer Academic, New York.

King, C.J. 1982. *Separation Processes*. McGraw-Hill, New York.

Maroulis, Z.B. and Saravacos, G.D. May 9, 2003. *Food Process Design*. Marcel Decker, New York.

Mullin, J. 1993. *Crystallization*, 3rd edn. Butterworths, London, U.K.

Perry, R.H. and Green, D. 1997. *Chemical Engineers' Handbook*, 7th edn. McGraw-Hill, New York.

Prausnitz, J.M., Lichtenthaler, R.N., and de Azevedo, E.G. 1999. *Molecular Thermodynamics of Fluid Phase Equilibria*, 3rd edn. Prentice-Hall, Englewood Cliffs, NJ.

Reid, R.C., Prausnitz, J.M., and Poling, B.E. 1987. *The Properties of Gases and Liquids*, 4th edn. McGraw-Hill, New York.

Saravacos, G.D. and Moyer, J.C. 1968. Volatility of some aroma compounds during vacuum-drying of fruit juices. *Food Technology*, 22(5): 89–95.

Saravacos, G.D. 2005. Mass transport properties of foods. In: *Engineering Properties of Foods*, 3rd edn., M.A. Rao, S.S.H. Rizvi, and A.K, Datta, eds. CRC Press, New York, pp. 327–373.

Saravacos, G.D. and Maroulis, Z.B. 2001. *Transport Properties of Foods*. Marcel Dekker, New York.

Schweitzer, P.A., ed. 1988. *Handbook of Separation Techniques for Chemical Engineers*, 2nd edn. McGraw-Hill, New York.

Treybal, R. 1980. *Mass Transfer Operations*, 3rd edn. McGraw-Hill, New York.

Van Winkle, M. 1967. *Distillation*. McGraw-Hill, New York.

Walas, S. 1988. *Chemical Processing Equipment*. Butterworths, London, U.K.

14 Novel Food Process Operations

14.1 INTRODUCTION

Conventional food process operations, such as thermal separations (evaporation, distillation) and thermal processing (pasteurization, sterilization), may cause undesirable heat damage to the quality of sensitive foods, due to the high temperatures and long times used. Advances in novel separation technologies, achieved in chemical engineering and chemical process industries, may find new applications in food processing (Saravacos and Kostaropoulos, 2002; Perry and Green, 1997).

Various novel food processing and preservation methods have been developed recently, and a few of them have been applied commercially, such as membrane separations and supercritical solvent extraction. Some other processes, such as food irradiation and high-pressure processing (HPP), are still in the development and evaluation stages, and various technical, economic, and public acceptance difficulties must be resolved, before large-scale application.

In the area of food preservation, the novel processes are at different stages of development, with food irradiation being closer to commercial application, followed by HPP and electrical pulse processing. All these methods are more expensive than conventional preservation processes, but they may become competitive, when large units are installed and operated.

An important advantage of all novel food processes is the use of less energy per unit mass of product processed, which is an important consideration in view of rising energy costs and anticipated worldwide energy shortages in the future. The most important novel food processes are reviewed in this chapter, emphasizing commercial applications, such as membrane separations (Grandison and Lewis, 1986).

14.2 MEMBRANE SEPARATION OPERATIONS

14.2.1 Introduction

Membrane separations are rate-controlled processes, based on the preferential permeation or rejection of various solute molecules in semi-permeable polymer or inorganic membranes. Mechanical pressure is basically the driving force through the thin membranes, but concentration or ionic charge differences may be utilized in some membrane separations. Membrane separations require low energy in their operation, and they are less expensive than conventional thermal separations. Typical examples of membrane separations are ultrafiltration (recovery of proteins) and reverse osmosis (desalination of water) (Sourirajan, 1977).

Membrane separations are classified according to the size of the molecule or the particle being separated from a solution or suspension. The following membrane separation processes are applied to the food processing industry:

1. Ultrafiltration, macromolecules 1–100 nm (protein recovery)
2. Nanofiltration, molecules 0.5–10 nm (whey desalting)
3. Microfiltration: particles 0.5–3 μm (removal of microorganisms)
4. Reverse osmosis, ions and molecules 0.1–1 nm (water desalting)
5. Pervaporation: molecules 0.2–1 nm (separation of ethanol)
6. Electrodialysis: ions 0.1–0.5 nm (water desalting)

14.2.2 Mass Transfer Aspects

Mass transfer through membranes depends on the physical structure of the membrane, which is usually a synthetic polymer or a porous ceramic material. In separations of suspended particles and macromolecules (ultrafiltration, nanofiltration, microfiltration) porous membranes are used, in which the driving force is a pressure gradient, and the permeability equation is applicable. Separation of small molecules and ions is based on molecular diffusion (Fick equation) and a concentration gradient as the driving force (Matsuura and Sourirajan, 1995).

In most food applications, the concentration of the solute molecules is low and the Henry law and the permeability equation are applicable. The transport rate of a solute through a membrane of thickness z (m) is given by the permeability (mass flux) equation, which is a combination of the Fick law and Henry law

$$J = \left(\frac{DS}{z}\right)(\Delta P) \tag{14.1}$$

where

J [kg/(m^2 s)] is the mass flux
ΔP (Pa) is the pressure drop
DS (kg/m s Pa) is the mass permeability of the membrane

Here, $DS = D/H$, where

D (m^2/s) is the diffusivity of the solute in the membrane
S (kg/m^3) is the solubility of the solute in the membrane
H (kg/m^3 Pa) is the Henry constant
z (m) is the membrane thickness

The pressure drop is defined as $(\Delta P) = (P_f - P_p)$, where (P_f, P_p) are the feed and product pressures.

In membrane separations, the permeate flux (J_p) is usually expressed in volumetric, instead of mass, units, i.e., (m^3/m^2 s) or (m/s), according to the equation,

$$J_p = A(\Delta P) = \frac{(\Delta P)}{R_m} \tag{14.2}$$

where

A [m/(s bar)] is the membrane permeability

$R_m = 1/A$ [(s bar)/m] is the overall membrane resistance to mass transfer

The following conversion factors are useful for flux units, 1 U.S. gallon/(ft^2 day) = 1.7 L/(m^2 h), and 1 L/(m^2 h) = 1.0 mm/h.

When the osmotic pressure (Π) of the solution is important, as in reverse osmosis, and when two additional resistances to flow may be important, such as fouling (R_f) and concentration polarization (R_p), the flux equation becomes

$$J_p = \frac{(\Delta P - \Delta \Pi)}{(R_m + R_f + R_p)} \tag{14.3}$$

where

$\Delta \Pi = (\Pi_f - \Pi_p)$

(Π_f, Π_p) are the osmotic pressures of the feed and product (penetrant) stream, respectively

Concentration polarization is caused by a significant increase of solute concentration near the membrane surface in the retentate side, particularly at high flow rates of the penetrant. Fouling can be caused by salts or macromolecules, precipitating on the membrane surface. Both resistances are reduced by high cross-flow rates or mixing of the retentate.

In membrane separations, especially ultrafiltration and microfiltration, the volumetric flux of permeate J_p (m^3/s) is also given by the Darcy equation,

$$J_p = \frac{K \Delta P}{(\eta z)} \tag{14.4}$$

where

η (Pa s) is the viscosity of the liquid

K (m^2) is the (Darcy) permeability

z (m) is the membrane thickness

Equation 12.4 can be written as

$$J_p = \frac{\Delta P}{(\eta r_m)} \tag{14.5}$$

where $r_m = z/K$ (1/m) is the flow (Darcy) resistance. It should be noted that the overall [R, (s Pa/m)] and flow (r_m, 1/m) resistances differ by the viscosity term (η), i.e. $R = \eta\, r_m$.

The volumetric flux in membrane separations systems with negligible osmotic pressure is given by the following modification of Equation 14.3:

$$\mathrm{J} = \frac{\Delta P}{\eta (r_m + r_f + r_p)} \tag{14.6}$$

The rejection parameter (R_j) of a solute in a membrane separation is calculated from the equation,

$$R_j = 1 - \left(\frac{C_p}{C_f} \right) \tag{14.7}$$

where C_p, C_f are the solute concentrations in the product and feed, respectively.

14.2.3 Membrane Modules

Commercial polymeric membranes include cellulose acetate, polyamides, polysulfones, polyacrylonitriles, polyethersulfones, and polypropylenes. The method of preparation determines the physical structure and the selectivity of the membrane. Polymeric membranes should have a precise pore size distribution and a thin skin on the surface about 0.5–1 μm thick. Two general types of membranes are used commercially, i.e., flat sheets and hollow fibers (capillary tubes).

Polymeric membranes are sensitive to temperature (T) and (pH), and the following limits should be used: Cellulose acetate $T < 30°C$, polyamides, polysulfones $T < 80°C$, and ceramic membranes $T < 130°C$. A narrow pH range is applied for cellulose acetate (pH 3–8), and polysulfones can be used over the range (pH 3–11).

Efficient commercial membrane modules include

1. Hollow fiber-capillary systems with outer diameter of the capillaries about 100 μm, used mainly in reverse osmosis. The liquid flows from the shell into the capillary. In ultrafiltration, the liquid flows from the interior to the exterior of tubes, which have diameters 0.25–6 mm.
2. Tubular modules of 12–25 mm diameter, supported on perforated or porous walls of stainless steel tubes. The liquid flows from the interior of the tube outward.
3. Spiral wound modules, providing large separation surface. Many membrane layers are wound to produce module of diameter up to 400 mm. The liquid feed flows from the outside and it is separated into the permeate, which flows in an internal tube, and the retentate, which is removed as side product.
4. Plate and frame modules, resembling the mechanical pressure (plate and frame) filters.
5. Ceramic modules with tubular capillaries (monoliths) of channel size 3 mm. Many monoliths are incorporated into a modular housing. Ceramic membranes have pore sizes in the range of 0.5–1.0 μm. Commercial ceramic membrane modules have length of 90 cm, module diameter of 15 cm, and channel diameter of 2.5 mm.

The separation capacity of UF membranes is characterized by the molecular weight cutoff (MWCO), which is the maximum molecular weight that can pass through the membrane. The size of the macromolecules is expressed by their molecular weight in Daltons (Da). Thus, a dextrin molecule of 5 kDa has a molecular weight of 5000, which corresponds to about 5000/180 = 28 glucose molecules (mol. weight of glucose = 180). The MWCO of UF membranes ranges from 2 to 200 kDa.

14.2.4 Membrane Separation Systems

14.2.4.1 Pressure-Driven Separations

Most membrane separations, such as ultrafiltration and reverse osmosis, are pressure driven; they utilize single units, with limited application of the multistage systems. Figure 14.1 shows the operating principle of a separation unit, consisting of the membrane module, the feed pump, and the accompanying piping for separating the feed into a permeate and a concentrate. The feed pump can be centrifugal for low pressures or positive displacement (piston type) for high pressures. The main product may be the permeate (e.g., desalted water in reverse osmosis) or the concentrate or retentate (e.g., recovered protein in ultrafiltration) (Maroulis and Saravacos, 2003).

The operating pressure of the separation process depends on the system, varying in the ranges shown in Figure 14.2.

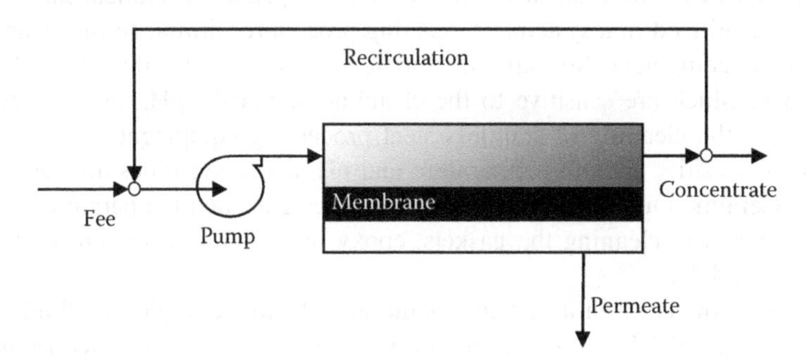

FIGURE 14.1 Diagram of a membrane separation system (pressure driven). (Modified from Maroulis, Z.B. and Saravacos, G.D., *Food Process Design*, CRC Press, Boca Raton, FL, May 9, 2003, fig. 10.2.)

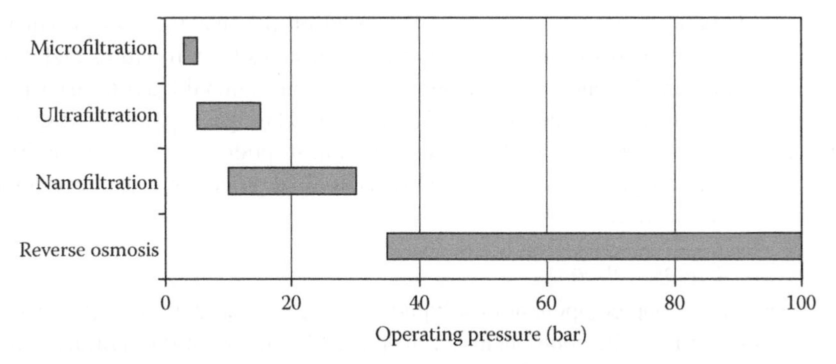

FIGURE 14.2 Operating pressures for membrane separation processes. (Modified from Maroulis, Z.B. and Saravacos, G.D., *Food Process Design*, CRC Press, Boca Raton, FL, May 9, 2003, fig. 10.3.)

Recirculation of the liquid concentrate is necessary in most applications to maintain high flow rate and increase the efficiency of the separation.

Flux reduction of permeate may be caused by membrane compaction, concentration polarization, or membrane fouling. Fouling is caused by plugging the membrane pores or by cake (gel) formation on the membrane surface.

Fouling can be prevented by selection of the proper membrane, e.g., cellulose acetate is a low-fouling polymer. Process configuration and flow rate can also control fouling. High shear rates, caused by high flow rates at the membrane surface reduce fouling by removing any deposits formed. An economic balance should be made to determine the optimum flow rate for minimum fouling and cleaning cost of the membrane system.

Membrane fouling, especially in ultrafiltration and microfiltration, is caused by suspensions of proteins and other biopolymers, which are unstable in the system.

Membrane fouling can be reduced by pulsating flow (periodic increase of pressure), backflushing, and washing of the active membrane surface.

Cleaning and sanitizing of fouled membranes involves physical removal of fouling substances, chemical removal of foulants, and hygienic (sanitary) treatment to remove all viable microorganisms. Detergents, heating, and mechanical energy (high velocity) are utilized in a systematic cleaning procedure, similar to the cleaning of food process equipment. Special care should be taken in cleaning the polymeric membranes, which are sensitive to the cleaning chemicals, pH, and temperature, compared to the cleaning of stainless steel processing equipment. Cellulose acetate is very sensitive to both temperature and pH, and polyamides are sensitive to chlorine. Ceramic membranes are resistant to chemicals and temperature, but care should be taken in cleaning the gaskets, epoxy resins, etc., which are part of the membrane module.

Turbulent flow is essential during membrane cleaning, employing fluid velocities in the range of 1.5–2.0 m/s in tubular systems. Hollow fibers, used in ultrafiltration, operate in the laminar region, but the high shear rates developed facilitate cleaning.

14.2.4.2 Special Membrane Separations

Special membrane separation processes are based on the selectivity of polymeric membranes to the diffusion and flow of various molecules from mixtures and solutions. They are simpler and more economical than conventional mass transfer processes, such as distillation, extraction, and absorption. Thus, gas separations (oxygen/nitrogen) by membranes can replace distillation, absorption, and adsorption. Two special membrane separations are of importance to food processing, i.e., pervaporation and electrodialysis.

14.2.4.2.1 Pervaporation

In pervaporation, one component of a liquid mixture is separated through a permselective membrane. The component is transported by diffusion through the membrane, and desorbed into the permeate space. In food systems, pervaporation is used either to remove water from liquids through hydrophilic membranes, or organic components through hydrophobic membranes.

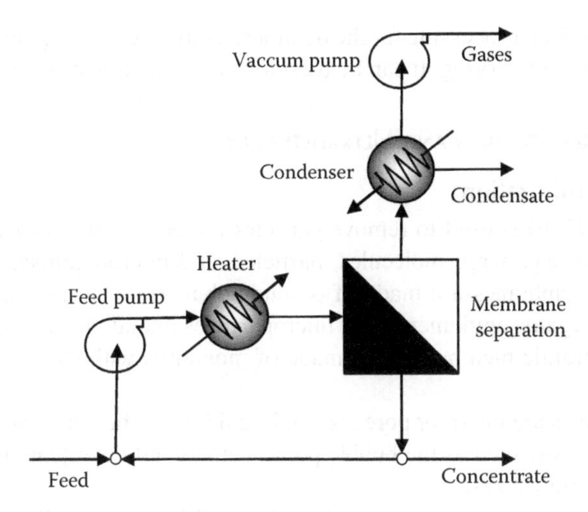

FIGURE 14.3 Diagram of a pervaporation system. (Modified from Maroulis, Z.B. and Saravacos, G.D., *Food Process Design*, CRC Press, Boca Raton, FL, May 9, 2003, fig. 10.4.)

The pervaporation membrane system includes heating and recirculation of the feed, condensation of the volatile component, and a vacuum pump (Figure 14.3).

An application of pervaporation is the separation of water (5% by volume) from the ethanol azeotrope mixture, using a hydrophilic membrane. Pervaporation can be used for the concentration of fruit juices, by removing only water through an appropriate membrane, while the volatile components are retained in the juice concentrate. The aqueous essence (aroma solution) of fruit juices can be concentrated, removing the water through a hydrophilic membrane.

Pervaporation is used for the de-alcoholization of wine, instead of the thermal distillation (spinning cone column) process, described in Chapter 13. Hydrophobic (special rubber) membranes are used to remove ethanol to a level below 0.5%, near room temperature.

14.2.4.2.2 Electrodialysis

Electrodialysis (ED) is a membrane-separation process of electrolytes from nonelectrolytes, based on the migration of ions through permselective membranes. The ED membranes consist of gels of polymers with fixed ionic charges. Cation exchange membranes consist of sulfonated polystyrene copolymerized with divinylbenzene. Anion exchange membranes consist of quaternary anions attached on polystyrene–divinylbenzene polymers.

The positive and negative ion exchange membranes are placed alternately in a stack, containing several membranes, similar to the plate and frame filtration system. The ions are induced to migrate by an electric potential, the anions passing the anion exchange membrane, and the cations, the cation exchange membrane.

Electrodialysis is competitive to RO in desalting brackish water (salt concentrations 500–5000 ppm TDS). However, it is not economical for desalination of seawater (above 25,000 ppm TDS).

ED is used in food processing for the demineralization (de-ashing) of whey before evaporation and spray drying. It can be used for the de-acidification of citrus juices.

14.2.5 Microfiltration and Ultrafiltration

14.2.5.1 Microfiltration

Microfiltration (MF) is used to remove particles larger than 0.2 μm from solutions and suspensions, e.g., large molecules, particles, and microorganisms (yeasts and bacteria). MF membranes are made of porous polymeric materials of sponge-like structure, or inorganic agglomerates of microparticles (metal, metal oxide, graphite, or ceramic). Ceramic membranes are made of monoliths with channels of 2–6 μm diameter.

MF membranes are rated for pore size and liquid flux. The pore size is characterized by special tests, such as the bubble point method and the retention of particles or bacteria of certain sizes.

Microfiltration of dilute suspensions (e.g., below 0.5%) is normally carried out by "dead-end" filtration, i.e., flow perpendicular to the membrane surface. Suspensions of higher concentration are treated by cross-filtration (parallel flow to the membranes), which reduces membrane fouling.

MF and other separation membranes are asymmetric, i.e., they have a tight layer (skin) on the top of a more porous structure. The membranes are usually operated with the skin in contact with the feed liquid. However, liquids with high solids loads are better filtered with inverted membranes, which retain the large particles in their pores. Periodic flushing of membranes with water is necessary to maintain an acceptable level of liquid flux.

Spiral wound modules are preferred in MF, because they are cheaper, but they are difficult to operate (fouling) with particles larger than 200 μm. Mechanical filters may be used for removing (pre-filtering) the large particles before MF is applied. Tubular MF modules foul less and they can be cleaned more easily than the spiral wound or hollow fiber systems.

Ceramic (inorganic) membranes are commonly used in MF. They consist of two layers, i.e., a porous support with pores 10 μm or larger, and an active coat of smaller pores. They are used as tubular monoliths, and they are particularly suited for high-temperature and high-pH applications.

Operating conditions of the MF units are, pressure 1–15 bar, and temperature 50°C–90°C. Temperature has a positive effect on the flux of juices and sugar solutions, since the viscosity decreases sharply at higher temperatures. However, temperature may have a negative effect on the flux of MF membranes treating protein suspensions, due to the precipitation (coagulation) of proteins at high temperatures.

Microfiltration is applied to several food processing operations, similar to those of ultrafiltration. Clarification of fruit juices, wine, and beer by MF can replace conventional mechanical filtration, eliminating the use of filter aids and reducing the process time and, finally, the process cost. Fouling of MF membranes (plugging with large particles) is a problem, which limits their widespread application.

MF can be used to remove bacteria and other spoilage microorganisms from liquid foods, such as milk, wine, and beer. The milk concentrate can be sterilized separately and recombined with the sterile milk permeate. The use of MF membranes for sterilization is governed by strict public health regulations (FDA). Fouling of MF membranes is caused mainly by protein components of the foods. The bovine serum albumin (BSA) is normally used as a model protein in fouling studies. Membrane fouling is reduced by backflushing, pulsing, and washing.

14.2.5.2 Ultrafiltration

Ultrafiltration (UF) is used to separate macromolecules and small particles (concentrate) from solvents (water) and ions and small molecules (permeate). It can replace evaporation, which may damage heat-sensitive food products. The most important application of UF is in the dairy industry where milk and whey proteins are separated from lactose and other solutes. UF finds applications in biotechnological processing and in the treatment of wastewater (Cheryan, 1998).

14.2.5.2.1 UF Systems

UF membranes include cellulose acetate, polyamides, polysulphones, polyvinylidene fluoride, and polyvinyl alcohol-polyethylene copolymers. Ceramic membranes, based on porous alumina and carbon substrates, are also used.

UF membranes are characterized by their permeability and retention properties. The membrane permeability is determined by the pore size distribution and the thickness of the active layer. Permeability data are reported at standard conditions, e.g., at $\Delta P = 3.43$ bar and $T = 25°C$. New membranes may have permeabilities of 0.1–1 mm/s, which are reduced considerably during processing.

The separating capacity of the UF membranes is characterized by the molecular weight cut-off (MWCO), the maximum size of the molecules that will pass through. The size of polymer molecules is expressed by their molecular weight, e.g., Dextran 250 is a dextrin of molecular weight 250,000 or 250 kDa, where 1 kDa = 1000 Da. The separation of polymer molecules is not sharp, since the pore size of UF membranes is not uniform but dispersed, following the normal or log normal distribution models. The MWCO of UF membranes is determined experimentally, using dextrins of known molecular weight in the range of 2–300 kDa.

UF modules used include tubular (diameters 5–25 mm), hollow fiber/capillary (diameters 0.25–1 mm), spiral wound, and plate and frame. The hollow fibers require clean feeds, and the tubular systems are used when the feed contains significant amounts of suspended particles, such as cloudy fruit juices.

Concentration polarization is a difficult problem of UF and RO membranes, reducing significantly the flux of permeate at higher operating pressures. At low pressures, the flux increases linearly with the applied pressure. However, above a certain pressure the flux increases slowly or becomes independent of the applied pressure. Equilibrium is established between concentration of the solute on the membrane surface and dispersion into the flowing liquid. Polarization is reduced by cross-flow of the liquid, parallel to the membrane surface, which removes the accumulated solute. Flow geometry and flow rate control concentration polarization. Most of the mechanical energy in UF is consumed to reduce concentration polarization.

UF separation systems operate at temperatures in the range 5°C–45°C. In general, the flux of permeate increases with increasing temperature, mainly because of a reduction of the liquid viscosity. The viscosity of concentrated sugar solutions drops sharply as the temperature is increased (high activation energy). However, temperature has a smaller effect (low activation energy) on the (apparent) viscosity of non-Newtonian (pseudoplastic) solutions/suspensions.

Rejection (R) of macromolecules and suspended particles by UF membranes is between 0.9 and 1.0, while for small molecules and ions $R = 0$–0.1. Flux rates in UF vary over the range of 5–500 L/m^{2}h. Pumping energy consumption is in the range of 0.5–5 kWh/m^3 permeate.

Diafiltration is a special case of UF, in which water is added to the concentrate side of the membrane system, facilitating the removal (permeation) of unwanted solutes or ions from the concentrated product. Water is added either batch-wise or continuously. Typical applications of diafiltration are the removal of lactose from the protein concentrate of UF-treated whey, and the fractionation of proteins with the removal of low-molecular weight components, e.g., ions, sugars, and ethanol.

14.2.5.2.2 Food Applications of UF

Commercial applications of UF include dairy technology, juice technology, water technology, and wastewater treatment. The whey in cheese manufacture is separated into a protein concentrate and an aqueous permeate containing lactose and minerals, which may be further concentrated by RO or evaporation. UF concentration of whole milk (twofold to fivefold) is used in the preparation of special types of cheese (e.g., feta) and yogurt.

UF is used in combination with RO and ion exchange in the cleaning and desalting of drinking and process water. Hollow-fiber UF membranes of 20 kDa (MWCO) are used at pressures about 3 bar with water fluxes 15–75 L/m^{2}h.

UF is used to clarify fruit juices, replacing conventional juice clarification technology (enzyme treatment, filter aids, cake filtration), with significant operating and economic advantages. Polymeric membranes of 30 kDa (MWCO) are used with fluxes 35–300 L/m^{2}h.

UF ceramic membranes are suitable for juice clarification, because they are more stable and more easily cleaned than polymeric membranes. Typical UF ceramic membranes, used for the clarification of apple juice, have 0.2 μm pores, tube diameter 4 mm, and length 85 cm. High flux rates of 400–500 L/m^{2}h are obtained with pressures 1–7 bar.

Combination of UF clarification and RO concentration can yield clarified juice concentrates up to about 35°Brix. Higher concentrations require thermal evaporation.

UF is used in combination with ion exchange to de-bitter citrus juices (removal of bitter components limonin and naringin). The cloudy juice is passed through a UF membrane system to remove suspended particles and colloids, which would plug the ion exchange resins. The bitter components from the clarified juice are adsorbed on resins, e.g., divinylbenzene copolymer, and the de-bittered juice is mixed with the separated pulp.

Wine and beer can be clarified by UF membranes, replacing chemical treatment and cake filtration. UF membranes of 0.45 μm pores will retain yeasts and other microorganisms, eliminating the need for thermal pasteurization. UF and RO separations can be used in the treatment of wastewater of food-processing plants, reducing pollution, and recovering water for reuse.

14.2.6 Reverse Osmosis and Nanofiltration

14.2.6.1 RO Systems

Reverse osmosis (RO) is used to remove water from aqueous solutions and suspensions, employing tight membranes, which retain the dissolved molecules, ions, and suspended materials. The applied pressure must be higher than the osmotic pressure of the solution in order to overcome the membrane resistance and the resistances of concentration polarization and membrane fouling, according to the flux equation. Nanofiltration (NF) is a more selective separation process than RO, retaining the dissolved molecules and the polyvalent ions, but allowing the permeation of the monovalent ions, along with the water. RO operates at pressures 35–100 bar and NF at 10–30 bar (Cheryan, 1998).

The osmotic pressure of aqueous solutions and food liquids depends on the size and concentration of the dissolved molecules (% total solids in water), as shown in the following examples:

1. Sodium chloride: 1%, 8 bar; 3%, 35 bar
2. Sucrose: 25%, 27.2 bar; 53%, 107 bar; 65%, 200 bar
3. Milk, nonfat: 9%, 7 bar
4. Whey: 6%, 7 bar
5. Orange juice: 11%, 16 bar
6. Apple juice: 15%, 20 bar
7. Coffee extract: 28.5%, 35 bar

Very high osmotic pressures are obtained in concentrated solutions of small molecules and (monovalent) ions, e.g., salt (sodium chloride) and sucrose solutions. The elevated osmotic pressures limit the application of RO in the production of concentrated salt, sugar solutions, and fruit juices.

RO membranes are rated for (water) flux and % rejection of the solutes. RO membranes, used in water desalting, may have rejections up to 99.7% (basis of rejection of sodium chloride). The rejection of nanofiltration membranes is based on magnesium sulfate.

Membrane fouling can be prevented by pre-filtration, pH adjustment, and chlorination.

14.2.6.2 Food Applications of RO

Desalting of brackish water and desalination of seawater to produce potable and process water are the most important applications of RO. The RO process competes with evaporation in small and medium installations, while evaporation is favored in large-scale applications, using multi-effect systems. Potable water should contain less than 500 ppm of total dissolved solids (TDS), while the concentrations of the brackish and seawater are about 1,500 and 30,000 ppm

TDS, respectively. Water recovery in the desalting of brackish water is high (70%–90%), but low in seawater (25%–30%). Energy requirements for brackish water desalting are low (about 1 kWh/m^3), compared to the high-energy requirements for seawater desalting (about 6 kWh/m^3).

Demineralized water, suitable for steam boilers and other specialized applications, is produced by RO in combination with ion exchange (Chapter 13). The water is first filtered from the suspended substances, then it is treated in an RO unit to remove most of the dissolved solids, and finally it is treated in an ion exchange system to remove any residual salt ions.

RO is used as a pre-evaporation step in the concentration of milk and fruit juices, removing most of the water, before the final concentration in the evaporator. Milk is concentrated by a factor of 2–3, using RO. Concentration of the milk by 1.5 times is used for the preparation of yogurt. Whey is concentrated from 6% to 24% TS. Due to the high osmotic pressure of sugar solutions, most applications of RO are in preparing juice concentrates up to about 25°Brix–30°Brix.

Nanofiltration is used to separate by preferential permeation monovalent ions (sodium chloride, hydrogen) from polyvalent ions and other solids of whey and citrus juices. A typical application of membrane separations in the dairy industry is shown in Figure 14.4. Cheese whey is separated into its components (fat, protein, lactose, salts, and water) by successive application of MF, UF, and RO.

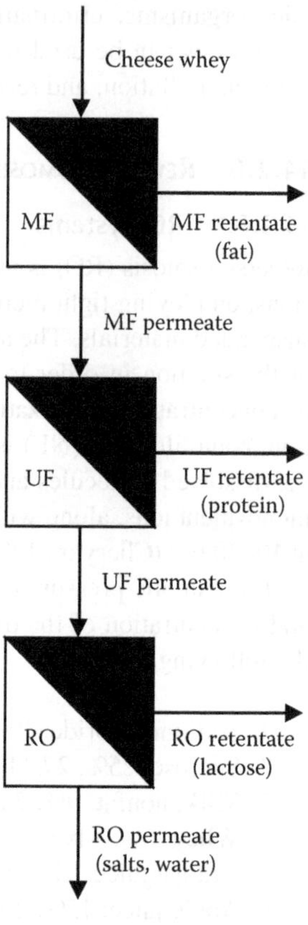

FIGURE 14.4 Membrane separations of cheese whey components. MF = microfiltration, UF = ultrafiltration, RO = reverse osmosis. (Modified from Maroulis, Z.B. and Saravacos, G.D., *Food Process Design*, CRC Press, Boca Raton, FL, May 9, 2003, fig. 10.7.)

14.3 SUPERCRITICAL FLUID EXTRACTION

Supercritical fluid (SCF) extraction separates the components of a mixture, using a SCF, i.e., a fluid operating above its critical temperature and pressure. Carbon dioxide is a preferred SCF fluid in food processing, because of its advantages (nontoxic, moderate operating pressures and temperatures, easily removed, inexpensive) (McHugh and Krukonis, 1994).

SCF extraction is used commercially in the decaffeination of coffee and the extraction of flavor components from plant materials.

14.3.1 SUPERCRITICAL FLUIDS

Values of critical properties are given in tables and data banks of the chemical engineering literature. Table 14.1 shows the critical pressure (P_c), temperature (T_c), and volume (V_c) of carbon dioxide and water. In practice, the operating pressures and

TABLE 14.1
Critical Properties of Carbon Dioxide and Water

Fluid	P_c, bar	T_c, °C	V_c, L/kmol
Carbon dioxide	73.8	31.1	73.9
Water	221.2	374.1	57.1

temperatures of SCF extraction are considerably higher than the critical values. The high temperatures and pressures of supercritical water prevent its commercial use in SCF extraction of heat-sensitive food products (Brun and Ely, 1991).

SCFs behave as dense gases, but they cannot be condensed as liquids by increasing the pressure. The density of the SCFs depends on the pressure and temperature, varying in the range of 400–700 kg/m^3, i.e., it is significantly lower than the density of liquids (water, 1000 kg/m^3), an important advantage in extraction operations.

The viscosity of SCF carbon dioxide is about 0.5 mPa s, significantly lower than, e.g., the viscosity of liquid hexane (3 mPa s). The molecular diffusivity of SCF carbon dioxide at 40°C is about 10×10^{-10} m^2/s, i.e., about one order of magnitude higher than the diffusivity of CO_2 in the liquid state (1×10^{-10} m^2/s). The favorable transport properties are desirable in extraction operations, since low viscosities facilitate the penetration of the SCF into the particulate beds and reduce power requirements in transferring the fluid through the system. Higher diffusivities increase mass transfer rate and extraction efficiency.

Equilibrium between the SCF solvent (e.g., carbon dioxide) and the food component to be extracted is expressed by the solubility of the component at a given temperature and pressure, which is usually determined experimentally. In general, the solubility increases when the pressure is increased and the temperature is reduced.

The CO_2 solubility of low molecular weight and low polarity organic compounds, such as hydrocarbons, alcohols, carboxylic acids, esters, and aldehydes is very high (complete miscibility). Macromolecules and highly polar molecules (sugars, starch, proteins, salts) are insoluble in CO_2. The solublility of some insoluble components can be increased by the addition of entrainers, such as ethanol, acetone, and ethyl acetate. Thus, β-carotene is solubilized in carbon dioxide–ethyl acetate mixtures. Water has a positive function in the SCF extraction of food solids. It is sorbed by the dry food materials, which are expanded, facilitating the transport of solvent and solute in the mixture.

14.3.2 SCF Extraction Operations

Most of the pilot plant and commercial supercritical extractions are batch operated, because of the difficulty of continuous feeding of solid materials into high-pressure vessels. Expensive alloy materials are used for the pressure vessels, such as stainless steel 316 and Hastelloy. Special attention should be given to the safety aspects of the high-pressure equipment.

SCF extraction of components of different solubility can be achieved by increasing the pressure progressively. Equilibrium is approached at each pressure, when the rate of removal of the component levels off. Fractionation of various components can also be accomplished by dissolving all extractants at high pressure and temperature (of carbon dioxide), and reducing the pressure successively, separating each component at the corresponding equilibrium pressure.

14.3.3 Food Applications of SCF Extraction

SCF extraction with carbon dioxide has found two major applications in food processing, i.e., decaffeination of coffee and extraction of hop flavor components. Other potential applications of SCF, which are expensive at the present time, include oilseed extraction (e.g., soy beans, replacing hexane), lecithin purification, fractionation of fish oils, and reducing cholesterol in egg yolk and butterfat.

The decaffeination of coffee and tea by SCF is economically competitive with the chemical extraction processes, and it has the advantage of no chemical residues in the food product. Coffee contains about 1% caffeine, while tea may contain up to 3%. The extraction of caffeine is facilitated by wetting the coffee beans with water, which tends to dissolve and desorb caffeine from the sold materials. The operating conditions of a decaffeination plant are, pressure 300 bar and temperature 40°C. Figure 14.5 shows a diagram of a coffee decaffeination operation.

Fresh and recirculated carbon dioxide at a supercritical pressure and temperature are introduced into the extractor. The extract is transferred through an expansion valve to the separator, which operates at a lower pressure, and separates it into two phases, the aqueous caffeine extract, and the carbon dioxide, which is recycled. The caffeine can be recovered, e.g., by adsorption on an activated carbon column. Water is introduced into the separator to facilitate the separation of caffeine from the carbon dioxide. Caffeine is more soluble in water (e.g., 3% at 25°C) than in carbon dioxide (0.1%), while the solubility of carbon dioxide in water at 25°C is 0.16%.

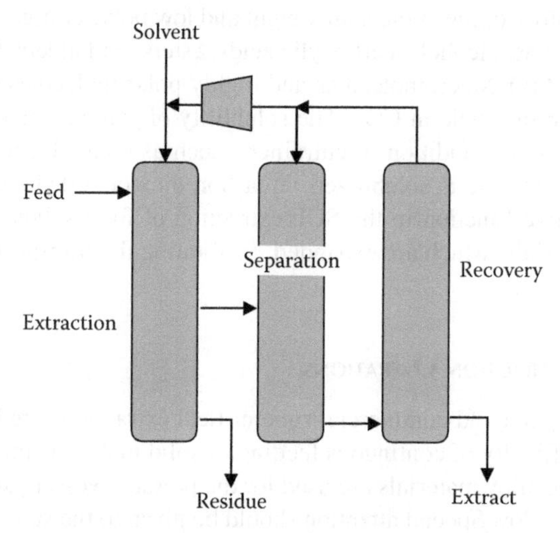

FIGURE 14.5 Diagram of a SCF decaffeination process for coffee.

14.4 FREEZE CONCENTRATION

The freeze concentration of aqueous food liquids is based on the crystallization of ice and its mechanical separation from the concentrated solution. It has the advantages of low-energy requirements and better quality retention than conventional evaporation, but it has not found wide applications yet due to unfavorable economics.

The freeze concentration process consists of two basic operations, i.e., crystallization and separation, as shown in the diagram of Figure 14.6 (Hartel, 1998).

14.4.1 CRYSTALLIZATION OF ICE

Crystallization of ice from water or aqueous solutions involves nucleation and ice growth. Most of the nucleation (formation of small ice nuclei) takes place in the scraped surface heat exchanger, used to cool the feed solution below its freezing point. Controlled crystallization (crystal growth) is carried out in a stirred crystallizer with recirculation of the solution through the cooler.

The freezing point depression (ΔT_f), i.e., the temperature difference between the freezing points of water and solution (°C) in dilute solutions, is a linear function of the molality (m) of the solution (moles of solute per kg of solution), and it can be estimated from the simplified equation,

$$\Delta T_f = k\,m \tag{14.8}$$

where the constant (k) depends on the molecular weight, the latent heat of fusion, the freezing point of water, and the gas constant (R).

Typical values of ΔT_f for apple juice are 1°C at 11% TS, and 10°C at 50% TS (total solids). The freezing-point depression is similar to the boiling-point elevation during the evaporation of aqueous solutions (Chapter 10). It is higher in solutions of low molecular weight solutes, such as sugars, than in solutions of macromolecules, such as starch and proteins.

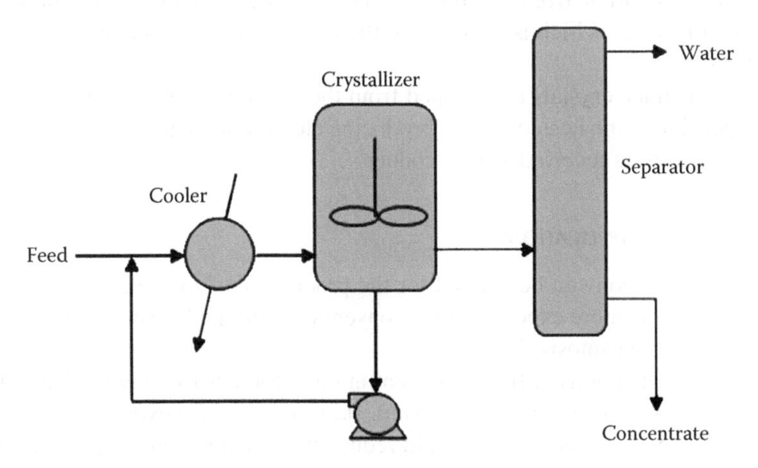

FIGURE 14.6 Diagram of a freeze concentration system.

The water fraction (X, kg/kg solution), which must be removed for concentrating a solution from C_o to C (kg/m^3) is given by the equation,

$$X = \frac{1-(C)}{(C_o)} \tag{14.9}$$

Most of the water is removed from the solution during the initial stages of the concentration process.

The mechanism of ice crystallization and crystal growth is analogous to the mechanism of crystallization of solutes from solution. Nucleation is predominantly heterogeneous and it takes place on the walls of the scraped surface heat exchanger, used for subcooling the feed, and in recirculating the solution. In general, large crystals are desired, without solute inclusions, which can be separated easily from the concentrated solution.

Crystal growth involves diffusion of water molecules to the crystal surface and transfer of the heat of ice formation away from the crystal. The rate of crystal growth increases linearly with increasing subcooling. However, the increased viscosity of the concentrated solution tends to decrease the growth rate.

Increasing the residence time in the crystallizer results in "ripening" of the crystals, increasing the mean crystal size. Large spherical crystals can be obtained by long residence time in the agitated crystallizer, which are separated easier in the wash column.

14.4.2 Ice Separation

Separation of the ice crystals from the concentrated solution is carried out in wash columns, which are preferred over the conventional filters and centrifuges.

The wash column consists of a vertical cylinder, which is fed at the bottom with the slurry of ice/concentrated solution from the crystallizer. The slurry is pushed upward with a perforated piston, which allows the draining and removal of the concentrated solution from the bottom. The ice bed, moving slowly upward, is washed with water, which is returned to the column from the ice melter, installed over the top.

The washed ice crystals are scraped from the top of the column and transported as a suspension to the heated melter, producing clean water, part of which is recirculated and the rest is received as the product.

14.4.3 Food Applications

Freeze concentration can be applied for the production of freshwater from the sea, but this process is more expensive than conventional thermal evaporation and newly developed reverse osmosis (RO).

Potential applications of freeze concentration in food processing include concentration of citrus juices (orange, grapefruit), skim milk, coffee extracts, vinegar, and beer. Improvement of the process could reduce the cost and make the process more competitive. A three-stage countercurrent system is more efficient than single-stage crystallizers, because the ice is separated at the lowest solute concentration (lowest viscosity), and the ice crystals can grow faster in lower concentrations.

14.5 NONTHERMAL FOOD PRESERVATION

Established thermal preservation methods, such as pasteurization and sterilization, may cause significant thermal damage of food quality (nutritive and organoleptic) during processing. Alternative nonthermal preservation processes have been investigated during the recent years, especially irradiation, high-pressure, and pulsed electrical field processing. Food irradiation is applied in small scale to some food and packaging materials. HPP needs further development before wider application. Electrical pulse processing is still in the early stages of testing and development (Barbosa-Canovas et al, 2000).

The new food preservation methods reduce the levels of the vegetative microbial cells by about 5 log cycles, without destroying the more resistant spores. They must be subjected to the required food safety tests and evaluation used in conventional thermal processing.

14.5.1 Food Irradiation

14.5.1.1 Radiation Dosages

Ionizing radiations can inactivate spoilage and pathogenic microorganisms in an analogous mechanism with heat. However, enzymes are more resistant to radiation than heat, so that a combination of heat and irradiation may be indicated in some cases. The ionizing radiations include gamma rays, x-rays, and high-energy electrons.

The penetration of radiation in the food depends on its energy, measured usually in MeV (Mega electron Volts). Table 14.2 shows the penetration of various kinds of radiation in food and water, which is the main component of most foods.

The efficiency of irradiation of a product is related to the energy absorbed by the material, which is measured in Gray (Gy), which is the absorption of 1 J/kg or 100 erg/g material (1 J = 10^7 erg). In the past, the unit rad was used (1 Gy = 100 rad).

TABLE 14.2
Penetration of Ionizing Radiations

Radiation	Material	Penetration
Ultra violet	Milk	0.01–0.02 cm
Gamma rays	Air	19–20 m/MeV
X-rays	Water	10–20 cm/MeV
High energy	Air	30 cm/MeV
Electrons	Water/food	0.5 cm/MeV

TABLE 14.3
Applications of Food Irradiation

Area of Application	Dosage (kGy)	Examples
Sprout inhibition	0.05–0.15	Potatoes, onions
Pest disinfestations	0.15–0.75	Cereals, nuts, fish
Inactivation of non-sporogenic	0.23–1.00 m.o.	Fresh fruits, and vegetables
Food preservation	1–3	Fruits, fish
Control of pathogenic m.o.	3–10	Spices, nuts
Food packages sterilization	10–30	Food packages
Food sterilization	30	Beef, pork, poultry

In food applications, the irradiation dosage is usually expressed in kGy (equivalent to 10^5 rad or 0.1 Mrad). In food preservation, three dosage regions are distinguished:

Low irradiation 0.05–1.00 kGy
Medium irradiation 1–10 kGy
High irradiation 10–75 kGy

The reduction of a microbial population follows a first-order reaction rate equation, similar to the thermal inactivation of microorganisms (Chapter 9).

In general, the smaller the species (microorganism, spores of microorganisms, enzymes), the larger the dose required for its inactivation. On the average, spores and enzymes require three to seven times more radiation energy than microbial cells for their inactivation.

Table 14.3 gives the radiation dosages for various food preservation processes.

Food irradiation has found some applications at low and intermediate dosages. However, high-dose sterilization of meat products has not been approved yet by Health Authorities.

14.5.1.2 Food Irradiation Equipment

In food preservation, only gamma rays and high-energy electrons can be used practically. The gamma rays are usually produced by irradiators consisting of packaged Cobalt 60 (^{60}Co) or Cesium 137 (^{137}Cs) radionuclides of strength up to about 1 MCi (MegaCurie), where $1\,Ci = 3.7 \times 10^{10}$ Bq; 1 Bq (Bequerel) is equal to 1 fission per second. Cobalt 60, with a half-life of 5.3 years, is produced by irradiation of non-radioactive Cobalt 59 with neutrons. Cesium 137, with a half-life of 30 years, can be produced from nuclear fission by-products. In both cases (Cobalt 60 and Cesium 137), the energy of radiation is not larger than 1.3 MeV, which is the upper limit for gamma radiation, that will not induce secondary radiation. The advantage of using gamma rays in food irradiation is that the irradiated energy is constant and that no complicated mechanical installations are required. Their disadvantages are the dependence on a relatively restricted number of radiation materials and the continuous radiation, even when the source is not used.

The high-energy electrons are produced by linear accelerators or Van de Graaff electrostatic generators. The penetration of the electrons is a function of their energy and the density of the material. The penetration of electrons is lower than gamma radiation, and it has a nonuniform distribution, with a maximum within the material. The penetration depth d (mm), which is defined as the depth at which the dose is equal to the dose at the surface, is given by the following equation:

$$d = \left(\frac{(4E - 10)}{\rho} \right) \tag{14.10}$$

where
E (MeV) is the electron energy
ρ (g/cm^3) is the density of the material

Thus, for an accelerator of $E = 5$ MeV, the penetration in water ($\rho = 1$ g/cm^3) will be $d = 10$ mm. Because of the uneven distribution, irradiation from both sides of the material will have an effective penetration of $2.4d = 24$ mm. To avoid induced radiation to foods, the energy of electrons should be lower than 10 MeV.

In irradiation installations, the cost of sheltering (usually concrete walls) of the units is about 50% of the total investment. In Cobalt-60 installations, the product is loaded in racks and is conveyed into the irradiation chamber through a labyrinth-like corridor, and after receiving the required dose leaves the system.

14.5.2 High-Pressure Processing

HPP can preserve foods by inactivating spoilage and pathogenic microorganisms without heating. Pressures in the range 1–8 kbar (100–800 MPa) can inactivate vegetative cells, but they are ineffective for spores and enzymes. Combined HPP and heat treatment can sterilize various food products with heat damage to sensitive products.

HPP is based on the same principles of thermal processing technology, i.e., kinetics of microbial inactivation, enzyme and nutritional changes, and process design and evaluation, discussed in Chapter 9.

HPP is suitable for inactivation of foodborne pathogens in some raw foods and fruit juices, which are not thermally pasteurized. For example, a 5-log inactivation of pathogenic *Escherichia coli* 0157:H7 can be achieved with HPP treatment of unheated apple and other fruit juices.

HPP treatment of fruit juices can be accomplished in continuous pressure systems, which consist of piston pressure pumps (modified homogenizers), at pressures up to 3 kbar.

However, most of the HPP applications are hydrostatic batch systems, due to the difficulty of continuous operation at very high pressures. High-pressure steel vessels, resembling autoclaves, are loaded with the food product and pressurized hydraulically (isostatic operation). Water and liquid foods are the pressure transfer media. After the required process time, e.g., 1–10 min, the vessel is depressurized, the product is removed, and the cycle is repeated. Batch sterilization in pressure vessels (e.g., 215 L capacity) can be achieved at 6.9 kbar and 100°C for a process time of 1.5 min and cycle times 6–8 min. The processing capacity of a 5-vessel system is about 90 ton/day.

HPP in the presence of carbon dioxide has been suggested to pasteurize and stabilize the cloud of orange juice (inactivation of enzymes), e.g., 3.5 kbar and 8.6 min residence time. The food products entering the HPP vessels should be at a low temperature, e.g., 5°C, since pressurization above 1 kbar increases the temperature by about 20°C.

HPP technology can be applied, alone or in combination with other processes, to the preservation of minimally processed fruits and vegetables.

14.5.3 Pulsed Electric Field Processing

Pulsed electric field (PEF) processing is used for the inactivation of spoilage and pathogenic microorganisms of food products. The food material is placed between electrodes and microbial inactivation is achieved through dielectric breakdown of the bacterial membranes. Similar microbial inactivation can be affected by other nonthermal processes being investigated, e.g., oscillating magnetic fields, light pulses, and ultrasound (Barbosa-Canovas et al, 2001).

Microbial inactivation is a function of the particular electric field (maximum discharge rate/electrode distance), and the number of pulses applied (total treatment time). Typical PEF conditions for pasteurization of fruit juices are electric fields of 30–60 kV/cm and pulse duration of about 1 μs.

Factors affecting the inactivation of microorganisms by PEF include (a) the electric field, (b) the wavelength applied, (c) the ionic concentration of food material, (d) the food temperature, (e) the microbial cell concentration, and (f) the growth stage of the cells.

PEF processing is at the development and testing stage, and it is suitable for batch or continuous nonthermal pasteurization of food products, improving their quality and safety.

APPLICATION EXAMPLES

Example 14.1

The osmotic pressure of a solution is given by the van't Hoff equation of Thermodynamics. Estimate the osmotic pressure of two sugar solutions of 30% glucose and sucrose at 20°C, respectively.

Solution

The van't Hoff equation relates the osmotic pressure Π (Pa) to the molecular weight M (kg/kmol), the concentration c (kg/m^3), the temperature T (K), and the gas constant R (8.31 kJ/mol K), $\Pi = (CRT)/M$. Note that 1 J = 1 Pa m^3.

The concentration (C) of a 30% sugar solution (about 30°Brix, density 1130 kg/m^3) will be $c = 0.30 \times 1130 = 339$ kg/m^3.

For glucose $M = 180$ kg/kmol, for sucrose $M = 342$ kg/kmol $R = 8.31$ kJ/(kmol K).

Osmotic pressure glucose $\Pi = (339 \times 8.31 \times 293)/180 = 4585$ kPa = 45.8 bar.

Osmotic pressure sucrose $\Pi = (339 \times 8.31 \times 293)/342 = 2413$ kPa = 24.1 bar.

The van't Hoff equation is applicable to relatively low solute concentrations, for instance, in sugar solutions lower than 50°Brix. Experimental measurements of osmotic pressure are used at higher sugar concentrations.

Example 14.2

Membrane separations have found applications in the processing of cheese whey, a difficult-to-dispose liquid by-product of the dairy industry.

Table 14.4 shows typical milk and cheese whey composition, while Figure 14.7 shows the component particle sizes, and a schematic of potential membrane separations.

A dairy plant produces 2 ton/h of whey waste, which is to be separated into protein and lactose solutions, using ultrafiltration and reverse osmosis.

(a) Ultrafiltration

Assuming a density of the whey $\rho = 1000\,\text{kg/m}^3$, the volumetric feed flow rate to the UF plant will be $V_f = 2\,\text{m}^3/\text{h} = 0.56 \times 10^{-3}\,\text{m}^3/\text{s} = 0.56\,\text{L/s}$.

An ultrafiltration membrane of mean pore diameter 5 nm will be used, which will separate almost quantitatively the protein and fat from the lactose and salt

TABLE 14.4
Typical (%) Composition of Milk and Cheese Whey

Component	Milk	Cheese Whey
Fat	3.5	0.1
Protein	3.3	0.6
Lactose	4.9	5.0
Salts	0.7	0.6
Total solids	12.4	6.3

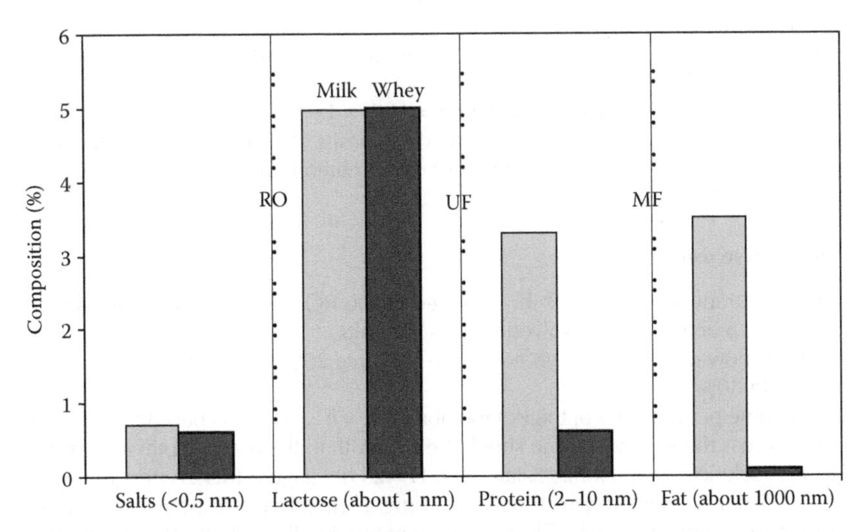

FIGURE 14.7 Particle sizes of milk and cheese whey components. MF, microfiltration; UF, ultrafiltration; RO, reverse osmosis. (Modified from Maroulis, Z.B. and Saravacos, G.D., *Food Process Design*, CRC Press, Boca Raton, FL, May 9, 2003, fig. 10.6.)

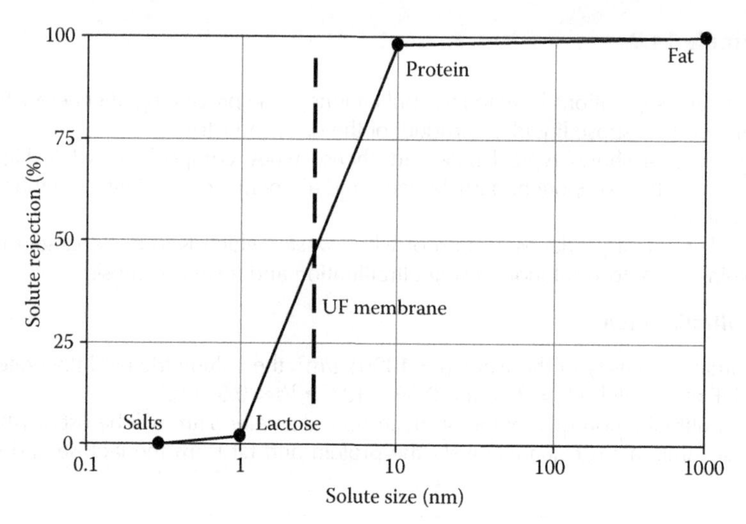

FIGURE 14.8 Solute rejection in UF of whey. (Modified from Maroulis, Z.B. and Saravacos, G.D., *Food Process Design*, CRC Press, Boca Raton, FL, May 9, 2003, fig. 10.9.)

solution (Figure 14.8). The volumetric permeability of the UF membrane is taken from the literature as $A = 8 \times 10^{-6}$ m/(s bar). The pressure drop through the membrane is assumed to be $\Delta P = 1$ bar.

The volumetric feed flux to the UF module will be $J = 2\,\text{m}^3/(\text{m}^2\ \text{h}) = 0.56$ mm/s.

In steady-state operation, the total feed flux (J) will be equal to the permeate (J_p) and retentate (J_r) fluxes, $J = J_p + J_r$.

The volumetric flow rate (V) is equal to the flux times the membrane surface area (S), $V = S\ J$. It is assumed that the retentate flow rate is 5% of the feed rate, i.e., $V_r = 0.05 \times 2 = 0.1$ m³/h. Therefore, the penetrant flow rate will be $V_p = 2.0 - 0.1 = 1.9$ m³/h, and the flow flux, $J_p = A\Delta P = 8 \times 10^{-6}$ [m/(s bar)] (1 bar) $= 8 \times 10^{-6}$ m/s.

The membrane surface area will be $S = V_p/J_p = 1.9/(3600 \times 8 \times 10^{-6}) = 66\,\text{m}^2$.

The UF membrane module is assumed to consist of hollow fibers 1 mm diameter and 2 m long. The required number of membrane fibers will be $N = 66/(2 \times 3.14 \times 0.001) = 10{,}510$.

(b) Reverse osmosis

The penetrant of the UF membrane is fed to the RO membrane module, which separates practically all dissolved lactose and salts.

Feed flow rate $V_f = 1.9\,\text{m}^3/\text{h}$. Retentate flow rate 20% $V_r = 0.38\,\text{m}^3/\text{h}$, penetrant flow rate 80% $V_p = 1.52\,\text{m}^3/\text{h}$.

Assume permeability of the RO membrane $A = 4 \times 10^{-7}$ m/(s bar). The pressure drop across the RO membrane should be higher than the osmotic pressure of the lactose solution, which for 20% lactose is $\Pi = 25$ bar (Figure 14.9). A pressure drop $\Delta P = 50$ bar is selected, which will overcome all membrane resistances and provide a reasonable flow rate. Thus, the volumetric flux through the RO membrane will be, $J_p = A\Delta P$ or $J_p = 4 \times 10^{-7} \times 50 = 2 \times 10^{-5}$ m/s.

The membrane surface area will be $S = V_p/J_p = 1.52/(3600 \times 2 \times 10^{-5}) = 21.1\,\text{m}^2$.

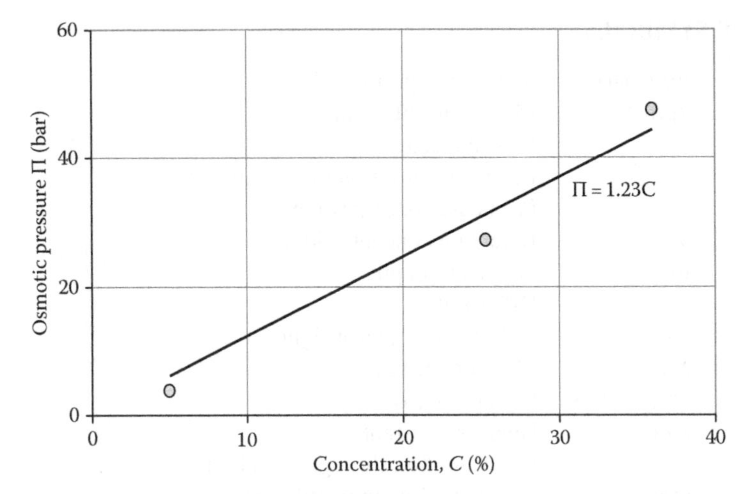

FIGURE 14.9 Osmotic pressure of lactose. (Modified from Maroulis, Z.B. and Saravacos, G.D., *Food Process Design*, CRC Press, Boca Raton, FL, May 9, 2003, fig. 10.17.)

A hollow fiber membrane can be used of 1 mm diameter and 2 m length. The required number of hollow fibers will be $N = 21.1/(2 \times 3.14 \times 0.001) = 3360$.

The power requirement for pumping in RO operations is much higher than in UF systems, due to the high pressure through the membrane, e.g., 50 bar (RO) versus 1 bar (UF).

PROBLEMS

14.1 Literature data of osmotic pressure for aqueous solutions of sucrose and sodium chloride are given in the Table 14.5.

a. Plot on semilog paper the data log(Π) versus % C (% concentration) for both solutions.
b. Estimate the regression equations of log(Π) versus % C for both solutions.
c. Explain the differences between the (Π) of the two solutions.

14.2 In RO separations, the permeate flux is affected considerably by the resistances due to fouling and concentration polarization.

Discuss engineering design and operating procedures to reduce these resistances for typical food liquids containing dissolved small molecules and fouling suspensions.

TABLE 14.5

Aqueous Solution	Osmotic Pressure (Π), Bar
Sucrose 25%	27
Sucrose 50%	100
Sucrose 65%	200
Sodium chloride 1%	8
Sodium chloride 3%	35
Sodium chloride 5%	100

LIST OF SYMBOLS

A	m/(s bar)	Membrane permeability
C	kg/m^3	Mass concentration
C_f	%	Feed concentration
C_{Fj}	%	Feed concentration of j component
C_p	%	Permeate concentration
C_r	%	Retentate concentration
d	m	Diameter, depth
D	m^2/s	Diffusivity
Da	—	Dalton molecular weight
E	MeV	Electron energy
Gy	Gray (J/kg)	Radiation dose
H	$kg/(m^3$ Pa)	Henry's constant
J_p	m/s	Penetrant flux through membrane
J_r	m/s	Retentate flux on membrane
J	m/s	Total flux to the membrane
K	m^2	Darcy permeability
M	mol/kg	Molality
M	kg/mol	Molecular weight
MF	—	Microfiltration
MWCO	—	Molecular weight cut-off
N	—	Number of hollow fibers
P	bar	Pressure
R	kJ/(kmol K)	Gas constant (8.31)
RO	—	Reverse osmosis
r_m, r_f, r_p	1/m	Darcy flow resistance (membrane, fouling, polarization)
R_m, R_f, R_p	(s Pa)/m	Flow resistance (membrane, fouling, polarization)
R_j	—	Rejection parameter
S	m^2	Membrane surface area
S	kg/m^3	Solubility
SCF	—	Supercritical fluid (extraction)
UF	—	Ultrafiltration
V_f	m^3/s	Feed flow rate
V	m^3/s	Permeate flow rate
V	m^3/s	Retentate flow rate
X	kg/kg	Mass fraction
Z	m	Thickness
ΔT	K	Temperature difference
ΔP	bar	Pressure drop
η	Pa s	Viscosity
Π	bar	Osmotic pressure
ρ	kg/m^3	Liquid density

REFERENCES

Barbosa-Canovas, G.V., Pothatehakamury, V.R., Palou, E., and Swanson, B.G. 2000. *Nonthermal Preservation of Foods*. Marcel Dekker, New York.

Barbosa-Canovas, G.V. and Zhang, Q.H. 2001. *Pulsed Electric Fields in Food Processing*. Technomic Publ., Lancaster, PA.

Bruno, J.T. and Ely, J.F. 1991. *Supercritical Fluid Technology: Reviews of Modern Theory and Applications*. CRC Press, New York.

Cheryan, M. 1992. Concentration of liquid foods by reverse osmosis. In: *Handbook of Food Engineering*, D.R. Heldman and D.B. Lund, eds. Marcel Dekker, New York, pp. 393–436.

Cheryan, M. 1998. *Ultrafiltration and Microfiltration Handbook*. Technomic Publ., Lancaster, PA.

Grandison, A.S. and Lewis M.J. 1996. *Separation Processes in the Food and Biotechnology Industries*. Technomic Publ., Lancaster, PA.

Hartel, R.W. 1992. Evaporation and freeze concentration. In: *Handbook of Food Engineering*. D.R. Heldman and D.B. Lund, eds. Marcel Dekker, New York, pp. 341–192.

Maroulis, Z.B. and Saravacos, G.D. May 9, 2003. *Food Process Design*. CRC Press, Boca Raton, FL.

McHugh, M. and Krukonis, V. 1994. *Supercritical Fluid Extraction: Principles and Practice*, 2nd edn. Butterworths, London, U.K.

Matsuura, T. and Sourirajan, S. 1995. Physicochemical and engineering properties of food in membrane separation processes. In: *Engineering Properties of Foods*, 2nd edn. M.A. Rao and S.S.H. Rizvi, eds. Marcel Dekker, New York, pp. 311–388.

Perry, R.H. and Green D. 1997. *Chemical Engineers' Handbook*, 7th edn. McGraw-Hill, New York.

Saravacos, G.D. and Kostaropoulos, A.E. 2002. *Handbook of Food Processing Equipment*. Kluwer Academic/Plenum Publ., New York.

Saravacos, G.D. and Maroulis, Z.B. 2001. *Transport Properties of Foods*. Marcel Dekker, New York.

Sourirajan, S. 1977. *Reverse Osmosis and Synthetic Membranes*. National Research Council, Ottawa, Canada.

15 Elements of Food Packaging Operations

15.1 INTRODUCTION

Food packaging consists mainly of mechanical operations, which include package preparing, product preparing, package filling, package closing (sealing), processing/control of filled packages, and storage/shipment (Brody, 1997).

The equipment used in filling food into packages depends on the nature and the properties of the food (liquids, granulates, or small pieces, larger pieces).

The packages are either ready to fill, partially ready-to-fill, or they are constructed in the food processing plant, shortly before filling. Typical ready-to-fill packages are glass and metallic containers, woven textile package (bags), and some plastic containers. Packages partially ready-to-fill are mainly cartons, cardboard, and some special kinds of materials. Packages constructed before filling include blown mold, thermo-formed and tube-formed packages. Packages that come directly into contact with food are usually sterilized before filling, either chemically (e.g., hydrogen peroxide) or by hot air/steam (Bureau and Multon, 1996; Robertson, 1993).

Figure 15.1 shows the main packaging operations applied to food processing. In addition to the mechanical operations involved, food product characteristics are of paramount importance in designing and operating food packaging equipment. Specific operations and equipment are required for various food products to meet the strict safety and quality requirements of packaged foods (Soroka, 1998).

Recent improvements in food packaging are based on a combination of research and development in food science, process engineering, and practical experience.

Good packages provide protection to the product from external factors, such as moisture, biological (microorganisms, insects, rodents), oxygen, odors, and adulteration. They should be appealing to the consumers, provide information on its contents, and be economical.

15.2 PACKAGING MATERIALS

The main packaging materials used for foods are metal (tin plate, aluminum), glass, paper and its products, and plastics. In addition, combined packaging materials, such as film laminates, and plastic-coated glass or metal, are used (Hernandez, 1997).

15.2.1 METALS

Metals are used to fabricate cans, tubes, aerosol containers, and larger containers, such as drums, thin aluminum sheets, trays, and bottle and jar caps. Metal cans are

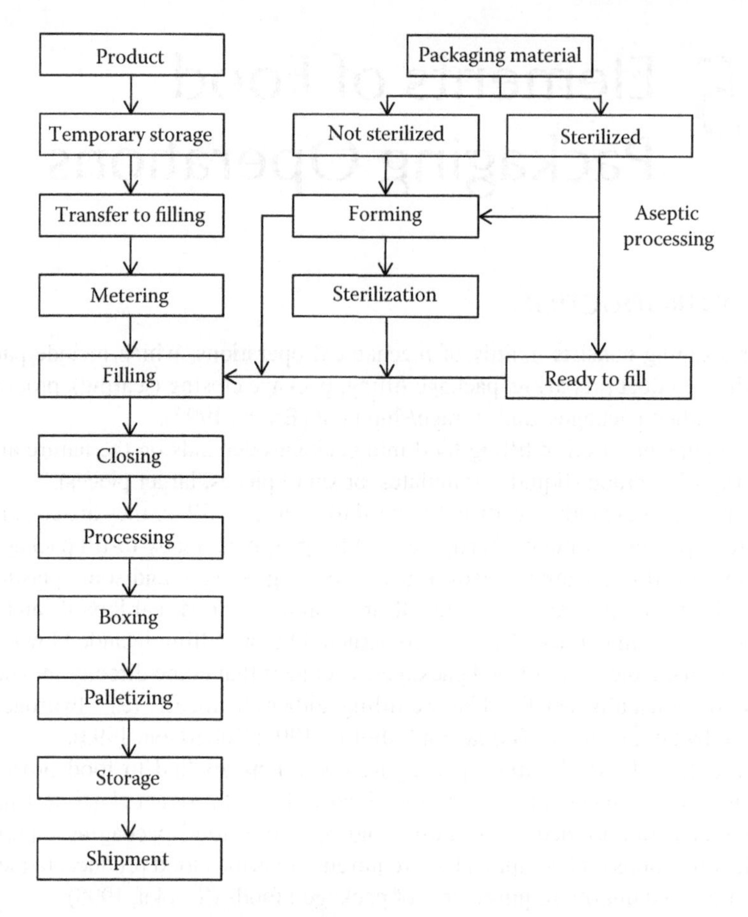

FIGURE 15.1 Food packaging operations.

the most common type of metal containers. Tinplate and aluminum cans are the most common types of metal cans used in a wide range of foods.

Aluminum and smaller tinplate containers are made of two pieces, while larger tinplate containers and a part of the smaller cans are made of three pieces. Tinplate cans are made of 0.15–0.50 mm thick low-carbon mild steel. Thinner cans are constructed by the drawn and ironed method, and the use of corrugated tinplate. The containers are enameled with, e.g., phenolic, epoxy-phenolic, acrylic, oleoresin, and vinyl enamels (4–12 μm) to reduce interaction between the food and the metal container. Most containers are enameled before being constructed, but if the tinplate is stressed significantly during the construction of the container, the container is enameled after its construction.

Aluminum cans are not enameled externally since they are corrosion-resistant, due to the aluminum oxides covering the surface of the containers. However, they are coated internally, in the absence of the protective oxide layer, when there is no oxygen. The aluminum cans are lighter, but they require about 35% more material to achieve the same strength as the tinplate cans.

Larger containers are used for packaging liquids and food pulps. They are also used as a protection for other packages, such as aseptically filled bags. Thin aluminum films, 0.015–0.200 mm thick, are used as gas barriers in laminated films and in the packaging of products, such as chocolate.

Aluminum food trays are made by compressing 0.02–0.25 mm thick aluminum sheets into molds, and they are used in the baking and packaging of ready meals and frozen food. Metal caps for bottles are made of aluminum, while the larger lids are made of steel. In all cases, the caps and the lids are coated, or they have cork or rubber-like additions. The twist caps are usually made of aluminum.

15.2.2 Glass

Glass is a good material for food packaging as it does not influence the food, it cannot be corroded, it can be sterilized, it is gas and moisture impermeable, transparent, and can be cleaned up relatively easily. However, glass is brittle, it is heavy, and it cannot withstand sharp temperature changes. Bottles are manufactured by the double-blow method, and the wide mouth jars by the press-blow method. In both cases, the dimensions and the homogeneity of glass packages cannot be as accurate as metal containers. The reduced dimensional accuracy of glass bottles e.g., can cause problems when using high speed bottling equipment, and the inhomogeneity of the wall thickness increases its breaking possibility. An abrupt change of temperature must not exceed 40°C.

Leakage of glass closures is possible, if it is not properly constructed, because it is made of other materials (e.g., metal). The popular crown cap and the traditional cork are good air-tight closures of bottles. In addition, aluminum twist caps are used especially in connection with one-way and other hermetic closures. The caps and lids consist of an inside metallic coated shell (steel or aluminum) and an impermeable sealing material, which may be natural or synthetic rubber, polyvinyl chloride material, or other appropriate plastic. There is quite a wide variety of closing systems, which are usually patented. Capping can be achieved, e.g., by screw on, crimp on, roll on, and push on closing equipment.

Better designed glass containers, coated externally with 0.2% polyurethane-based plastic film and of better design, have increased the strength against impact forces by 50% and reduced the weight of bottles by 40%. It also reduces attrition between bottles during conveyance in the filling station.

15.2.3 Paper and Cardboard

Paper is widely used in food packaging to wrap products and to create bags, cartons, cardboard, and their products. Paper is a relatively low cost material, it can be used in combination with other materials (e.g., in laminates), it has light weight, it has a neutral behavior against food, and that it is environmentally friendly. The disadvantages of paper are its low strength against mechanical stress, low moisture resistance, inability to be sealed, and vulnerability to attacks by insects and microorganisms.

Two main paper categories are used in packaging; fine sulfite paper (greaseproof, glassine, and tissue paper) and coarse sulfate paper (kraft and parchment paper).

Sulfite paper is used for small bags and pouches, for foil laminating, and for making waxed papers. Kraft paper is used in the wrapping and manufacturing of multiply papers, corrugated board, cardboard, and their products (bags, sacks, boxes, etc.). Kraft paper is strong heavy-duty paper, weighing 70–300 g/m^2.

Vegetable parchment paper has greater wet and oil strength than kraft paper, it weights 12–75 g/m^2, and it has almost the same tensile strength as kraft paper. Sulfite paper weights 35–300 g/m^2 and it is used for grocery bags and labels since it can be printed on effectively. Greaseproof paper, weighing 40–150 g/m^2, is used in packaging baked and fatty products. It is fat-resistant, but it loses this property when wetted. The surface of glassine paper is smooth, it is fat-resistant, and about half as heavy and strong as kraft paper. Tissue paper is light (20–50 g/m^2), its strength is low, and it is used in wrapping fruits to protect them from dust and dirt.

Papers can also be waxed, if water resistance and sealability is required. Waxed papers are used in bread wrapping and for the inner lining of cereal cartons. However, since wax sealing is not so strong and wax also tends to crack when the paper temperature is low, wax coating is often replaced by plastic coating. This is, for example, the case with cartons that are used for frozen food.

Paperboard consisting of 3–6 plies of kraft paper, weighing 65–114 g/m^2 each, is used in manufacturing paper bags. Paperboard is stiff and it creases (wrinkles) without cracking.

Cartons for food are made from white board of bleached pulp, which weighs 250 g/m^2.

Cardboard is thicker and heavier than paperboard, consisting of 2–5 plies of kraft paper about 1.00–3.00 mm thick, and weighing about 560–1800 g/m^2. There are two main forms of cardboard; solid cardboard used in paper drums and boxes, and corrugated cardboard used for wrapping. Sealable cardboard is coated with polyethylene or polyvinyl chloride.

Simple corrugated board is made of fluted kraft paper, and it is used as a cushion against impact stresses. Corrugated cardboard, used for boxes, is a multiply material, consisting of plain and corrugated kraft paper. The height and the number of flutes of the corrugated configuration determine the type of cardboard. For heavier packaging materials, more layers of corrugated and plain paper are used.

15.2.4 Plastics

Plastic materials are widely used in food packaging, in the form of bags, pouches, and containers, for wrapping, and as components of compound and laminated packaging units. They are relatively low cost materials and easy to process. Their properties can be modified, thermally sealed, and used in combination with other packaging materials. However, they are not resistant to high temperatures, their use in foods may be restricted for safety reasons, and they may pollute the environment. Plastics are used in flexible and rigid food packages. Their functional properties can be influenced by molecular size, the method of preparation, the addition of substances, and radiation.

The plastics that are commonly used in food packaging are, polyethylene (PE), polyvinyl chloride (PVC), polypropylene (PP), poly ethylene terephthalate (PET), cellulose, and rubber hydrochloride. There are two basic forms of PE—high density

PE (HDPE) and low density PE (LDPE). Both types can be heat-sealed (sealing temperature: 121°C–170°C), and are suitable for shrink-wrapping. HDPE has a 95% crystalline form and can be used up to 120°C, while LDPE has a 60% crystalline form and can be used at temperatures up to 90°C. PE is cheaper than other plastic films and has relatively good mechanical properties. The lowest temperature at which PE can be used is −60°C (Miltz, 1992).

HDPE is stronger and more brittle than LDPE. In general, stiffness, tensile strength, and chemical resistance increase as the PE density increases, while at the same time, gas and vapor permeability, elongation, and crack resistance decrease.

Polyvinylidene chloride (PVdC) has low moisture, gas, fat, and alcohol permeability. It is a little less elastic than PE and it is an odor barrier. Due to its low moisture, gas, and odor permeability, it is used in wrapping many foods, or as a gas and moisture barrier in compound packages. It is heat sealed with other materials or with itself and withstands hot filling and retorting (steam sterilization).

PP has about the same elasticity as PVdC. Its tensile strength is four times that of PE, and it is widely used in making rigid plastic packages and in flexible packaging. Its moisture and fat permeability is low, but it is permeable to many gases and air. Low density PP is used in forming double seam seals, and in making caps and closures for bottles and thin-walled pots by the blow-mold method.

PET has significant tensile strength in a wide range of temperatures (−60°C to 150°C). It has relatively low moisture, gas, and aroma permeability and chemical resistance. It is often used for beverage containers, except milk. It is also used for bags that contain frozen products and are subsequently put in boiling water. PET is used for creating bottles by the blow-mold method and as external film in laminates with PE.

Polyamide (PA) is a strong plastic material and its tensile strength is three times that of PE and 1/3 that of PET, and it is 10 times harder than PE. PA, which is also known as "Nylon," has a low water permeability (40 times less than PET), and a low gas permeability (comparable to that of PET). It can be used at temperatures between −40°C and above 100°C (melting point of Nylon is −6, 215°C). Therefore, it is used for boil-in-bags or for packed products to be processed thermally.

Polytetrafluorethylene (PTFE), known as Teflon, is very heat-resistant, has a very low coefficient of friction and nonadhesive properties, and is chemically inert. It can be used at temperatures up to 230°C. However, since it is expensive, it is used only in connection with thermal resistant coatings of containers that are reused, and when stickiness must be avoided. In packaging technology, heat-sealing elements may be coated by PTFE.

Rubber hydrochloride has the tensile strength of PVC, the elasticity of PVdC, and almost the same shrink properties. It can be used up to 90°C, but it becomes brittle at low temperatures.

Two types of cellulose are used in packaging—cellulose acetate (CA) and regenerated cellulose (RC). The tensile strength of both types is three times that of PE and the hardness of CA is half of that of PA. The water permeability of both types is very high. However, cellulose is often used in food packaging, since it is cheap, environmentally friendly, very transparent, resistant to oil if not wetted, and can be used at temperatures between −50°C and 100°C, RC is used in packaging of dry and other nonmoist materials. CA is mainly used to protect fruits from dust and in "transparent windows" of carton packages.

15.3 PREPARATION OF FOOD CONTAINERS

Food packages are either supplied by container manufacturers as ready-to-use containers (metal, glass), or they are formed in place (plastic bags) during the packaging operation.

15.3.1 Forming of Packages

15.3.1.1 Metal Containers

Metal containers are manufactured in special factories that deliver them to the canning plant either "just in time," or the containers are stored until the next processing period.

Large canning companies manufacture the metal containers themselves. The construction of rigid containers (cans, bottles, drums, etc.) is not covered in this book. However, the construction of paper and plastic containers during the packaging operations in the food processing plants is briefly discussed here.

15.3.1.2 Cartons and Cardboard Packages

Cartons, consisting of laminated cardboard, 0.3–1 mm thick, are used in the packaging of frozen and dried products, and as an additional package of paper or plastic bags. Cardboards are made of corrugated kraft paper, and they are used for packing grouped smaller containers.

The cardboard blanks, which are supplied folded in packs, are placed in the packaging equipment magazines. Each folded cardboard blank is shaped automatically.

In food packaging, the cartons are preferably heat-sealed or glued, protecting them against microorganisms and adulteration, rather than formed mechanically. In addition, gluing or sealing requires half the time of the mechanical locking of carton parts.

In gluing packages of frozen products, the glue must be stable at low temperatures.

15.3.1.3 Film-Based Packages

Plastic packages are formed by using plastic film or by extruding plastic granulate. Film packages are made by (a) the tube method (bags and rectangular packages), (b) the pouch method, and (c) the thermoforming method (cups).

15.3.1.3.1 Tube Packages

Tube packages are used in packaging liquid foods and in aseptic packaging. Laminated materials are used consisting of polyethylene, aluminum, paper, and 2–3 other materials. Aluminum is used for gas sealing, polyethylene for direct contact with food and for heat sealing, and paper for rigidity. Aluminum film thicker than 9 μm is an excellent gas barrier.

Plastic bags are formed by pulling the film out from a reel, downward. The film is wound around a metallic tube and both its ends are sealed vertically together, forming a plastic tube. Horizontal jaws seal the plastic tube at exact distances, as it slips downwards.

The plastic tube moves vertically by pulling the film down each time the jaws seal the plastic tube, or by movement of the plastic tube by roll-belts. The filled and sealed bags are separated from the plastic tube by knives, incorporated in one of the two jaws. The tube system can form various shapes, such as the rectangular package.

15.3.1.3.2 *Pouch Packages*

Plastic pouches are formed from film that comes from two reels and is seamed on four sides. The bags are formed by (a) sealing the two vertical sides of a folded film, or (b) sealing the films rolled out of the two reels. In both types, the films are sealed on three sides, while the fourth side is sealed after the pouch is filled with the product through a tube. This system is similar to the tubular packaging system, but it has a higher capacity.

15.3.1.3.3 *Thermoformed Packages*

Thermoformed plastic cups or trays are used for packaging milk products, such as yogurt and ice cream, juice, jam, dressings, and ready meals. Cups are formed by pressing the laminated film that comes stretched over a mold (die), which has the shape of the package that will be formed. The film comes from a roll and it can be preheated to about 155°C–160°C, either in an infrared (IR) radiant panel heater, or in the mold.

The thickness of the laminated plastic material used in thermoforming is 75–200 μm, depending on the package to be produced (physical properties, strength).

The plastic cups can be formed, filled with the product, and sealed in one operation.

15.3.1.3.4 *Blow-Mold Packages*

Plastic containers are formed by extruding powder plastics in molds consisting of two parts. Pressurized air is used to blow the hot plastic material and to help it to take the form of the mold. After cooling, the two parts of the mold go apart and the ready package is removed.

The formation of plastic packages can be combined with the filling of liquid products in one simultaneous operation.

Blow-mold equipment can produce and fill low viscosity products; about 400 bottles/h (1 L bottles).

15.3.2 Filling of Packages

Before filling, the product is transferred from the short storage to a metering device and then to the filler, while the package is prepared to receive the product (e.g., by air evacuation of the container). Filling can be done under atmospheric conditions, in vacuum, in modified atmosphere, or aseptically.

Filling of food affects the shelf life of the packed products. If food is not filled properly, contamination of fresh or frozen products, or recontamination of pasteurized or sterilized food, can occur. Furthermore, the package must contain the right quantity of product, or the right combination of constituents of food, according to the labeling laws.

The following factors are important in the filling of food in packages: (1) hygienic (sanitary) conditions, (2) coordination with other related packaging steps, (3) high capacity, (4) no waste of product, (5) precision, and (6) flexibility.

The coordination of packaging operations with other food processing operations in a food plant, is essential. A delay in any of the processing steps (forming, filling, sealing, etc.) may offset the whole packaging line. The equipment of a packaging line should be able to "cooperate" with equipment of other processing operations. Thus, bottlenecks could result if excessive product reaches packaging, or if the removal of packaged food is not fast enough. Such problems can be faced by the use of buffer round tables, accumulating packed or nonpacked food for a short time, or diversion to other packaging lines.

The production rate of filling equipment is expressed usually as the instantaneous output, which is the rate under steady state conditions, guaranteed by the equipment manufacturer. Usually, packaging lines are fully automated resulting in high capacities.

The design and operation of filling and packaging equipment require high accuracy, in contrast to other food processing equipment. Filling machines must be flexible, especially in food plants manufacturing seasonal products, when adjustment is necessary in processing different foods (Saravacos and Kostaropoulos, 2002).

Flowability of food products is very important in filling operations. Low viscosity liquid foods, such as beverages and oil, most granulates, and some powders, have a high flowability and they can be transferred by gravity. High-density liquid foods, such as concentrates and some granulates and powders, may need additional force to flow, e.g., positive or negative pressure (vacuum). Increasing the temperature can increase the flowability of viscous liquid food by reducing the viscosity. The flowability of some sticky products, e.g., raisins, is increased if they are dipped in oil and then dried.

The required quantity of food material added to a package can be controlled by (a) volumetric filling, (b) level, (c) overflow, (d) weight filling, (e) time-controlled filling, and (f) counting-filling (Figure 15.2).

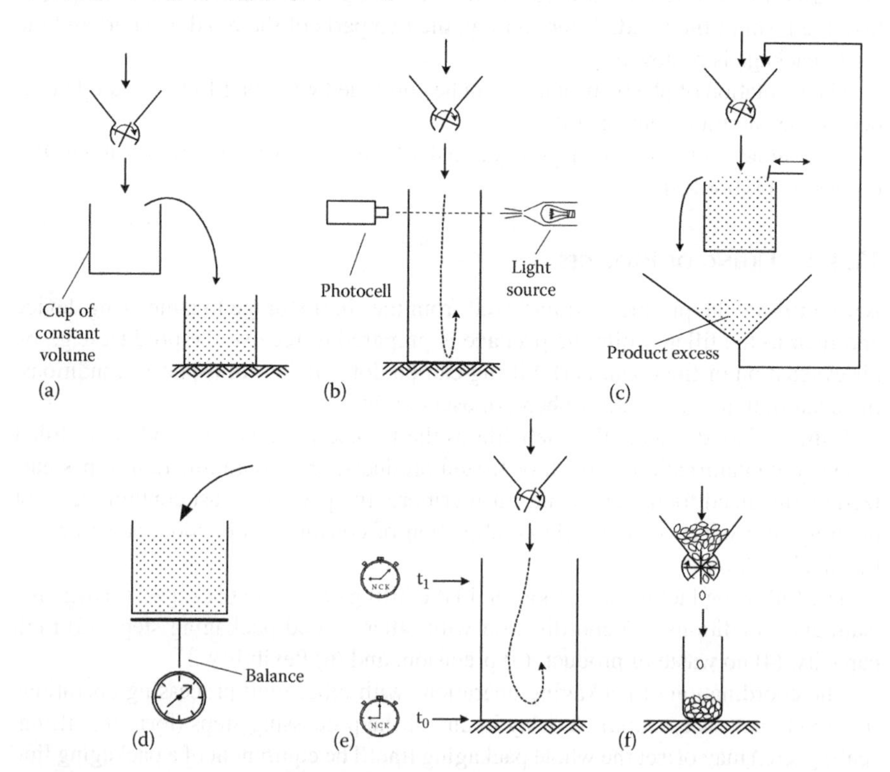

FIGURE 15.2 Methods of filling foods in packages. (a) volumetric, (b) level, (c) overflow, (d) weight, (e) timing, and (f) counting.

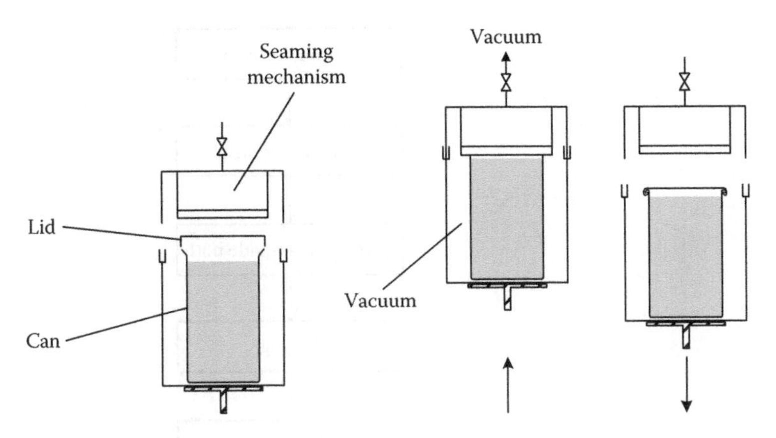

FIGURE 15.3 Vacuum seaming of cans.

The capacity of filling equipment can vary over a wide range, e.g., 5,000–25,000 jars/h (jams), 2,000–25,000 cups/h (yogurt), 8,000–36,000 bottles/h (wine).

15.3.3 Closing of Packages

Metal cans are seamed and jars or bottles are closed by using several types of caps. Cartons are heat-sealed or glued, while cardboard boxes are closed by adhesive plastic or paper tapes. Bags are stitched or closed by clips. Plastic bags are heat-sealed or closed mechanically by clips. Cartons are either sealed by glue, or they are closed mechanically.

Two-piece metal cans, in which only one double seam exists, have replaced the traditional three-piece can. The closing equipment closes a can in two steps by using two different roller die profiles. During closing, steam is often blown through jets on the surface of the can contents, creating a vacuum in the headspace of the can after steam is condensed. Mechanical vacuum can also be applied continuously (Figure 15.3) (Downing, 1996).

The forming, filling, and sealing operations of plastic and paper packages can be combined in a complex machine, such as the Monoblock, which has a higher capacity than conventional separate operations. The Monoblock system (Figure 15.4) is used in aseptic packaging, in vacuum, or in modified atmosphere packaging.

15.3.4 Processing of Packages

15.3.4.1 Conventional Thermal Processing

The conventional thermal processing of food packages is discussed in Chapter 9 (Thermal Process Operations). Sealed food packages (mainly metal and glass containers) are pasteurized or sterilized in retorts (autoclaves) at time-temperature combinations according to the food product and the container size. The quality of

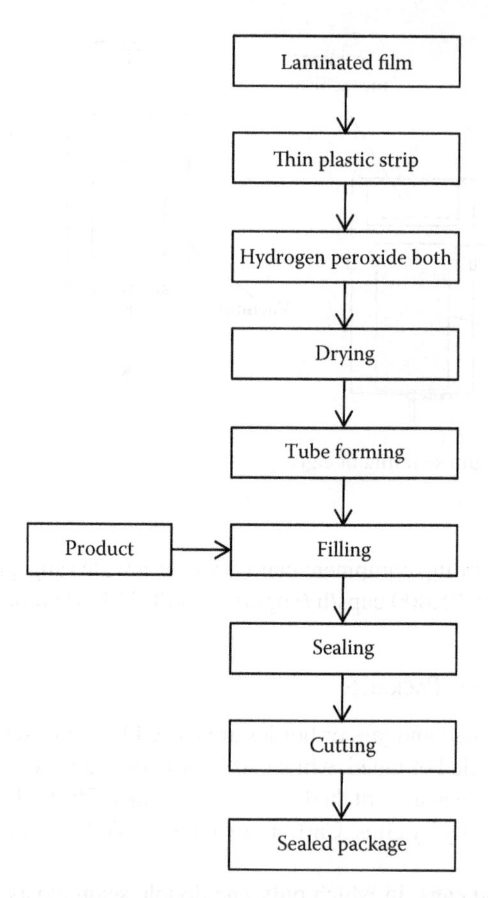

FIGURE 15.4 Diagram of Monoblock forming, filling, sealing (FFS) packaging.

thermally packaged foods may be damaged particularly in large containers and at long processing times.

15.3.4.2 Aseptic Packaging

Aseptic packaging is based on the fast separate sterilization of product and container and sealing the package aseptically, resulting in reduced heat damage to the product. Fast sterilization of liquid foods is achieved in continuous heat exchangers, while the containers are sterilized by steam or hydrogen peroxide. Aseptic processing is also used in the packing of sterilized liquid foods in large and bulk containers.

Aseptic processing is particularly effective when plastic or paper packages are used, which are difficult to process in conventional retorts (Reuter, 1988).

The operations of forming, filling, and sealing (FFS) of plastic packages can be combined in integrated operations, such as the Monoblock system (Figure 15.4). Aseptic packaging of paper cartons can be achieved by combining all necessary operations in an integrated operation, such as the Combiblock system (Figure 15.5).

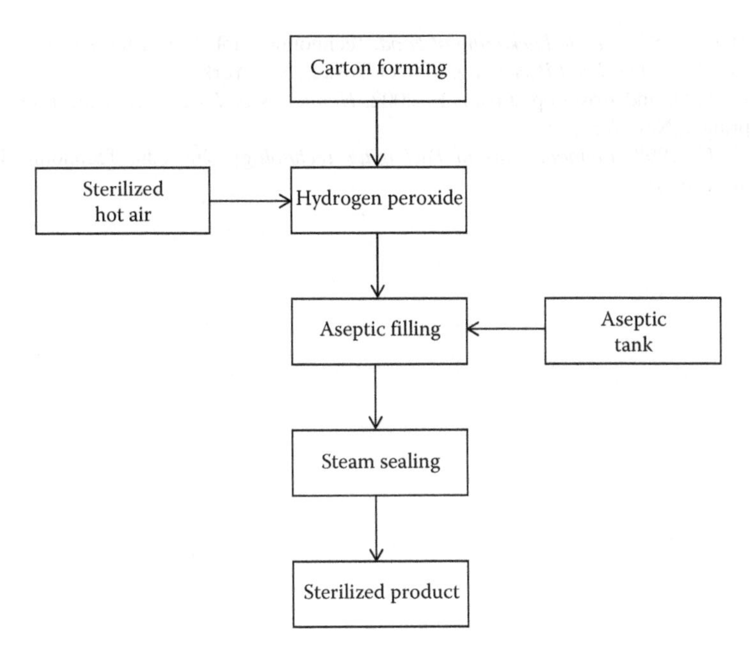

FIGURE 15.5 Diagram of Combiblock aseptic packaging.

15.4 HANDLING OF FOOD PACKAGES

The processed metal cans, after cooling to room temperature, are labeled in special equipment and then packed in carton boxes of standard capacity, e.g., 24 No. 21/2 cans per box. The can boxes are palletized mechanically and transported to a storage room.

Similar labeling, boxing, and palletizing operations are used for glass or jar bottles.

Plastic cups and paper cartons are not labeled, since all labeling information is printed on the outside of the package during forming. They are packed in special carton boxes.

REFERENCES

Brody, A. and Morsh, K.S. 1997. *Wiley Encyclopedia of Packaging Technology*, 2nd edn. Wiley, New York.

Brown, W.E. 1992. *Plastics in Food Packaging*. Marcel Dekker, New York.

Bureau, G. and Multon, J.L., eds. 1996. *Food Packaging Technology*, Vols. 1 and 2. VCH, New York.

Downing, D.L. ed. 1996. *A Complete Course in Canning*. CTI Publ., Timonium, MD.

Hernandez, R. 1997. Food packaging materials, barrier properties, and selection. In: *Handbook of Food Engineering Practice*. K.J. Valentas., E. Rotstein, R.P. Singh., eds. CRC Press, New York.

Miltz, J. 1992. Food packaging. In: *Handbook of Food Engineering*. D.R. Heldman and D.B. Lund, eds. Marcel Dekker, New York.

Reuter, H. ed. 1988. *Aseptic Packaging of Food*. Technomic Publ., Lancaster, PA.
Robertson, G.L. 1993. *Food Packaging*. Marcel Dekker, New York.
Saravacos, G.D. and Kostaropoulos, A.E. 2002. *Handbook of Food Processing Equipment*. Springer, New York.
Soroka, W.G. 1998. *Fundamentals of Packaging Technology*, 2nd edn. Technomic Publ., Lancaster, PA.

16 Spreadsheet Applications

16.1 INTRODUCTION

General purpose spreadsheet software can be used effectively for engineering calculations. For example, Microsoft Excel® with Visual Basic® for Applications is an effective tool for process design. Spreadsheets offer sufficient process model "hospitality." They are connected easily and online with charts and graphic objects, resulting in powerful and easy-to-use graphical interfaces. Excel also supports mathematical and statistical tools. For instance, Solver is an excellent tool for solving sets of equations and performing optimization. Databases are effectively and easily accessed. In addition, Visual Basic for Applications offers a powerful object-oriented programming language capable of constructing commercial graphics interfaces.

Our book *Food Process Design* (Maroulis and Saravacos, 2003) presents a systematic approach to solve engineering problems in a spreadsheet environment. In particular, it shows integrated procedures for robust process design by analyzing the following topics:

- Modeling and spreadsheets
- Analyzing the Solver
- Sensitivity analysis using Excel tables
- Controls and Dialog boxes to input data
- Graphics to get results
- Databases
- Visual Basic as a programming language

In the present book, a more simplified concept is adopted. It is a step from classic hand calculations toward more sophisticated spreadsheet calculations.

The following topics are introduced and applied to problems similar to those of the previous chapters of this book.

- Name variables
- Insert data
- Insert equations using names
- Use "Goal Seek" to solve an equation
- Automate using Visual Basic
- Assign a macro to a button
- Use Excel tables
- Use Excel charts
- Use the scroll bars

- Construct a simple database
- Use Combo Boxes
- Use matrix operations
- Use Solver

16.2 SHELL AND TUBES HEAT EXCHANGER

In this example, the following Excel operations are introduced:

- Name variables (Ctrl + Shift + F3)
- Insert data
- Insert equations using names

16.2.1 Problem Formulation

Calculate the appropriate shell and tube heat exchanger for a tomato paste heating process using the available heating steam.

The design specifications are

$F = 1$ kg/s	Feed flow rate
$T_1 = 50°C$	Feed temperature
$T_2 = 100°C$	Target temperature
$T_s = 120°C$	Steam temperature

The required data of the tomato paste and the heating steam are

$\rho = 1130$ kg/m^3	Fluid density
$\lambda = 0.55$ W/(m K)	Fluid thermal conductivity
$\eta = 0.27$ Pa s	Fluid apparent viscosity
$C_p = 3.50$ kJ/(kg K)	Specific heat
$\Delta H_s = 2200$ kJ/kg	Latent heat of steam condensation

The following design variables are also assumed:

$u = 1$ m/s	Fluid velocity in tubes
$d = 0.01$ m	Tube diameter
$dx = 5$ mm	Tube thickness
$n = 4$	Number of passes in tubes

16.2.2 Problem Solution

The thermal load is calculated by the equation

$$Q = FC_p\left(T_2 - T_1\right)$$

The required steam flow rate Q is calculated by the equation

$$Q = F_s \Delta H_s$$

The mean temperature difference ΔT_m

$$\Delta T_m = \frac{(T_s - T_1) - (T_s - T_2)}{\ln\left[(T_s - T_1)/(T_s - T_2)\right]}$$

The total number of tubes N

$$u = \frac{F}{\rho (N/n)(\pi d^2/4)}$$

The surface heat transfer coefficient outside tubes h_o

$$h_o = 2750\left(\frac{Nd}{F_s}\right)^{1/3}$$

The Pr number

$$Pr = \frac{C_p \eta}{\lambda}$$

The Re number

$$Re = \frac{\rho d u}{\eta}$$

The Nu number

$$Nu = 1.86\left(RePr\frac{d}{L}\right)^{1/3}\left(\frac{\eta}{\eta_w}\right)^{0.14}$$

The surface heat transfer coefficient inside tubes h_i

$$Nu = \frac{h_i d}{\lambda}$$

The overall heat transfer coefficient U

$$\frac{1}{U} = \frac{1}{h_i} + \frac{1}{h_o}$$

	A	B	C	D	E	F
1	Variable	Text Name	Excel Name	Value	Units	Equation
2	Feed flow rate	F	F	1	kg/s	
3	Feed temperature	T_1	T1.	50	°C	
4	Target temperature	T_2	T2.	100	°C	
5	Steam temperature	T_s	Ts	120	°C	
6	Fluid specific heat	C_p	Cp	3.50	kJ/kgC	
7	Latent heat of steam condensation	ΔH_s	dHs	2200	kJ/kg	
8	Fluid density	ρ	p_	1130	Kg/m³	
9	Fluid thermal conductivity	λ	k_	0.55	W/mK	
10	Fluid apparent viscosity	η	n_	0.27	Pa s	
11	Fluid velocity in tubes	u	u.	1.00	m/s	
12	Tube diameter	d	d	0.01	m	
13	Tube thickness	δx	dx	5	mm	
14	Number of passes in tubes	n	m	4	—	
16	Thermal load	Q	Q	175	kW	:=F*Cp*(T2.−T1.)
17	Steam flow rate	F_s	Fs	0.08	kg/s	:=Q/dHs
18	Mean temperature difference	ΔT_m	dTm	40	°C	:=((Ts−T1.)−(Ts−T2.))/LN((Ts−T1.)/(Ts−T2.))
19	Total number of tubes	N	M.	45	—	:=F/p_/(u./m)/(3.14*d^2/4)
20	Surface heat transfer coefficient outside tubes	h_o	ho	4.90	kW/m²K	:=2.75*(M.*d/Fs)^(1/3)
21	*Pr* Number	Pr	Pr	1718	—	:=Cp*n_/k_*1000
22	*Re* Number	Re	Re	42	—	:=u.*p_*d/n_
23	*Nu* Number	Nu	Nu	9.28	—	:=1.86*(Re*Pr*d/10)^(1/3)*1.2
24	Surface heat transfer coefficient inside tubes	h_i	hi	0.51	kW/m²K	:=Nu*k_/d/1000
25	Overall heat transfer coefficient	U	U	0.46	kW/m²K	:=(1/(1/ho+1/hi))
27	Heat transfer area	A	A	9.48	m²	:=Q/U/dTm
28	Shell diameter	D	D.	0.20	m	:=(d+2*dx/1000)*(M./0.319)^(1/2.142)
29	Tube length	L	L	6.70	m	:=A/M./3.14/d

FIGURE 16.1

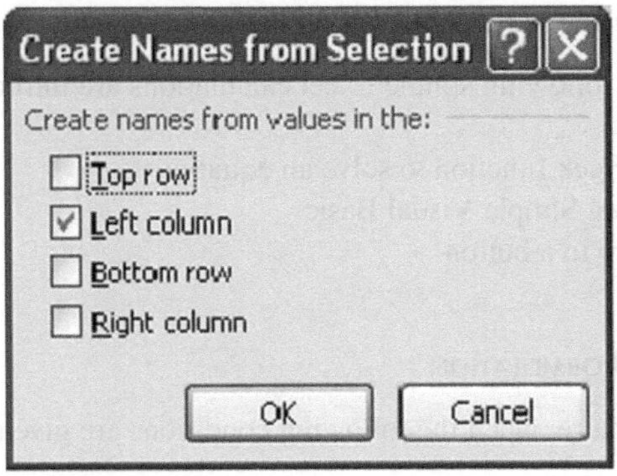

Create Names from Selection

Create names from values in the:

- [ ] Top row
- [x] Left column
- [ ] Bottom row
- [ ] Right column

OK Cancel

FIGURE 16.2

The total heat transfer area A

$$Q = AU\Delta T_m$$

The shell diameter D

$$N = 0.319\left(\frac{D}{d}\right)^{2.142}$$

The tubes length L

$$A = N\pi dL$$

16.2.3 Excel Implementation

In an Excel spreadsheet and in the range A2:A29, type the names of the process variables incorporated in the previous solution as shown in Figure 16.1. The range A2:A14 is used for the data while the range A16:A29 for the results. In the next column and in the range B2:B29, type the symbols for these variables as used in the solution provided earlier.

In the range C2:C29, type the corresponding names in Excel. Use names similar to their symbols in the text if possible. In the range E2:E29, type the corresponding units.

By selecting the range C2:D29 and by simultaneously pressing the buttons Ctrl + Shift + F3, the names in column C are assigned to the cells of column D (Figure 16.2).

Enter the data for the given variables in the range D2:D14. Enter the equations according to the previous solution procedure into the range D16:D29. These equations are presented for information in the range F16:F29.

The problem has been solved. Any change of the data will change the result. The problem is solved with just 13 equations in the range D16:D29 by using 13 data in the range D2:D14. All other text in the spreadsheet is for information (Figure 16.1).

16.3 PSYCHROMETRIC CALCULATIONS

In this example, the following simple Excel calculations are introduced:

- Use the Goal Seek function to solve an equation
- Automate Using Simple Visual Basic
- Assign a macro to a button

16.3.1 Problem Formulation

For an air/water mixture, when the following conditions are given:

$P = 1$ bar	Pressure
$T = 65°C$	Temperature
$Y = 0.035$ kg/kg db	Total humidity (liquid + vapor)

Calculate the following properties:

P_d	bar	Dew pressure
P_s	bar	Vapor pressure at temperature T
P_w	bar	Vapor pressure at temperature T_w
T_b	°C	Boiling temperature
T_d	°C	Dew temperature
T_w	°C	Wet bulb temperature
Y_L	kg/kg db	Humidity in liquid
Y_V	kg/kg db	Humidity in vapor
Y_s	kg/kg db	Saturation humidity at temperature T
Y_w	kg/kg db	Saturation humidity at temperature T_w
a_w	—	Water activity
H	kJ/kg db	Enthalpy of humid air
C_P	kJ/kg K db	Specific heat of humid air
ΔH_S	kJ/kg	Latent heat of condensation of water vapor at temperature T

Suppose that the following data are valid:

$m = 0.622$	Air/water molecular weight ratio
$C_{PA} = 1.00$ kJ/kg K	Specific heat of air
$C_{PV} = 1.90$ kJ/kg K	Specific heat of water vapor
$C_{PW} = 4.20$ kJ/kg K	Specific heat of liquid water
$\Delta H_0 = 2.50$ MJ/kg	Latent heat of water evaporation at 0°C
$a_1 = 1.19 \times 10^1$	Antoine equation constants for water
$a_2 = 3.99 \times 10^3$	
$a_3 = 2.34 \times 10^2$	

16.3.2 Problem Solution

A simple psychrometric model has been presented by Maroulis and Saravacos (2003) as follows:

$P_S = \exp\,[a_1 - a_2/(a_3 + T)]$	(E01)
$Y_S = m\,P_S/(P - P_S)$	(E02)
$Y_V = \min\,(Y, Y_S)$	(E03)
$Y_L = Y - Y_V$	(E04)
$Y_V = m a_w P_S/(P - a_w P_S)$	(E05)
$H = C_{PA}T + Y_V\,(\Delta H_0 + C_{PV}\,T) + Y_L C_{PL} T$	(E06)
$T_b = -\,a_3 + a_2/(a_1 - \ln P)$	(E07)
$T_d = a_2/(a_1 - \ln[Y\,P/(m + Y)]) - a_3$	(E08)
$P_d = \exp[a_1 - a_2/(a_3 + T)]$	(E09)
$P_w = \exp[a_1 - a_2/(a_3 + T_w)]$	(E10)
$Y_w = mP_w/(P - P_w)$	(E11)
$C_P = C_{PA} + Y_V\,C_{PV}$	(E12)
$\Delta H_S = \Delta H_0 - (C_{PL} - C_{PV})T$	(E13)
$(Y_V - Y_W)/(T - T_W) = -\,C_P/\Delta H_S$	(E14)

Based on these equations, the following algorithm can be used:

(E01)	→	P_s		
(E02)	→	Y_s		
(E03)	→	Y_V		
(E04)	→	Y_L		
(E05)	→	a_w		
(E06)	→	H		
(E07)	→	T_b		
(E08)	→	T_d		
(E09)	→	P_d		
		T_w	Trial value	←
(E10)	→	P_w		
(E11)	→	Y_w		
(E12)	→	C_P		
(E13)	→	ΔH_S		
(E14)	→	T_w	Corrected value	→

16.3.3 Excel Implementation

Using the same conventions as in the previous example, the following spreadsheet is constructed (Figure 16.3):

As the algorithm suggests, wet bulb temperature is calculated by trial and error, which means that we assign a trial value to cell D23 and examine the resulting cell

	A	B	C	D	E	F
1	Variable	Text Name	Excel Name	Value	Units	Equation
2	Antoine equation constant for water a1	a_1	a1.	1.19E+01	—	
3	Antoine equation constant for water a2	a_2	a2.	3.99E+03	—	
4	Antoine equation constant for water a3	a_3	a3.	2.34E+02	—	
5	Air/water molecular weight ratio	m	m	0.622	—	
6	Latent heat of water evaporation at 0	ΔH_0	dHo	2.50	MJ/kg	
7	Specific heat of liquid water	PW	Cp1	4.20	kJ/kg K	
8	Specific heat of water vapor	PV	Cpv	1.90	kJ/kg K	
9	Specific heat of air	PA	Cpa	1.00	kJ/kg K	
10	Pressure	P	P	1.0	bar	
11	Temperature	T	T	65.0	°C	
12	Total humidity (liquid + vapor)	Y	Y	0.035	kg/kg db	
13	Vapor pressure at temperature	P_s	Ps	0.25	bar	:=EXP(a1.–a2./(a3.+T))
14	Saturation humidity at temperature	Y_s	Ys	0.206	kg/kg db	:=IF(m*Ps/(P–Ps)<0,10,m*Ps/(P–Ps))
16	Humidity in vapor	Y_V	Yv	0.035	kg/kg db	:=MIN(Y,Ys)
17	Humidity in liquid	Y_L	Yl	0.000	kg/kg db	:=Y–Yv
18	Water activity	a_w	aw	0.214	—	:=(P/Ps)/(1+m/Yv)
19	Enthalpy of humid air	H	H	156.8	kJ/kg db	:=Cpa*T+Yv*(dHo*1000+Cpv*T)+Yl*Cpl*T
20	Boiling temperature	T_b	Tb	99.9	°C	:=a2./(a1.–LN(P))–a3.
21	Dew pressure	P_d	Pd	4.7	Bar	:=(m/Y+1)*Ps
22	Dew temperature	T_d	Td	34.0	°C	:=a2./(a1.–LN(Y*P/(m+Y)))–a3.
23	Wet bulb temperature (trial value)	T_w	Tw	39.0	°C	
24	Vapor pressure at temperature	P_w	Pw	0.07	bar	:=EXP(a1.–a2./(a3.+Tw))
25	Saturation humidity at temperature	Y_w	Yw	0.047	kJ/kg db	:=m*EXP(a1.–a2./(a3.+Tw))/(P–EXP(a1.–a2./(a3.+Tw)))
27	Specific heat of humid air	C_P	Cp	1.07	kJ/kgK db	:=Cpa+Yv*Cpv
28	Condensation heat of water vapor at	ΔH_s	dH	2.35	MJ/kg	:=(dHo*1000–(Cpl–Cpv)*T)/1000
29	Wet bulb temperature (corrected value)		Twc	39.0		:=T–(Yw–Yv)*dH/Cp*1000
30	Wet bulb temperature (corrected minus trial value)		dTw	9.57E-05		:=Twc–Tw

wbt

FIGURE 16.3

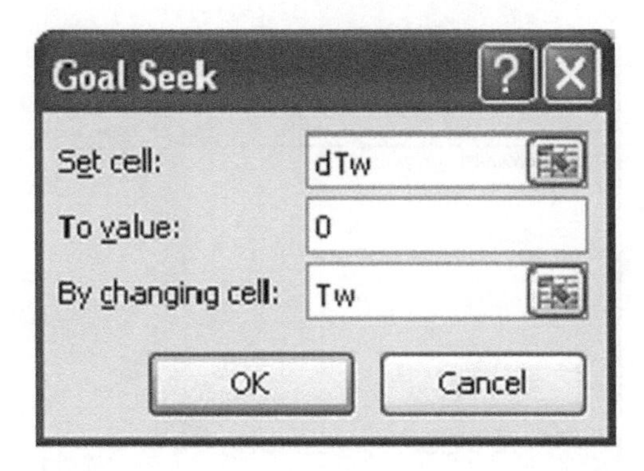

FIGURE 16.4

D29. We try different values in cell D23 until we obtain the same value in cell D29. The difference between the trial and corrected values is calculated in cell D30.

Alternatively, we can use the "Goal Seek" function of Excel to automatically set cell D30 to equal 0 by changing cell D23. Goal Seek can be found in the "Data" menu by selecting the "What-if Analysis" submenu (Figure 16.4).

Furthermore, we can automate the procedure as follows:

- From the "Developer" menu and the "Insert" submenu select a Button (Form Control) and insert it into the spreadsheet.
- By right clicking on the button, select "Edit text" and insert a caption e.g., wbt (Wet Bulb Temperature).
- By right clicking on the button, select "Assign Macro" (Figure 16.5) and then "Edit."
- Type the following subroutine within the Visual Basic Editor.
  1. Sub wbd()
  2. Range("Tw").Value = Range("Td").Value
  3. Range("wbt").GoalSeek Goal: = 0#, ChangingCell:=Range("Tw")
  4. End Sub

The calculations are automatically happened every time the button "wbt" is clicking.

16.4 PSYCHROMETRIC CHART

In this example, the following Excel functions are introduced:

- Use a 2D Excel table
- Use an Excel chart

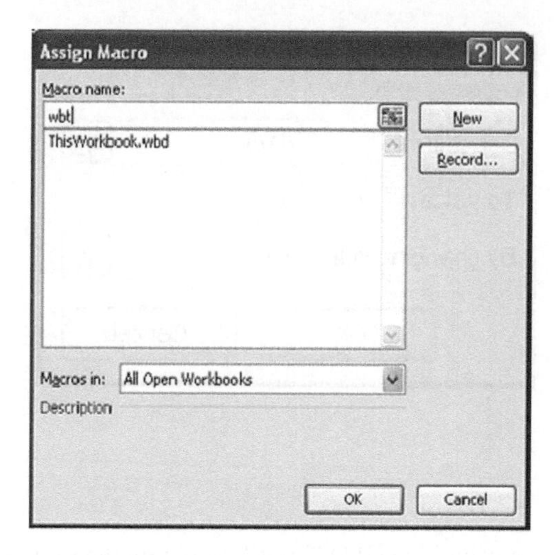

FIGURE 16.5

16.4.1 Problem Formulation

Construct a psychrometric chart, i.e., plot the humidity in vapor Y_V (kg/kg db) versus the temperature T (°C) for some constant values of water activity a_w (–) for an air water-vapor mixture at any pressure P (bar).

Suppose the following parameters are given:

$m = 0.622$	Water molecular weight ratio
$a_1 = 1.19 \times 10^1$	Antoine equation constants for water
$a_2 = 3.99 \times 10^3$	
$a_3 = 2.34 \times 10^2$	

16.4.2 PROBLEM SOLUTION

The humidity is calculated from the equation

$$Y_V = \frac{m\, a_w P_S}{(P - a_w P_S)}$$

where the vapor pressure is calculated by the Antoine equation

$$P_S = \exp\left[\frac{a_1 - a_2}{(a_3 + T)}\right]$$

Thus, for any given values of P, T, and a_w, the Y_V is obtained.

16.4.3 Excel Implementation

The solution just presented is provided in the following spreadsheet (Figure 16.6):

	I	J	K	L	M	N
1	Variable	Text Symbol	Excel Name	Value	Units	Equation
2	Antoine equation constant for water a1	a_1	a1.	1.19E+01	—	
3	Antoine equation constant for water a2	a_2	a2.	3.99E+03	—	
4	Antoine equation constant for water a3	a_3	a3.	2.34E+02	—	
5	Air/water molecular weight ratio	m	m	0.622	—	
6	Water activity	a_w	aw.	1.00	—	
7	Pressure	P	P.	1.00	bar	
8	Temperature	T	T.	65.00	°C	
9	Vapor pressure at temperature	P_s	Ps.	0.249	bar	:=EXP(a1.–a2./(a3.+T.))
10	Humidity in vapor	Y_V	Yv.	0.206	kg/kg db	:=m*aw.*Ps./(P.–aw.*Ps.)

FIGURE 16.6

A 2D Excel table is used to calculate the humidity for various values of temperature and water activity as follows (Figure 16.7):

1. In the range R4:R99, type the desired values for temperature and in the range S3:W3, the desired values for water activity.
2. In cell R3 insert the equation "=Yv."
3. Select the range R3:W25, and from the menu "Data" and submenu "What-if Analysis" select "Data Table." The Data Table dialogue will appear (Figure 16.8).
4. In Data Table Dialogue assign cell "T" for the Row Input Cell and cell "aw" for the Column Input Cell. The corresponding values for humidity are automatically filled in the range S4:W25 (Figure 16.7).

Select the range R3:W25 shown in Figure 16.7 and from the menu "Insert" select the "Scatter" diagram. The psychrometric chart in Figure 16.7 is constructed.

16.5 CALCULATIONS ON PSYCHROMETRIC CHART

In this example, the following Excel function is introduced:

- Use the scroll bar to change the values of a variable

16.5.1 Problem Formulation

In the psychometric chart of Example 3, select a point of given temperature and humidity (T, Y) and construct the corresponding dew point (T_d, Y), the saturated point (T, Y_s), and the wet bulb point (T_w, Y_w).

P	Q	R	S	T	U	V	W
1							
2	Yv(T, aw)						
3	0.206	0.20	0.40	0.60	0.80	1.00	
4	0	0.001	0.002	0.002	0.003	0.004	
5	5	0.001	0.002	0.003	0.004	0.006	
6	10	0.002	0.003	0.005	0.006	0.008	
7	15	0.002	0.004	0.006	0.009	0.011	
8	20	0.003	0.006	0.009	0.012	0.015	
9	25	0.004	0.008	0.012	0.016	0.020	
10	30	0.005	0.011	0.016	0.022	0.028	
11	35	0.007	0.014	0.022	0.029	0.037	
12	40	0.009	0.019	0.029	0.039	0.050	
13	45	0.012	0.025	0.038	0.052	0.066	
14	50	0.016	0.032	0.050	0.068	0.087	
16	55	0.020	0.042	0.065	0.089	0.116	
17	60	0.026	0.054	0.084	0.117	0.154	
18	65	0.033	0.069	0.109	0.155	0.206	
19	70	0.041	0.088	0.142	0.205	0.279	
20	75	0.052	0.112	0.185	0.275	0.386	
21	80	0.065	0.144	0.244	0.375	0.552	
22	85	0.081	0.185	0.326	0.527	0.836	
23	90	0.100	0.240	0.445	0.779	1.418	
24	95	0.125	0.314	0.629	1.265	3.215	
25	99	0.150	0.394	0.864	2.145	19.447	
27							

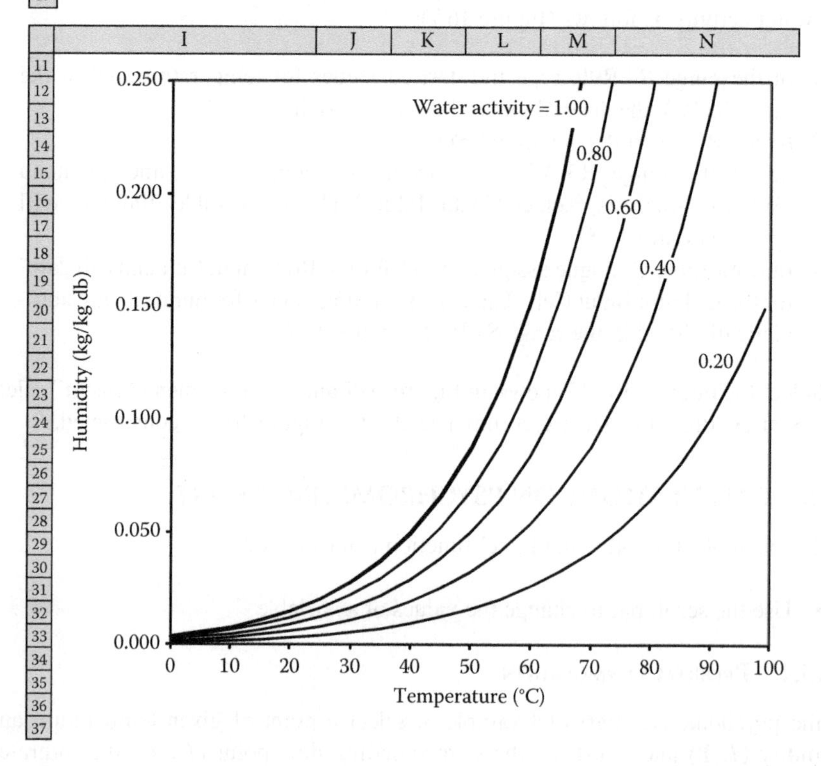

FIGURE 16.7

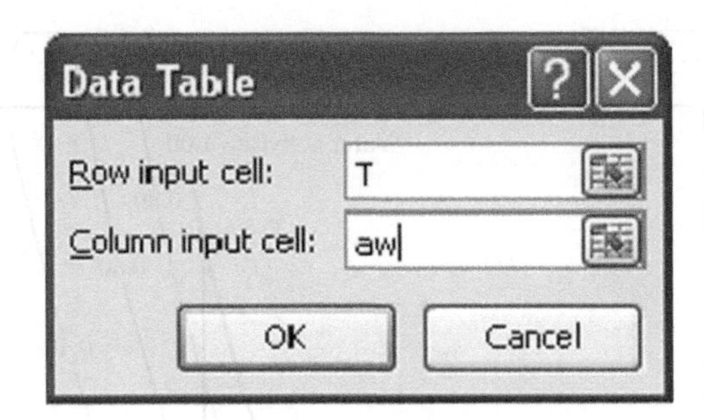

FIGURE 16.8

16.5.2 Problem Solution

When P, T, and Y are given, all the corresponding psychrometric properties are calculated according to Example 2.

Thus, if we combine Examples 2 and 3 in one spreadsheet, the problem is solved by plotting the results of Example 2 in the chart of Example 3.

16.5.3 Excel Implementation

In the range AC27:AD33, insert the equations presented in the range AH29:AI31. In the psychrometric chart insert two more curves, AC27:AD31 and AC32:AD33, and the icon appears in Figure 16.9.

If the values of cells T and Y change, the above points are moved in the chart and the new conditions appear. Obviously, the button "wbt" described in Example 2 should be pressed (Figure 16.3).

More automation can be applied by using "scroll bars":

From the "Developer" menu and the submenu "Insert" insert a "Scroll Bar." By right clicking on the Scroll Bar select "Format Control" and select the temperature cell T (F11) for the "Cell Link" (Figure 16.10).

In order to automatically calculate the wet bulb temperature (without clicking the "wbt" button) right click the scroll bar and select "Assign Macro." Assign the macro "wbt" to the scroll bar.

The Scroll Bar is now able to automatically change the temperature and move the point (T, Y) left or right. All other points (T_d, Y), (T, Y_s), and (T_w, Y_w) are automatically redrawn.

Scroll bar can also be inserted for pressure and/or humidity.

16.6 ISOTHERMS

In this example, the following Excel functions are introduced:

- Construct a simple properties database
- Use a Combo Box to present data from a database

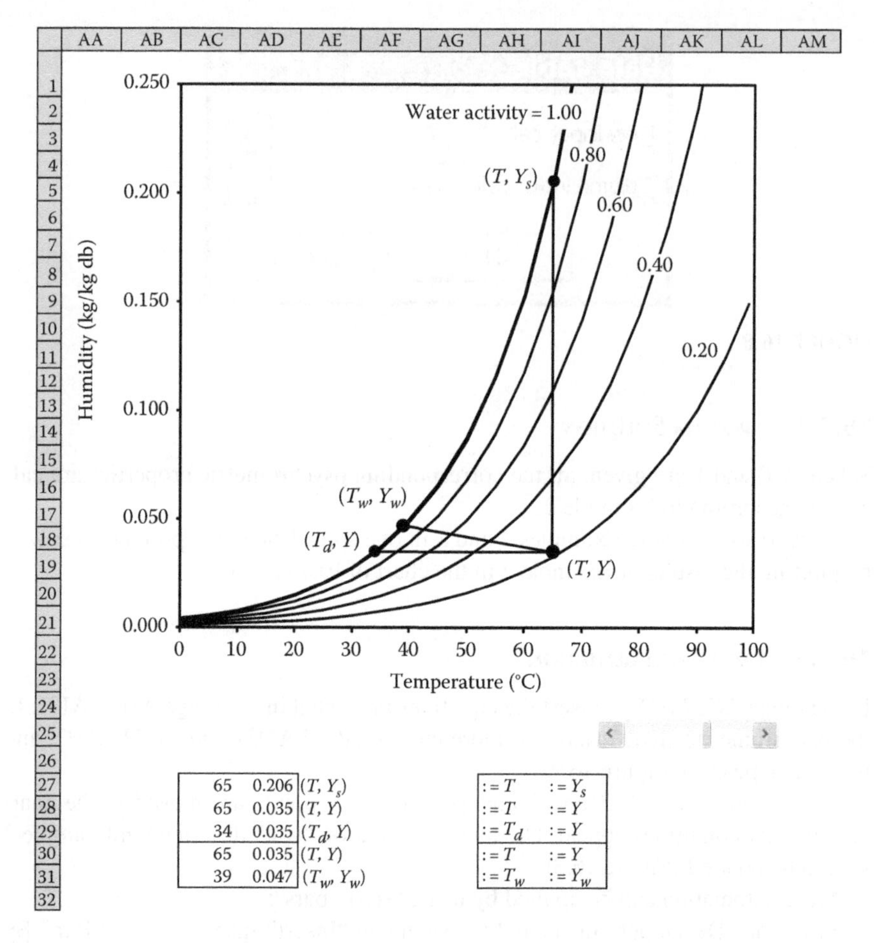

FIGURE 16.9

16.6.1 Problem Formulation

The equilibrium material moisture content X_e depends on air temperature, T and water activity, a_w.

Various empirical or semitheoretical models have been proposed in the literature, but a modified Oswin model seems to be the most appropriate in process design calculations:

$$X_e = b_1 \exp\left[\frac{b_2}{(273+T)}\right]\left[\frac{a_w}{(1-a_w)}\right]^{b_3}$$

where b_1, b_2, and b_3 are adjustable empirical constants, depending on the material characteristics.

The purpose of this example is to construct a database in an Excel environment to incorporate the Oswin constants for some materials and by using the various Excel

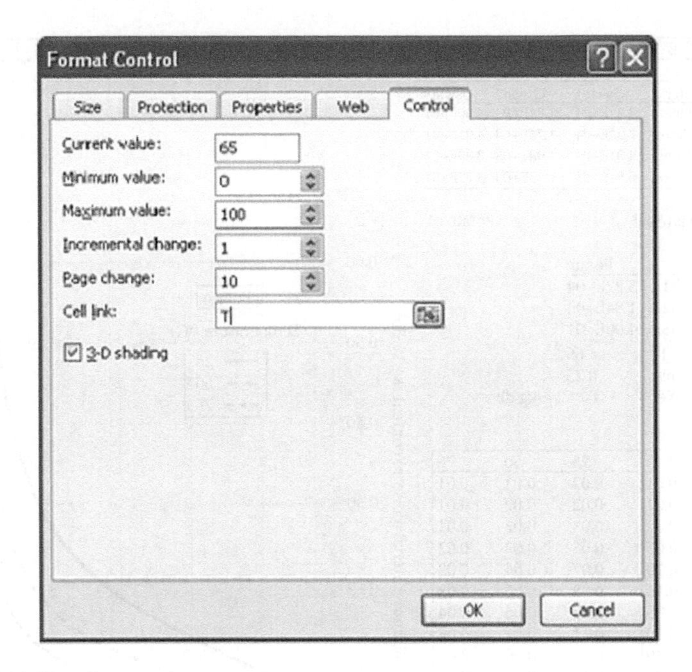

FIGURE 16.10

utilities to construct an environment in which the moisture isotherms for various materials will be examined automatically.

16.6.2 Excel Implementation

Retrieve some values for the Oswin constants from the literature (e.g., Maroulis and Saravacos 2003) and insert these values in to an Excel spreadsheet as shown in the range B2:E6 in Figure 16.11.

Insert initial data values for the variables as follows:

jMaterial	Material identification number
T	Temperature
Aw	Water activity

Insert the following equations for the following corresponding variable:

Name	:=INDEX(Material,jMaterial)
b1.	:=INDEX(Oswin1,jMaterial)
b2.	:=INDEX(Oswin2,jMaterial)
b3.	:=INDEX(Oswin3,jMaterial)
Xe	:=b1.*EXP(b2./(273 + T))*(aw/(1-aw))^b3.

The INDEX function selects a specific value from an array.

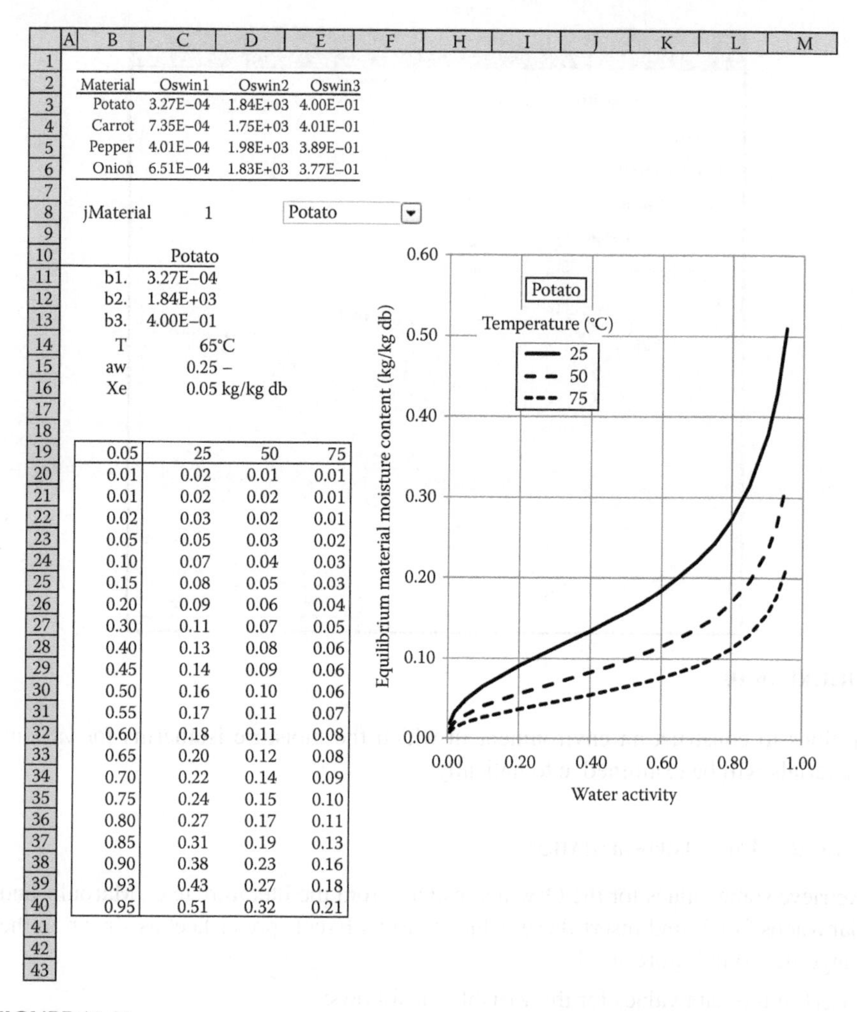

Material	Oswin1	Oswin2	Oswin3
Potato	3.27E−04	1.84E+03	4.00E−01
Carrot	7.35E−04	1.75E+03	4.01E−01
Pepper	4.01E−04	1.98E+03	3.89E−01
Onion	6.51E−04	1.83E+03	3.77E−01

jMaterial 1 Potato

	Potato
b1.	3.27E−04
b2.	1.84E+03
b3.	4.00E−01
T	65°C
aw	0.25 –
Xe	0.05 kg/kg db

0.05	25	50	75
0.01	0.02	0.01	0.01
0.01	0.02	0.02	0.01
0.02	0.03	0.02	0.01
0.05	0.05	0.03	0.02
0.10	0.07	0.04	0.03
0.15	0.08	0.05	0.03
0.20	0.09	0.06	0.04
0.30	0.11	0.07	0.05
0.40	0.13	0.08	0.06
0.45	0.14	0.09	0.06
0.50	0.16	0.10	0.06
0.55	0.17	0.11	0.07
0.60	0.18	0.11	0.08
0.65	0.20	0.12	0.08
0.70	0.22	0.14	0.09
0.75	0.24	0.15	0.10
0.80	0.27	0.17	0.11
0.85	0.31	0.19	0.13
0.90	0.38	0.23	0.16
0.93	0.43	0.27	0.18
0.95	0.51	0.32	0.21

FIGURE 16.11

The following Excel names should be assigned in the following cells:

Name	Cell
aw	C15
b1.	C11
b2.	C12
b3.	C13
jMaterial	C8
Material	B3:B6
Oswin1	C3:C6
Oswin2	D3:D6
Oswin3	E3:E6
T	C14
Xe	C16

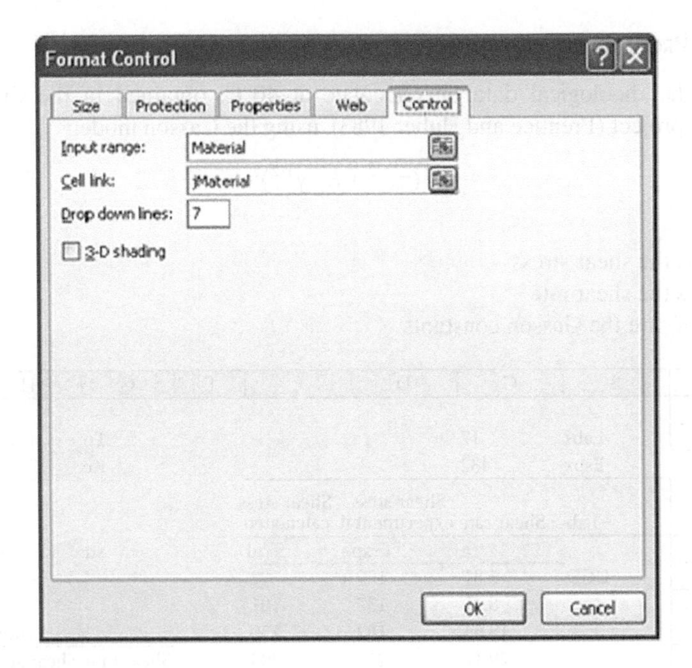

FIGURE 16.12

Insert a 2D Excel table using water activity values as the Column Input Cell and Temperature values as the Row Input Cell. The resulting table is shown in Figure 16.11.

Insert a Scatter Graph, as also shown in Figure 16.11, to graphically represent the results of the 2D Excel table.

Automate the material selection using a "Combo Box":

- Insert a "Combo Box" from the menu "Developer" and submenu "Insert."
- Right click on the Combo Box and select "Format Control."
- In the "Format Control Dialogue," assign the "Material" name to the "Input Range" and the "jMaterial" name to the "Cell Link" (Figure 16.12).

The spreadsheet is now fully automated:

By clicking the Combo Box, the available material appears in the Combo Box List, and when a material is selected from the list, the material isotherms are automatically plotted in the Chart (Figure 16.11).

16.7 FITTING RHEOLOGICAL DATA FOR CHOCOLATE

In this example, the following Excel functions are introduced:

- Use matrix operations (Alt + Ctrl + Shift + ok)
- Excel Solver

16.7.1 Problem Formulation

Analyze the rheological data on chocolate at 40°C, obtained in the COST 90 European project (Prentice and Huber 1983), using the Casson model:

$$\tau = (\tau_o^{1/2} + K_c\,\gamma^{1/2})^2$$

where

- τ (Pa) is the shear stress
- γ (1/s) is the shear rate
- τ_o and K_c are the Gasson constants

A	B	C	D	E	F	G	H
1							
2	Labs	17				To	27.0
3	Espx	432				Kc	1.68
4	Lab	Shear rate	Shear stess experimental	Shear stess calculated			
5		g	Texp	Tcal		sd	1446
6	UCD	6.81	109	92			
7		8.41	137	101			
8		13.64	182	130			Model
9		29.11	217	203		Shear rate	Shear stress
10		35.96	307	233		0.00512	28
11		58.33	254	325		0.05	31
12		72.20	462	379		0.1	33
13		89.18	367	443		0.2	35
14							41
15							47
16							57
17							80
18							110
19							161
20							291
21							483
22							838
23							1827
24							2777
25							
26							
27							
28							
29							
30							
31							
32							
33							
34							
35							
36		0.34	50	38			
37		0.43	52	40			

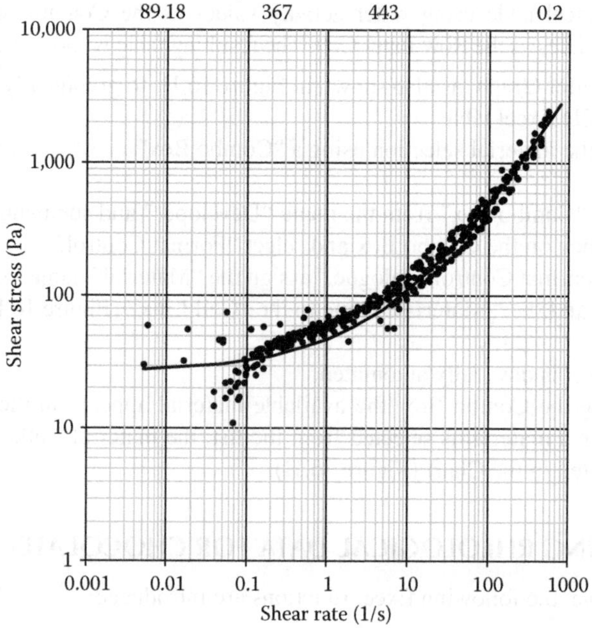

Shear stress (Pa)
10,000
1,000
100
10
1
0.001 0.01 0.1 1 10 100 1000
Shear rate (1/s)

FIGURE 16.13

16.7.2 Excel Implementation

In the spreadsheet in the Figure 16.13, 423 experimental data from 17 different laboratories are introduced in the range C6:D428.

Initial values for the Gasson constants are also introduced in the range H2:H3.

In the range E6:E428, the calculated values using the Gasson model are introduced using the equation:= (To^0.5 + Kc*g^0.5)^2 in each cell.

The following Excel names should be assigned in the following cells:

Name	Cell
g	C6:C428
Kc	H3
sd	H5
Tcal	E6:E428
Texp	D6:D428
To	H2

The standard deviation between the experimental and calculated values is estimated in cell H5 by using the equation:

:={SQRT(SUM((Texp-Tcal)^2))}

Since Texp and Tcal are arrays, the mathematical operation in the cell refers to the matrix operation.

This is denoted by pressing Alt + Ctrl + Shift simultaneously when ok is clicked to end the equation editing. If the brackets {} exist in the equation, the matrix operation has been accepted.

A comparison between the experimental and calculated values can be obtained by introducing the appropriate chart.

Curve fitting means finding the best values for the parameters To and Kc to minimize the standard deviation sd between the experimental and calculated values.

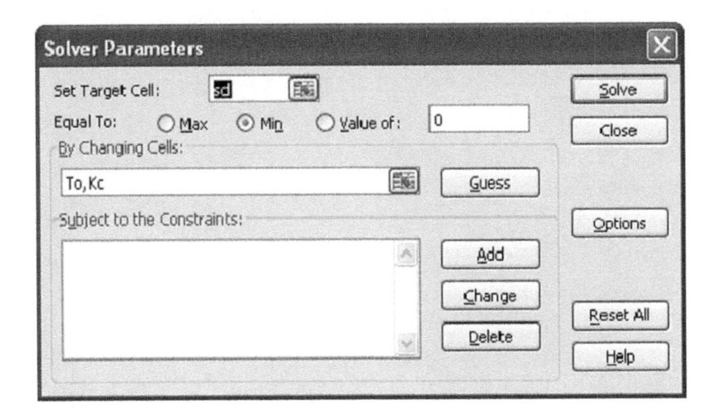

FIGURE 16.14

This operation can be performed using the Excel Solver:

- From the menu "Data" select "solver" and the Solver Parameters Dialogue will appear (Figure 16.14).
- Select the "min" operation, insert the variable "sd" in the "Set Target Cell" parameter, and the variables "To; Kc" in the "By Changing Cells" parameter.
- By pressing the button "Solve", the optimization procedure is executed, and after some iteration, the optimal parameter estimates are obtained.

REFERENCES

Maroulis, Z.B. and Saravacos, G.D. 2003. *Food Process Design.* Marcel Dekker, New York.

Prentice, J.H. and Huber, D. 1983. Results of the collaborative study on measuring rheological properties of foodstuffs. In: *Physical Properties of Foods.* R. Jowitt, F. Escher, B. Halstrom, H.F.Th. Meffert, W.E.L. Spiess, G, Vos, eds. Applied Science Publ., London, pp. 123–183.

Saravacos, G.D. and Maroulis, Z.B. 2001. *Transport Properties of Foods.* Marcel Dekker, New York.

Appendix A

A.1 CONVERSION TO SI UNITS

From	To (SI) Units	Multiply By
Acre	m^2	4046
Angstrom	m	1×10^{-10}
atm (760 Torr)	bar	1.013
bar	Pa	1×10^5
barrel (oil)	m^3	0.159
barrel (liquors)	m^3	0.117
box	kg	41
Btu	kJ	1.055
Btu/h	W	0.293
Btu/(h ft^2)	W/m^2	3.154
Btu/(h ft °F)	W/m K	1.729
Btu/(h ft^2 °F)	W/m^2 K	5.678
Btu/lb	kJ/kg	2.326
Bushel	m^3	0.036
cP	Pa s	0.001
cu ft	m^3	0.0284
cu ft/min (CFM)	m^3/s	0.5×10^{-3}
cu ft/lb	m^3/kg	0.0624
cwt (hundredweight)	kg	50.8
dyn	N	1×10^{-5}
ft	m	0.305
ft/min (FPM)	m/s	0.0051
ft of water	Pa	2990
ft lb	J	1.355
gallons (U.S.)	m^3	3.785×10^{-3}
gallons (Imperial)	m^3	4.543×10^{-3}
gallons/min (GPM)	m^3/s	0.063×10^{-3}
hectare	m^2	10×10^3
HP	kW	0.745
HP (boiler)	kW	9.80
in. (inches)	m	0.0254
in. Hg	Pa	3386
kcal	kJ	4.18
kg force (kp)	N	9.81
kW h	MJ	3.6
L (lit, l)	m^3	0.001
lb mass	kg	0.454

(continued)

(continued)

From	To (SI) Units	Multiply By
lb force	N	4.45
lb/cu ft	kg/m^3	16.02
lb/(ft s)	kg/(m s)	1.488
lb/ft^2	Pa	47.9
$lb/(ft^2 h)$	$kg/(m^2 s)$	0.00133
$lb/in.^2$ (psi)	Pa	6894
miles	km	1.609
mm water	Pa	9.81
oz (ounce)	kg	0.028
P (poise)	Pa s	0.10
Pa	N/m^2	1.00
Pa s	kg/(m s)	1.00
RPM (rpm)	1/s	0.0167
sq ft (ft^2)	m^2	0.093
sq in. ($in.^2$)	m^2	0.645×10^{-3}
therm	MJ	105.5
ton (metric)	kg	1000
ton (U.S., short)	kg	907.2
ton-refrigeration	kW	3.51
Torr (mm Hg)	Pa	133.3
W	J/s	1.00
K = °C + 273, °C = (°F − 32)/1.8		

A.1.1 Multipliers of SI Units

dk (deco)	× (10)
h (hecto)	× (100)
k (kilo)	× (1,000)
M (mega)	× (1,000,000)
G (giga)	× (1,000,000,000)
d (deci)	× (0.1)
c (centi)	× (0.01)
m (milli)	× (0.001)
μ (micro)	× (0.000001)
n (nano)	× (0.000000001)

Note that in the United States, the following technical multipliers are sometimes used, M (thousand) × (1,000), and MM (million) × (1,000,000).

A.2 SATURATED STEAM TABLE (ABBREVIATED)

T (°C)	P (bar)	H_L (kJ/kg)	H_V (kJ/kg)	ΔH_{VL} (kJ/kg)
0.01	0.0061	0.00	2500	2500
1	0.0066	4.18	2502	2498
2	0.0071	8.40	2504	2496
3	0.0076	12.61	2506	2493
4	0.0081	16.82	2508	2491
5	0.0087	21.02	2510	2489
10	0.0123	42.00	2519	2477
15	0.0170	63.00	2528	2465
20	0.0234	83.84	2537	2453
25	0.0317	104.75	2546	2441
30	0.0424	125.67	2555	2430
35	0.0523	146.60	2564	2418
40	0.0738	167.50	2573	2406
45	0.0959	188.42	2582	2394
50	0.1234	209.33	2591	2382
55	0.1575	230.24	2600	2370
60	0.1993	251.15	2609	2358
65	0.2502	272.08	2617	2345
70	0.3118	293.91	2626	2333
75	0.3856	313.96	2635	2321
80	0.4737	334.93	2643	2308
85	0.5786	355.92	2651	2295
90	0.7012	376.93	2660	2283
95	0.8453	397.98	2668	2270
100	1.0132	419.06	2676	2257
102	1.0877	427.51	2679	2251
104	1.1667	435.95	2682	2246
106	1.2503	444.41	2685	2241
108	1.3388	452.87	2688	2235
110	1.4324	461.34	2691	2230
112	1.5313	469.61	2694	2224
114	1.6358	478.29	2697	2219
116	1.7461	486.78	2700	2213
118	1.8623	495.28	2703	2208
120	1.9848	503.76	2706	2202
125	2.3201	525.07	2713	2188
130	2.7002	546.41	2720	2174
135	3.1293	567.80	2727	2159
140	3.6119	589.24	2734	2145
150	4.7572	632.32	2746	2114
160	6.1766	675.65	2758	2082
170	7.9147	719.28	2768	2049
180	10.0190	753.25	2778	2015
374	221	2086	2086	0 (critical point)

T (°C), saturation temperature; P (bar), saturation pressure; H_L (kJ/kg), enthalpy of liquid; H_V (kJ/kg), enthalpy of vapor; ΔH_{VL} (kJ/kg), heat of evaporation of water (Haar et al, 1984; Perry and Green, 1997).

The density of saturated liquid water (ρ_L, kg/m^3) changes from 999.85 (1°C) to 958.39 (100°C) and 887.06 (180°C). The density of saturated water vapor (ρ_V, kg/m^3) varies from 0.0052 (1°C) to 0.5975 (100°C) and 5.134 (180°C).

A.3 SUPERHEATED STEAM

Superheated steam is produced by heating saturated steam at constant pressure to temperatures above saturation. The properties of superheated steam at various temperatures and pressures can be found in special steam tables of the literature.

Of particular importance in process engineering is the enthalpy of superheated steam (H_s), which can be estimated from the enthalpy of the saturated steam (H_V), using the equation $H_s = H_V + C_p(T_s - T_o)$, where C_p is the specific heat of superheated steam, and T_s and T_o are the superheated and saturation temperatures, respectively. For typical process applications of superheated steam, approximately $C_p = 2\,\text{kJ/(kg K)}$.

As an example, the enthalpy of superheated steam at an absolute pressure of 1 atm (1.013 bar) and 150°C is estimated as $H_s = 2676 + 2 \times (150 - 100) = 2776\,\text{kJ/kg}$.

A.4 ENGINEERING PROPERTIES OF FOOD SYSTEMS

A.4.1 Properties of Water and Air

Property (20°C)	Water	Air (Dry)
Density (ρ), kg/m^3	1000	1.3
Specific heat (C_p), kJ/(kg K)	4.18	1.1
Viscosity (η), mPa s	1.0	0.018
Thermal conductivity (λ), W/(m K)	0.60	0.025
Thermal diffusivity (α), m^2/s	1.4×10^{-7}	1.7×10^{-5}
Mass diffusivity (D), m^2/s		
Water/water	1×10^{-9}	—
Air/water	—	1.9×10^{-9}
Air/air	—	1.8×10^{-5}

Source: Saravacos, G.D. and Maroulis, Z.B., Transport Properties of Foods, Marcel Dekker, New York, 2001.

A.4.2 Engineering Properties of Some Liquid Foods (20°C)

Liquid Food	ρ, kg/m³	C_p, kJ/kg	η, mPa s	λ, W/(m K)
Milk	1030	3.8	2	0.48
Orange juice 12°Brix	1040	3.5	1.5	0.66
Apple juice (clear) 65°Brix	1350	2.2	70	0.35
Vegetable oil	920	1.8	90	0.17
Honey	1380	2.0	8000	0.34

Source: Rahman, M.S., *Food properties Handbook*, 2nd edn., CRC Press, New York, 2009; Rao, M.A., Rizvi, S.S.H., and Datta, A.E., Engineering Properties of Foods, 3rd edn., Taylor & Francis, New York, 2005.

A.4.3 Engineering Properties of Some Solid Foods (20°C)

Food Material	X, kg/kg dm	ε,—	C_p, kJ/(kg K)	λ, W/(m K)	D, ×10⁻¹⁰ m²/s
Potato, raw	3.00	0.10	3.6	0.45	5
Potato, dried	0.05	0.15	2.0	0.10	1
Corn	0.20	0.10	2.1	0.20	0.5
Bread	0.80	0.50	2.8	0.40	2
Pasta	0.15	0.10	2.1	0.20	0.3
Raisins	0.25	0.10	2.2	0.20	1
Minced beef	0.75	0.15	2.8	0.40	1.5
Codfish	0.50	0.20	2.5	0.30	2

Source: Data from Saravacos, G.D. and Maroulis, Z.B., *Transport Properties of Foods*, Marcel Dekker, New York, 2001.

Note: X, moisture content; ε, porosity; C_p, specific heat; λ, thermal conductivity; D, water diffusivity.

A.5 ENGINEERING DIAGRAMS

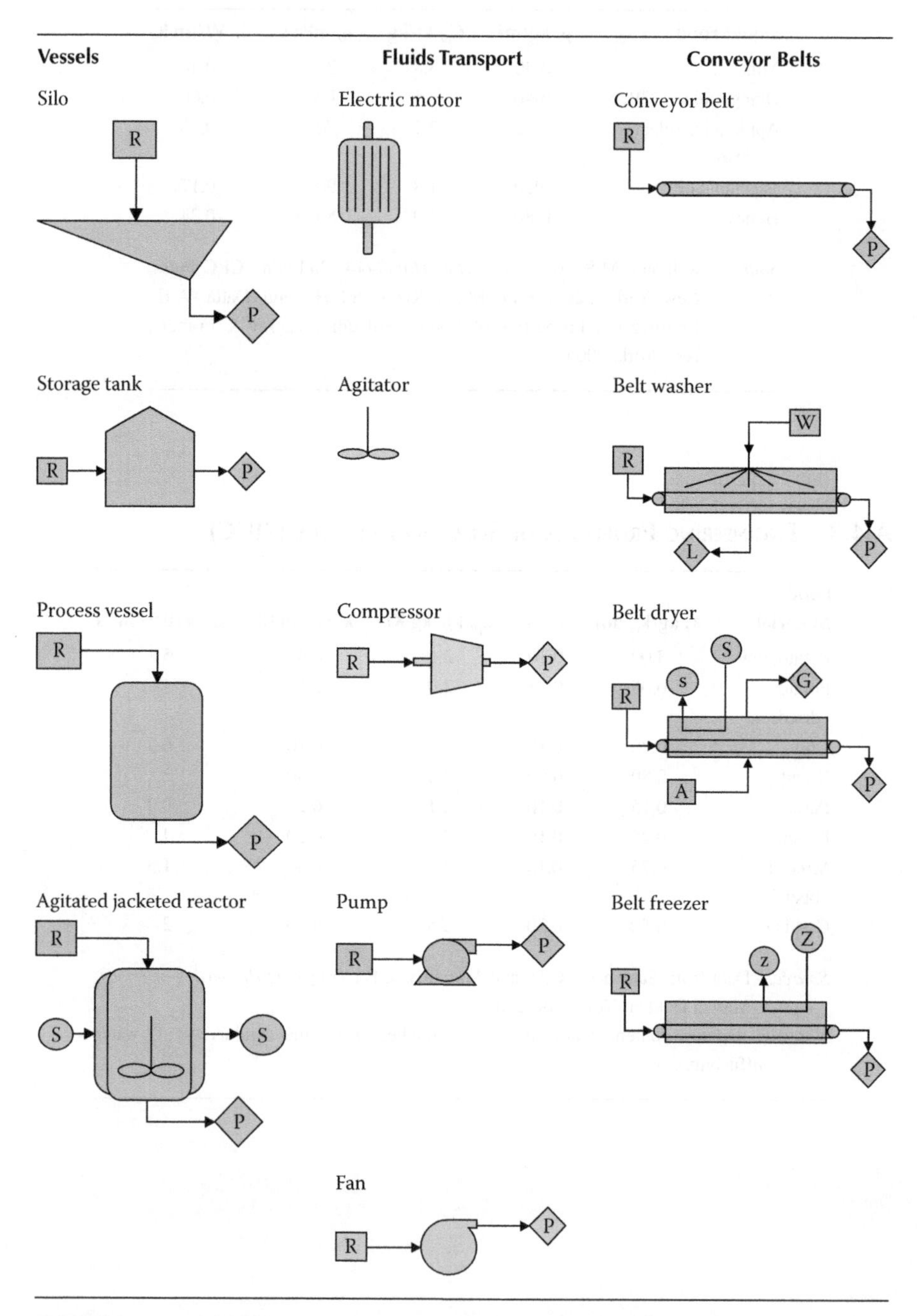

FIGURE A.1 Symbols of process equipment. (Modified from Maroulis, Z.B. and Saravacos, G.D., *Food Plant Economics*, CRC Press, New York, August 2, 2007, fig. 5.4.)

Filters

Vacuum drum filter

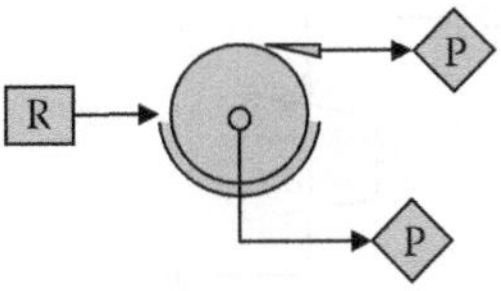

Plate filter

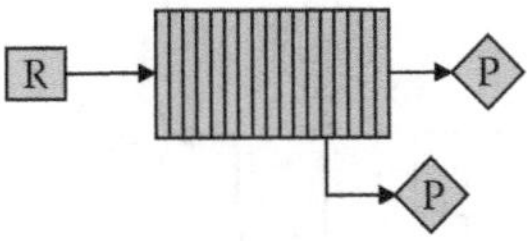

Vibrating screen

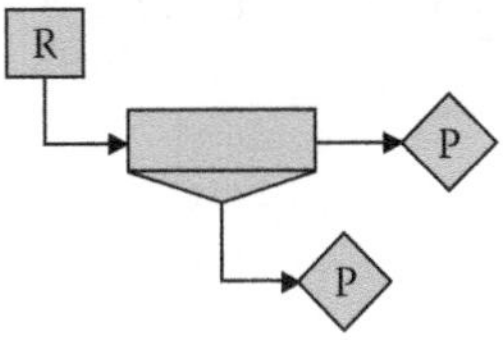

Heat Exchangers

Scraped surface heat exchanger

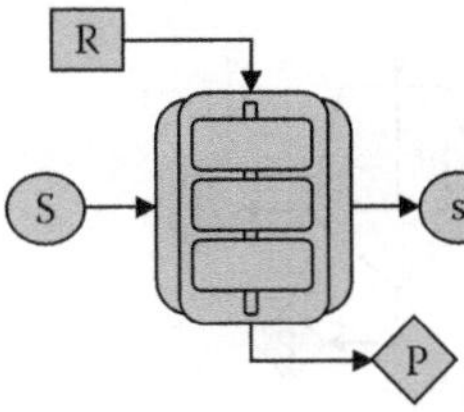

Shell and tubes heat exchanger

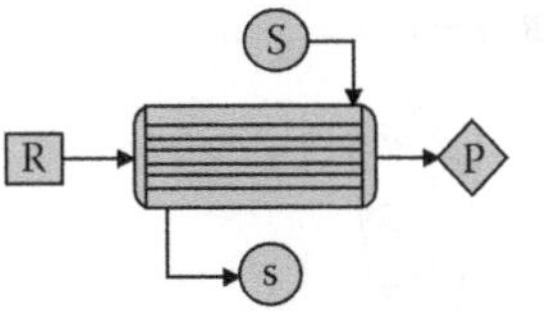

Plate heat exchanger

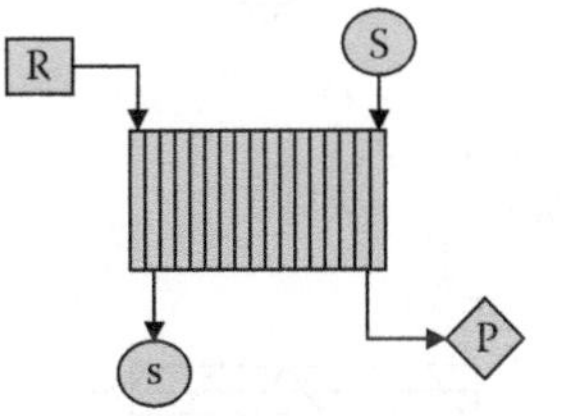

Tubular evaporator

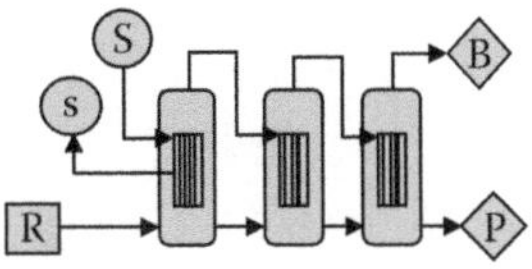

Forced circulation evaporator

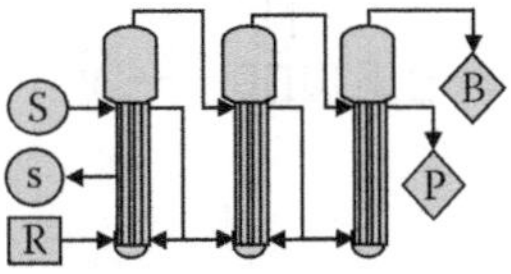

Dryers

Tray dryer

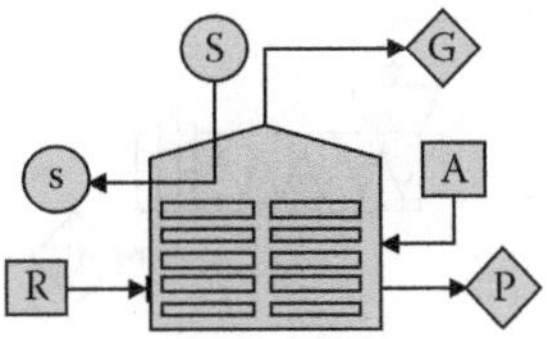

Vibratory conveyor dryer

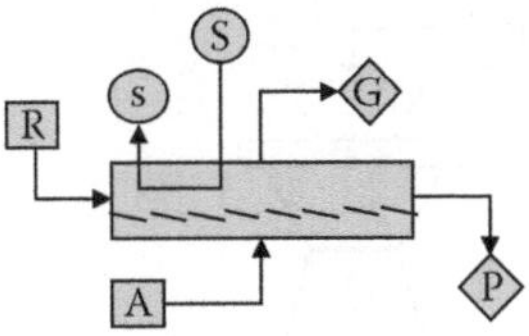

Rotary dryer

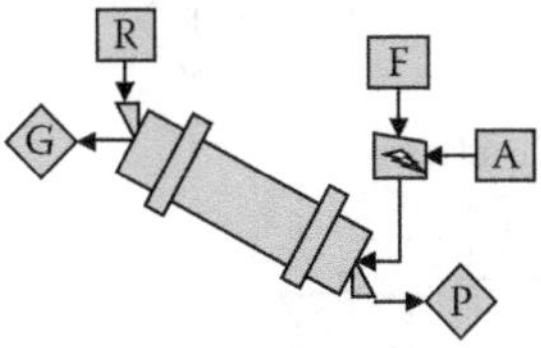

Fluidized bed dryer

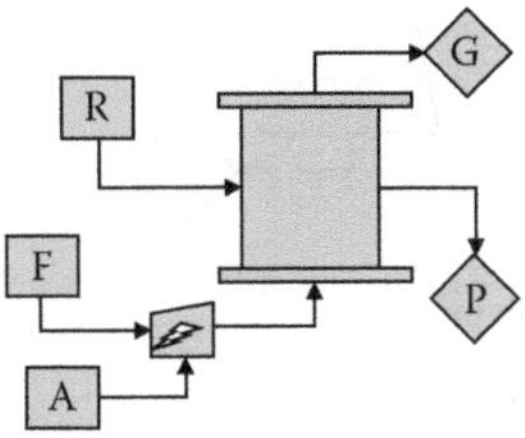

FIGURE A.1 (continued)

(continued)

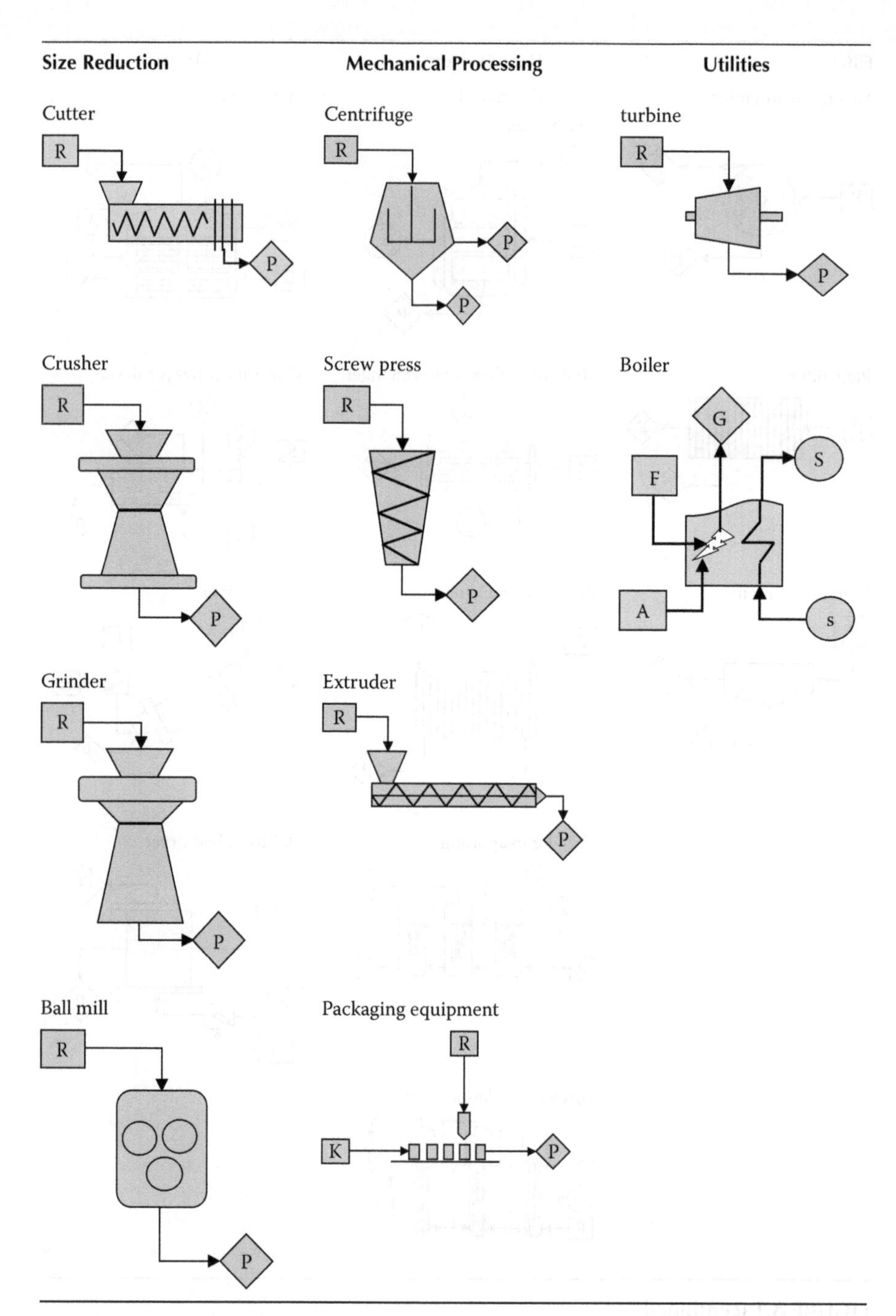

FIGURE A.1 (continued)

List of Equipment

Vessels

Silo
Storage tank
Process vessel
Agitated jacketed reactor

Fluids transport

Electric motor
Agitator
Compressor
Pump
Fan

Conveyor belts

Conveyor belt
Belt washer
Belt dryer
Belt freezer

Heat exchangers

Scraped surface heat exchanger
Shell and tubes heat exchanger
Plate heat exchanger
Tubular evaporator
Forced circulation evaporation

Filters

Vacuum drum filter
Plate filter
Vibrating screen

Dryers

Tray dryer
Vibratory conveyor dryer
Rotary dryer
Fluidized bed dryer

Size reduction

Cutter
Crusher
Grinder
Ball mill

Mechanical processing

Centrifuge
Screw press
Extruder
Packaging equipment

Utilities

Turbine
Boiler

Nomenclature

System input streams

R Row material
K Packaging material
X Auxiliar material
F Fuel
W Process water
A Ambient air

System internal recycled streams

S Steam
s Steam condensate
C Cooling water
c Cooling water return
Z Refrigerant
z Refrigerant return
M Compressed air

System output streams

Product

Byproduct

Liquid or solid waste

Gas waste

FIGURE A.1 (continued)

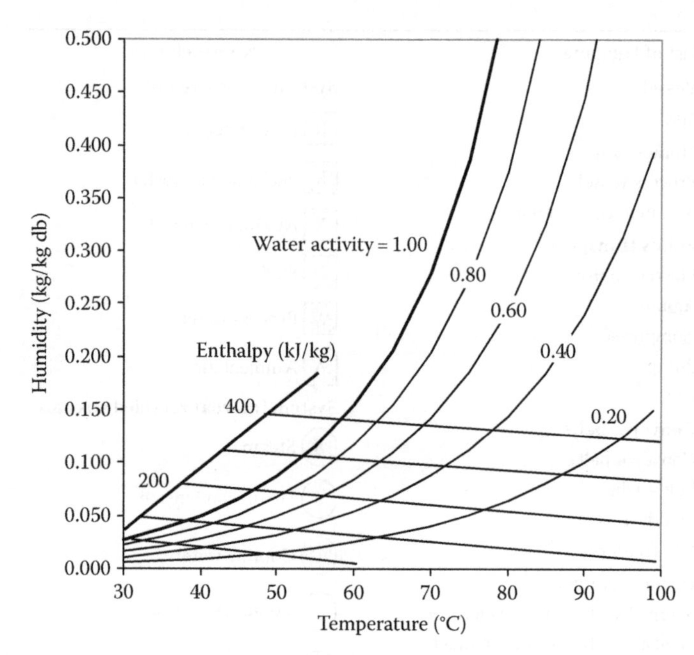

FIGURE A.2 Psychrometric chart at pressure 1 bar and temperatures 30°C–100°C (calculated from the psychrometric model of Chapter 11). Air water activity a_w = (% relative humidity)/100.

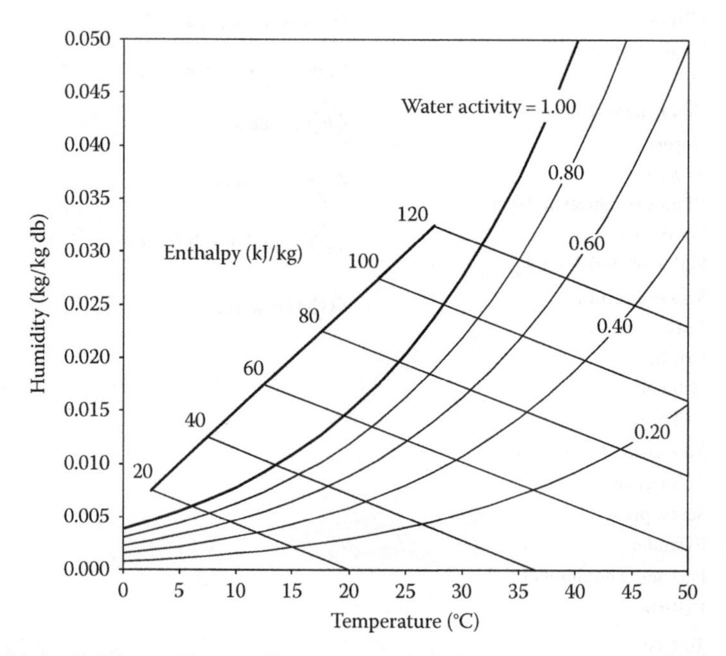

FIGURE A.3 Psychrometric chart at pressure 1 bar and temperatures 0°C–50°C (calculated from the psychrometric model of Chapter 11). Air water activity a_v = (% relative humidity)/100. (Maroulis, Z.B. and Saravacos, G.D. *Food Process Design*, CRC Press, Boca Raton, FL, 2003.)

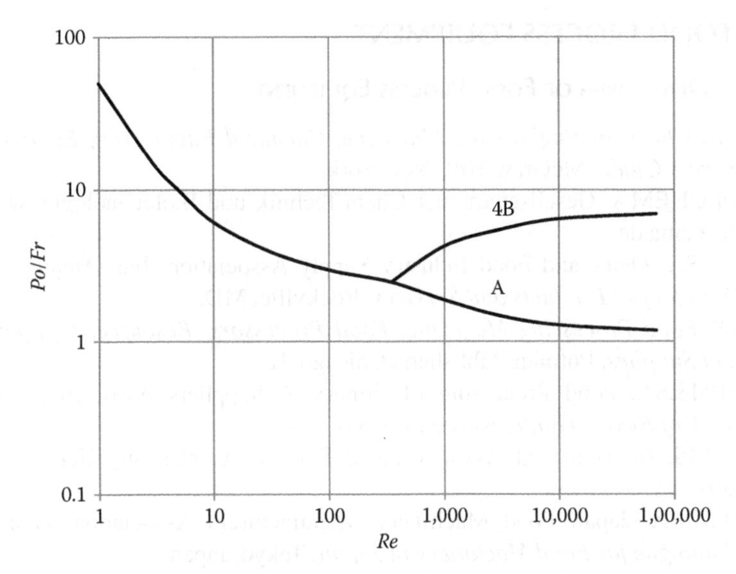

FIGURE A.4 Power number (*Po*) in agitated vessels, Froude number (*Fr*), Reynolds number (*Re*). (4B) four baffles, (A) un-baffled vessel. (Saravacos, G.D. and Kostaropoulos, A.E., *Handbook of Food Processing Equipment*, Springer, New York, 2002.)

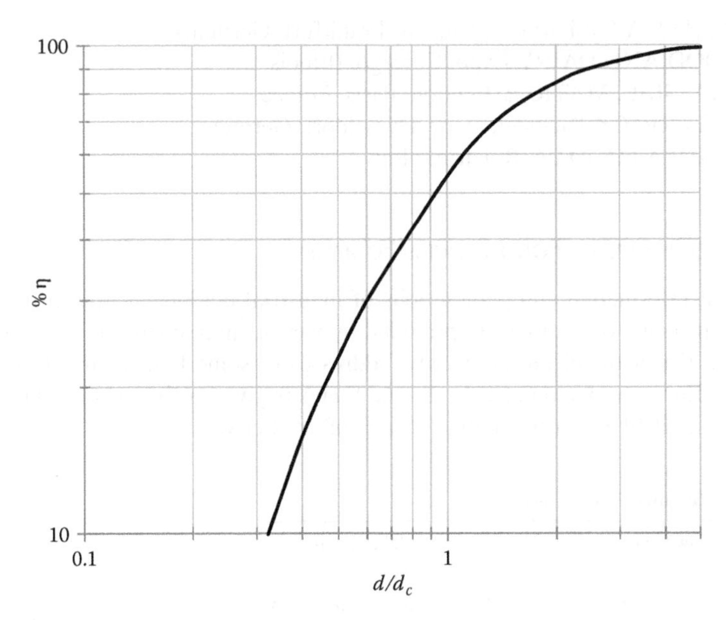

FIGURE A.5 Efficiency (η) of cyclone collectors (Lapple diagram), particle diameter (d), cut particle diameter (d_c).

A.6 FOOD PROCESS EQUIPMENT

A.6.1 Directories of Food Process Equipment

- CE, *Chemical Engineering Magazine, Chemical Engineering Equipment Buyers Guide*, McGraw-Hill, New York.
- DECHEMA, Gesellschaft fuer Chem.Technik und Biotechnologie, www.dechema.de
- DFISA, Dairy and Food Industry Supply Association, Inc., *Membership Directory of Products and Services*, Rockville, MD.
- FP, *Food Processing Magazine, Food Processors' Resource, Equipment and Supplies*, Putman Publishers, Chicago, IL.
- FPM&SA, Food Processing Machinery & Suppliers Association, *Blue Book of Buyers Guide*, Alexandria, VA.
- IAFIS, International Association of Food Industries Suppliers, www.iafis.org
- JFMMA, Japan Food Machinery Manufacturers Association, *General Catalogue for Food Machinery in Japan*, Tokyo, Japan.
- PPMA, Processing and Packaging Machinery Association,www.ppma.co.uk
- TEMA, Tubular Manufacturers Association, Inc., www.tema.org

A.6.2 Exhibitions of Food Process Equipment

- ACHEMA Exhibition–Congress, Frankfurt, Germany
- FOOD AND DAIRY Expo, Chicago, Illinois
- GIA, SIEL, MATIC Exhibition, Paris, France
- INTERPACK Packaging Fair, Düsseldorf, Germany
- PARMA FOOD FAIR, Parma, Italy

A.6.3 Suppliers of Food Process Equipment

Technical information and photographs of industrial equipment used in food process engineering operations are provided by various manufacturers and suppliers. A partial list of the electronic (www) addresses of some known suppliers mainly from the European Union and the United States is given is this section. The equipment is classified according to the chapters of this book.

Chapter 6 (Pumps, Fans, Conveyors)

Alfa Laval (Sweden)	alfalaval.com
APV (UK)	apv.com
Mono Pumps Ltd (UK)	mono-pumps.com
Moyno Pumps (USA)	moyno.com
Waukesha Cherry-Burell (USA)	gowcb.com

Chapter 7 (Size Reduction, Mixers, Homogenizers, Extruders)

Amandus Kahl (Germany)	amanduskahl-group.de
Buhler (Switzerland)	buhlergroup.com
Clextral (France)	clextral.com
Fitzpatrick (USA)	fitzmill.com
Hosokawa Micron Corporation (Japan)	hmc.hosokawa.com
Urschel (USA)	urschel.com
Werner & Pfeiderer (Germany)	wpib.de
Wenger (USA)	wenger.com

Chapters 8 and 9 (Heaters, Sterilizers, Blanchers)

Alfa Laval (Sweden)	alfalaval.com
APV Crepaco (UK)	apv.com
Cabinplant (Denmark)	cabibplant.com
Dixie Canner (USA)	dixiecanner.com
FMC Corporation (USA)	fmc-foodtech.com
Lagarde (France)	lagarde-autoclaves.com
Stock America (USA)	stockamerica.com
Stork (Netherlands)	stork-food-dairy.com
Tetrapak (USA)	tetrapak.com
Waukesha Cherry-Burrell (USA)	gowcb.com

Chapter 10 (Evaporators)

Alfa Laval (Sweden)	alfalaval.com
APV Systems (UK)	apv.com
Fenco (Italy)	fenco.it
GEA-Wiegand (Germany)	gea-wiegand.com
Unipektin (Switzerland)	unipektin.com

Chapter 11 (Dryers)

GEA Niro Atomizer (Denmark)	niro.dk
National Drying Equipment (USA)	nationaldrying.com
Pavan Mabimpianti (Italy)	pavan.com
Wolverine Proctor (USA)	cpmwolverineproctor.com

Chapter 12 (Refrigerators, Freezers)

GEA Grasso (Netherlands)	grasso.nl
Jackstone Froster Ltd (UK)	jackstonefroster.com
York International Corporation (USA)	york.com

Chapter 13 (Distillation, Extraction)	
Flavourtech (Australia)	flavourtech.com
GEA-Wiegand (Germany)	gea-wiegand.com
Koch Modular Process Systems (USA)	modular-process.com

Chapter 14 (Membrane Separations, HP Processing)	
Avure Technology (USA)	avure.com
Dow Chemical (USA)	dowmembranes.com
Millipore (USA)	millipore.com
PCI Membrane Systems (UK)	pcimem.com

Chapter 15 (Canning, Aseptic, Plastic, Paper)	
Angelus Sanitary Can Machine (USA)	angelusmachine.com
Tetrapak (Sweden)	tetrapak.com
Lubeca-Scholz (Germany)	scholz-mb.de
Packaging Machinery (Canada)	pfmusa.com
R. Bosch Verpackungsmaschinen (Germany)	boschpackaging.com
SIG Holding (Switzerland)	sig-group.com

REFERENCES

Haar, L., Gallagher, J., and Kell, G. eds. 1984. *NBS/NRC Steam Tables.* Hemisphere Publications, New York.

Maroulis, Z.B. and Saravacos, G.D. 2003. *Food Process Design.* CRC Press, Boca Raton, FL.

Maroulis, Z.B. and Saravacos, G.D. 2007. *Food Plant Economics.* CRC Press, New York, August 2, 2007.

Perry, R. and Green, D. eds. 1997. *Perry's Chemical Engineer's Handbook*, 7th edn. McGraw-Hill, New York.

Rahman, M.S. 2009. *Food Properties Handbook*, 2nd edn. CRC Press, New York.

Rao, M.A., Rizvi, S.S.H., and Datta, A.E. 2005. *Engineering Properties of Foods*, 3rd edn. Taylor & Francis, New York.

Saravacos, G.D. and Kostaropoulos, A.E. 2002. *Handbook of Food Processing Equipment.* Springer, New York.

Saravacos, G.D. and Maroulis, Z.B. 2001. *Transport Properties of Foods.* Marcel Dekker, New York.

Index

G

H

Q

R